Praise for *Moscow 1812*

"Zamoyski elegantly delivers gripping storytelling, bold revisionism, and poignant suffering. . . . Adam Zamoyski has reexamined the evidence and created a modern account that . . . takes a giant step closer to how it really was."

—Simon Sebag Montefiore, *The Evening Standard*

"Zamoyski's achievement rests on firmer foundations than previous studies. No historian has dug so deeply into eyewitness accounts. . . . Significant [and] convincing." —Alan Palmer, *Literary Review*

"This is a towering history—a thoroughly enjoyable read that is worthy of the monumental scope of its subject. Zamoyski's writing is vivid and, perhaps more important, he knows when to let his sources speak for themselves." —*Seattle Times*

"A magnificent book. . . . No review can do justice to the scholarly integrity and human sensitivity of this book, or to the horrors it describes. I could not sleep after reading his description of the last days of the French retreat."

—Christopher Woodward, *The Spectator*

"It is one of history's great stories of hubris and tragedy, not to mention a titanic clash of personalities and men—and it's all here in Adam Zamoyski's *Moscow 1812*. Told with vigor, sweep, and insight, *Moscow 1812* brings this epic moment to life i
fascinating way." —Jay Winik, author

"Adam Zamoyski's *Moscow 1812* is a brilliant, chilling
Although the muscle of this book is historical detai
comes from the fact that it is about both history a
quences." —Ed Vulliamy,

"This is a well-paced and stylishly written narrative, executed on a scale to match the multinational experience of war in the dramatic 175 days that separated the French crossing of the Niemen from their return." —Ian Beckett, *Times Literary Supplement*

"This massive study of Napoleon's famous Russian campaign may rank as the best recent study in English. . . . The author spares none of the harrowing details of cold, storm, starvation, and the vigorous efforts of the Russians to turn defeat into disaster."

—*Publishers Weekly*

"Penetrating. . . . As well as mastering the huge geopolitical and strategic issues at stake in the campaign, Zamoyski is brilliant at explaining what it must have been like to be a foot soldier marching through thousands of miles of Russian pine forests."

—Andrew Roberts, *The Mail on Sunday*

"Napoleon's invasion of Russia in 1812 is a dark and starry epic, about which hundreds of books have been written and then some. Now, almost two centuries after that drama, Adam Zamoyski's *Moscow 1812* is more than one of the best; it is perhaps the best."

—John Lukacs, author of *Five Days in London*

"Zamoyski displays not only narrative ability but also persuasive interpretive skill when he turns to events in the Russian camp."

—*Booklist* (starred review)

"Magnificent. . . . Zamoyski is both a superb storyteller and a meticulous historian. . . . His book springs to life on every page. . . . Deft and elegant. . . . The fullest and most objective popular account in English. . . . It is narrative history at its best." —*New York Sun*

Roderick Field

About the Author

ADAM ZAMOYSKI was born in New York and educated at Oxford. He is the author of *Holy Madness: Romantics, Patriots, and Revolutionaries, 1776–1871*. He lives in London.

Moscow 1812

By the same author

CHOPIN: A BIOGRAPHY

THE BATTLE FOR THE MARCHLANDS

PADEREWSKI

THE POLISH WAY

THE LAST KING OF POLAND

THE FORGOTTEN FEW

HOLY MADNESS: ROMANTICS, PATRIOTS, AND REVOLUTIONARIES, 1776–1781

Moscow 1812

Napoleon's Fatal March

ADAM ZAMOYSKI

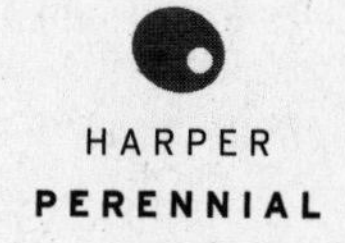

HARPER
PERENNIAL

HARPER PERENNIAL

First published as *1812: Napoleon's Fatal March on Moscow*
in Great Britain in 2004 by HarperCollins Publishers.

A hardcover edition of this book was published in 2004 by HarperCollins Publishers.

FIRST HARPER PERENNIAL EDITION PUBLISHED 2005.

The Library of Congress has catalogued the hardcover edition as follows:

Zamoyski, Adam.
[1812]
Moscow 1812 : Napoleon's fatal march / Adam Zamoyski.—1. ed.
p. cm.
Originally published: 1812: Napoleon's Fatal March on Moscow. London :
HarperCollins, 2004.
Includes bibliographical references and index.
ISBN 0-06-107558-2
1. Napoleonic Wars, 1800–1815—Campaigns—Russia. 2. Napoleon I,
Emperor of the French, 1769–1821—Military leadership. 3. France—History,
Military—1789–1815. 4. Russia—History, Military—1801–1917. I. Title

DC235.Z35 2004
940.2'742'0947—dc22 2004047575

ISBN-10: 0-06-108686-X (pbk.)
ISBN-13: 978-0-06-108686-1 (pbk.)

05 06 07 08 09 ❖/RRD 10 9 8 7 6 5 4 3 2 1

Contents

Illustrations

Maps

Select Glossary of Place-Names in the Former Polish Lands of the Russian Empire

Babinovitse: Babinowicze (Polish), present-day Babinavičy (Belarus)
Berezina: Berezyna (Polish), present-day Bjarezina (Belarus)
Beshenkoviche: Bieszenkowicze (Polish), present-day Bešankovičy (Belarus)
Bobr: Bóbr (Polish), present-day Bobr (Belarus)
Borisov: Borysów (Polish), present-day Barysau (Belarus)
Brest: Brześć (Polish), present-day Brèst (Belarus)
Dnieper: Dniepr (Polish), present-day Dnjapro (Belarus)
Drissa: Dryssa (Polish), present-day Verhnjadzvinsk (Belarus)
Dubrovna: Dubrowna (Polish), present-day Dubrovno (Belarus)
Dunaburg: Dźwińsk (Polish), present-day Daugavpils (Latvia)
Glubokoie: Głębokie (Polish), present-day Glybokae (Belarus)
Grodno: Grodno (Polish), Grodna (Belarus)
Kobryn: Kobryń (Polish), present-day Kobryn (Belarus)
Kovno: Kowno (Polish), present-day Kaunas (Lithuania)
Ladi: Lady (Polish), present-day Liadi (Belarus)
Loshnitsa: Łosznica (Polish), present-day Lošnica (Belarus)
Miedniki: Miedniki (Polish), present-day Medininkai (Lithuania)
Minsk: Mińsk (Polish), present-day Minsk (Belarus)
Mogilev: Mohylów (Polish), present-day Magilev (Belarus)

Molodechno: Mołodeczno (Polish), present-day Maladzečna (Belarus)
Mstislav: Mścisław (Polish), present-day Mscislav (Belarus)
Niemen (river): Niemen (Polish), present-day Nemunas (Lithuania)
Nieshviezh: Nieśwież (Polish), present-day Njasviž (Belarus)
Orsha: Orsza (Polish), present-day Orša (Belarus)
Oshmiana: Oszmiana (Polish), present-day Ašmjany (Belarus)
Ostrovno: Ostrowno (Polish), present-day Astrovna (Belarus)
Pleshchenitse: Pleszczenice (Polish), present-day Plescanicy (Belarus)
Polotsk: Połock (Polish), present-day Polack (Belarus)
Ponary: Ponary (Polish), Panarai (Lithuania)
Shvienchiany: Święciany (Polish), present-day Svencionys (Lithuania)
Smorgonie: Smorgonie (Polish), present-day Smarhon' (Belarus)
Studzienka: Studzienka (Polish), present-day Studenka (Belarus)
Tolochin: Tołoczyn (Polish), present-day Talačyn (Belarus)
Troki: Troki (Polish), Trakai (Lithuania)
Vesselovo: Weselowo (Polish), Veselovo (Belarus)
Vilia: Wilja (Polish), present-day Neris (Lithuania)
Vilna: Wilno (Polish), present-day Vilnius (Lithuania)
Vitebsk: Witebsk (Polish), present-day Vicebsk (Belarus)
Volkovisk: Wołkowyski (Polish), present-day Vavkavysk (Belarus)
Zakrent: Zakręt (Polish)
Ziembin: Ziembin (Polish), present-day Zembin (Belarus)

Introductory Note

Napoleon's invasion of Russia in 1812 was one of the most dramatic episodes in European history, an event of epic proportions, etched deeply in the popular imagination. I only had to mention the subject of this book for people to come to life, stirred by recollections of Tolstoy's *War and Peace*, by the scale of the tragedy, by some anecdote that had lodged itself in their memory, or just a mental image of snowbound Napoleonic tragedy. But the flash of recognition was almost invariably followed by an admission of total ignorance of what had actually happened and why. The reasons for this curious discrepancy are fascinating in themselves.

No other campaign in history has been subjected to such overtly political uses. From the very beginning, studies of the subject have been driven by a compulsion to interpret and justify that admits of no objectivity, while their sheer volume – over five thousand books and twice as many articles published in Russia alone in the hundred years after 1812 – has helped only to cloud the issue.[1]

This was to be expected, considering what was involved. There were great reputations at stake: those of Napoleon, of Tsar Alexander, of Field Marshal Kutuzov, to name only the obvious ones. There was also a need to make sense of the whole business, for this war, unprecedented in the history of Europe in both scale and horror, was not easy to assess in military terms. The action was often confused. Both sides claimed victory in every engagement. And if the French had lost the campaign, the Russians could hardly be said to have won it. At the same time, people on both sides had behaved with a savagery that neither nation wished to contemplate.

In France, early attempts at a balanced study were complicated by political factors: the regime which replaced Napoleon's soon after the events required anything to do with him to be represented in the most negative terms. Censorship also played a part in Russian assessments, for more complex reasons. The events of 1812 and their aftermath raised questions about the very nature of the Russian state and its people, and, as the historian Orlando Figes nicely puts it, "the nineteenth-century quest for Russian nationhood began in the ranks of 1812."[2]

This quest was innately subversive of the Tsarist system, and led in the first place to the Decembrist Rising of 1825. It was pursued, along divergent paths, by those who sought a more modern Russia integrated into the mainstream of Western civilisation, and by the slavophiles, who rejected the West and all it stood for, seeking instead a truly "Russian" way. The events of 1812 were used by both sides to back up their arguments, rapidly attaining mythological status and becoming increasingly distorted as a result. This dualism was only complicated with the advent of Marxism.

The first French historians to write about 1812 were either hostile to Napoleon or motivated by a desire to ingratiate themselves with the post-Napoleonic regime, and therefore laid all blame at the feet of the demon Bonaparte. But most French writers on the events of the campaign, whether they were participants or later academic historians, have followed a more measured, and broadly similar, path. While often displaying a degree of embarrassment over such an apparently imperialist venture and the misery France inflicted on the Russian people, not to mention her own and her allies' soldiers, they have tried to redeem Napoleon's reputation and the honour of French arms by a generous representation of the doughtiness of the Russian soldier and of the implacable nature of the Russian climate. They have also clutched at the straw of comfort held out by the Romantic imagination of the 1820s and 1830s, which turned the picture of sordid disaster into a vision of greatness in adversity.

In the last decades of the nineteenth century, distance as well as a growing cordiality between the two nations made it possible for French historians to approach the subject more objectively. The centenary, coming as it did just before the Great War and at a time when the two nations were allies, saw cooperation between the historical commissions of the French and Russian staffs, and led to the publication of much primary source-material. But French historians continue to show a certain reluctance to deal with the war, and have come up with no satisfactory general study of it.

The first Russian account of the events, by a colonel on the general staff, was produced with such alacrity that it was published, in English, as far afield as Boston within a year. It was undoubtedly a piece of propaganda, intended to pave the way for Russia's future role in the affairs of Europe, but it did reflect the perceptions of large sections of Russian society. It depicted Alexander as the catalyst who rallied a gallant patriotic nobility and a loyal peasantry eager to defend Faith, Tsar and Fatherland.

Dmitri Petrovich Buturlin, himself a participant, who wrote the first detailed account of the war, added a couple of new elements. One was the idea of Russia as the innocent victim of aggression. The other was the image of Kutuzov as the quintessential Russian hero, simple but wise. A.I. Mikhailovsky-Danilevsky, whose four-volume history came out in 1839, depicted Alexander as a moral beacon awakening the spiritual as well as the physical forces of the Russian people in defence of their fatherland. It was he who first called it the "*Otechestvennaia Voina*," the "Patriotic War." Underlying much of their writing was the view that it was the Almighty, acting through the Tsar and the Russian people, who had confounded the evil one. This being so, French assertions that they had been defeated by the Russian winter rather than the Russians themselves were dismissed as irrelevant.

It was against the backdrop of this fundamentally spiritual

interpretation that in the summer of 1863 Tolstoy began work on his novel *War and Peace*, in which he was to add his own highly personal gloss on the events.

Tolstoy had originally been enthusiastic about the programme of liberal reforms launched by Tsar Alexander II when he came to the throne in 1855. He had even tried to pre-empt them by offering the serfs on his estate a deal that would free them from their obligations and give them the land they worked. But the serfs were suspicious and rejected his offer. Instead of turning Tolstoy against the peasants, this helped to put him off liberalism in general. He embraced the slavophile view that the liberals would destroy Russia by imposing foreign ideas and institutions alien to the Russian character. He also reacted against the wave of self-abasement among intellectuals who had turned the recent defeat in the Crimean War into a paradigm for Russian backwardness. In his depiction of the events of 1812, Tolstoy traces a metaphor of the penetration of Russia by foreign influences: Napoleon is the harbinger of an "alien" order, which some of Alexander's "contaminated" entourage favour. But it is rejected by the Russian nation. Yet this is no glorification of the Russian common people – the hero of Tolstoy's novel is a deferential peasantry led by the minor nobility who, unlike the Frenchified aristocrats, have remained true to Russian values. But Tolstoy's work is not all fantasy, and he did his homework.

The first sentence of *War and Peace* expresses outrage at French doings in Genoa and Lucca in 1799, while on the next page one of the protagonists affirms that Russia will be "the saviour of Europe." With this opening, Tolstoy firmly dismissed the notion of the French invasion of 1812 as an act of gratuitous aggression: to him it was clear that it was merely part of a prolonged struggle between France and Russia for hegemony over Europe. Yet it was to be some time before a Russian historian would mention this.

The second half of the nineteenth century saw the publication of a great many diaries, recollections and letters of participants in the events; of staff documents giving troop numbers and dispositions;

and of official documents, orders and letters. It also brought forth a number of very useful studies of specific aspects of the campaign, of individual battles, and of social reactions to the events.

The next generation of Russian historians to deal with the subject in any depth were influenced by the works of Marx and Engels and took a more pragmatic view. It is true that Aleksandr Nikolaevich Popov, writing in 1912, idealised Kutuzov and "Russian society," but his more down-to-earth contemporary Vladimir Ivanovich Kharkievich admitted that Kutuzov had faults, and rejected the image of Russia as innocent victim. Konstantin Adamovich Voensky took a similar line, and related Russian military failures in 1812 to the shortcomings of the country's constitution and social structure. A number of other historians produced studies of specific aspects, in which they turned up evidence of a less glorious response by Russian society than had been represented hitherto, and confirmed that logistics and climate had been largely responsible for the outcome.

Perhaps the most forceful of this generation of historians, and the one who reacted most vigorously against the old pieties, was Mikhail Nikolaevich Pokrovsky. According to him, the Tsarist state was bent on extending Russian hegemony beyond its borders in order to guarantee the survival of an essentially feudal system at home. He went so far as to say that Napoleon's invasion of Russia was "an act of necessary self-defence" on his part. He was deeply critical of Kutuzov and other Russian generals. He stressed the role of the weather in the defeat of the French, and belittled the role of "Russian society," questioning the myth of the patriotic peasants. Those who did resist the invader did so, according to him, in defence of their chickens and geese, not their fatherland.[3]

This view was endorsed by Lenin, and held sway during the first two decades of Soviet rule. The war was not referred to as the "Patriotic War" during this period, as it concerned only the respective economic interests of the Russian imperialists and the French bourgeoisie. The Russian army had made a mess of defending the country precisely because it was commanded by nobles, and the government's fear of arming the peasants prevented the development of a guerrilla war against the French.

At Stalin's prompting, a ruling of the Central Committee on 16 May 1934 recommended a fresh approach to the study of history, aimed at engaging the masses. How that would affect the representation of the events of 1812 was not immediately clear. Writing in 1936, the historian Evgenii Viktorovich Tarle affirmed that the Russian people had played no part in the war, dismissing evidence of peasant guerrilla activity as no more than the opportunistic murder of French stragglers. The following year he published an account of the war in which he said almost exactly the opposite, representing it as the triumph of the patriotic Russian people. After a certain amount of hair-splitting argument couched in the language of Marxist dialectics, it was once again dubbed the "Patriotic War," but only in inverted commas. Tarle also admitted that the weather might have had something to do with the French débâcle, but was later accused of retailing the ideas of what one writer termed "the Trotskyite-Bukharinite counter-revolutionary enemies of the people" and "the lying inventions of foreign authors" – even the account by the great military theorist Karl von Clausewitz, himself a participant in the campaign on the Russian side, was dismissed as "lies."

Tarle adopted a traditional spiritual view of the events, representing the French victory at Borodino as "a moral victory" for the Russians and the war itself as the crucible of all that was best in Russian history over the next decades. He also built up the image of Kutuzov, as a kind of metaphysical emanation of the Russian people, their true leader in every sense.[4] But it was his colleague P.A. Zhilin who made the obvious connection between Kutuzov and Stalin as saviours of the fatherland.

Hitler's invasion of Russia in 1941 and the titanic struggle that followed added substance to this connection, and the events of 1812 provided a wonderful source of propaganda material. The "Patriotic War of 1812," as it would henceforth be known, could be viewed as a dress rehearsal for what became the "Great Patriotic War." Tarle's book was translated and published widely in the West, in order to help make the point that a peace-loving Russia had been attacked for no reason at all – deftly burying the embarrassing fact that, just as in 1812, Russia had been a complicitous ally of the other side up to the very outbreak of hostilities – but that her people and the great leaders that sprung from their bosom were invincible.

For a brief period following Stalin's death in 1953 an element of objectivity entered Russian historiography, and a number of solid studies on the economic, political and diplomatic background, the military preparations and other aspects saw the light of day. But the advent of Brezhnev put the lid on this. Historians such as L.G. Beskrovny plugged the old patriotic nostrums and shamelessly repeated obvious falsehoods. French numbers were regularly inflated and those of the Russian forces scaled down. The persona of Kutuzov took on a life of its own. The luxury-loving prince was transformed into a kind of peasant leader who was in some mysterious way "in conflict" with the Tsar and the system. Every blunder he made was represented as a piece of cunning, the actual effect of which was not defined, and every failure to act as a brilliant strategic ploy.

This kind of interpretation went unchallenged until the late 1980s, when a new generation of historians, such as A.A. Abalikhin, V.G. Sirotkin, S.V. Shvedov, Oleg Sokolov and N.A. Troitsky brought a freshness and honesty to the subject never known before. But it will probably be some time before a satisfying synthesis emerges from this.

The handful of Western historians who have written on the subject have made modest use of available Russian primary sources, relying instead on the works of their Russian colleagues. Not surprisingly, they have accepted the facts and figures they found in these. More surprisingly, most have also accepted some of the interpretations and ingested, albeit unconsciously, a dose of their emotional and political flavour.

Virtually all the extant documentary material concerning the political and military events covered by this book has been published and available for decades. It would be interesting, and possibly worthwhile, to investigate further areas such as the question of how the episode impacted on the structures of the Russian state, its economy and attitudes to authority. It might also be profitable to go back to the manuscript originals of some of the printed sources, particularly where these have been translated from French into Russian. But it is highly improbable that new documents of any significance will come to light, or that further detailed research in any ancillary field will throw up fresh evidence on the causes of the war, its conduct, the numbers involved, the extent of the losses or any of the other vital aspects.

The ground has therefore been thoroughly prepared, and now that the nationalist passions and the political imperatives have ebbed away, the task of writing about the events of 1812 should present a less daunting prospect. But it remains a formidable one. For this was not just any war. It was the climax in a protracted struggle between Napoleon and Alexander, between France and Russia, and between the ideological inheritance of the European Enlightenment and the French Revolution on the one hand, and a reactionary combination of Christianity, monarchism and traditionalism on the other. It involved the whole of Europe, and its repercussions were therefore widespread as well as long-lasting. Its scale was unprecedented, and raised a number of issues hitherto unknown in military history. It was also the first modern war, in that the entire Russian people were forced by their own government to participate actively, and popular feeling became an element of military strategy. It is impossible to isolate any of these elements from the others, as the conflict does not make sense without at least an awareness of the depth and breadth of the issues involved.

To do justice to such a subject would take many years, and a book at least twice the length of this one, which is not intended to be in any way a definitive work. It is not a full record of the military operations, which involved dozens of engagements and ranged over a vast area. Nor does it aspire to be any more than an outline of the diplomatic relationship between France and Russia. My principal aim in writing this book has been to tell an extraordinary story, of which everyone has heard but very few have any real knowledge. I have attempted to place it in its wider context and to touch on its deeper significance. Above all, I have tried to convey what these events meant for those concerned, at every level – for this is *par excellence* a human story, of hubris and nemesis, of triumph and catastrophe, of glory and squalor, of joy and suffering.

I have therefore drawn heavily on the first-hand accounts of participants, of which there are a remarkable number. There is much variation between them, in terms of both accuracy and literary quality: some are original letters or diaries; others are memoirs written from diaries; memoirs written from memory, some of them composed within a year or two, others decades later; accounts based on personal experience and documentation; and regular histories written by participants, some of whom were in key positions, others only witnesses. I have taken these factors into account in making use of them, and I have avoided basing myself too heavily on, for instance, the much-quoted Ségur, who was not a central figure but who wrote as though he had been, and subjected his writing to exalted literary pretensions; I also avoid his main critic, Gourgaud, who was himself not as well placed to know what was going on as he suggests, and who subjects his account to uncritical worship of Napoleon.

I have been driven by the same desire to reproduce the human experience in my choice of illustrations, and in this I have been aided by another unique aspect of this war. This was the only campaign before the age of photography to have been graphically recorded by a number of participants, some of whom were distinguished artists. Not until the American Civil War, half a century later, would the realities of war be conveyed through such vivid insights. In view of this, I decided to dispense with the array of pompous and largely meaningless battle scenes that usually adorn this kind of book, and to concentrate on providing something more akin to a photographic record of life on campaign. Apart from the small number of portraits of the main protagonists, almost every image was drawn or painted by a participant, either on the spot or from memory, and the few exceptions to this were executed under the direction of participants.

Where I feel a quotation or a statement of fact needs it, I have given references to other first-hand accounts which support it. But in order to keep down the quantity of reference numbers on the page, I have often lumped together several quotations or a series of linked facts contained in a single paragraph under the same one. The translations are all mine, except in cases where the book was only available to me in another's translation, and in the case of the translations from German, in which I was helped by others.

There are several methods of transliterating Russian words and names, none of which is, to my mind, completely satisfactory. This is mainly because they attempt consistency where none is possible, and also because every new scheme necessarily outlaws words based on previous methods that have already become familiar. I have therefore followed my own instinct and what I believe to be common sense. I realise that specialists may find this irritating.

I have transliterated Russian names as they are pronounced, preferring Yermolov to Ermolov, or as they have been known in the West for decades, sticking to Tolstoy rather than Tol'stoi, Galitzine rather than Golitsuin. I have stuck to the –sky ending for Russian names as opposed to the –skii, for the same reasons. But I have observed the universally accepted new spelling in the bibliography, since that is how the names appear in (most) library catalogues. In the cases of non-Russians serving in the Russian army, I use the original spelling, as I see no reason to turn a Wittgenstein into Vitgenshtain, a Czaplic into a Chaplits, or a Clausewitz into a Klausevits, except in the cases of Baggovut (Swedish Baggehuffwudt) and Miloradovich (Serbian Miloradovič), for purely pragmatic reasons.

Perhaps the most difficult question is that of place names. The action of the campaign unfolds over territories which had recently passed from one sovereignty to another, and sometimes back again, and which are now in entirely new countries. I have used German names for what was then East Prussia. I have used Russian names for Russia beyond Smolensk (except for St Petersburg and Moscow). I have used Polish forms and spellings for places in the Grand Duchy of Warsaw, and transliterated Polish names on the territory which had been Polish in the decades before 1812. I have done this because the French used these forms in their accounts (albeit with curious spellings). The Russian forms of these names hardly differ from the Polish, while the present-day Lithuanian or Belarussian ones would be confusing (I have enclosed a brief glossary of these names with their present-day forms, for purposes of identification). I have made an exception of the capital of Lithuania: the French mostly spelt it Wilna, the Russians called it Vilna, and the Polish Wilno would seem out of place, particularly if transliterated into Vilno – so I have opted for Vilna. For similar reasons, I have preferred the Russian Glubokoie to the Polish Głębokie, which would be transliterated as Gwembokie.

All dates are given in the new style, according to the Gregorian calendar.

I should like to thank Professors Isabel de Madariaga, Janet Hartley, Lindsey Hughes, Dominic Lieven and Alexander Martin for their advice and assistance. My thanks must also go to Mirja Kraemer and Andrea Ostermeier for reading a number of German texts for me, and to Galina Babkova for the speed and efficiency with which she ferreted out, copied and despatched to me whatever I required from libraries in Russia. I am grateful to Dr Dobrosława Platt, Laurence Kelly, Artemis Beevor and Jean de Fouquières for their help in tracking down illustrations. Shervie Price was, once again, a long-suffering reader and an invaluable critic of my typescript, and Robert Lacey was an exceptionally meticulous and sensitive editor. Trevor Mason deserves a medal for his patience with me over the maps and the diagram.

I should also like to thank Ambassador Stefan Meller for his assistance during my trip to the theatre of operations, and Mikołaj Radziwiłł for being such a good driver on the roads of Russia, Lithuania and Belarus, and companion in Vilnius, Orša and Smolensk, on the battlefield of Borodino and the banks of the Berezina.

Above all I want to thank my wife Emma, for everything.

Moscow 1812

1

Caesar

As the first cannon shot thundered out from the guns drawn up before the Invalides on the morning of 20 March 1811, an extraordinary silence fell over Paris. Wagons and carriages came to a standstill, pedestrians halted, people appeared at their windows, schoolboys looked up from their books. Everyone began to count as the discharges succeeded each other at a measured pace. In the stables of the École Militaire, the cavalry of the Guard were grooming their horses. "Suddenly, the sound of a gun from the Invalides stopped every arm, suspended every movement; brushes and curry-combs hung in the air," according to one young *Chasseur*. "In the midst of this multitude of men and horses, you could have heard a mouse stir."[1]

As news had spread on the previous evening that the Empress had gone into labour, many *patrons* had given their workmen the next day off, and these swarmed expectantly in the streets around the Tuileries palace. The Paris *Bourse* had ceased dealing that morning, and the only financial transactions taking place were bets on the sex of the child. But the excitement was just as great among those who had nothing riding on it.

"It would be difficult to imagine with what anxiety the first cannon shots were counted," recalled one witness: everyone knew that twenty-one would announce the birth of a girl, and one hundred that of a boy. "A profound silence reigned until the twenty-first, but when the twenty-second roared forth, there was an explosion of congratulation and cheering which rang out simultaneously in every part of Paris."[2] People went wild, embracing total strangers and shouting "*Vive l'Empereur!*" Others danced in the streets as the remaining seventy-eight shots thundered out in a rolling barrage.

"Paris had never, even on the greatest holidays, offered a picture of more general joyfulness," noted another witness; "there was celebration everywhere."[3] A balloon went up, bearing into the sky the celebrated aeronaut Madame Blanchard with thousands of printed notices of the happy tidings, which she scattered across the countryside. Messengers galloped off in all directions with the news. That evening there were fireworks and the capital was illuminated, with candles in the windows of even humble mansard rooms. Theatres staged special performances, printmakers began churning out soppy images of the imperial infant borne on celestial clouds with crowns and laurels hovering over him, and poets set to work on commemorative odes. "But what one will never be able to convey adequately," wrote the young Comte de Ségur, "is the wild intoxication of that surge of public rejoicing as the twenty-second cannon shot announced to France that there had been born a direct heir to Napoleon and to the Empire!"[4]

The twenty-year-old Empress Marie-Louise had felt the first pains at around seven o'clock on the previous evening. Dr Antoine Dubois, *Premier Accoucheur* of the Empire, was on hand. He was soon joined by Dr Corvisart, the First Physician, Dr Bourdier, the Physician-in-Ordinary to the Empress, and Napoleon's surgeon Dr Yvan. The Emperor, his mother and sisters, and the various ladies of the Empress's household brought to twenty-two the number of those attending her, either in her bedroom or in the next chamber.

Beyond that, the salons of the Tuileries were filled with some two hundred officials and dignitaries, who had been summoned at the first signs of the Empress going into labour and stood about awkwardly in full court dress. Every now and then, one of the ladies-in-waiting on duty would come out and give them a progress report. As the evening wore on, small tables were brought in and they were served a light supper of chicken with rice washed down with Chambertin. But the banter was subdued: things were clearly not proceeding smoothly in the Empress's bedroom. At about five in the morning the Grand Marshal of the Empire came out and informed them that the pains had ceased and the Empress had fallen asleep. He told them they could go home, but must remain on call. Some went, but many of the exhausted courtiers stretched out on benches or rolled up carpets into makeshift mattresses and lay down on them in all their finery to snatch some sleep.

Napoleon had been with Marie-Louise throughout, talking to her and comforting her with all the solicitude of a nervous father-to-be. When she fell asleep Dubois told him he could go and take some rest. Napoleon could do without sleep. His preferred means of relaxation was to lie in a very hot bath, which he believed in as a cure for most of his ailments, be it a cold or constipation, from which he suffered regularly. And that is what he did now.

He had not been luxuriating in the hot water for long when Dubois came running up the concealed stairs that led from his apartment to the Empress's bedroom. The labour pains had started again, and the doctor was anxious, as the baby was presenting itself awkwardly. Napoleon asked him if there was any danger. Dubois nodded, expressing dismay that such a complication had occurred with the Empress. "Forget that she is Empress, and treat her as you would the wife of any shopkeeper in the rue Saint Denis," Napoleon interrupted him, adding: "And whatever happens, save the mother!" He got out of his bath, dressed hastily and went down to join the doctors at his wife's bedside.

The Empress screamed when she saw Dubois take out his forceps, but Napoleon calmed her, holding her hand and stroking her while the Comtesse de Montesquiou and Dr Corvisart held her still. The baby emerged feet first, and Dubois had a job getting the head clear. After much pulling and easing, at around six in the morning he delivered it. The baby appeared lifeless, and Dubois laid it aside as he and the others attended to the mother, who seemed to be in danger. But Corvisart picked up the child and began to rub him briskly. After about seven minutes of this he came to life, and the doctor handed him to the Comtesse de Montesquiou, with the comment that it was a boy. Napoleon, who could see that Marie-Louise was by now out of danger, took the baby in his arms and, bursting into the adjoining room where all the senior officers of the Empire were gathered, expecting the worst, exclaimed: "Behold the King of Rome! Two hundred cannon shots!"

But when his sister-in-law, Queen Hortense, came up to congratulate him a moment later, he replied: "I cannot feel the happiness – the poor woman has suffered so much!"[5] He meant it. They had been married for just one year, and the arranged match had quickly turned into an almost cloyingly loving relationship. One of thirteen children of the Austrian Emperor Francis II, Marie-Louise had been her father's favourite, his "*adorable poupée.*" She had been brought up to hate Napoleon and to refer to him as "the Corsican," "the usurper," "Attila" or "the Antichrist." But, when diplomacy demanded it, she bowed to her father's will. And once she had tasted the pleasures of the marital bed there was no restraining her enthusiasm for the Emperor. Napoleon, who had been thrilled at the idea of having in his bed "a daughter of the Caesars," as he referred to her – and one half his age – quickly became moonstruck, and their marriage turned into a middle-class idyll.

That evening, as the capital celebrated, the child was baptised according to the age-old rites of the French royal family. The next day Napoleon held a grand audience, seated on the imperial throne, to receive formal congratulations. The entire court then accompanied him to see the infant, who lay in a superb silver-gilt cradle presented by the city of Paris. It had been designed by the artist Pierre Prudhon and represented a figure of Glory holding a triumphal crown and a young eagle ascending towards the bright star which symbolised Napoleon. The chancellors of the Légion d'Honneur and of the Iron Cross laid the insignia of both orders on cushions beside the sleeping child. The painter François Gérard set to work on a portrait.

Over the next days homage of every kind poured in, and cities throughout the country joined Paris in celebrating as the news reached them, each in turn sending a delegation to deliver its congratulations. The same process was repeated as the news rippled out to the more far-flung parts of the Empire and to other countries. Such expressions were to be expected in the circumstances. But there was a great deal more to the celebrations and congratulations than just loyal humbug – to most Frenchmen the birth of a boy heralded a period of peace and stability, and much more besides.

France had been at war virtually without interruption for nineteen years. She had been attacked, in 1792, by a coalition of Prussia and Austria. Over the next years these were joined by Britain, Spain, Russia and other lesser powers, all of them bent on defeating revolutionary France and restoring the Bourbon dynasty. It was not a fight over territory. It was an ideological struggle over the future order of Europe. Atrocities aside, revolutionary France had brought into public life all the ideals of the Enlightenment, and her very existence was seen by the monarchical powers as a threat to theirs. She had made ample use of this weapon in order to defend herself, by exporting revolution and subverting provinces belonging to her enemies. She had gradually turned from victim to aggressor, but she was nevertheless fighting for survival. Revolutionary France could not secure a lasting peace, as virtually every other power in Europe would not reconcile itself to the survival of the republican regime, and felt a necessity to destroy it.

General Napoleon Bonaparte's seizure of power in Paris in November 1799 should have broken this vicious circle of fear and aggression. He reined in the demagogues, closed the Pandora's box opened by the revolution and tidied up the mess. Being a child of the Enlightenment as well as a despot, he mobilised the energies of France and harnessed them to the task of building a well-ordered, prosperous and powerful state, the "*état policé*" of which the *philosophes* of the

Enlightenment had dreamed. He was following in the footsteps of rulers such as Frederick the Great of Prussia, Catherine the Great of Russia and Joseph II of Austria, who had introduced social and economic reforms while strengthening the framework of the state, and who were universally admired for this. But to their successors, Bonaparte was but a grotesque upstart, a malignant outgrowth of the evil revolution.

By 1801, following a series of resounding victories, Bonaparte was able to force peace on all the powers of the European continent. France's security was guaranteed by expanded frontiers and the creation of a series of theoretically autonomous republics in northern Italy, Switzerland and Holland which were in fact French provinces. In March 1802 Bonaparte even concluded the Peace of Amiens with Britain. But this was not likely to last.

To Britain, France's hegemony in Europe was intolerable. To France, Britain's superiority at sea was a constant threat. French designs on Malta, Egypt and India were a hazy but nevertheless haunting nightmare to Britain, while Britain's ability to use allies on the European mainland to make war by proxy was a source of continuing anxiety to France. Hostilities between the two resumed in May 1803.

During the following year Bonaparte himself revived opposition to his rule throughout Europe. In March 1804 he ordered the young Bourbon Duc d'Enghien to be seized at Ettenheim in the state of Baden just outside the borders of France and brought to Paris. He was convinced that the Duke was involved in a conspiracy to overthrow him and restore the monarchy, and had him executed after a summary judgement. This violation of every accepted law and principle horrified Europe. It confirmed the opinion of those who saw Bonaparte as the devil incarnate, and reinforced the notion of a fight to the death between the sanctified order as embodied in the *ancien régime* and the forces of evil in the form of revolutionary France.

France was in fact no longer exporting revolution. She had become little more than a vehicle for the ambitions of Bonaparte, who a couple of months later proclaimed himself Emperor of the French under the name of Napoleon I. What exactly these ambitions consisted of has perplexed and divided historians over two centuries, for Napoleon was never consistent in anything. His utterances can at best be taken to illustrate some of his thoughts and feelings, while his actions were often erratic and contradictory. He was intelligent and pragmatic, yet he allowed himself to indulge the most far-fetched fantasies; he was the ultimate opportunist, yet he could get caught up in his own dogma; he was a great cynic, yet he pursued romantic dreams. There was no grand idea or master project.

Napoleon was in large measure driven by nothing more complicated than the lust for power and domination over others. Attendant on this was an often childish set of reactions at being thwarted in any way. Having no sense of justice and no respect for the wishes of others, he took any objection to his actions as gratuitous rebellion, and responded with disproportionate vehemence. Instead of ignoring a minor setback or turning an obstacle, he would unleash bluster and force, which often involved him in unnecessarily costly head-on collisions.

He was also driven by a curious sense of destiny, a self-invented notion of a kind often affected by young men brought up on Romantic literature (his favourite reading had been the poems of Ossian and *The Sorrows of Young Werther*), which he came to believe in himself. "Is there a man blind enough not to see," he had declared during his Egyptian campaign in 1798, "that destiny directs all my operations?"[6] Napoleon was also a great admirer of the plays of Corneille, and there is reason to believe he saw himself as acting out some great tragedy in their mould.

This sense of living out a destiny was to lead him repeatedly into acting against his better judgement in pursuit of nebulous dreams. His triumphs in Italy, followed by his spectacular victories at Austerlitz and Jena, only confirmed him in this fantasy, which communicated itself to his troops. "The intoxication of our joyful and proud exaltation was at its height," wrote one young officer after

Napoleon's triumph over Prussia. "One of our army corps proclaimed itself 'the Tenth Legion of the New Caesar!,' another demanded that Napoleon should henceforth be known as 'The Emperor of the West!' "[7]

But Napoleon was also the ruler of France. As such, he was inevitably driven by the same political, cultural and psychological motors which had dictated the policies of French rulers of the past such as François I and Louis XIV, who had striven for French hegemony over Europe in order to achieve lasting security.

France had always sought to impose a balance in central Europe that would prevent a major mobilisation of German forces against her, and she had achieved this by the Treaty of Westphalia back in 1648, in which she and Austria, jointly with a number of other powers, had put in place a whole series of checks and balances. This system had been undone in the second half of the eighteenth century by the rise of Prussian power and the emergence of Russia as a player in European affairs, manifested most critically in huge shifts of power in Germany, the partition and disappearance of Poland, and the race for control of the Balkans. In view of this, it was quite natural that Napoleon should seek to reassert French interests, and in doing so he was pursuing a traditional vision of a "French" Europe as much as his own personal ambition. It was a vision that appeared to have history on its side.

In the eighteenth century France had become the cynosure of Europe in terms of culture and political thought. Her paramountcy in these spheres was consolidated by the revolution, whose fundamental message and ideas were admired and accepted by élites all over the Continent. The French political and military classes saw themselves as "*la Grande Nation*," the first nation in Europe to have emancipated itself, and considered themselves to be armed with a mission to carry the benefits of what they had achieved to other peoples. This was the age of neo-Classicism, and they began to see France as the next Rome, the fount from which this new ideological civilisation radiated, the capital of the modern world.

Napoleon was not immune to the enthusiasms of his age. As befitted the most powerful individual since the days of the Caesars, he issued decrees ordering the cleaning of the Tiber and the Forum Romanum, and the preservation of its monuments. Shortly after the birth of the King of Rome, he set in motion plans for an imperial palace on the Capitol. But he also intended to build one for the Pope in Paris, arguing that this was where he should move, just as St Peter had moved to Rome from the Holy Land.[8]

As early as the mid-1790s, the French revolutionary armies began to bring home to Paris not only valuables and works of art, but also libraries, scientific instruments and entire archives. This epic bout of looting was not the product of mere greed. The idea was that everything most useful to the development of civilisation should be concentrated at the heart of the Empire, and not allowed to benefit only a few in outlying provinces. "The French Empire shall become the metropolis of all other sovereignties," Napoleon once said to a friend. "I want to force every king in Europe to build a large palace for his use in Paris. When an Emperor of the French is crowned, these kings shall come to Paris, and they shall adorn that imposing ceremony with their presence and salute it with their homage." It was not so much a question of France "*über alles*." "European society needs a regeneration," Napoleon asserted in conversation in 1805. "There must be a superior power which dominates all the other powers, with enough authority to force them to live in harmony with one another – and France is the best placed for that purpose." He was, like many a tyrant, utopian in his ambitions. "We must have a European legal system, a European appeal court, a common currency, the same weights and measures, the same laws," Napoleon once said to Joseph Fouché. "I must make of all the peoples of Europe one people, and of Paris the capital of the world."[9]

France's claim to the mantle of Imperial Rome seemed to gain validation when, in 1810, Napoleon married Marie-Louise, daughter of the last Holy Roman Emperor Francis II. His father-in-law, now Emperor of Austria under the name of Francis I, appeared to acquiesce in this transference of power. When Napoleon produced an heir, Francis ceded to the child the title of King of Rome, which had traditionally been that of the son of the Holy Roman Emperor.

France's position on the Continent was by then one of unprecedented power; her political culture and the new system were imposed over vast areas of Europe. But to the average Frenchman this was of less interest than the benefits the past decade had brought him at home. All the most positive gains of the revolution had been salvaged, but order, prosperity and stability had been guaranteed, and a general amnesia if not amnesty had allowed those divided by the struggles of the revolution to put the more unpleasant aspects of the past behind them. Whether this new order would survive depended not only on Napoleon's ability to defend it, but on his ability to guarantee its continuance by cancelling out the possibility of a Bourbon restoration. A return of the Bourbons would mean not only a return to the *ancien régime*; it would also raise the prospect of much score-settling.

In this respect, the birth of the King of Rome was crucial. Most of Napoleon's subjects believed that their ruler, who had recently turned forty, would henceforth be inclined to spend more time with his family than with his armies, that Napoleon the Great would in time be succeeded by Napoleon II, and that the rest of Europe would accept that the Bourbons had been consigned to history. That was why they rejoiced. "People sincerely anticipated a period of profound peace; the idea of war and occupations of that sort were no longer entertained as being realistic," wrote Napoleon's chief of police, General Savary, adding that the child appeared to all as the guarantor of political stability.[10]

Napoleon himself rejoiced for much the same reasons. "Now begins the finest epoch of my reign," he exclaimed. He had always been keenly aware that a man who seizes the throne can never rest easy on it, and that he could only achieve security of tenure by means of the dynastic principle. "With the birth of my son, there is a future in my destiny," he told one of his diplomatic agents. "I am now founding a legitimacy. Empires are created by the sword and are conserved by heredity."[11]

But he was not yet ready to lay down his arms. He had managed to destroy the unity of purpose which had fed the coalitions against France for so long. Austria, Russia and Prussia were now as ready to fight each other as to fight France, the original repugnance to treat with "the Corsican upstart" had largely evaporated, his imperial title was recognised across the Continent, and the Bourbon pretender Louis XVIII was beginning to look like an anachronism. Yet Napoleon was keenly aware of his continuing vulnerability, for nothing had been finally settled.

Over the past decade he had turned France into an empire which included the whole of Belgium, Holland and the North Sea coast up to Hamburg, the Rhineland, the whole of Switzerland, Piedmont, Liguria, Tuscany, the Papal States, Illyria and Catalonia, and ruled directly over some forty-five million people. The French Empire was surrounded by a number of dependent states – the Kingdom of Westphalia, the kingdoms of Saxony, Bavaria, Württemberg and other states grouped in the Confederation of the Rhine, the Grand Duchy of Warsaw, the kingdoms of Italy, Naples and Spain, ruled by Napoleon's siblings, relatives or devoted allies. The only part of his vast imperium where there was open unrest was in Spain, where armed opposition to his brother King Joseph was being supported by a British army. This was not in itself a major problem, and could be dealt with by a concerted operation under his own direction.

The real problem facing Napoleon was how to achieve some kind of finality and to fit all his conquests into a system that would guarantee his and his successors' position. While others regarded him as a megalomaniac bent on conquering all, he saw his wars as defensive, aimed at guaranteeing France's security as well as his own. "To leave my throne to my heirs," he told one of his chamberlains, "I will have had to be master of all the capitals of Europe!" In the written instructions to one of his diplomatic envoys, he explained that although France

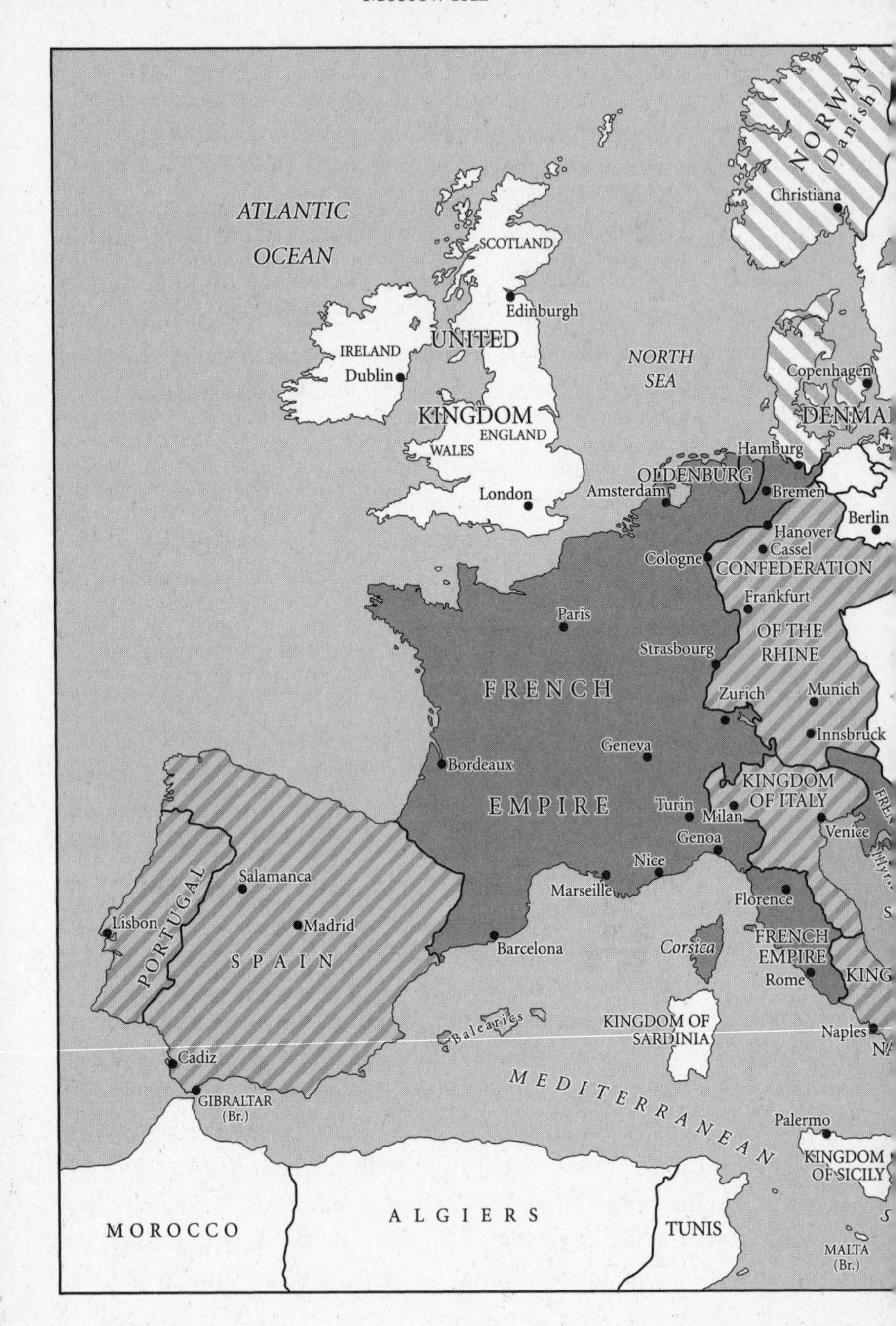
ATLANTIC
OCEAN
SCOTLAND
Edinburgh
IRELAND
Dublin
UNITED
KINGDOM
ENGLAND
WALES
London
NORTH
SEA
NORWAY
(Danish)
Christiana
Copenhagen
Hamburg
OLDENBURG
Amsterdam
Bremen
Berlin
Hanover
Cassel
Cologne
CONFEDERATION
Frankfurt
OF THE
RHINE
Paris
Strasbourg
FRENCH
EMPIRE
Zurich
Munich
Innsbruck
Geneva
Bordeaux
KINGDOM
OF ITALY
Turin
Milan
Venice
Genoa
Nice
Marseille
Florence
PORTUGAL
Salamanca
Lisbon
Madrid
SPAIN
Barcelona
Corsica
FRENCH
EMPIRE
Rome
Balearics
KINGDOM OF
SARDINIA
Naples
Cadiz
GIBRALTAR
(Br.)
MEDITERRANEAN
Palermo
KINGDOM
OF SICILY
MOROCCO
ALGIERS
TUNIS
MALTA
(Br.)

EUROPE IN 1811
St Petersburg
Stockholm
Nizhny Novgorod
Riga
BALTIC SEA
Moscow
Tilsit
Smolensk
Konigsberg
Vilna
Friedland
RUSSIAN EMPIRE
Minsk
GRAND DUCHY OF WARSAW
Warsaw
Kiev
Cracow
EMPIRE
Pest
Odessa
HUNGARY
MOLDAVIA
TRANSYLVANIA
Brassó
WALLACHIA
Belgrade
Bucharest
Ruschuk
MONTENEGRO
Sofia
Constantinople
OTTOMAN EMPIRE
Corfu
Ionian (Br.) Islands
Athens
Crete
Key
French Empire
French Satellites
French Allies

was at the height of her power, "if she cannot fix the political constitution of Europe now, she may tomorrow lose all the advantages of her position and fail in her enterprises."[12]

But a final settlement that would secure his gains for the future eluded him, partly because he kept expanding its scope, meaning it to be comprehensive, and partly because war was his element; he could not see his way to achieving his ends by other means. That was why all his treaties to date were no more than truces, and all his arrangements remained fluid pending the elusive ultimate peace settlement. The Empire was a work in progress.

At the time of the birth of his son, Napoleon was forty-two years old. He was five feet two inches tall, which was small even at the time, but he had a well-proportioned figure. "His complexion had never had much colour; his cheeks were of a matt white, giving him a full, pale face, but not of the kind of pallor that denotes a sick person," wrote his secretary Baron Fain. "His brown hair was cut short all over and lay flat over his head. His head was round, his forehead was large and high; his eyes were grey-blue, with a gentle look in them. He had a handsome nose, a graceful mouth and beautiful teeth." But he had recently begun to put on weight. His body filled out, his neck, which was short anyway, thickened, and he developed a paunch. Those close to him noted that his eyes grew less piercing. He spoke more slowly and took longer to make decisions. His phenomenal powers of concentration diminished, and those used to his fits of fury were surprised to find him growing more pensive and hesitant. Something was eating away at the vital force of this Promethean creature. It has been convincingly suggested that his pituitary gland failed as he reached the age of forty, causing dystrophia adiposogenitalis, a condition that leads to weight gain and loss of energy.[13]

It is impossible to say whether Napoleon himself was aware of any decline. His enemies had certainly noted that his victories were no longer as resounding as they had been, and he must have realised this as well. Even if this was not quite the twilight of his life, the end of his active career could not be that far off, so the final battles would have to be fought soon and a permanent settlement put in place in the near future.

The principal obstacle to such a settlement was Britain, with which France had become locked in a self-perpetuating duel. With her control of the seas, Britain could cripple French trade and support resistance anywhere on the European mainland, as she was currently doing in Spain. After the annihilation of his fleet at Trafalgar in 1805 Napoleon could not hope to confront the British navy in battle. He had therefore decided to ruin her economically, by closing the whole Continent to her trade.

The idea was not new. It was one of the fundamental French beliefs that Britain's wealth came not from herself but from her colonies, which supplied commodities she could sell on to Europe at vast profit. Every conflict between Britain and France over the past century had included a tariff war, and the revolutionary government and the Directory inherited this tradition. As there was widespread commercial jealousy of Britain, this was a popular policy. Napoleon carried on this tradition, setting ever higher tariffs and eventually banning all British trade from the Continent.

In theory, the French policy was bound to bring about economic hardship in Britain that would undermine support for her war effort. The Whigs, currently in opposition, had sympathised with the revolution in France and opposed the waging of war against her, and many admired Napoleon himself. Although they were in a minority, their calls for peace with France might well have carried the day if British trade had really begun to suffer. But in the long run, France probably suffered more than Britain. And Napoleon's Continental System, as he called it, was in effect unenforceable. Smuggling and corruption holed it even in French ports, while some of France's dependent states and allies were hardly enforcing it at all.

Worse, it imposed real hardships on the populations of subject and allied states. Nowhere more so than in the very area France most needed to control. Germany was feeling the cost very keenly, and political discontent was mounting. Although most of the sovereigns who ruled there were strongly attached to the French cause, the mood of their people might make them think twice if an alternative became possible. Such a situation might arise if French power were challenged, but there were only two powers capable of mounting such a challenge – Britain, which could not gain a serious foothold on mainland Europe, and Russia, which was an ally of France.

But Russia was not a happy ally, and nobody realised better than Napoleon that if she were to break out and challenge his authority, Britain could never be brought to the negotiating table, and the whole of Germany would be destabilised. Russia was therefore the key, and she would have to be brought back on side before any final settlement could be achieved. What he could not appreciate was that it was already far too late for that, and that even as French society was looking forward to a golden age of peace, Russia was coming to see war with France as unavoidable, desirable even, while her ruler was entertaining dreams of his own for the regeneration of Europe.

2

Alexander

When Catherine the Great came to the throne, exactly half a century before 1812, Russia had been of little significance outside the immediate area of eastern Europe. Peter the Great had done much to modernise his kingdom, and he put it on the map by building a fancy new capital at St Petersburg. In 1721 he even awarded himself the title of Emperor. But he was succeeded by a series of largely ineffectual monarchs, most of them ushered in through disreputable palace revolutions. They were feared by their subjects but generally despised by the other rulers of Europe, none of whom recognised the imperial title Peter had assumed.

Catherine changed all that. She worked hard at organising the state, involved herself in the affairs of Europe, and initiated an aggressive foreign policy which over the next fifty years was to add the whole of Finland, what are now Estonia, Latvia, Lithuania, Belarus and Ukraine, most of Poland, the Crimea, some of what is now Romania, the Kuban, Georgia, Kabardia, Azerbaidjan, parts of Siberia, Chukchi and Kamchatka to her dominions, as well as part of Alaska and a military settlement just north of San Francisco. This not only increased the size and population of Russia, it also brought her frontiers six hundred kilometres further into Europe and her rulers into European affairs. By 1799 Russian armies were operating in Switzerland and Italy. In a memorandum to Catherine's successor Paul I, the

ICELAND
FAROE ISLANDS
(Danish)
NORWAY
(Danish)
ATLANTIC
OCEAN
SCOTLAND
Edinburgh
IRELAND
Dublin
GREAT
BRITAIN
ENGLAND
WALES
London
NORTH
SEA
Copenhagen
DENMA
Hamburg
Amsterdam
Hanover
Cologne
Paris
Frankfurt
Strasbourg
FRANCE
BAVARIA
Munich
Innsbruck
Geneva
Bordeaux
Turin
Milan
Genoa
Venice
Nice
Marseille
Florence
PORTUGAL
Lisbon
Madrid
SPAIN
Barcelona
Corsica
Rome
Balearics
SARDINIA
Naples
Cadiz
MEDITERRANEAN SEA
GIBRALTAR
(Br.)
Palermo
Oran
Algiers

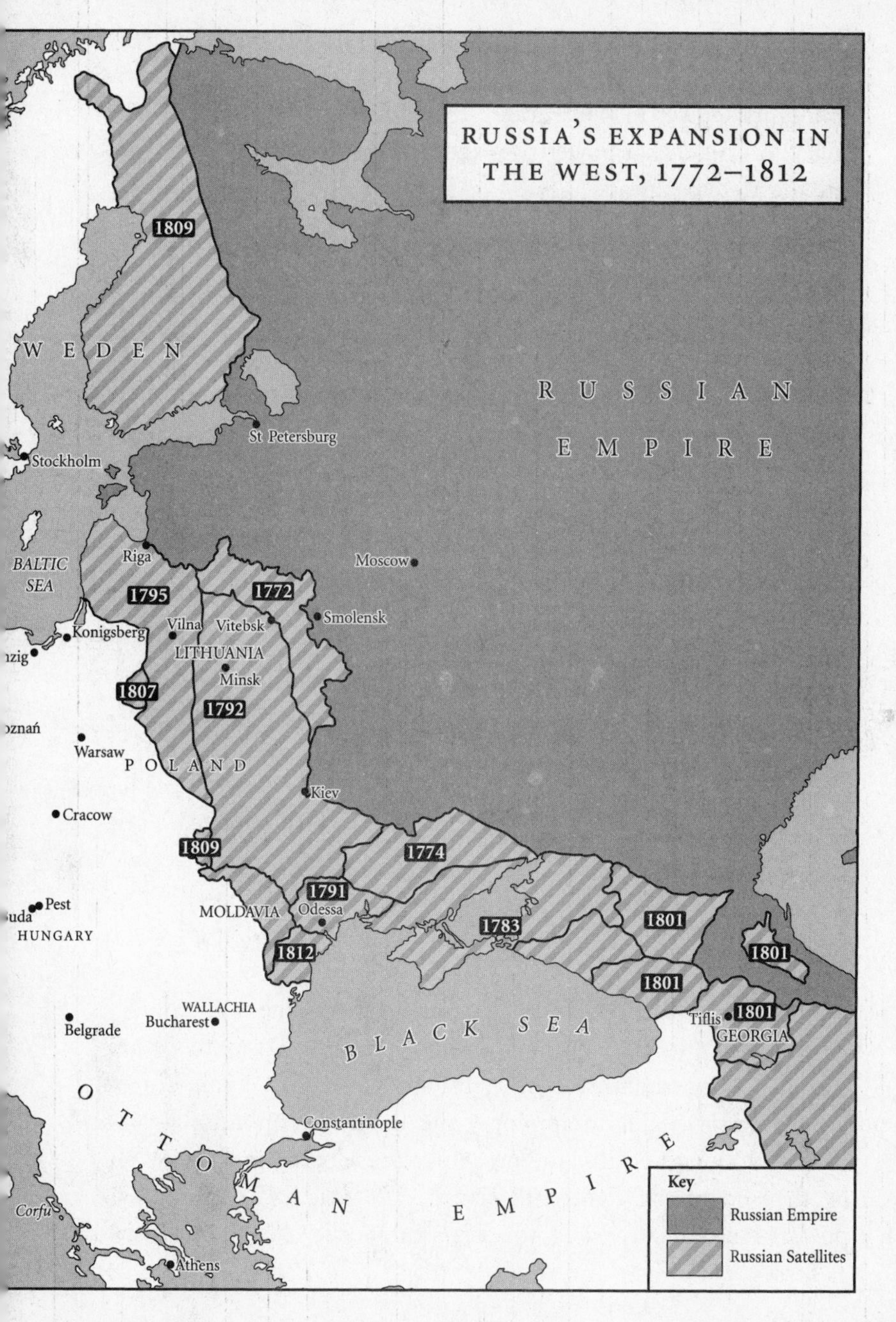
RUSSIA'S EXPANSION IN THE WEST, 1772–1812
RUSSIAN EMPIRE
1809
WEDEN
St Petersburg
Stockholm
BALTIC SEA
Riga
Moscow
1795
1772
Smolensk
Konigsberg
Vilna
Vitebsk
LITHUANIA
Minsk
1807
1792
Poznań
Warsaw
POLAND
Kiev
Cracow
1809
1774
1791
Odessa
MOLDAVIA
Pest
HUNGARY
1812
1783
1801
1801
1801
1801
WALLACHIA
Bucharest
Belgrade
BLACK SEA
Tiflis
GEORGIA
Constantinople
OTTOMAN EMPIRE
Corfu
Athens
Key
Russian Empire
Russian Satellites

Russian Chancellor Fyodor Vasilievich Rostopchin wrote: "Russia, as much by her position as by her inexhaustible resources, is and must be the first power in the world."[1] It was a constant aim of Russian policy to extend that power over the Balkans, Ottoman Turkey and into the Mediterranean.

Many in Europe were alarmed at this seemingly inexorable onward march of Russian power. There was talk of ravening Asiatic hordes and some fear, particularly after the first partition of Poland in 1772, that Russia might engulf the whole of Europe as the barbarians had done with ancient Rome. "Poland was but a breakfast . . . where will they dine?" Edmund Burke wondered, echoing the fears of many.[2] Diplomats were struck by the single-mindedness and ruthlessness of Russia's foreign policy: she did not play by the same rules as others. What few appreciated was the extent to which Russia saw herself as a special case.

When Ivan IV, popularly known as the Terrible, was crowned in the Uspensky cathedral in the Kremlin in 1547, he took the title of Tsar (Caesar) and laid claim to the legacy of Byzantium. "Ivan was claiming not only sovereignty, independence from other powers," in the words of Geoffrey Hosking, "but the actual superiority of his realm, as the universal Christian monarchy, to all others on earth."[3] He used the regalia of Byzantium and had himself depicted alongside Roman emperors. His successors and their political servants remained faithful to this legacy and the mission it imposed. It was not for nothing that Catherine had named her two eldest grandsons Alexander and Constantine.

France had traditionally kept a string of allies in the east – Sweden, Poland and Ottoman Turkey – whose purpose it was to contain the then dominant threat of Habsburg power in central Europe. When Russia began to impinge, she depended on this "*barrière de l'est*" to guard against the new threat developing in the east. But by the end of the eighteenth century, Sweden had declined as a power, Poland had ceased to exist, and Turkey had been pushed out of the Crimea and Moldavia, and was in a state of political decay. France would have to look elsewhere for allies.

In 1801 General Bonaparte, who was then First Consul, decided to make an ally of Russia herself. When, during negotiations on the exchange of prisoners, the British and Austrians refused to accept seven thousand of their Russian allies taken prisoner by the French in Switzerland in exchange for French prisoners they were holding, Bonaparte offered them free to Tsar Paul. He even volunteered to clothe and arm them. Paul, who had previously held everything to do with revolutionary France in abhorrence, was as disarmed by this chivalrous gesture as he was annoyed by the mean-mindedness of his Austrian and British allies. Bonaparte, who knew how much the Russians lusted for a harbour in the Mediterranean, followed this up by offering Paul the island of Malta (which was about to be captured by the British anyway). He would at this stage even have contemplated awarding Constantinople to Russia in order to enlist her support against Britain. He was well on the way to achieving this when, on the night of 23–24 March a group of generals and court officials forced their way into Paul's bedroom in the Mikhailovsky Palace in St Petersburg and murdered him.[4]

Paul had been mentally and emotionally unstable, if not actually mad, and there was an open sense of relief in Russia at his death. Whenever his son and successor Alexander showed himself in public in the first weeks of his reign, he was mobbed by people kissing his hands and clothes, and Pushkin later wrote of "the magnificent dawn of Alexander's days." But while he stands out among the monarchs of his day by his generous nature, his lack of vindictiveness and his hatred of injustice and cruelty, Alexander was also marked by severe psychological problems.

Though not unintelligent, he suffered from an inability to think through the consequences of his words and actions. This need not have mattered much had it not been for the education his grandmother, Catherine the Great, had devised for him. She was a despot who admitted no liberal ideas in or near her dominions. Yet alongside mathematicians and priests, she engaged the services of the Swiss republican *philosophe* Frédéric César de La Harpe as tutor for her grandson. The child was subjected to a regime of moral education which consisted of the study of improving stories drawn from the scriptures, history and mythology, as well as a whole canon of secular Enlightenment morality. His limited mind could hardly have been expected to square the religious precepts with the profane, or to accommodate within the despotic reality the radical concepts preached by La Harpe. "This little boy is a knot of contradictions," Catherine commented, somewhat disingenuously, after a few years of this diet.[5]

Alexander's principal failings – vanity, weakness and laziness – also need not have mattered much, had it not been for the brand of moral education to which he was subjected, and which expanded his perceived duties well beyond his capacities. He had to keep notebooks, "archives of shame," in which he jotted down every failing, every piece of bad behaviour, every loss of temper or lack of diligence in study. "I am an idler, given over to irresponsibility, incapable of true thought, speech and action," the twelve-year-old notes on 19 July 1789. "Egoism is one of my shortcomings, and vanity its main cause; it is easy to see to what they might lead me if I give them a chance to develop," on 27 August.[6] This continuous self-flagellation only aggravated an innate sense of inadequacy.

When he came to the throne at the age of twenty-three, Alexander was a young man of great charm, burning with desire to improve the world. But as he struggled to live up to what he thought was expected of him, he was undermined by a terrible moral canker. His father's murderers had naturally made him a party to their plans, since it was in order to put him on the throne that they had decided to act. He would claim that he made them swear they would not kill Paul, but he was nevertheless an accomplice in the crime of parricide. He could hardly penalise them, so they continued to hold high office at court and rank in the army. Alexander was racked with guilt for the rest of his life for the part, however passive, he had played in the murder.

He was indeed a mass of contradictions. He claimed to despise the principles of hereditary monarchy, and recoiled before the necessity of assuming power. "My plan is to settle with my wife on the banks of the Rhine, where I shall live peacefully as a private person finding happiness in the company of friends and in the study of nature," he confided to one of his friends at the age of nineteen. But he soon fell out of love with his wife and with the notion of a tranquil, private life. He also used to hold forth on the liberal constitutions he was going to introduce. But once he had gained power, he grew jealous of letting anyone else have any say in how things should be done, and notoriously took offence whenever privileges and rights he had granted were actually invoked.[7]

Alexander wanted to bring an element of professionalism into the governance of the Russian empire through the introduction of institutional structures. He reorganised the civil service, making entrance into the higher grades dependent on a university degree or a written exam (which did not endear him to the nobility). He set up ministries and a State Council, which were supposed to help run the country. What he would have liked to introduce was something along the lines of the system Bonaparte was creating in France – authoritarian government mobilising the whole nation in an efficient way along rational and liberal lines. But this would have required emancipating the serfs and breaking down the entire social structure of Russia, and he lacked the nerve to implement it.

Absorbed as he was by internal reforms, Alexander paid little attention to foreign policy. He was horrified by Bonaparte's abduction and judicial murder of the Duc d'Enghien, and joined every other ruler in Europe in robust condemnation of the act. It offended every fibre in his chivalrous nature, and he felt the outrage personally: the Duke of Baden, on whose territory Enghien had been seized, was his father-in-law. He therefore couched his condemnation in grandiloquent terms. But he was made to regret it. The French response was to remind the world that Paul's assassins had not only never been punished, but actually held high office at his son's court, thereby putting in question Alexander's right to point the finger at anyone, in view of the part he had played in the murder of his own father. Alexander was stung, and hated Bonaparte for it. When Bonaparte took the title of Emperor a few months later, Alexander's hatred turned to indignant rage, and the bearer of Peter the Great's invented title denounced that taken by the upstart Corsican.

Alexander believed that Europe had reached a crisis, moral as well as political, and wrote to the British Prime Minister William Pitt suggesting a reorganisation of the Continent into a league of liberal states founded on the sacred rights of humanity. Pitt was not interested in the scheme, but he pandered to Alexander, and, allowing him to dream of greater things, in 1805 managed to enrol him into the third coalition against Napoleon: Austria and Russia were to attack France, and Britain would pay for it.

Russia had no reason for going to war with France, as none of her interests were threatened, and France was Russia's cultural beacon. Russian society was divided on the matter. While those who regarded Napoleon as an evil being who had to be crushed were probably in the majority, there were plenty who thought otherwise. The former Chancellor Count Rostopchin was vociferous in his criticism, propounding the view that Russia was being used by Britain; his future successor, Count Nikolai Rumiantsev, regarded France as Russia's natural ally. Napoleon had many admirers in Russia, particularly among the young – some of whom would be drinking his health even after the war had begun.[8]

But Alexander had come to see the whole question as part of a wider moral issue. He had assumed the role of knightly defender of a Christian monarchical tradition against the onslaught of the new barbarism as represented by Napoleon. An element of emulation also came into it, for he longed to distinguish himself on the battlefield. He had inherited his father's love of parades and the minutiae of military life – he was always checking details of uniforms and drill –

and believed that a Tsar's place was at the head of his troops. He therefore insisted on setting off to war in person, although he gave overall command of his armies to the only experienced general to hand, the fifty-eight-year-old Mikhail Ilarionovich Kutuzov.

Kutuzov had first seen action against Polish insurgents, and subsequently distinguished himself in several campaigns against Turkey. In 1773 in the Crimea he had received a bullet in the head which severed the muscles behind his right eye, causing it to sag in a grotesque way; eventually he lost sight in it. Kutuzov had been military governor of St Petersburg at the time of the murder of Tsar Paul, so he knew a thing or two about that. This was not the least of the reasons for which Alexander feared and resented him, and as a result he dismissed him and exiled him to his country estate. There, Kutuzov relieved his boredom and his rheumatic pains with drink and whatever sexual solace the rural retreat could provide his notorious appetite. And it was there, in the summer of 1805, that he suddenly received the order to take command of the army and join forces with the Austrians.

The army was not ready, so Kutuzov set off with an advance guard to reinforce the Austrian General Mack. Napoleon acted with speed and surrounded Mack, forcing him to surrender at Ulm while Kutuzov was still on the march. Massively outnumbered, Kutuzov was obliged to fall back and join up with the Russian main army, led by Alexander, and the remainder of the Austrian forces under the Emperor Francis.

Napoleon had never seen any good reason for France and Russia to fight, and was convinced that Alexander had been manipulated by Britain into joining the coalition. He therefore sent General Savary to the Tsar with the suggestion that they get together and sort out any differences amicably. But Alexander haughtily declined, famously addressing his reply to "the Head of the French Government," as he could not stomach acknowledging Napoleon's imperial title.

Kutuzov wanted to retreat further, but Alexander was determined to fight, and obliged him to give battle at Austerlitz on 2 December.

Like a subaltern playing at being commander, Alexander overruled Kutuzov's suggestions and made him adopt a plan devised by one of the Austrian generals. On the day, he bossed and chivvied Kutuzov for the slowness of his deployment, and then watched in horror as the allied army was routed. Forced to flee from the battlefield, Alexander was mortified. "He was himself even more thoroughly defeated at Austerlitz than his army," according to the French diplomat Joseph de Maistre.[9] The Tsar now resented Kutuzov all the more, and dismissed him from his command, giving him the minor post of Governor of Kiev.

Austria sued for peace, but the war went on, as Prussia joined the coalition. The thirty-five-year-old King Frederick William III had sat on the fence, until his beautiful and spirited wife Louise had finally induced him to come out against Napoleon. But in a whirlwind campaign in October 1806 his renowned army was routed at Jena and Auerstädt, and he had to flee his capital of Berlin. Napoleon entered the city and pursued Frederick William, who took refuge in East Prussia at the side of the Russian army, now under the command of General Lev Bennigsen.

Alexander showed remarkable determination in adversity. He raised more troops, and in 1807 called up a peasant militia. But he had to take precautions to ensure that these serfs would remain loyal to a system that kept them enslaved. News of the revolutionary happenings in France over the past fifteen years was slow to spread among the uneducated peasants of central and eastern Europe. But that very slowness meant that it often mingled with local legend and even religious millenarian longings as it went, with the result that the figure of Napoleon was sometimes confused with a number of mythic folk heroes, lending him the attributes not only of a liberator, but of a messiah as well. The Russian authorities were well aware of this, and prepared accordingly as the French armies drew close to the boundaries of the empire.

While calling on a high official in 1806 the writer Sergei Glinka had been intrigued to see a civil servant clutching a copy of the

Apocalypse. There was a long tradition in Russia of associating the enemy with the Antichrist in order to raise the fighting spirit of the soldiers, and now the authorities had hit on the idea of substituting Napoleon for the rulers of the abyss, Abaddon and Apollyon. In November 1806 the Holy Synod of the Russian Orthodox Church issued a thundering denunciation of Napoleon, accusing him of taking on the role and the name of Messiah and conspiring with Jews and other evil people against the Christian faith. The clergy also made much of the fact that when in Egypt Napoleon had declared his regard for Islam – it must be remembered that the Russians had been in a semi-permanent state of war with Muslim Tatars and Turks, which they saw as a kind of crusade. Thus the average soldier and peasant was given the impression that Napoleon was in league with all the devils of hell.[10]

But the crusade against him was cut short. In January 1807 Bennigsen lost 25,000 men in a fierce engagement at Eylau, and he was routed by Napoleon at Friedland in June. Alexander faced a stark choice. He could either fall back and try to regroup, which would involve letting the enemy into his own empire, or he could come to terms with Napoleon. His army was unpaid, unfed and badly officered, and the territory he would be falling back through, which had only been taken from Poland ten years before, was full of potential partisans.

On 24 June 1807 Alexander sent General Lobanov-Rostovsky to Napoleon's headquarters at Tilsit on the river Niemen with a personal message saying that he would be delighted to make not just peace but an alliance with him. "An entirely new system must replace the one which has existed up to now, and I flatter myself that we will easily reach an understanding with the Emperor Napoleon, provided that we meet without intermediaries," he wrote.[11]

Negotiations began the next day. A tented pavilion was constructed for the purpose on a raft moored in the middle of the Niemen. Alexander turned up in his most fetching uniform, determined to charm Napoleon and get himself out of the desperate straits he was in. For his part, Napoleon wanted to seduce Alexander in order to break up the coalition once and for all, and in the process gain a useful pawn in his struggle against Britain.

Alexander may have had great charm, but Napoleon was the better manipulator of men. He flattered Alexander shamelessly, treating him as an equal. He also spared no occasion of driving a wedge between him and his Prussian ally. Frederick William III had not been allowed on to the raft, and on the day negotiations opened he could be seen watching from the Russian bank, even at one stage edging his horse forward until it had water up to its chest, as though trying to eavesdrop. On the next day Napoleon relented and allowed Alexander to present Frederick William to him, but he was curt and did not invite him to the dinner he was giving for the Tsar that evening. He repeatedly told Alexander that he was only leaving the wretched King on his throne in deference to his, Alexander's, wishes. However much he might have been pained or shocked by such insults to a brother monarch, Alexander could not fail to be flattered at the difference in the status accorded to the two by Napoleon.

While the foreign ministers of both states negotiated the actual treaties, Napoleon and Alexander assisted at parades, went out for walks, drives and rides; they sat up together after dinner, talking far into the night. Napoleon would let drop the odd phrase about how Russia's frontier really ought to be on the Vistula, about a possible mutual carve-up of Turkey, about the two of them resolving all the problems of Europe together. He pandered to Alexander's dreams of reforming the world. He would unfold maps of Europe and Asia, and together they would speculate on ideal solutions to the world's ills through some monumental territorial rearrangement. Napoleon told of how he had modernised France, giving Alexander the impression that he too could achieve great things, that all the self-flagellation he had been obliged to perform before his tutor would finally be vindicated by some magnificent act.[12]

Alexander had grown up hating Napoleon and all he stood for, as did his family and court. On the day of the first meeting on the raft, his sister Catherine wrote to him vehemently denouncing Napoleon as a liar and a monster, urging him to have no truck whatever with him. But there is no doubt that the flattery of the conqueror of Europe, however monstrous he might be, had worked its magic. For Alexander, unsure of himself, aware of his inadequacies, brought up to think of himself as a failure, to be treated as an equal by a man who had achieved so much, whose very name made Europe tremble, was strong liquor. The subaltern sat at the table of the most successful general in history. "Just imagine my spending days with Bonaparte, talking for hours quite alone with him!" he wrote back to Catherine. "I ask you, does not all this seem like a dream?"[13]

And while Napoleon had set out with the most cynical attitude, he too seems to have fallen for Alexander's boyish charm and enjoyed being with him in an elder-brotherly way. They were also to some extent carried away by the epic nature of the proceedings. Their meeting on the raft, in full view of two great armies drawn up in parade uniforms on either bank; the banquets at which the two most powerful men in Europe drank each other's health and embraced, pledging to build a better world; the grenadiers of both armies mingling to drink the health of the emperors of the Orient and the Occident; the touching scenes as Napoleon, having asked the Russians to name their bravest ranker, pinned the order of the Légion d'Honneur on his breast, a gesture reciprocated by Alexander with the Cross of St George – were all so much playacting. But it was grand spectacle, and actors are notorious for being taken in by their own histrionics.[14]

In the treaties signed on 7 July at the conclusion of these three weeks of posturing, Russia ceded the Ionian Islands to France, but received a small part of Poland in return. She agreed to pull her troops out of the Danubian Principalities, while France negotiated a settlement with Turkey on her behalf. Most importantly, she allied herself with France in the war with Britain, promising to close her ports to all British trade unless Britain made a speedy peace with France by the end of the year.

The obvious loser at Tilsit was Prussia. Frederick William was only just allowed to keep his throne, in deference to the wishes of Alexander. He had to give up most of the territory Prussia had taken from Poland in the past decades, to pay France a huge indemnity for having made war on her, to reduce his army to a symbolic force, and to accommodate French garrisons all over his kingdom. With the Polish lands taken from Prussia, Napoleon formed the Grand Duchy of Warsaw, a new French satellite.

Considering he had been obliged to sue for peace, Tilsit was a triumph for Alexander: he had managed to avoid being treated as a defeated party. But a closer look at the treaty revealed it to be not a peace settlement, but the initiation of a new war and the foundation of a partnership, one which bound Russia more than it bound France. All the exciting nocturnal talk remained vapour hanging in the air, while Russia had committed herself to make economic war on Britain. And while the stationing of French troops on her territory was a humiliation and an expense for Prussia, it was clear to all but the most naïve that they were there to keep Russia in check and to shore up the newly-founded Grand Duchy of Warsaw. This in itself was an open challenge to Russia. It was tiny, but it was a potential kernel for the resurrection of the Polish state which had been wiped off the map only a decade before, and a chunk of which currently formed the whole western belt of the Russian empire.

Whatever else he had managed to save, Alexander did not have to wait long to find out that he had not, as far as his subjects were concerned, saved his face or Russia's honour. His sister Catherine called the treaty a humiliating climbdown, and his mother refused his embrace when he returned to St Petersburg. The court, already disapproving of his desertion of the popular Empress Elizabeth for his mistress Maria Antonovna Naryshkina, sensed a betrayal. The traditionalist aristocracy opposed any negotiation with the despised "upstart" and saw the treaty as a sell-out. Many felt Alexander had been made a fool of by Napoleon. The playwright Vladislav Aleksandrovich Ozerov wrote *Dmitry Donskoi*, a play whose historical heroics, applauded frantically by full houses, made Alexander look ineffectual.

Although the Russian army had been beaten by Napoleon, the younger officers felt a new confidence and entertained dreams of fighting on to ultimate victory, and consequently felt betrayed. The soldiers could not understand why their Tsar was suddenly embracing as an ally the man they had been told was the Antichrist. A vigorous whispering campaign against Alexander's conduct of policy was initiated by General Wilson, a British adviser formerly attached to the Russian army. Rumours of plots to depose or assassinate the Tsar were rife. "Take care, sire! You will end up like your father!" one of his courtiers warned him. As there had been so many palace revolts in the past century, many people assumed that the dissatisfied courtiers would reach for this "Asiatic remedy," as one diplomat called it. "I saw this prince enter the cathedral preceded by the assassins of his grandfather, surrounded by those of his father, and followed, no doubt, by his own," wrote a French émigré after attending Alexander's coronation. Such fears were probably exaggerated, but the possibility could not be discounted.[15]

Matters only got worse when, Britain having failed to make peace with France, Russia had to honour her undertaking and declare war on her. This went against the grain, and revealed the true implications and consequences of the Tilsit settlement. "Russia's alliance with Your Majesty, and particularly the war with England, has upset the natural manner of thinking in this country," Napoleon's ambassador reported from St Petersburg in December. "It is, one could say, a complete change of religion."[16] Alexander had difficulty in finding ministers whom he could trust to implement his policy. The only one wholeheartedly in favour of the French alliance was Count Nikolai Rumiantsev, who now became Foreign Minister.

It is difficult to know what Alexander really thought of Napoleon and of the arrangement reached at Tilsit, as he was learning to be more secretive and devious. But outwardly he had to pretend that he stood by the treaty and his friendship with the Emperor of the French. Feeling rejected by society, Alexander withdrew into himself, and, as he steeled himself against public opinion, he bandaged his hurt pride and swathed his vulnerable convictions in such spiritual scraps as had been left behind by his strange upbringing.

Ironically, the treaty signed at Tilsit also bore the germ of Napoleon's undoing. On the face of it, he had achieved a great deal. He had broken up the coalition and set up the Grand Duchy of Warsaw as a French marcher outpost, an ambiguous piece on the diplomatic chessboard, to be used aggressively against one or all of his potential enemies, or traded for something. It was a powderkeg laid under one of the bastions of Russia's position in central Europe, as well as a threat to Austria. The treaty had neutered Prussia, and left a strong French military presence in the area ready to intervene at the slightest sign of trouble. Above all, it was an affront to Britain, whose shipping was excluded from even more ports, and who could now find no allies on the European mainland. Napoleon felt the moment draw near when Britain would be obliged to negotiate with him. Shortly after signing the treaty he turned his attention to excluding Britain from the Iberian peninsula, and in November 1807 French troops entered Lisbon.

The crucial element in the Tilsit treaty was that it was meant to embody an alliance, a real *entente*, between the two emperors. Yet Napoleon did not know how to treat allies: he was used to vassals. And this alliance was a particularly unnatural one. It dispelled Russia's primal dream of continued expansion at the expense of Turkey; it placed a question mark over her possession of Poland; and it forced her to penalise herself by making economic war on Britain. Those Russians who did not care about the stain on their country's honour would feel the pinch in their purses. Russia had been pushed into a loveless and unequal marriage with France, and soon adopted the sullen resentment of the unhappy wife. Sooner or later, she would be unfaithful, and Napoleon would have to go to war again in order to bring her back to heel. And it is much easier to defeat and even dispossess countries than to force them to do one's bidding.

Napoleon had made Russia the cornerstone of his strategy. "The affairs of the whole world will be decided there . . . the general peace is to be found in Petersburg," he said to the special envoy he sent there after Tilsit.[17] For this crucial mission he chose one of his most trusted officers, General Armand de Caulaincourt, Master of the Horse. Caulaincourt was only thirty-four years old, but he had come a long way. The scion of an old noble family of Picardy, he had been brought up partly at the court of Versailles, which made him a little more acceptable to supporters of the *ancien régime*. He knew Russia, as he had already been sent to St Petersburg once by Napoleon, to negotiate with Paul. His brief was to keep the special relationship between Napoleon and Alexander, the "mood of Tilsit," alive by every possible means.

As Napoleon's ambassador extraordinary, Caulaincourt appeared in public at Alexander's side, sat at his table and enjoyed a position which singled him out from the rest of the diplomatic corps in the Russian capital. He spent lavishly on balls and dinners, and while Russian society avoided him at first, he soon seduced even the most obdurate. In an effort to replicate this situation in the French capital, Napoleon bought his brother-in-law Murat's Paris residence – furniture, silver, bedlinen and all – for an astronomical sum so that Alexander's ambassador, Count Tolstoy, should be comfortable on his arrival.[18] But Tolstoy remained cool, hardly able to conceal his disdain and dislike of Napoleon. His successor, Prince Aleksandr Borisovich Kurakin, a caricature of the boundlessly wealthy and profligate Russian grandee, nicknamed "*le prince diamant*," was hardly more amenable.

Feeling the atmosphere grow cool, Napoleon decided to dangle another bauble before Alexander. In a long letter on 2 February 1808 he laid before him a grandiose plan for a joint attack on the British in India, holding out a prospect of empire in the east. It was an old idea. As early as 1797 General Bonaparte had declared that the surest way to destroy Britain was by throwing her out of India, and when he sailed for Egypt in May 1798 he took with him atlases of Bengal and Hindustan. He wrote to Tippoo Sahib, the Sultan of Mysore, who was then fighting the British, promising to come to his aid.

"I was full of dreams, and I saw the means by which I could carry out all that I had dreamed," he confided two years later. "I saw myself founding a religion, marching into Asia, riding an elephant, with a turban on my head and in my hand the new Koran that I would have composed to suit my needs. In my undertakings I would have combined the experiences of the two worlds, exploiting for my own profit the theatre of all history, attacking the power of England in India, and, by means of that conquest, renewing contact with the old Europe. The time I spent in Egypt was the most beautiful of my life, for it was the most ideal." He felt that the East offered a grander stage on which to act out his destiny. "There has been nothing left to achieve in Europe over these last two centuries," he declared a couple of years later. "It is only in the East that one can work on a grand scale." Napoleon would far rather have emulated Alexander the Great than Charlemagne.[19]

In 1801 he had sold the idea of a joint march on India to Paul, who had actually begun moving troops towards the Caucasus as a preliminary, and he had touched on it again at Tilsit. Circumstances were now inviting. The ruler of Persia, Shah Fath Ali, whose recent capture of Kabul and Kandahar brought her armies closer to the British outposts in India, greatly admired Napoleon and wanted French arms and officers to modernise his army. He had sent an ambassador, who reached Napoleon's headquarters early in 1807, and in May a treaty of alliance was duly signed. General Gardane was sent to Persia as ambassador with a seventy-man military mission and instructions to survey the routes to India and map out convenient halting points. He came up with a route through Baghdad, Herat, Kabul and Peshawar.[20]

"If an army of 50,000 men, Russian, French, and perhaps even partly Austrian were to set off from Constantinople into Asia it would need to get no further than the Euphrates to make England tremble and fall at the feet of the continent," Napoleon wrote to Alexander on 2 February 1808. Caulaincourt noticed the Tsar's expression change and grow animated as he read the letter. "This is the language of Tilsit," Alexander exclaimed. He thrilled at the grandeur of the concept and seemed keen to participate.[21] But there would be no talk of the East at their next meeting, a few months later, as in the short term Napoleon needed his ally for another purpose.

A revolt had broken out against French rule in Madrid on 2 May 1808, and although this had been crushed with severity, insurrection had spread through the whole of Spain. A blow was dealt to French military prestige on 21 July when a force of some 20,000 men under General Dupont was cut off by a Spanish army and obliged to capitulate at Bailén. Exactly a month later, General Junot was defeated by the British at Vimiero in Portugal. Napoleon concluded that he must go to Spain and conduct operations in person. But he suspected that the moment he was fully engaged on the other side of the Pyrenees Austria would take the opportunity to make war on him. He therefore needed to make sure that his Russian ally was going to cover his back.

The two emperors agreed to meet at Erfurt in Thuringia. They arrived in the city on 27 September 1808 and spent the next two weeks in each other's company. Alexander was treated to the spectacle of Napoleon as the master of Europe, surrounded by the kings of Westphalia, Württemberg, Bavaria and Saxony, the Duke of Weimar and a dozen other sovereign princes, all doing obeisance. He sat through bombastic performances of classics by Corneille, Racine and Voltaire performed by the best actors of Paris, brought along specially for the purpose. Among them were some of the most celebrated beauties, whom Napoleon apparently tried to introduce into Alexander's bed. Napoleon had his troops parade before the Tsar, spent hours talking to him about administrative reforms, new buildings, the arts, and all the things he knew interested him. He took him off to visit the battlefield of Jena, and on the knoll from which he had commanded the action he gave a dramatic account of the battle. After this they sat down to a bivouac dinner, as though they were on campaign. Outwardly, Alexander appeared to be duly impressed. When the line "The friendship of a great man is a gift of the gods" rang out during the performance of Voltaire's *Oedipe* one evening, Alexander rose from his seat and ostentatiously took Napoleon's hand, while the whole audience applauded.[22] But it was all sham.

When Alexander had announced his intention of going to Erfurt, most of his entourage begged him not to go, knowing only too well his weakness and fearing that he would be forced into some new agreement. There was also a latent fear that he might never come back: only a few months earlier, Napoleon had invited the Spanish King Charles IV and his son to a meeting at Bayonne, and had promptly deposed and imprisoned them. The underlying fears are best expressed in a long letter the Tsar's mother wrote to him just as he was setting off. In measured tones that nevertheless betray a sense of despair, she implored him not to go, saying that his attendance on Napoleon would insult the dignity of every Russian and lose him their confidence. "Alexander, the throne is but poorly secured when it is not based on that strong sentiment," she wrote. "Do not wound your people in all that they hold most sacred and dear in your august person; recognise their love in their present anxiety and do not go voluntarily to bow your forehead adorned with the most beautiful diadem before the idol of fortune, an idol accursed of present and future humanity; step back from the edge of the precipice!" Again and again she came back to her real fear. "Alexander, in the name of God avoid your downfall; the esteem of a people is easily lost but not so easily regained; you will lose it through this meeting, and you will lose your empire and destroy your family . . ."

Alexander's reply was calm, well reasoned and Machiavellian in its clear-sightedness. He poured cold water on the enthusiasm aroused by Bailén and Vimiero, pointing out that they were of no significance, and that Napoleon was strong enough to conquer Spain and beat Russia, even if Austria were to come to her aid. The only course of action was to work at mobilising the power of Russia and wait patiently for the moment when that power, along with that of Austria, could be brought to bear in a decisive way. "But it is only in the most profound silence that we must work towards this aim, not by boasting of our armaments and preparations in public, or in loudly denouncing him whom we wish to defy," he explained. He pointed out that France would always prefer alliance with Russia to a state of conflict, and this meant that Napoleon would not harm him and would not move against Russia if she did not provoke him. He was afraid Austria might be tempted into going to war too soon, thereby sealing her own downfall and putting back for years the moment at which they could stand up to Napoleon effectively. He believed that by going to Erfurt and appearing to be ready to support France against her, he might make Austria think twice before launching an attack that was doomed to failure. "If the meeting were to have no other result than that of preventing such a deplorable calamity, it would compensate with interest for all the unpleasantness involved in it," he concluded. To his sister Catherine, he replied more succinctly. "Napoleon thinks that I'm just a fool," he wrote, "but he who laughs last laughs longest."[23]

Napoleon could have had no inkling of these thoughts, but he was unpleasantly struck by the change that had taken place in Alexander. He found him more self-possessed and annoyingly steadfast, and their interviews were nothing like those of Tilsit – so much so that one day Napoleon grew so heated in the discussion that he tore his hat from his head, threw it on the floor and stamped on it.[24]

Alexander had come to Erfurt looking for some advantage or concession with which he could justify his apparent subjection to Napoleon to sceptics at home. But Napoleon was not in a giving mood. He deflected Alexander's plans for expansion in the direction of Constantinople, as he had come to the conclusion that any division of the Ottoman Empire would benefit Russia far more than France. He allowed Alexander to hang on to Moldavia and Wallachia, and to take Finland from Sweden. He agreed to withdraw French troops from the Grand Duchy of Warsaw and to start evacuating his garrisons in Prussia. But that was the sum total of his concessions.

Alexander did not openly challenge the basis of the alliance, and agreed to act out the role of faithful ally with respect to the Austrian threat. "The two emperors parted relatively satisfied with their arrangements, but, at bottom, dissatisfied with each other," in the words of Caulaincourt.[25]

Having, as he thought, secured a degree of support from Alexander, Napoleon turned his attention to Spain, where he went in November. On 4 December he was in Madrid, and from there he set about pacifying the country. Just as he had anticipated, Austria seized the opportunity of his back being turned, and in April 1809 invaded the territory of his Bavarian and Saxon allies.

Napoleon recrossed the Pyrenees and marched to their defence. On 21 May he confronted the Austrian army at Essling. The battle was little short of a defeat for Napoleon, dimming the aura of invincibility that hung about him and giving heart to all his enemies. On 6 July he won the decisive battle of Wagram and dictated a treaty with Austria. But he was far from satisfied. Alexander, on whose assistance he had called as soon as he heard of the Austrian attack, had been slow to respond, and his army had taken an eternity to reach the theatre of operations. When it did so, it began executing a series of military minuets aimed at avoiding the Austrian forces until all was over. It was so successful that it suffered just one casualty during the entire campaign.

Napoleon had taken Alexander for granted, and was now paying the price. He would henceforth have to make more of an effort to bring his ally back on side, and he began to consider what concessions he might make to him. But he had no idea of how far Alexander had strayed from his influence. He certainly did not know that his own Foreign Minister, Talleyrand, had been involved in secret talks with the Tsar at Erfurt. "It is up to you to save Europe and you will only achieve this by standing up to Napoleon," Talleyrand claimed to have told Alexander. What Talleyrand probably did not know was that the Tsar had already come to see himself as being locked in a personal contest with Napoleon. Instead of acquiring a useful ally, Napoleon had helped to create a formidable rival, one who was already working at supplanting rather than merely defeating him. "There is no room for the two of us in Europe," Alexander had written to his sister Catherine before setting off for Erfurt; "sooner or later, one of us will have to bow out."[26]

3

The Soul of Europe

That Alexander could be beginning to think of himself as a counterweight or even an alternative to Napoleon on the international stage is eloquent testimony to what a mess the Emperor of the French had made of his dealings with the other nations of Europe, and with the Germans in particular.

France's had long been the dominant intellectual and cultural influence on the Continent, and by the end of the eighteenth century progressives and liberals of every nation fed on the fruits of her Enlightenment. The fall of the Bastille on 14 July 1789, followed by the abolition of privilege, the declaration of the Rights of Man, the introduction of representative government and other such measures elicited wild enthusiasm among the educated classes in every corner of Europe. Even moderate liberals saw revolutionary France as the catalyst that would bring about the transformation of the old world into a more equitable, and therefore more civilised and peaceful one.

The horrors of the revolution put many off, and others were offended by France's high-handed behaviour with regard to areas, such as Holland and Switzerland, caught up in her military struggle against the coalitions lined up against her. But the French were convinced that they were engaged on a mission of progress, bringing happiness to other nations. So, in a more pragmatic way, was Napoleon, who used to say that "What is good for the French is good for everyone." Liberals everywhere clung to the view that a process of transformation and human regeneration was under way, and that casualties were only to be expected. Those suffering foreign or aristocratic oppression continued to look longingly at the example set by France. With some justification.

The political boundaries criss-crossing much of Europe at the end of the eighteenth century and the constitutional arrangements within them were largely the legacy of medieval attempts at creating a pan-European empire. Germany was broken up into more than three hundred different political units, ruled over by electors, archbishops, abbots, dukes, landgraves, margraves, city councils, counts and imperial knights. What is now Belgium belonged to the Habsburgs and was ruled from Vienna; Italy was divided up into eleven states, most of them ruled by Austrian Habsburgs or French and Spanish Bourbons; the Holy Roman Empire of the German Nation included Czechs, Magyars and half a dozen other nationalities; and Poland was cut up into three and ruled from Berlin, Vienna and St Petersburg.

Every time a French army passed through one of these areas, it disturbed a venerable clutter of archaic law and regulation, of privilege and prerogative, of rights and duties, releasing or awakening a variety of pent-up or dormant aspirations in the process. And every time France annexed a territory she reorganised it along the lines of French Enlightenment thought. Rulers were dethroned, ecclesiastical institutions were abrogated, ghettos were opened, guild rights, caste privileges and other restrictions were abolished, and serfs and slaves were freed. Although this was often accompanied by cynical exploitation of the territory in the French cause and shameless looting, the net effect was nevertheless a positive one in the liberal view. As a result, significant sections, and in some cases the majority, of the politically aware populations of such countries as Belgium, the Netherlands, Switzerland, Italy, Poland and even Spain ranged themselves in the camp of France against those seeking to restore the *ancien régime*, even if they resented French rule and decried the depredations of French troops. Nowhere more so than in Germany.

The Holy Roman Empire, founded by Charlemagne a thousand years before, included almost all the lands inhabited by German-speaking people, but it did not bring them together or represent them. The absurd division of the territory into hundreds of political units inhibited cultural and economic, as well as political life. German eighteenth-century thought was cosmopolitan rather than nationalist, but most educated Germans nonetheless longed for a more coherent homeland.

Between 1801 and 1806, following his victories over Austria and Prussia, Napoleon thoroughly transformed the political, social and economic climate throughout the German lands. He secularised ecclesiastical states and abolished the status of imperial cities, swept away anachronistic institutions and residues of gothic rights, in effect dismantling the Empire and emancipating large sections of the population in the process. In 1806, after his defeat of the Emperor Francis at Ulm and Austerlitz, he forced him to abdicate and to dissolve the Holy Roman Empire itself. In a process known as "mediatisation," hundreds of tiny sovereignties were swept away as imperial counts and knights lost their lands, which were fused into thirty-six states of varying size, bound together in the Confederation of the Rhine. With them went all the nonsensical borders and petty restrictions that had made life so difficult. In their place came institutions moulded on the French pattern.

The ending of feudal practices gave agriculture a boost, the abolition of guild and other restrictions encouraged industry and trade, the removal of tolls and frontiers liberated trade. The confiscation of Church property was followed by the building of schools and the development of universities. Not surprisingly, all this made Napoleon popular with the middle classes, with small traders, peasants, artisans and Jews, as well as with progressive intellectuals, students and writers. Johan Wilhelm Gleim, a poet more used to singing the glories of Frederick the Great, wrote an ode to Napoleon, Friedrich Hölderlin also immortalised him in verse, and Beethoven dedicated his "Eroica" symphony to him.

Although many were put off by his decision to take the imperial crown and some even felt betrayed by the act, German intellectuals continued to be fascinated by Napoleon, whom they saw as a figure in the mould of Alexander the Great. Some hoped he would revive the old German empire like a latter-day Charlemagne. To others, he appeared as some kind of avatar. The young Heinrich Heine imagined Christ riding into Jerusalem on Palm Sunday as he watched Napoleon making his entry into his native Düsseldorf. Georg Wilhelm Friedrich Hegel famously identified him as "the world-spirit on horseback."

But that moment, just after the victories of Jena and Auerstädt, in which Napoleon destroyed the Prussian army and shook the Prussian state to its core, was to be something of a turning point. The Prussians were shocked and insulted by the French victories, but they also saw them as proof of the superiority of France and her political culture. When Napoleon rode into Berlin he was greeted by crowds which, according to one French officer, were as enthusiastic as those that had welcomed him in Paris on his triumphant return from Austerlitz the previous year. "An undefinable feeling, a mixture of pain, admiration and curiosity agitated the crowds which pressed forward as he passed," in the words of one eyewitness.[1] Napoleon won the hearts of the Berliners as well as their admiration over the next weeks.

But he treated Prussia and her King worse than he had treated any conquered country before. At Tilsit he publicly humiliated Frederick William by refusing to negotiate with him, and by treating Queen Louise, who had come in person to plead her country's cause, with insulting gallantry. He did not bother to negotiate, merely summoning the Prussian Minister Count Goltz to let him know his intentions. He told the Minister that he had thought of giving the throne of Prussia to his own brother Jérôme, but out of regard for Tsar Alexander, who had begged him to spare Frederick William, he had graciously decided to leave him in possession of it. But he diminished his realm by taking away most of the territory seized by Prussia from Poland, so that the number of his subjects, which had grown to 9,744,000, was reduced to 4,938,000. Napoleon would brook no discussion, and Frederick William had to submit.[2]

Having done so, he wrote to the Emperor on 3 August 1807 entreating him to accept Prussia as an ally of France, addressing him as "the greatest man of our century." Napoleon ignored the request. The reason he did not wish to encumber himself with such an ally was that he intended to despoil the country. In the treaty he had foisted on Prussia, he had undertaken to evacuate his troops, but only after all the indemnities agreed upon had been paid. But the level of the indemnities was never agreed, and while vast amounts of money did pour out of the Prussian treasury into French coffers, some 150,000 French troops continued to live off the land, happily helping themselves to everything they required. French military authorities virtually supervised the administration, while the economy plummeted. The Prussian army had been reduced to 42,000 men, with the result that hundreds of thousands of disbanded soldiers and even officers wandered the land begging for their subsistence.[3]

Napoleon did consider abolishing Prussia altogether. The kingdom had only emerged as a major power sixty years before (as a result of a French defeat), but it was efficient and expansive, and might one day rally the rest of Germany, which was something he wanted to avoid at all costs. But while he continued to exploit and humiliate it in every way, he did not get around to dismantling it. In effect, Napoleon's treatment of Prussia is paradigmatic of his whole mishandling of the German issue, for which his successors were still paying in 1940.

If Frederick William had every reason to feel aggrieved, most of the other rulers in Germany, grouped in the Confederation of the Rhine, had much to thank Napoleon for. For one thing, they were relieved to be rid of the heavy-handed Habsburg overlordship. Although they were now subjected to Napoleon through a series of alliances, they had grown in power within their own realms. Several had even been promoted, and most had gained in territory, becoming proper sovereigns with their own armies.

Landgrave Ludwig of Hesse-Darmstädt had seen the size of his fief swell, and became a grand duke; the tiny Landgravate of Baden had also become a grand duchy, and its ruler Frederick Charles willingly married his grandson to Napoleon's stepdaughter Stephanie de Beauharnais. The Elector of Saxony had seen his realm expand and turn into a kingdom. Bavaria too was enlarged and turned into a kingdom, and in 1809 King Maximillian I acquired more territory, making his realm larger than Prussia. Württemberg, which had been a mere duchy, was extended with every Napoleonic victory and its elector Frederick was promoted to the rank of king in 1806. He was only too happy to see his daughter marry Napoleon's brother Jérôme.

Jérôme himself ruled over the Kingdom of Westphalia, created by Napoleon at the heart of Charlemagne's Germany with its capital at Cassel, extended again in 1810 to include Hanover, Bremen and part of the North Sea coast. "What the people of Germany desire impatiently is that individuals who are not noble but have talents should have an equal right to your consideration and to employment, that all kinds of servitude and all intermediary links between the sovereign and the lowest class of the people should be entirely abolished," Napoleon wrote to Jérôme as he took up the throne of Westphalia. "The benefits of the *Code Napoléon*, transparency of procedures and the jury system will be the distinguishing characteristics of your monarchy. And if I have to be quite open with you, I count more on their effect for the extension and consolidation of your monarchy than on the greatest victories. Your people must enjoy a liberty and equality and a well-being unknown to the other peoples of Germany," he continued, making it clear that the security of his throne and that of France were better served by this great benefit she was able to bestow than by any number of armies or fortresses.[4]

Some of the other rulers did follow the French example and adopted the *Code Napoléon*. King Maximillian of Bavaria even brought in a constitution. Most of them, however, only introduced those French laws which gave them greater power over their subjects, sweeping away in the process venerable institutions and hard-won privileges. But whether they were enlightened liberals or authoritarian despots like the King of Württemberg, their subjects were immeasurably better off in every way than they had been before they had heard of Bonaparte.

Causes for discontent nevertheless began to pile up. The most vociferous opponents of the new arrangements were, unsurprisingly, the horde of imperial counts and knights who had lost their estates and privileges. More liberal elements were disappointed that the changes wrought by Napoleon had not gone far enough. The old free cities and some of the bishoprics, which had been havens of German patriotism, had been awarded to one or other of the rulers Napoleon had favoured. Along with their independence they lost some of their freedoms. Many were disappointed that the old aristocratic oligarchy had not been replaced by republics, and some would have liked to see the creation of one great German state.

The high-handedness of the arrangements, with Napoleon callously shunting provinces from one state to another, could not fail to offend Germans at every level. French became the official language in some areas. French officials were placed in key posts, and the higher ranks in the armies of the various sovereigns were reserved for Frenchmen. The large-scale official looting was also highly offensive. French military impositions and the Continental System, which actually had the effect of stimulating the coalmining and steel industries in Germany, became a cause for everyday grumbling by the very classes that naturally supported the changes brought in by Napoleon.

Cultural factors also played a part. Cosmopolitan and outward-looking as the Germans were, they were generally, whether they were Catholics or Protestants, very pious, and they found the godlessness of revolutionary and Napoleonic France shocking. In Lutheran circles, the ribbon of the Légion d'Honneur was even referred to as "the sign of the Beast." Napoleon was more popular amongst Catholic Germans, until June 1809, when he dispossessed the Pope and imprisoned him in Savona, drawing upon his head the Pontiff's excommunication. The Germans also nurtured an age-old sense of their "otherness," a vision of themselves as "true" and "pure" in contrast to the French, whom they viewed as essentially flighty and artificial, if not actually false and corrupt.[5]

It was not long before these feelings began to have practical consequences. Her catastrophic defeat in 1806 had prompted Prussia to embark on a far-ranging programme of reform and modernisation. Those in charge of carrying it out realised that a real revolution was required, both in the army, where the soldier was transformed from a conscript motivated entirely by ferocious eighteenth-century discipline into a professional inspired by love of his country, and in society as a whole, where an edict passed in 1807 swept away the remnants of feudalism and emancipated the peasantry.

This was to be a revolution from above, carried out, in the words of Frederick William's Minister Count Karl August von Hardenberg, "through the wisdom of those in authority" rather than by popular impulse. It was also to be a spiritual revolution. One of its chief architects, Baron vom Stein, a mediatised knight, wanted "to reawaken collective spirit, civic sense, devotion to the country, the feeling of national honour and independence, so that a vivifying and creative spirit would replace the petty formalism of a mechanical apparatus."[6]

The process was largely carried out by German nationalists from other parts of the country. Baron vom Stein was from Nassau, Count Hardenberg was from Hanover, as was General Gerhard Johann Scharnhorst; Gebhart Blücher was from Mecklemburg, August Gneisenau was a Saxon. They were inspired by the example of revolutionary France in their determination to infuse a national spirit into every part of the army and administration. But their reforms aimed not so much at emancipating people as at turning them into efficient and enthusiastic servants of the state. Many of them believed that only a strong Prussia would be able to liberate and unite the German lands, and then go on to challenge French cultural and political primacy. A powerful tool in this was to be education, and Wilhelm von Humboldt was put in charge of a programme of reform of the system that culminated in the opening of a university in Berlin in 1810.

At a popular level, the urge to seek regeneration through purification manifested itself through the formation of the *Tugendbund*, or League of Virtue, by a group of young officers in Berlin. Its aims were non-political in principle, consisting of self-perfection through education and moral elevation, but since this included the fostering of national consciousness and the encouragement of love of the fatherland, they were deeply so in practice. The membership never exceeded a few hundred, and all they did was sit around talking of insurrection, guerrilla war and revenge. But it is in the very nature of secret societies to appear more powerful and threatening than they actually are, and the *Tugendbund* had profound symbolic significance.

It also acted as an inspiration and a focus to disaffected elements in other parts of Germany. The German nation's impotence in the face of the arrogance of the French was underlined as the cost of the Continental System made itself felt. Wounded pride turned into grim determination in the minds of many German patriots, and it received its first encouragement with the news of Bailén in the summer of 1808. "The events in Spain have had a great effect and show what can be done by a nation which has force and courage," Stein wrote to a friend.[7]

Napoleon was well aware of the new spirit at work in Germany. He was not particularly concerned by it, but he did, during his stay in Erfurt and Weimar at the time of the meeting with Alexander in 1808, make a desultory effort to garner some popularity, inviting professors from the university of Jena to lunch with him. He decorated Goethe with the Légion d'Honneur. He had the poet Christoph Martin Wieland brought to Weimar, and spent upwards of two hours discussing German literature with him during a ball, while a circle of astonished guests looked on. He then walked over to Goethe and engaged him in conversation. The event was commented on in the court bulletin, which explained that "the hero of the age thereby gave proof of his attachment to the nation of which he is the protector, and that he esteems its language and literature, which are its national binding force." But the next day he visited the battlefield of Jena, on which he had made the Germans build a small temple to commemorate his triumph over them.[8]

In 1802, the German philosopher Friedrich Schlegel had gone to Paris with the intention of founding an international institute of learning in this new Rome. Now he was looking more to Germany. Goethe, who wore his Légion d'Honneur with pride and used to refer to Napoleon as "my Emperor," was also beginning to complain of the shameful state of submission into which Germany had been forced. The philosopher Johann Gottlieb Fichte, the writer Ernst Moritz Arndt and the theologian Friedrich Daniel Schleiermacher were among those who called for a national German revival and a rejection of the French hegemony. Many of those who had seen Napoleon as a liberator now saw in him nothing but an oppressor.

There had been a predictable surge of national feeling in Austria following her defeat by Napoleon in 1806, with papers and pamphlets calling for a united German front against the French. Austria's natural desire to avenge the humiliating defeat and regain some of her losses had been powerfully reinforced by the many disgruntled mediatised counts and knights, the deposed north Italian and particularly Piedmontese nobles and the many German patriots from the Confederation of the Rhine who had taken refuge and in many cases service there. In January 1808 the Emperor Francis married for a third time. His bride, Maria Ludovica of Habsburg, was the daughter of the Captain-General of Lombardy, who had been thrown out by Napoleon, and this was not the least of her reasons for loathing the French.

The new government under Count Philip Stadion appointed by Francis in 1808 began preparing for a confrontation with France, instituting, amongst other things, a national militia, the *Landwehr*. This war of revenge, Stadion made clear, was to be a national, German one, aimed at expelling France and her influence from central Europe altogether. While Maria Ludovica and a poet and fitness enthusiast called Caroline Pichler reinvented a supposedly traditional German form of dress, the *Tracht*, the historian Johannes von Müller, the publicist Friedrich von Gentz and others underpinned the anti-French arguments with facts. They made much of what they saw as the struggle of the Spanish people against foreign domination, holding it up as an example for the Germans to follow. Authors of every kind were invited to police headquarters, where they would be asked to use their pens in the national cause, and publishers of periodicals were instructed to print patriotic poems and articles, on pain of having their publications closed down. Unbidden, the poet and dramatist Heinrich von Kleist published "*Die Hermannschlacht*," a poetic appeal to Germans and Austrians alike to rise up against the French and to punish all pro-Napoleonic "traitors."[9]

In April 1809, judging Napoleon to be bogged down in Spain, Austria invaded Bavaria and launched a war for the "liberation" of Germany. "We fight to assert the independence of the Austrian monarchy, to restore to Germany the independence and national honour that belong to her," declared Stadion in his manifesto. The commander-in-chief Archduke Charles issued a proclamation penned by Friedrich Schlegel which dwelt on the pan-German character of the war, representing it as an opportunity for the redemption and regeneration of the nation.[10]

Their call did not go unanswered. A Prussian officer, Frederick Charles de Katt, attempted to seize Magdeburg with a gang of partisans, but failed and was forced to take refuge in Austrian Bohemia. Colonel Dornberg, a Hessian serving in King Jérôme's royal guard who had been plotting with Stein, Gneisenau and Scharnhorst, intended to seize Jérôme and call the population to arms. In the event, he only managed to raise six to eight hundred men and was easily defeated. This was bad news for Major Schill, a Prussian officer who had distinguished himself in 1806–1807 by his determined defence of Kolberg. On 28 April 1809 he marched out of Berlin with his regiment, telling his men that he was going to invade Westphalia and evict the French from Germany. He was expecting to link up with Dornberg, who should by then have seized Jérôme, but he soon found himself facing superior forces and was obliged to retreat to the Baltic coast, where he hung on, vainly hoping for British seaborne support, until he was killed in a skirmish on 31 May.

An altogether more serious response came in the Tyrol, where resentment of the French ran much deeper. The area had traditionally been governed by the Habsburgs with much respect for local tradition and idiosyncrasies, but Bavaria, to which it was transferred by Napoleon in 1806, operated a more centralised administration. The locals were offended by higher rates of taxation and by enforced conscription. The parish clergy did not approve of the secularisation taking place in Bavaria, adding to the discontent. In January 1809 Andreas Hofer and a handful of other Tyrolese went to Vienna to prepare an insurrection to coincide with Austria's invasion of Bavaria. On 9 April beacons were duly lit and the Tyrol rebelled. A Bavarian corps of two thousand men was forced to capitulate, and Austrian forces occupied Innsbrück. But they were soon ejected from it by the French under Marshal Lefèbvre.

On 21–22 May Napoleon fought the twin battles of Aspern-Essling against the Austrians under Archduke Charles. Although technically a French victory, they reverberated through Europe as a defeat. Napoleon suffered a personal loss in the death of Marshal Lannes, and had to bring Lefèbvre back to join the main army. This allowed the revolt in the Tyrol to erupt with renewed vigour, under the slogan "God and the Emperor," which had enemies of Napoleon all over Europe rubbing their hands at what they thought was a new Spain.

At this point, the Duke of Brunswick-Oels appeared on the scene. His father had been ignominiously defeated at Auerstädt in 1806, and he had vowed eternal hatred to the French. He had gone to Vienna, where he obtained a subsidy in order to raise a 20,000-strong "Legion of Vengeance" with which he intended to liberate northern Germany. He now sallied forth, defeated the Saxons at Zittau, seized Dresden on 11 June and Leipzig ten days after that. On 21 July he marched on, through Brunswick and Hanover, but he met with little enthusiasm, and was eventually forced to take refuge on a British man-of-war in the Baltic.

In the meantime, Napoleon had won the conclusive battle of Wagram, and Austria was forced to sign the Treaty of Vienna, which reduced it to a state of powerlessness. Her image as a potential liberator of Germany was shattered, and she settled down meekly within the Napoleonic system. Francis was only too happy to pay tribute by giving his favourite daughter to the Corsican ogre, and the marriage was hailed as a happy event by his people.

Austria's failure stemmed in large measure from her inability to engage the support of Russia, and above all to draw Prussia into the war against the French. The pan-German plotters had been active in this respect, and Vienna had been in close touch with Stein, Hardenberg, Scharnhorst and the other Prussian reformers, who were doing everything to bounce Frederick William into declaring war. But the pusillanimous Prussian King was afraid. He was afraid of the French, and he was afraid of starting a "national" war that might end up by costing him his own throne. It was only when, with Schill marching into Westphalia and popular opinion at a high pitch of excitement, he thought he might lose his throne if did not act that he considered going to war.

One of the stipulations of the Treaty of Vienna was that Francis had to banish all French émigrés, Piedmontese, and Germans from other states who had settled or taken service in Austria. A number, including Karl von Grolmann, a Prussian officer who had come to fight for Austria, now headed for Spain, where they could carry on their crusade against Napoleonic France. Many more took the St Petersburg road, already trodden by some of those German patriots who had gone to Prussia in the hope that she might become the champion of Germany. With both Prussia and Austria discredited, Alexander was beginning to look like the only alternative. He was still an ally of Napoleon, and had acted as such by sending an army to threaten Austria during the recent war. But he had done only the minimum demanded of him. Assisted by large doses of wishful thinking, many of those opposed to Napoleon and French hegemony had begun to see in Alexander a tutelary angel of their own particular cause.

One of the first to fall for Alexander, when she had met him in 1805, was Frederick William's Queen, Louise, who saw him as "a Schiller hero come down to earth." "In you, perfection is incarnate," she wrote to him; "one must know you to know perfection." The feeling did not go unrequited, which was probably what had saved Prussia from extinction at Tilsit. But there was little more that Alexander could do for her and her dismal husband. In January 1809 he invited them to St Petersburg, where he honoured and fêted them, thereby sending out a strong signal to all Napoleon's enemies in Europe. The mutual esteem between Alexander and Louise grew. When he heard of her death in Prussia in July 1810, he saw her as a victim of Napoleon's barbaric oppression and reacted with requisite chivalry. "I swear to you that I shall avenge her death and shall make certain that her murderer pays for his crime," he is alleged to have said to the Prussian Minister in St Petersburg.[11]

Alexander was also viewed as a potential saviour by other humiliated or dispossessed monarchs and nobles, including the kings of France, Sardinia, the Two Sicilies, Spain, the Grand Master of the Order of Malta, a gaggle of dispossessed Germans, as well as hordes of French, Piedmontese, Spanish and other émigrés.

This did not prevent elements more or less violently opposed to the *ancien régime* from looking to him as well. Many of the Germans who placed their hopes in Alexander were republican or at least liberal nationalists in conflict with the Prussian monarchy. The same went for *Tugendbunders* and even the Freemasons, who were regarded by Frederick William as dangerous subversives. Among the Spaniards and Italians who placed their hopes in the Tsar were liberals who would in time be clamped in irons by their own monarchs.

Other unlikely members of the club were French liberal opponents of Napoleon, such as Benjamin Constant and Madame de Staël, who, while subscribing to most of the achievements of the French Revolution, hated him for his despotic tendencies and for the cultural arrogance with which he treated Europe. Her bestselling novel *Corinne*, published in 1807, was a thinly veiled criticism of French doings in Italy, while her treatise on German literature, *De l'Allemagne*, was so implicitly critical that the first printing was confiscated on Napoleon's orders.

His behaviour and his policies were rapidly losing Napoleon the dominion over hearts and minds he had enjoyed in earlier years, while a great *internationale* of alienated people all over Europe was gathering, bound together only by their detestation of him. Even Wellington was beginning to see and to portray his war against Napoleon in Spain as a part of some kind of moral crusade.[12]

None of this was of any immediate consequence, and Napoleon's position in Europe was still paramount. He controlled his vast imperium through a web of loyalty, beginning with his crowned brothers. He had created all over Europe a new international aristocracy beholden to him, endowed with fiefs which had fallen vacant through the mediatisation of the Holy Roman Empire or through conquest, and by 1812 the imperial almanac listed four princes, thirty dukes, nearly four hundred counts and over a thousand barons, not including the titles Napoleon had given to members of his own family.

It is also worth noting that he had many natural allies bound to him by self-interest of one sort or another. Frederick William and Hardenberg feared the social upheaval that might result from any national revival more than they resented Napoleon. Others in Germany and Europe as a whole feared the relentless onward march of Russian expansion and believed that a weakening of French influence would entail Russian hegemony, and distrusted Alexander's motives.[13]

Napoleon's spies nevertheless kept a close watch on potential subversives all over Germany and on the support they were receiving from Russia. By the summer of 1810 he was growing irritated by the numbers of Russians visiting European courts and capitals trying to incite people against France. In Vienna, the former ambassador Count Razumovsky, the *salonière* Princess Bagration and Napoleon's old Corsican enemy Pozzo di Borgo, now in Russian uniform, made up a real propaganda network between them. Others were rallying anti-Napoleonic sentiment in the watering places of Germany. He asked Alexander to recall them all to Russia, but received scant satisfaction.

In November 1810 Talleyrand's successor as Foreign Minister, Jean-Baptiste de Champagny, reported to Napoleon that "a vast revolution" was brewing in Germany, fuelled by national hatred of France. This gathered in strength as tension between France and Russia mounted, and as the Continental System began to bite. Although he tended to make light of the threat, Napoleon was beginning to take more serious note of it, and declared his intention of "uprooting the German national spirit." And the only way he would be able to "uproot" this burgeoning growth was by cutting off its chief source of nourishment, which came from Russia.[14]

4

The Drift to War

At Erfurt, Napoleon had in an offhand way asked Caulaincourt what he thought Alexander might think of a dynastic union between the two empires. He did not seem to attach much importance to the matter, but came back to it a couple of times. This did not surprise Caulaincourt. Ever since Napoleon had assumed the imperial crown, the question of an heir had presented itself, and as the Empress Josephine was no longer of childbearing age, there had been much talk of a divorce. After Tilsit, gossip had it that he might marry one of the Tsar's sisters to cement the new *entente*.

Alexander had two unmarried sisters, the Grand Duchess Catherine, who was charming, witty and highly regarded, and the Grand Duchess Anna, who was only fourteen years old. Before his assassination, their father had issued a special *ukaz* giving his consort, now the Dowager Empress, absolute power to decide whom their daughters married. She loathed Napoleon, and in 1808, no doubt alarmed by the gossip, quickly found a husband for Catherine. Shortly after Alexander's return from Erfurt, the Grand Duchess was married to Prince George of Holstein-Oldenburg.

This did not bother Napoleon, who had anyway been thinking of the younger sister. He was in no hurry, and he wanted to keep his options open. It also suited Alexander, permitting him to express a degree of enthusiasm for the idea of the marriage, knowing that he need not commit himself for some years.

But as the cracks in the alliance began to show, Napoleon decided to paper them over with a dynastic union. At the end of November 1809 he instructed Caulaincourt to approach Alexander with a request for his sister's hand. Alexander's response was positive, but he took the matter no further. When Caulaincourt pressed him for a definite answer, he asked for two weeks to consider the matter and to gain his mother's approval. At the end of the two weeks, he asked for another ten days. Then another week. At the beginning of February 1810 he was still stalling, saying that his mother objected on the grounds that Anna was too young. Napoleon, who felt insulted by the lack of enthusiasm he sensed in Alexander and was beginning to suspect that he would never agree to the match, decided to pre-empt the humiliation of a refusal by turning to Austria instead.

He had sounded out the Austrian court on the subject in a vague way in the previous year, so he could now act with speed. After reading the despatches from Caulaincourt describing Alexander's negative response on the morning of 6 February, he summoned Prince Karl von Schwarzenberg, the Austrian ambassador in Paris, and pinned him down for a binding decision straight away. Schwarzenberg seized what he believed was a historic chance, and, overstepping his powers, gave him the reply he wanted. The courier bearing Napoleon's letter to Alexander notifying him of the change of plan crossed with one from the Tsar bearing a letter in which he in effect refused Napoleon's offer, explaining that his mother felt there could be no question of marrying off Anna for at least two years.

When he heard the news of Napoleon's betrothal to Marie-Louise, Alexander assumed that he had been carrying on parallel negotiations with Austria all along, and was stung by the apparent duplicity. It is almost certain that Napoleon would have preferred to marry the Grand Duchess Anna, as it would have had tremendous resonance as a symbolic marriage of East and West reuniting the two halves of the Roman Empire. But as his marriage to Marie-Louise went ahead, absorbing the attention of Europe with its pomp and *éclat*, Alexander was made to look ridiculous in front of his own people. He had stood up for the *entente* with Napoleon in the face of almost universal opposition at home, only to end up in the role of jilted party. And the marital junketing going on in Paris appeared to hide a deeper threat.

At the marriage feast the Austrian Chancellor Count Metternich, who was representing his imperial master, stood up and raised his glass "To the King of Rome!," thereby expressing the hope that Napoleon would produce an heir, and ceding the old imperial title to the house of Bonaparte. From St Petersburg it looked very much as though France and Austria were entering into an alliance even closer than the special relationship forged at Tilsit. An unpleasant sign of the way public opinion was swinging was that a Russian loan which Alexander was trying to float on the Paris exchange in order to raise much-needed funds suddenly found no subscribers. And the new situation had other implications.[1]

When Alexander was approached on the subject of the marriage of his sister, he had let it be known that he would make his agreement conditional on a convention ruling out forever the restoration of a Kingdom of Poland. Napoleon had responded positively, quite happy to trade Poland for Anna. But now Alexander had lost his main bargaining counter in this matter of crucial importance.

Napoleon's creation of the Grand Duchy of Warsaw in 1807 had, in effect, introduced the first material conflict of interest between France and Russia. The new political unit inevitably raised the possibility of a restoration of the Kingdom of Poland. Such a restoration would entail the loss by Russia of some if not all of her acquisitions at the expense of Poland in the partitions – an area of 463,000 square kilometres with a population of some seven million.

The creation of the Grand Duchy of Warsaw had also raised the spectre of another threat to Russia through the introduction of the *Code Napoléon*. This transformed social relationships and would lead to the full emancipation of the peasants. The landowners of Russia,

95 per cent of whose population were serfs, could not look on such a neighbour with equanimity.

The Poles, whether they were citizens of the Grand Duchy or not, certainly saw it as the nucleus of a restored Kingdom of Poland, and there was much dreaming and plotting in provinces still under Russian or Austrian rule. When Austria went to war with France in 1809, one of her armies had seized Warsaw but had then been obliged to fall back, pursued by the Poles, who proceeded to march into Galicia, a part of Poland annexed by Austria.

In the peace settlement, Napoleon had allowed the Poles to incorporate a small part of the liberated territory into the Grand Duchy of Warsaw. This alarmed the Austrians, who feared they might in time have to give up the rest, and annoyed the Poles, who felt they should have been allowed to take back the whole area they had liberated. They were, moreover, outraged by the fact that Napoleon had also ceded a piece of it to Russia as a sweetener.

But the Russians were not mollified. Caulaincourt reported that everyone in St Petersburg, from the Tsar down, was adamant that no part of Galicia should be added to the Grand Duchy, as this would set a dangerous precedent. "All the news coming from Moscow and the provinces confirms that this agitation is universal: it is necessary to take up arms and die, they are saying, rather than suffer the reunification of Galicia with the Grand Duchy," he wrote. And the issue transcended loyalty to the Tsar, whom many did not trust.* "There is not the slightest restraint in the allegations being made against the Emperor Alexander; there is open talk of assassinating him," reported Caulaincourt. "I have not seen minds so agitated since my arrival in Petersburg."[2]

Napoleon never had any intention of restoring Poland – all his

* While many assumed that Napoleon's creation of the Grand Duchy owed much to his infatuation at the time with his mistress Maria Walewska, now many in Russia suspected Alexander's mistress Maria Antonovna Naryshkina, also a Pole, of exercising a similar influence.

statements to the contrary date from later, when he was trying to keep the Poles on his side or to pluck straws of self-justification from the wind. He therefore proposed to Alexander that they sign a joint convention binding themselves not to encourage the Poles in their dreams. As a sign of his discouragement of Polish aspirations, he sent the best units of the army of the Grand Duchy, the Legion of the Vistula, to fight in Spain.

But Alexander produced a draft convention which would excise the words "Poland" and "Poles" from all official correspondence, ban the wearing of Polish decorations and forbid the use of Polish emblems in the Grand Duchy. He wanted Napoleon to pledge that he would never allow the restoration of Poland, and that he would take up arms against the Poles if they attempted it. Napoleon replied that while he could declare his opposition to such a revival, he would not and could not undertake to hinder it. The wording suggested by Russia was nonsensical, as it bound France to pledges she would be in no position to carry out. He pointed out that he could have re-established Poland if he had wished to in 1807, and added the whole of Galicia to the Grand Duchy in 1809, but had not done so because he had no intention of doing so. Nevertheless, tens of thousands of Poles had fought loyally alongside the French for over a decade, inspired by their hopes of a free motherland and by France's sympathy to their cause. To sign the text suggested by Russia would "compromise the honour and dignity of France," as Napoleon put it to Champagny.[3]

Alexander continued to insist on his draft rather than the more general one proposed by Napoleon. Hoping to put pressure on the Emperor to acquiesce, he dropped hints that he might not find it so easy to keep up the blockade against Britain without wholehearted support from him. But his increasingly urgent insistence with regard to the convention, as well as the suspicions he voiced, revealed how little he trusted Napoleon, who began to wonder what lay behind it all. "One cannot conceive what aim Russia might have in mind in refusing a version which accords her everything she wants in favour of one which is dogmatic, irregular, contrary to common prudence, and which, ultimately, the Emperor cannot subscribe to without dishonouring himself," he wrote to Champagny on 24 April 1810.[4]

On 30 June, when Champagny brought a communication from St Petersburg with a list of complaints and a renewed demand that he sign the Russian draft of the convention, Napoleon lost his temper. He summoned the Russian ambassador, Prince Kurakin. "What does Russia mean by such language?" he demanded. "Does she want war? Why these continual complaints? Why these insulting suspicions? If I had wished to restore Poland, I would have said so and would not have withdrawn my troops from Germany. Is Russia trying to prepare me for her defection? I will be at war with her the day she makes peace with England." He then dictated a letter to Caulaincourt in St Petersburg telling him that if Russia was going to start blackmailing him and using the Polish question as an excuse to seek a *rapprochement* with Britain, there would be war.[5]

It was the first time Napoleon had mentioned war, and it was a remark thrown out in the heat of the moment. The last thing he wanted was war with Russia. Russia, on the other hand, was increasingly looking forward to one. Russian society had been hostile to the French alliance from the start, and attitudes had only hardened over the years. The reasons were cultural and psychological rather than strategic.

Russia was a young society, and its upper echelons consisted of a rich social and ethnic mix. At court, in the administration and in the army old boyar families jostled with a new aristocracy whose origins lay in the political instability and culture of favouritism of the past century, which had produced grand aristocratic families such as the Razumovskys and the Orlovs, only a couple of generations from the servants' quarters or the barrack room. To this, conquest and annexations had added Germanic Baltic barons, Polish nobles, Georgian and Balkan princes, while the need for talent in the rapidly expanding state had sucked in immigrants from many lands. It was a mobile society, highly dynamic, but also beset by cultural insecurity.

Over the past hundred years educated Russians had drawn heavily on French culture. To them more than to any other European society, France was the fount of civilisation. The nobility were brought up by French tutors on French literature, and spoke French amongst themselves.* Few of them had any more Russian than was needed to give orders to servants. French books were as widely read in Moscow and St Petersburg as in Paris. Fluency in French was mandatory for anyone wishing to make a career in the army or the administration. The only senior officer in the Russian army in 1812 not to speak French fluently was General Miloradovich, who was of Serbian extraction, and Alexander prided himself on the fact that his French was better than Napoleon's.[6]

Underpinning this francocentrism was a huge colony of teachers, artists, musicians, tailors, dressmakers, cabinetmakers, jewellers, dancing masters, hairdressers, cooks and servants, some of whose parents or grandparents had settled in Russia and established dynasties. From the beginning of the revolution in France they were joined by thousands of French émigrés, some from the highest aristocracy, many of whom took service in the Russian army.†

Knowledge of French meant much more than just a linguistic skill. It implied familiarity with the literature of the past hundred years and with the ideas of the Enlightenment, as well as with all the pseudo-spiritual and occult fads of the day. Freemasonry had spread through the upper reaches of Russian society, and its more spiritual offshoots, such as Martinism,‡ were embraced with enthusiasm. Alexander himself had close relations with many Masons, and even founded a lodge consisting of himself, Rodion Kochelev, a follower of Saint-Martin and Swedenborg, and Aleksandr Nikolaevich Galitzine (whom he

* It is true that some teachers at Russian universities did use German or Latin.

† According to some sources, almost certainly apocryphal, even Captain Bonaparte applied for a post in the Russian army in 1795.

‡ Martinists were followers of the illuminist philosopher and mystic Louis Claude de Saint-Martin, whose obscure writings had a surprisingly wide influence.

had also appointed Procurator of the Holy Synod). This seemingly paradoxical situation was symptomatic of a wider malaise, for the dependence on French culture sat uneasily with a visceral attachment to traditional Orthodox values. And while French culture ruled and society spoke French, dressed French and aped the French in everything, there had always been a concurrent resentment of France herself. The revolution had magnified this resentment, and the events of 1805–1807 had turned it into something of a national movement.

For most young officers, military service had meant little more than attending parades (the non-commissioned officers did all the training, so all they had to do was lead their men) and court festivities. The rest of the time was given over to gaming, drinking and womanising. They underwent hardly any training or military instruction. "We had no sense of morality, an entirely false conception of honour, very little true education and, in almost every case, a surfeit of foolish high spirits which I can only call depraved," wrote Prince Sergei Volkonsky, a junior officer of the Chevaliergardes.[7]

They marched away to war in 1805 as though they were off to a hunting party. Some, like Tolstoy's Prince Andrei, dreamed of emulating Napoleon. They were routed at Austerlitz. They were defeated at Pułtusk and two other minor battles in the following year; in 1807 they lost the bloody battle of Eylau, and were finally vanquished at Friedland.

Most of the Russian officers took these defeats very badly. The campaign had been a sobering experience, and they had begun to grow up. Even the most depraved of the aristocratic layabouts felt a spark of patriotism flare inside them, and the valour of their soldiers had awakened a novel respect for these serfs in uniform. They felt humiliated at the apparent facility with which the French could inflict defeat on them however hard they fought, and their resentment of them was heavily tinged with an inferiority complex which shines through their writings on the subject. Lieutenant Denis Davidov and his brother officers were outraged when the Comte Louis de Périgord, the bearer of a letter from Marshal Berthier to General Bennigsen, did not remove his fur *kolpak* when ushered into the Russian general's presence. They saw it as an insult to Russia's honour, and developed a dogged determination to go on fighting until they finally won a battle against France. They regarded the peace of Tilsit as something akin to a betrayal.[8]

The wounded pride of these officers was reflected by a sense of humiliation felt by sections of the nobility back home. Once nations have embarked on the pursuit of great-power status they begin to develop a curious perspective on what represents a threat to their very existence. And the Russians were fast catching up with the French in this respect. "Our land was free, but the air had grown heavier, we walked about freely but could not breathe," complained Nikolai Grech after Tilsit. "Hatred of the French grew apace." But there was more to it than mere hatred. There were the beginnings of a sense of mission. Ordinary backwoods xenophobia came together with anti-Masonic paranoia and the first stirrings of Romanticism to create a conviction that Russia was somehow different from other European countries, more spiritually alive, and that she should reject the mainstream (i.e. French) culture of Europe and go her own way.[9]

There was a flurry of pamphlets, passionately argued, semi-religious, deeply anti-French, advocating a return to Russian values, and in 1808 Sergei Glinka founded a new periodical, *Russkii Viestnik*, which was to be "purely Russian," and would oppose the treacherous philosophy of the West with the manners and virtues of old Russia, an imagined culture of idyllic innocence. Defenders of the Russian language joined the fray, and a discussion club, *Biesieda*, was founded by a group including the poet Gavril Romanovich Derzhavin to combat foreign influences in literature. Patriots denounced the employment of French tutors and chambermaids as "gallomania," and the retired Admiral Aleksandr Semionovich Shishkov called for children to be brought up in traditional Russian ways. Alexander's sister Catherine, whose German husband George of Oldenburg had been given the post of Governor of Tver, Yaroslavl and Novgorod, somehow contrived to become the *belle idéale* of the most fervent champions of Russian culture. They included the former Chancellor Count Fyodor Rostopchin and the historian Nikolai Karamzin, who used to refer to her as "the demi-goddess of Tver."[10]

In these circumstances, the creation by Napoleon of the Grand Duchy of Warsaw was as a red rag to a bull. Russia had actually gained a piece of Polish territory in the operation, but territory was not the only consideration. Orthodox Russian traditionalists tended to regard the Catholic and unmistakably Western Poles as the rotten apples in the Slav basket. Now the Polish inhabitants of Russia's western provinces, some of whom had only become subjects of the Tsar a dozen years back, could potentially form a terrible fifth column of Western corruption inside the Russian empire.

This kind of thinking gave rise to a paranoid conviction, voiced by Sergei Glinka and others, that France under the satanic leadership of Napoleon was bent on the subjugation of Russia, and that Tilsit and indeed any peace concluded with her was but a truce putting off the terrible day. The sense of paranoia was only intensified when, at the end of May 1810, the Swedes elected Napoleon's Marshal and kinsman, Jean Baptiste Bernadotte, Prince of Ponte-Corvo, to the position of Crown Prince and *de facto* ruler of Sweden.

With its colony in Pomerania, Sweden still ruled over more than half of the entire coastline of the Baltic Sea. She had lost Finland to Russia in 1809 and a constitutional crisis resulted in the half-mad Gustav IV being toppled in favour of Charles XIII. The new King was senile and childless, and in their search for a successor the Swedes turned to Napoleon for advice. He declined to involve himself in their internal affairs, and in the end they chose a man they believed he might have nominated, and whom they considered to be agreeable to him. Their mistake was to have momentous consequences.

Bernadotte was an old colleague of Napoleon. When the two were no more than aspiring officers he had succeeded, and possibly supplanted, the future Emperor in the affections of the lovely Désirée Clary, whom he had subsequently married. Désirée's sister Julie had married Napoleon's brother Joseph, which might have made for a happy family. But it did not. Bernadotte was jealous of his colleague's meteoric rise. While he happily accepted the rank of Marshal of France and the princely title Napoleon had bestowed on him, he cloaked his resentment in righteous disapproval of Napoleon's assumption of the imperial purple and his pursuit of conquest. Napoleon for his part had a low opinion of Bernadotte and once said that he would have had him shot on at least three occasions had it not been for the bond of kinship.[11]

When Bernadotte became Crown Prince of Sweden, Napoleon realised that he might prove less than cooperative, but assumed he would behave as a Frenchman and as a Swede. Sweden had traditionally been a close ally of France, and her natural enemies had always been Russia and Prussia. Only the previous year Russia had invaded and forced her to give up Finland after a protracted war. The Swedes' friendly feelings towards France were put under a certain amount of strain by the Continental System, but their long coastline permitted them to breach it and trade with Britain, while their Pomeranian colony on the northern coast of Germany meant that they could sell on to the German market with profit.

The Russians could only view the combination of the creation of the Grand Duchy of Warsaw, Napoleon's marriage to the daughter of the Emperor of Austria and the recent developments in Sweden as aggressive encirclement, and Bernadotte's election was greeted with uproar.

All these feelings were given added poignancy by the economic hardships caused by the Continental System, which had turned into a regular tariff auction. Britain had responded to Napoleon's Berlin decree of 1806 banning her ships from all ports under his control by declaring that any ship trading with a port from which her ships were excluded was fair game for confiscation by the Royal Navy. French, Spanish, Dutch and German traders tried to get around this by using neutral American vessels to carry goods, but Britain decreed that no vessel could be considered neutral if it were carrying goods between hostile ports. In order to get around this, American ships would pick up their cargoes, take them to an American port, unload them, reload them and take them to a European port. Britain refused to accept this as legal. Napoleon retaliated in December 1807 by decreeing that any ship which had put in at a British port or paid British duty was automatically liable to seizure. On 1 March 1809 the United States closed its ports to all British and French shipping, but Napoleon managed to reach an agreement with the Americans to the detriment of Britain, which would ultimately lead to the outbreak of hostilities between Britain and the United States in 1812.

Russia had little industry, and was dependent on imports for a huge variety of everyday items. These now had to be smuggled in via Sweden or through smaller ports on Russia's Baltic coastline. Her exports – timber, grain, hemp and so on – were bulky and difficult to smuggle. The Russian rouble fell in value against most European currencies by some 25 per cent, which made foreign goods exorbitantly expensive. Between 1807 and 1811, the price of coffee more than doubled, sugar became more than three times as expensive, and a bottle of champagne went from 3.75 to twelve roubles. Russian noblemen had to pay through the nose not only for champagne, but for everything they did not produce at home, and they could not find a market for the produce of their own estates.[12]

This cocktail of hurt pride and financial hardship produced ever more violent criticism of Alexander's policy and of his State Secretary, Mikhail Mikhailovich Speransky, who was virtual prime minister. Speransky was the son of a priest, a very able man of lowly social background, ascetic and devoid of any social or financial ambition. A radical at heart, he believed autocracy to be incompatible with the rule of law, and would have liked to carry out far-reaching reform of the structure of the state. But he accepted the limitations imposed by his position and concentrated on modernising the administration. Soon after his appointment in 1807 he had promulgated reform of the legal system, which was never implemented, of government finances and of the administration.

The nobility, who sensed an enemy in him, did everything to undermine his position. There were soon rumours circulating to the effect that Speransky was a Freemason and revolutionary secretly in league with Napoleon, and that he meant to bring the whole social system crashing down.

The Tsar of Russia was theoretically an all-powerful autocrat, but his relationship with his people was a complex and ambivalent one. There was a mystical, sacred foundation to his power, since he was both his subjects' religious hierarch and the representative of God on earth. This imposed strong bonds of obedience to him on them. But if a Tsar was felt to have betrayed his divinely ordained purpose, he became something worse than just a wicked Tsar – he became a devil who must be destroyed. At the secular level, his position was just as ambiguous. The very fact that all power was concentrated in him meant that he had no instruments with which to impose his will. He was thus in a curious way dependent on the goodwill of the nobility, which staffed the army and all the organs of state, and therefore on public opinion. And public opinion was by now strongly against Alexander and his policies on virtually every point. He was seen by many as the author of Russia's shame, and he realised that the only way he could wipe away that shame was through war. The conquest of Finland had helped slightly, but it was not enough.

On 26 December 1809, while he was assuring Napoleon that he would do everything to make the marriage to his sister Anna possible and begging him to bury the Polish question forever, Alexander summoned Prince Adam Czartoryski, a close friend and a prominent Polish patriot who had ten years before elaborated a plan for the restoration of the Kingdom of Poland under Russia's protection. The Tsar told him that he would now like to put this plan into action, by "liberating" the Grand Duchy of Warsaw and uniting it with the Polish provinces currently under Russian rule, and asked Czartoryski to sound out the Poles on the subject. The Prince did not need to do much research. He knew that the plan could only have worked in 1805 or in 1809, during Napoleon's war with Austria. He nevertheless went to Warsaw and saw the man who would be the key figure in such a plan – Prince Józef Poniatowski, commander-in-chief of the Grand Duchy's army and nephew of the last King of Poland. Predictably, Poniatowski rejected the Russian proposal.[13]

Czartoryski reported back to Alexander personally in April 1810. He pointed out that many Poles had got wind of Alexander's negotiations with Napoleon to prevent the restoration of Poland, and that this hardly inspired confidence. But the Tsar clung to his view that the Poles could be won over. "We are now in April, so we could begin in nine months' time," he concluded.[14]

Caulaincourt noticed during the winter of 1809–1810 that Alexander was less and less amenable to French policies, and by the spring of 1810 he was finding the friendship he had built up with the Tsar increasingly at odds with his ambassadorial role. He began to hint to Napoleon that he would like to be recalled. But Napoleon paid no heed to his warnings or his wishes.

He had persuaded himself that Britain was suffering economically, and that a few more months would probably bring her to the negotiating table. He therefore adopted a more aggressive attitude to the application of the Continental System. His correspondence bristles with detailed instructions to the rulers and administrators of the coastal areas under his control on which ships and goods to impound and which to allow through. He suggests alternative sources of the supplies cut off and explains the principle behind his policy, exhorting all to enforce it with strictness.

Adding insult to injury, Napoleon decided to recoup some of the cost to France of the system at the expense of others. He took a leaf from the smugglers' book and licensed a number of merchants to import goods from Britain (for which they paid a hefty price to his treasury), and these goods were then exported overland, many of them to Russia. Such procedures left Alexander with little option but to defy the system openly. On 31 December 1810 he issued an *ukaz* opening Russian ports to American ships and at the same time imposing hefty tariffs on (French) manufactured goods imported overland into Russia. British goods were soon pouring into Germany from Russia. The Continental System was in tatters. Yet Napoleon refused to accept this. "The Continental System is uppermost in his mind, he is more taken up with it than ever," noted his secretary Baron Fain early in 1811; "too much so perhaps!"[15]

In his determination to control all points of import, Napoleon annexed the Hanseatic ports. In January 1811 he did the same with the Duchy of Oldenburg, whose ruler was the father of Alexander's brother-in-law. He did offer him another German province as compensation, but this was refused. Alexander was outraged, and felt personally insulted – his supposed ally was now dethroning members of his family, thereby reinforcing the view, widely held in Russia, that Tilsit was not an alliance but a subjection. He felt he had to act, if only to safeguard his position at home. "Blood must flow again," he told his sister Catherine.[16]

On 6 January 1811 he wrote once more to Czartoryski, asking him to try persuading the Poles to accept him as their liberator and restorer. His Minister of War General Barclay de Tolly was already drawing up plans for a strike into the Grand Duchy followed by an advance into Prussia to link up with the Prussian forces.* In a second letter to Czartoryski, Alexander detailed the troops he had already massed on the border to support the operation: 106,500 in the front line supported by a second line of 134,000 men, and a third army of 44,000 men supplemented by 80,000 recruits who had already finished their training. These forces could, in case of need, be supplemented with a few divisions from the army operating against the Turks in Moldavia. "There can be no doubt that Napoleon is trying to provoke Russia into a break with him, hoping that I will make the mistake of being the aggressor," he explained. "It would indeed be a mistake in the present circumstances, and I am determined not to make it – but everything would look different if the Poles were to rally to me. Reinforced with

* In February a similar plan was submitted by General Bennigsen, while Generals Bagration, Württemberg, d'Allonville and Saint Priest also contributed theirs.

the 50,000 men whom I could count on from them, by the 50,000 Prussians who could then join me without risk, and by the moral revolution which would unfailingly result in Europe, I would be in a position to reach the Oder without striking a blow."[17]

Alexander's troop movements could hardly be kept secret, and by the summer of 1811 the forthcoming war was being widely discussed all over Russia. His agitation in Poland, as well as the soundings his diplomats were taking in Vienna and Berlin, were no secret either. This has prompted some to conclude that he was in fact bluffing. But whether he meant to attack at this stage or not, he had taken a step which could not fail to lead to armed confrontation.[18]

Napoleon had to take the threat seriously. He had already been alerted by Poniatowski to Russian troop concentrations along the border of the Grand Duchy in the autumn of 1810, and he was desperately aware of the weakness of his forces in the area. He immediately instructed commanders on the spot to draw in exposed units and supply dumps against a surprise attack, and designated a fallback position along the Vistula while he set about strengthening his forces in Poland and Germany. He began bombarding Marshal Davout, in command of the French troops in northern Germany, with letters telling him to fortify strongpoints and put his men on a war footing. On 3 January 1811 he began regrouping his forces with the aim of strengthening the front line. "I considered that war had been declared," he later affirmed. Most people in France too considered it only a matter of time. "There is much talk of war here; sooner or later it must come to that, and now the time seems propitious," an officer of the Chevau-Légers of the Imperial Guard wrote from the depot at Chantilly to his sister on 9 April 1811.[19]

At the same time, Napoleon did everything he could to avert a conflict. In February he instructed Caulaincourt to demand an interview with Alexander and his Foreign Minister Rumiantsev, and to assure them that he wanted the alliance to continue, and that he would never make war on Russia unless she were to ally herself with Britain. In April he repeated this in his instructions to General

Marquis Jacques Law de Lauriston, the new ambassador he was sending to St Petersburg to replace Caulaincourt, who had finally been recalled. Napoleon also took every opportunity to tell Kurakin and any other senior Russian figure who passed through Paris that he wanted peace and friendship with their country. "I have no wish to make war on Russia," he declared to Prince Shuvalov during an interview at Saint Cloud in May 1811. "It would be a crime on my part, for I would be making war without a purpose, and I have not yet, thanks to God, lost my head, I am not mad." To Colonel Aleksandr Ivanovich Chernyshev, a trusted aide-de-camp whom the Tsar had sent to Paris a couple of times with letters for Napoleon, he repeatedly stated that he had no intention of fatiguing himself or his soldiers on behalf of Poland, and "he formally declared and swore by everything he held holiest in the world that the re-establishment of that kingdom was the very *least* of his concerns."[20]

But Alexander could not lay aside the Polish problem so easily. When he realised that he could not count on the Poles to undo Napoleon, he reverted to the idea of cementing his relationship with him over the body of the Polish question. Rumiantsev proposed to Caulaincourt just before the latter left Russia that they put the Duchy of Oldenburg and the Grand Duchy of Warsaw into a sack, shake it about, and see what dropped out. What he was suggesting was that Napoleon indemnify his uncle by marriage for the loss of Oldenburg with a piece of the Grand Duchy. Napoleon responded with anger to this proposal, and refused to consider it, although he did at one stage contemplate giving the throne of a restored Kingdom of Poland to Alexander's brother Constantine as a solution.[21]

When Caulaincourt's travelling chaise rolled into Paris on the morning of 5 June 1811, it drove straight on to Saint Cloud, where Napoleon was staying. Within minutes of it having trundled into the courtyard, Caulaincourt was ushered into Napoleon's presence, in which he spent the next seven hours. His account of the interview, noted down that very evening, provides an illuminating insight into Napoleon's thinking at this crucial stage.[22]

Caulaincourt told Napoleon that in his view Alexander desired peace but could not be expected to subject his people to the rigours of the Continental System, and needed reassurance on the subject of Poland. He also warned Napoleon that Alexander was no longer the malleable youth of Tilsit, and that he would not let himself be intimidated. Alexander had told him that if it came to war, he would go on fighting, in the depths of Russia if necessary, and would never sign a peace dictated to him in his capital, as the Emperor Francis and King Frederick William had done. Napoleon brushed this aside, saying that Alexander was "false and weak," and suggested that Caulaincourt had been taken in by him.

He himself was suspicious of the Tsar's intentions, believing that he would pounce on the Grand Duchy of Warsaw the moment his back was turned. He repeatedly affirmed that he was no Louis XV – referring to France's feeble response to the Russian partition of her Polish ally in the eighteenth century. The conversation went round in circles, with Napoleon eagerly asking Caulaincourt's opinion yet rejecting it when it was given. He was, in fact, probably right to think that Caulaincourt had been lulled into believing in Alexander's pacific intentions, yet he could not dismiss his arguments outright.

One thing that did seem to make a profound impression on Napoleon was one of the Tsar's statements as reported by Caulaincourt. "If fate decides against me on the field of battle," Alexander had said, "I would rather retreat as far as Kamchatka than give away provinces and sign in my capital any treaty which would only be a truce. The Frenchman is brave, but long privations and a bad climate tire him and discourage him. Our climate, our winter will fight for us. Prodigious victories are only achieved where the Emperor is, and he cannot be everywhere or spend years away from Paris."[23] Alexander had said that he was well aware of Napoleon's ability to win battles, and would therefore avoid fighting the French where they were under his command. He had also referred to the *guerrilla* in Spain, and said that the whole Russian nation would resist an invader. But on reflection Napoleon dismissed all this as bravado. He believed

Alexander was too weak a character to carry out such a plan, and that Russian society would not accept such sacrifices. He reasoned that the nobles would not want to see their lands ravaged for the sake of Alexander's honour, while the serfs would as likely revolt against their nobles and their Tsar as fight to the last man for a system of slavery.

When asked his opinion on what should be done, Caulaincourt came up with two alternatives. Napoleon should either give a significant part, if not the whole, of the Grand Duchy of Warsaw to Alexander, thereby cementing the alliance, or he should go to war with the aim of restoring the Kingdom of Poland. He pointed out that Austria could easily be compensated, and maintained that the cause of Poland was so universally recognised as a just one that even Britain would ultimately approve.[24] Asked which course of action he would adopt given the choice, Caulaincourt replied that he would give the Grand Duchy to Alexander, thereby guaranteeing a stable peace. Napoleon countered that he could not have peace without honour, and the abandonment of the Poles would dishonour him. At the same time, such appeasement of Alexander would inevitably lead to further Russian expansion into the heart of Europe.

Alexander's military ardour had in fact cooled by then. Memories of Austerlitz must have played their part, for, as Czartoryski noted, he was still "very afraid" of Napoleon. His mind was troubled by the uncertainties of his position at home, his heart was bruised by the public rejection of his policies and, at a more personal level, by the successive deaths, in 1808 and 1810, of two baby daughters. But perhaps the main consideration holding him back was that he did not want to be seen as the aggressor. In July 1811 he wrote to his sister that the best course to follow was to let time and circumstances destroy Napoleon. "It seems to me more reasonable to hope that this evil will be remedied by time and by its own sheer scale, for it is such that I cannot rid myself of the conviction that this state of affairs cannot last, that the suffering of all classes, both in Germany and in France, is so great that patience must necessarily run out."[25]

But it was Napoleon's patience that had run out. He viewed the

Russian abandonment of the Continental System as a betrayal, he saw her troop build-up as a threat and a provocation, and he was convinced that she was using the Polish question and the subject of trade as excuses to break out of the alliance. This seemed to be confirmed by the increased diplomatic activity of the Russians in Vienna, where they were quite openly trying to turn Austria away from France.

Napoleon needed to go and take charge of operations in Spain personally in order to throw out the British and pacify the peninsula, but he could not contemplate such a move with a Russian army hovering on the borders of the Grand Duchy of Warsaw and exciting German hopes of revenge. He was convinced that, just as the Austrians had done in 1809, Alexander would stab him in the back the moment he turned it.[26]

His exasperation erupted on 15 August 1811, his forty-second birthday. At midday he strutted into the throne room at the Tuileries, which was filled with the entire court and all the senior officers in Paris, all perspiring in full ceremonial and parade uniforms on what was an exceptionally hot day. He took his place on the throne to receive the good wishes of the dignitaries and the diplomatic corps. This part of the ceremony over, Napoleon stepped down from the throne and began to circulate among the guests.

When he reached the Russian ambassador Prince Kurakin, he mentioned Russian reports of a recent victory over the Turks at Ruschuk on the Danube, and queried why, if they had indeed won, the Russians had evacuated the town. Kurakin explained that the Tsar had been obliged to withdraw some troops from the Turkish front for financial reasons, and had therefore decided not to hold the town. At this Napoleon exploded, saying that the Russians had not won, they had been beaten by the Turks, and they had been beaten because they had withdrawn troops from the Turkish front not for any financial reasons, but because they were massing their armies on the frontiers of the Grand Duchy of Warsaw, and that all the so-called outrage over Oldenburg was but an excuse for their intention to invade the Grand Duchy in an open act of hostility to him, Napoleon. The unfortunate

Kurakin kept opening his mouth to reply, but could not get a word in edgeways and looked like a fish gasping for air, while perspiration poured down his face in the intense heat. Napoleon accused Russia of harbouring hostile intentions, and when Kurakin assured him of the contrary, he turned on the ambassador and asked whether he had powers to negotiate, for if he had, they could conclude a new treaty there and then. The answer was negative, so Napoleon merely walked away, leaving the ambassador in a state of shock.[27]

Napoleon was back at Saint Cloud late that evening, and on the following morning he locked himself up with the punctilious and hard-working Hugues Maret, Duc de Bassano, who had succeeded Champagny as Foreign Minister. Together they trawled through all the documentation concerning the Russian alliance since Tilsit. According to their analysis, the problems had started in 1809, when the Russians had hung back in the war against Austria instead of marching in loyally and capturing Galicia. Had they done so, they could have been allowed to keep it. As they did not, it was captured by the Poles, who could not be denied some of it. This caused panic in Russia and led the Tsar to demand slices of the Grand Duchy. France could never accede to such a request. Not just for the sake of her honour, but also because if Russia were to receive one piece of the Grand Duchy she would in time expect to get another, and would soon entrench her position on the Vistula if not the Oder. For similar reasons, France could not countenance any further Russian advance against Turkey.

In the memorandum summing up the situation, they stated France's position as follows: France wanted Russia's friendship and needed her as an ally in her struggle against Britain, which was the one remaining obstacle to a general peace. She did not want to fight Russia, as there was nothing that she wanted to take from her. Also, she had more pressing business in Spain, which required Napoleon's personal attention. But France could not go down the road of buying Russia's friendship through endless cession of Polish or Ottoman lands. France must therefore prepare for war in order to be in a position to dictate a peace. Lauriston was told that he had to make it clear that "we want peace, but we are prepared for war."[28]

Napoleon's sense of exasperation at not being able to bring Alexander back into a close alliance is obvious in a personal letter he had written him on 6 April. "The effect of my military preparations will make Your Majesty increase his own; and when news of his actions reaches me here, it will force me to raise more troops: and all this over nothing!" he wrote. They had been drawn into a spiral of mistrust and power politics that made it very difficult to arrive at a negotiated settlement. Napoleon later admitted that they had got themselves "into the position of two blustering braggarts who, having no wish to fight each other, seek to frighten each other."[29]

5

La Grande Armée

On the evening of 25 March 1811, as he was scouring the night sky from his makeshift observatory in Viviers, Honoré Flaugergues discovered a comet in the now defunct constellation of Argo Navis. He saw it again the following day and began to track its progress. The comet was low in the south and was moving northward and brightening. On 11 April it was spotted by Jean Louis Pons in Marseille, and on 12 May by William J. Burchell in Cape Town. The comet soon became visible to the naked eye, and by the late autumn it lit up the night sky from Lisbon to Moscow. People gazed up at it, some with interest, many more with a sense of foreboding.[1]

This seemed to increase the further east one went in Europe. "As they contemplated the brilliant comet of 1811," recalled a parish priest there, "the people of Lithuania prepared themselves for some extraordinary event." Another inhabitant of the province never forgot how everyone got up from dinner and went out to gaze on the comet and then talk of "famine, fire, war and bloodshed." In Russia, many linked the comet to a plague of fires which swept the land that summer and autumn, and a blind terror gripped them as they looked on it. "I remember fixing a long look on it on an autumn moonless night, and I was struck with childlike fear," wrote the son of a Russian landowner. "Its long, bright tail, which seemed to wave with the movement of the wind and to leap from time to time, filled me with such horror, that in the days that followed I did not look up at the sky at night, until the comet had disappeared."

In St Petersburg, Tsar Alexander himself became fascinated by the phenomenon, and discussed it with John Quincy Adams, then American ambassador at his court. He claimed to be interested only in the scientific aspects of the comet, and made fun of all those superstitious souls who saw in it a harbinger of catastrophe and war.[2]

But he was either being disingenuous or he was deluding himself, for the machinery of war had already clanged into gear, and its wheels were by now turning with such momentum that it would have taken a complete climbdown on the part of Napoleon or himself to stop them. Observing events from Vienna, Metternich was in no doubt that "the supreme struggle" between the *ancien régime* and what he termed Napoleon's revolutionary designs was now imminent. "Whether he triumphs or succumbs, in either case the situation in Europe will never be the same again," he wrote to his imperial master on 28 December 1811. "This terrible moment has unfortunately been brought on us by the unpardonable conduct of Russia."[3]

"I am far from having lost hope of a peaceful settlement," Napoleon wrote to his brother Jérôme on 27 January 1812. "But as they have adopted towards me the unfortunate procedure of negotiating at the head of a strong and numerous army, my honour demands that I too negotiate at the head of a strong and numerous army. I do not wish to open the hostilities, but I wish to put myself in a position to repulse them."[4] He therefore needed to field an army vast enough to intimidate Alexander or, failing that, to force him into submission with a rapid and shattering blow. There was an element of haste involved, as he had to count with the possibility of a Russian first strike at any moment. Fortunately, he was not starting from scratch.

Following the treaty of Tilsit, a body of French troops remained in the Grand Duchy of Warsaw while the local forces were being organised, along with garrisons in key fortresses in Prussia such as Danzig, Glogau, Stettin and Küstrin. After the 1809 war with Austria, Napoleon left further garrisons at Düsseldorf, Hanau, Fulda, Hanover,

Magdeburg, Bayreuth, Salzburg and Ratisbon. In May 1810 he strengthened all the forces on German soil and organised them into the Armée d'Allemagne, under Marshal Davout. In the autumn of 1810, following the Russian troop build-up along the border of the Grand Duchy of Warsaw, Napoleon reinforced this further. He also began moving units stationed in France closer to Germany, concentrating his artillery parks at Strasbourg, Metz, Wesel and La Fère, and withdrawing selected units from Spain.

In the spring of 1811, fearing a Russian invasion of the Grand Duchy of Warsaw, Napoleon ordered the Poles to mobilise 50,000 men. He had already ordered his stepson Prince Eugène de Beauharnais, Viceroy of Italy, to place the Army of Italy on a war footing. Now he instructed his brother Jérôme and other allied monarchs to mobilise the armies of Westphalia, Württemberg, Bavaria, Baden and the lesser German states. He meant to put together a force of half a million men with which to confront Russia. He began calling up men in France on a massive scale, and gendarmes combed the countryside for the tens of thousands of deserters who regularly sneaked away from the colours and went to ground. They would be rounded up and fed back into the army, along with the new recruits.

The French army was organised in divisions, which were usually made up of four regiments. A regiment of foot normally consisted of about 3800 men, with a hundred officers. It had up to five battalions, one of which was always at the depot, and these battalions consisted of six companies each, of which one would be a company of grenadiers, one of *voltigeurs* (skirmishers) and four of fusiliers. A company was supposed to number 140 men, including two drummer boys, and was commanded by one captain, one lieutenant, one sub-lieutenant, a sergeant major and a dozen other sergeants and corporals. To accommodate the new influx, Napoleon added a fifth and then a sixth battalion to existing regiments. The recruits were spread through the old battalions as well as the new ones, which were fortified with a sprinkling of veterans.

Napoleon attended personally to every detail. His correspondence in these matters reveals a staggering degree of familiarity with every brigade, regiment and battalion, where they were stationed, where they were due to move to, who commanded them, how many reinforcements they needed, where these could be drawn from, and how soon they could be made available. No detail was too insignificant for him. He attended to lettering on standards and badges, to the quality and calibre of arms and equipment, to numbers of horses and types of supply wagon required. To deal with the many rivers he would need to negotiate, he formed a bridging train equipped with pontoon boats and other necessaries at Danzig.

Curiously enough, the one thing he paid no attention to, now or at any stage in his military career, was the army's basic weaponry. The artillery still used the Gribeauval gun and gun carriage, designed fifty years before, while the footsoldier's weapon was a muzzle-loading flintlock musket of a design that had remained virtually unchanged for a hundred years. It was an extremely primitive instrument. To load it, a soldier would take a cartridge, consisting of a paper cylinder containing a measure of powder and a lead ball. He would bite off the end of the cartridge, keeping the ball in his mouth, sprinkle a little of the powder in the priming pan, and close the flap; he would then pour the remainder of the powder down the barrel, spit the ball in after it, screw up the paper into a wad, and ram the whole lot down to the bottom of the barrel with his ramrod. A trained soldier could reload and be ready to fire in one and a half minutes.

The musket was notoriously inaccurate even at short range and had a number of faults which could be dangerous. The black powder in the cartridges fouled the inside of the barrel, so that after a dozen or so shots it became increasingly difficult to ram anything down it, while the progress of the bullet being fired was also slowed. The powder in the pan might ignite, producing the usual plume of smoke, but the charge in the barrel might not go off – the proverbial "flash in the pan." In the din of battle, the soldier might not register that his charge had not gone off, and set about loading up with another cartridge. If the first one then went off, the barrel was likely to explode in his face. But that was considered just another of the hazards of war. Footsoldiers were expendable, and there were always plenty more where they came from.

The relentless build-up of forces continued through the autumn and winter of 1811 and into the spring of 1812. The twenty-year-old son of a wine-grower in Burgundy presented himself at seven o'clock on the morning of 3 January 1812 at the Préfecture in Lyon, and a couple of days later he was in the barracks of the 17th Light Infantry at Strasbourg. "The very morning after our arrival, we were uniformed and armed, and, without giving us time to breathe, the corporals set about inculcating in us the principles of our new trade," he remembered. "They were in a hurry . . ."[5]

Raising the troops was only part of the task: the men had to be fed, clothed and armed. On campaign, the French soldier was supposed to receive a daily ration of: 550 grams of biscuit, either thirty grams of rice or sixty grams of dried vegetables, 240 grams of meat or two hundred grams of salt beef and lard, some salt, a quarter of a litre of wine, a sixth of a litre of brandy and, in hot weather, a shot of vinegar. By January 1812 Napoleon had amassed fifty-day supplies of biscuit, flour, salted meat and dried vegetables for 400,000 men and forage for 50,000 horses at Danzig. This was on top of the million rations stored at Stettin and Küstrin.[6]

The enterprise also required the provision of hundreds of thousands of items of clothing, of boots of various kinds, and of small arms. It entailed the purchase of tens of thousands of horses for the cavalry, which had to be trained to carry a heavily armed rider and respond to his intentions as he wielded his sword, lance or carbine. They also had to be habituated to the roar of cannon and the clash of arms, by being led and then ridden, again and again, towards lines of men shouting, banging cooking pots and letting off guns in their direction, and to be rewarded each time with a carrot.

Napoleon prepared massive supplies of ammunition, setting up depots at Magdeburg, Danzig, Küstrin, Glogau and Stettin. By May 1812 he would have amassed 761,801 rounds of ammunition for his field artillery – over a thousand rounds per gun for some calibres of the more than eight hundred cannon he was putting into the field. This did not include the siege train of heavy guns which he had built up there so as to be able to reduce enemy fortresses. Such figures do not compare at all badly with the preparations made by a highly industrialised imperial Germany a hundred years later.

As he was expecting Russia to launch her attack at any moment, his first preoccupation was to secure the line of the Vistula and strengthen the garrisons of the fortresses at Modlin, Toruń and Zamość. This would allow his main forces to concentrate in the first couple of months of 1812. He hoped to have over 400,000 men in the area of northern Germany and Poland by the middle of March, which would allow him to deal with any Russian strike, even if it were accompanied by outbreaks of German national insurrection.[7]

The situation in Germany had been growing increasingly tense for some time, and patriots watched the preparations for war on both sides with mounting excitement. The Russian embassy in Vienna was orchestrating agitation throughout Germany. Colonel Chernyshev was recruiting disaffected Prussian officers and working on a plan to found a German Legion in Russia which, in the event of war, would enlist all prisoners of German nationality taken from Napoleon. He was also investigating the possibility of creating a fifth column of sympathisers all over Germany who would be ready to rise up when a Russian army marched in.[8]

Reports from French military commanders and diplomatic agents in Germany were full of stories of plots by secret societies, and warned Napoleon that the hardships imposed by the Continental System were driving people to desperation. In the autumn of 1811 Prussia appeared to be on the brink of revolt, with the King and his pro-French cabinet barely able to control the nationalists. The Prussian army was surreptitiously mobilising its reserves. In Westphalia, Jérôme was growing nervous. "The ferment has reached the highest degree, and the wildest hopes are fostered and cherished with enthusiasm," he reported to Napoleon on 5 December 1811. "People are quoting the example of Spain, and if it comes to war, all the lands lying between the Rhine and the Oder will be embraced by a vast and active insurrection." Napoleon did not believe the Germans had the stomach for popular insurrection and thought the secret societies ridiculous. But he instructed Davout to be ready to march on Berlin at a moment's notice in order to disarm the Prussian army.[9]

The army Napoleon was assembling would be the largest the world had ever seen.* It included soldiers from almost every nation of Europe. Its main body was made up of Frenchmen, Belgians, Dutchmen, Italians and Swiss from the areas incorporated into the Empire. This was supplemented by contingents from every vassal or allied state. The presence of such a wide variety of nationalities inevitably raised questions of cohesion, quite apart from motivation or loyalty. But with the exception of the Polish and the Austrian corps, all the contingents were commanded by French generals. And most were imbued with French military culture, and fortified by the reputation of French arms. "The belief that they were invincible made them invincible, just as the belief that they were sure to be beaten in the end paralysed the enemy's spirits and efforts," in the words of Karl von Funck, a German officer attached to the French imperial staff.[10]

"Three-quarters of the nations which were about to take part in the struggle had interests diametrically opposed to those which had decided the opening of hostilities," wrote Lieutenant Count von Wedel, a German serving in the 9th Polish Lancers. "There were many who in their heart of hearts wished the Russians success, and yet at the moment of danger, all fought as though they had been defending their own homes."[11] The urge to emulate was strong, and there was the magic presence of Napoleon.

"Anyone who was not alive in the time of Napoleon cannot imagine

* The term "*grande armée*" in French military parlance designated the main operational force in any given campaign, but in the popular imagination the two words are above all associated with the great force that marched on Moscow.

the extent of the moral ascendancy he exerted over the minds of his contemporaries," wrote a Russian officer, adding that every soldier, whatever side he was on, instinctively conjured a sense of limitless power at the very mention of his name. Wedel agreed. "Whatever their personal feelings towards the Emperor may have been, there was nobody who did not see in him the greatest and most able of all generals, and who did not experience a feeling of confidence in his talents and the value of his judgement . . . The aura of his greatness subjugated me as well, and, giving way to enthusiasm and admiration, I, like the others, shouted '*Vive l'Empereur!*' "[12]

The largest non-French contingent were the Poles, who numbered some 95,000. Many of them had been fighting under French colours since the late 1790s and were enthusiastic allies. In 1807 Napoleon created an élite regiment of Polish Chevau-Légers in the Imperial Guard as a token of how much he valued his Polish troops. In the same year the Grand Duchy of Warsaw began recruiting its own army, and raised the Legion of the Vistula, an auxiliary corps which was to fight for the French. These troops had distinguished themselves in various theatres, and had no difficulty in operating alongside the French. The only problem was that Napoleon's insistence on the Grand Duchy raising more troops than such a small state could support, either in human or economic terms, meant that the barrel had been scraped. Men who were physically unfit had been drafted, uniforms had been skimped on, training was inadequate, and nobody was paid after June 1812. But at least their loyalty to the cause and devotion to Napoleon were never in question.[13]

The next largest contingent were the Italians, grouped in the Army of Italy, commanded by Prince Eugène, and the Neapolitan army of Joachim Murat. The Army of Italy was a fine force of 45,000, which included 25,000 Italians organised on French lines, highly disciplined, with a strong *esprit de corps*, particularly in units such as the Royal Guard, and 20,000 Frenchmen, many from Savoy and Provence, stationed in Italy. It was also one of the more motivated contingents, inspired by national pride. As he looked at all the nationalities making up the

Grande Armée, one young Italian officer's mind drifted to the days of ancient Rome, whose legions were equally made up of disparate elements, and he felt a great sense of pride at being part of it.[14]

The same could not be said for the Neapolitan contingent. This was a largely worthless force, poorly trained and undermined by the existence of numerous rival secret societies. Whenever the troops were moved out of barracks they deserted in large numbers and formed bands of brigands who would terrorise the surrounding countryside.

Most of the German troops in the Grande Armée were of high quality. The 24,000 Bavarians were Napoleon's most reliable allies, having fought under his banner several times. The smaller Badenese forces, organised along French lines, had taken part in the campaign of 1805 against Austria and Russia, so they fitted relatively well into the composite army. The 20,000-strong Saxon contingent was disciplined and also marched quite comfortably in the ranks of the Grande Armée, to which it brought some of the best cavalry.[15]

The 17,000 men of the Westphalian contingent did not, according to Captain Johann von Borcke from Cassel, contain many Napoleonic enthusiasts. The Principal Minister of Westphalia reported that the men were loyal, but hated the idea of being sent far away more than they feared being killed. "An active resistance on their part seems impossible to me," he wrote to Maret in January 1812, "but the weight of their inertia could, in the first stages, cause trouble, mainly through large-scale desertion."[16]

On the whole, the German contingents were loyal to Napoleon. Many of the troops were fired by the idea of rolling the Russians back out of Europe, and felt a strong urge to prove the valour of German arms. Even if they had no love for the French, they tended to be more antagonistic to Germans from other parts of the country, with most of the troops from the Confederation of the Rhine showing a marked dislike of the Prussians. Finally there was military honour. "I know that the war we are fighting is contrary to the interests of Prussia," Colonel Ziethen of the Prussian Hussars said to a Polish officer, "but

I will, if necessary, let myself be hacked to pieces at your side, for military honour commands it."[17]

The Prussians were brought into the Grande Armée under the terms of the treaty signed between Napoleon and Frederick William on 24 February 1812, and made up an auxiliary corps of 20,000 men. There was also an Austrian contingent, under Prince Karl von Schwarzenberg, made up of 35,000 men. Most of them had last seen action against the French and the Poles, and while soldiers fight when and whom they are ordered to, they were not enthusiastic allies. Because of the political stance of their ruler and their commander, they were to play an insignificant part in the campaign.

Amongst the lesser contingents the four Swiss regiments should be singled out as being of very high quality and well tempered by a couple of years' service in Spain and Portugal. There were two battalions of Spanish volunteers from the Joseph-Napoléon Regiment, in distinctive white uniforms with green facings, which had spent the past year under Davout in Germany. They were commanded by Colonel Doreille, a Provençal who did not speak French. There were also many Spaniards, some three thousand of them, in the ranks of the second and third regiments of General d'Alorna's Portuguese Legion, which numbered around five thousand men in total, uniformed in brown with red facings and English-style shakos. "The men, who are highly motivated, make up a fine unit, on which I believe we can count," General Clarke, Napoleon's Minister of War reported. And finally there were two regiments of Croats, numbering just over 3,500 men.[18]

The worth of all these troops was hugely enhanced by the presence of Napoleon. Not only because he lent them the value of his reputation as a military genius, but also because he had the gift of drawing the best out of them. He was masterly in his treatment of soldiers, whom he captivated with his *bonhomie* and his sometimes brusque lack of ceremony. He always knew which regiments had fought where, and when he reviewed them, he would walk up to older rankers and ask them if they remembered the Pyramids, Marengo, Austerlitz, or wherever it was that particular unit had distinguished itself. They would swell with pride, feeling that he had recognised them, and they could feel the envy of the younger men all around them. With the younger soldiers Napoleon adopted a solicitous manner. He would enquire if they were eating enough, whether their equipment was up to scratch, sometimes asking to see the contents of their haversacks and engaging them in conversation. He was well known for tasting the soldiers' stew and bread whenever he passed a camp kitchen, so they felt his interest was genuine.

During a review shortly before the campaign, Napoleon stopped in front of Lieutenant Calosso, a Piedmontese serving in the 24th Chasseurs à Cheval, and said a few words to him. "Before that, I admired Napoleon as the whole army admired him," he wrote. "From that day on, I devoted my life to him with a fanaticism which time has not weakened. I had only one regret, which was that I only had one life to place at his service." Such a level of devotion was by no means rare, and transcended nationality. But Napoleon could not be everywhere, and the larger the army, the more diluted his presence would be.[19]

Napoleon's determination to assemble such a vast force was bound to have a negative effect on its quality. Louis François Lejeune, a senior officer on Berthier's staff, was detailed to inspect the troops already on the Oder and the Vistula in March 1812, and was bombarded with complaints from the commanders of the units he visited that half of the recruits they were receiving were useless.

He mentioned this to General Dejean, who was organising the cavalry in the area. Dejean told him that up to a third of the horses he had been sent were too weak to carry their burden, while nearly half of the men were too puny to wield a sabre. "I was not happy with the way the cavalry was being organised," echoed Colonel de Saint-Chamans, commanding the 7th Chasseurs à Cheval. "Young recruits who had been sent from depots in France before they had learnt to ride a horse or any of the duties of a horseman on the march or on campaign, were mounted on arrival in Hanover on very fine horses which they were not capable of managing." The result was that by the time they reached Berlin, the majority of the horses were suffering from lameness or saddle sores induced by the riders' bad posture or their failure to take care in saddling up. More than one officer noted that recruits were not taught about checking whether their saddle was rubbing or how to detect the early signs of saddle sores.[20]

Sergeant Auguste Thirion of the 2nd Cuirassiers had a rosier view. "Such fine cavalry has never been seen, never had regiments reached such high complements, and never had horsemen been so well mounted," he wrote, adding that their leisurely march through Germany had actually hardened the horses and men. But the cuirassiers were the élite of the French cavalry. And good horses could be a problem in themselves, according to Captain Antoine Augustin Pion des Loches of the Foot Artillery of the Guard. "Our teams were of the best, and the equipment left nothing to be desired, but everyone was agreed that the horses were too tall and too strongly built, and unsuited to supporting hardship and lack of abundant nourishment," he wrote on leaving the depot at La Fère on 2 March 1812.[21]

Napoleon was not particularly bothered by such a state of affairs. "When I put 40,000 men on horseback I know very well that I cannot hope for that number of good horsemen, but I am playing on the morale of the enemy, who learns through his spies, by rumour or through the newspapers that I have 40,000 cavalry," he told Dejean when the latter reported his findings. "Passing from mouth to mouth, this number and the supposed quality of my regiments, whose reputation is well known, are both rather exaggerated than diminished; and the day I launch my campaign I am preceded by a psychological force which supplements the actual force that I have been able to furnish for myself."[22]

The real strength of the French army was that all the men, even the lower ranks, were free citizens with a strongly patriotic education in the new public schools behind them. They could think as well as fight, and if they showed initiative as well as bravery they could gain promotion and rise very high. But Napoleon's habit of rewarding mere bravery with promotion eventually led to units being commanded by men who lacked the necessary competence. "Among the generals of rapid promotion," wrote Karl von Funck, "there were only a few who had the gift of leadership; many lacked even the most elementary military knowledge . . . In the madness of daring they had learnt how to fling their intrepid forces against the foe, but they had no notion of judging a position, of even the first principles of operations, of withdrawing in good order if the first onset should fail . . ." There was also a multitude of young officers drawn from the Parisian *jeunesse dorée* who had obtained promotion through string-pulling, who had mostly joined cavalry regiments because they liked the uniforms or the staff because they wanted to be close to Napoleon. Many were clearly not up to the job.[23]

Much of the revolutionary ardour that had fired the French armies of the 1790s and early 1800s had been quenched by 1812. "As the uniforms grew more embroidered and gathered decorations the hearts they covered grew less generous," as one observer put it. Napoleon himself sensed a lack of enthusiasm for the forthcoming campaign. "People have always followed him with excitement; he is surprised that they are not prepared to end their careers with the same dash with which they embarked on them," noted his secretary Baron Fain.[24] But it was he who had turned his commanders into what they were.

"From the moment that Napoleon came to power, military mores changed rapidly, the union of hearts disappeared along with poverty and the taste for material well-being and the comforts of life crept into our camps, which filled up with unnecessary mouths and with numerous carriages," in the words of General Berthézène. "Forgetting the fortunate experiences of his immortal campaigns in Italy, of the immense superiority gained by habituation to privation and contempt for superfluity, the Emperor believed it to be to his advantage to encourage this corruption."[25] He gave his marshals and his generals titles, lands and pensions on the civil list. He demanded of them that they keep palaces in Paris in which they were to entertain at the appropriate level. As members of his court, they must maintain a glittering entourage, as he increasingly did himself. This high living softened them up, and they became less and less willing to give up their warm beds and fine palaces in fashionable parts of Paris, not to mention their wives and mistresses, for the rigours of the bivouac and the uncertainties of war. This was particularly true of the marshals.

"They were most of them between the ages of thirty-five and forty-five when, after a stormy youth, a man begins to look for a settled domesticity," wrote von Funck. "They could hardly expect to win a higher degree of fame, but might well jeopardise the reputations they had made." A good example was Napoleon's chief of staff Marshal Berthier, Prince de Neuchâtel, a plump man of settled tastes in his mid-fifties with a magnificent *apanage* and an adored and adoring mistress in Paris.[26]

At the same time the lavish rewards given to those who distinguished themselves in battle were an irresistible incentive to soldiers and officers right up to the rank of general, who all saw in war the possibility of making their fortune. A simple soldier might obtain promotion, which would give him a higher salary and status, or the Légion d'Honneur, which meant pension rights. A general might obtain the coveted marshal's baton, which signified fame and fortune, and a ducal title to boot.

There was also the opportunity of making some money on the side through the more or less legitimate acquisition of precious items. Looting as such was not countenanced, but in the course of campaigns in distant lands it was possible to purchase things at knockdown prices and bring them home without paying any duties. Valuables found on the field of battle or in the enemy camp were fair game, as was anything that might be left masterless through the fortunes of war. As this great campaign to end all campaigns was being prepared, a good many felt that it would be their last opportunity to get rich.

There were also a great many for whom war furnished the prospect of adventure, the opportunity to distinguish themselves or a last chance to share in the glory. "At last I was going to find myself at some of those battles which are destined to change the course of history; I was going to fight under the eyes of so many of the illustrious warriors who filled the world with their renown, Murat, Ney, Davout, Prince Eugène, and so many others, and under the eyes of the greatest of them all, under the eyes of Napoleon!" remembered a Creole from Saint-Domingue. "There I would have my chance to distinguish myself, there I would be able to obtain decorations and promotion of which I would be proud and which I could hold up to the world! Before a year was out I would be *chef de bataillon*; I would be colonel by the end of the campaign, and after that . . ."[27]

Most of Napoleon's entourage, beginning with Caulaincourt and including close friends and collaborators such as General Duroc, repeatedly beseeched him not to go down the road of war. Many of them warned that Russia could not be defeated in conventional ways. Napoleon had read the accounts of Charles XII's disastrous foray into Russia almost exactly a hundred years before, beginning with the famous one by Voltaire, who wrote that "there is no ruler who, in reading the life of Charles XII, should not be cured of the folly of conquest." Napoleon dismissed such arguments with annoyance. "His capitals are as accessible as any others, and when I have the capitals, I hold everything," he snapped at one of his diplomats who had been pointing out the perils of going to war with Alexander. But he had a habit of making statements he did not believe in, as though he were trying to convince himself by convincing others. And the special nature of the forthcoming campaign was not lost on him. "If people think that I am going to make war in the old way, they are very much mistaken," he is alleged to have declared.[28]

He certainly prepared himself for the forthcoming campaign as he had prepared for no other. For one thing, realising that he would be operating in detached corps at some distance from each other, he decided to make an example of General Dupont, whose capitulation at Bailén had been such a humiliation. Before setting off for Russia, Napoleon had him retried and given a harsher sentence.[29]

He had officers who spoke German and Polish or Russian attached to every unit, while others were made to take Russian lessons so they could interrogate prisoners and gather intelligence, and he set up a network of intelligence agents fanning out from Poland into Russia's western provinces. He ordered his librarian to supply him with books on the Russian army, and on the topography of Lithuania and Russia, which he studied, paying particular attention to roads, rivers, bogs and forests. As early as April 1811 he commissioned the Dépot de la Guerre to engrave a series of large-scale maps of western Russia.[30]

Not the least daunting aspect of this campaign was that the enemy frontier, on the river Niemen, lay about 1,500 kilometres from Paris. Thus a length of time and a huge effort were required to move up men and supplies before the campaign could begin. And the Russian capital cities of St Petersburg and Moscow lay another 650 and 950 kilometres respectively beyond that. An army marching from Paris to Moscow without fighting would take up to six months to cover the distance. To make matters worse, the last three hundred of the 1500 kilometres up to the Niemen lay through poor, infertile areas of Prussia and Poland, while the first five hundred of the next 950 kilometres were in even less abundant country. It was, moreover, criss-crossed by rivers, mottled with bogs and forests, and contained large areas of wilderness.

Napoleon's tactic had always been to move fast, concentrate large numbers of men at the right point before the enemy knew what was happening, knock out their army with a decisive blow and force them to make peace on his terms. But he would have to work hard to achieve it in this theatre of operations.

His armies had in the past been able to move fast because they travelled light, a tradition originally forged of necessity. The French revolutionary armies of the 1790s had been hurriedly improvised and had not possessed a proper commissariat. Since they regarded enemy territory as belonging to tyrants and enemies of the revolution, they lived by looting. With time, they began to buy what they needed, but as they paid with largely worthless paper *assignats*, it amounted to the same thing. Napoleon disapproved of looting, and brought in administrators who would provide for the army's needs in a more methodical way. They bought what was needed, paying in real money or receipts that were generally honoured, when peace had been signed, by the government of the defeated country. But the fact remained that French armies lived off the fat of the land they moved through. And as they moved fast, they did not stay long enough to exhaust its resources.

Whenever the administrative machine broke down or failed to provide the necessities, the French reverted to the old system of "*la maraude*." Every so often a company or similar unit would send out eight or ten men under a corporal into the areas alongside the line of march. These little bands would fan out through local villages and farms, paying for what they took, and rejoin their company a few days later, their carts laden with grain, eggs, chickens, vegetables and other victuals, driving before them a small herd of cattle. From time to time the main force would halt in order to allow the foraging parties to catch up.

As the French armies were conscripted, a company usually contained a baker, a cobbler, a tailor's apprentice, a cooper, a blacksmith and a wheelwright, so not only could they bake their own bread, but also, given an occasional purchase of cloth, leather, iron and other raw materials, they could mend their uniforms, boots, equipment and wagons.

For everything else, there was the *cantinière* or suttler-woman, something of an institution, unique to the French army. "These ladies usually started out by following a soldier who had inspired tender feelings in them," explained Lieutenant Blaze de Bury. "You would see them first walking along with a cask of *eau-de-vie* slung round their neck. A week later, they would be comfortably seated on a horse someone had *found*, draped, to the right, to the left, in front, behind, with casks, saveloys, cheeses and sausages in precarious equilibrium. A month would not pass that a cart harnessed with a couple of horses and filled with provisions of every kind would not be testifying to the growing prosperity of their enterprise."[31]

In camp, the tent of the *cantinière* became the company café, where officers would come to sit around and play cards or gossip. It was also a bank, lending money and giving credit. On campaign, the *cantinière* went to untold lengths to stock up on all the little necessities which could transform a soldier's life by guaranteeing survival or just relief. She always had a little something for a soldier who had money or whom she trusted to pay her when he got some. She usually had a protector, sometimes a husband but mostly just a temporary mate who was able to provide her with security and help in return for being supplied with victuals, and sometimes a change of protector would entail a financial transaction between the two men involved.

These ladies saw themselves as part of the army, and despised anyone not in uniform. They viewed the dangers of war as part and parcel of their trade. If they fell prey to enemy marauders and lost everything, they would shrug it off as hard luck and start again. Some would even take kegs of brandy onto the battlefield and give free slugs to the men, and not a few were wounded in this act.

When a regiment moved out on campaign, it was followed by the *cantinière* and her small gang of purveyors, the servants of the officers, a dozen washerwomen and a horse thief or two. As it went, it picked up petty criminals for whom things had got too hot in the locality, young men looking for adventure, stray dogs and the odd whore. "While the regiment marched along the road in good order, or wherever it was sent, this mounted rabble – or to give it its proper name, this robber band – swarmed round it to left and to right, in front and behind, and used the regiment as a base," wrote Lieutenant von Wedel. "They all carried large and small haversacks and bottles in which to hide their plunder, and they were armed with swords, pistols, even carbines if they could lay hands on such a weapon. These bands often roamed far and boldly on the flank, and if they ever got back again, brought supplies for the troops. The work was dangerous and many lost their lives – in agony if they fell into the clutches of the infuriated peasants . . . This swarm of plunderers also formed a sort of flank patrol for the army, because if ever they bumped into enemy detachments they came flying back with great haste and loud shouts."[32]

Such tactics liberated French commanders from the necessity of hauling heavy stores along with them, which gave them the edge of speed over their more traditionally organised enemies. They worked well in the rich, densely populated, fertile and commercially developed areas of northern Italy and southern Germany, where small distances, good roads, frequent towns and an abundance of every kind of resource meant that a comparatively large army could indeed move fast and provision itself as it went. It even worked in the less populous and more arid expanses of Spain. It could not work in Russia, where the distances were huge, the roads primitive, towns few and far between, the countryside thinly populated and poor in resources. Nobody saw this more clearly than Napoleon. "One can expect nothing of the country, and we shall have to carry everything with us," he warned Davout.[33]

The commissariat he had founded, under the command of General Matthieu Dumas, was methodically stockpiling arms, munitions, uniforms, shoes, saddles, as well as food rations on a vast scale. But the problem of how to move these supplies about represented a logistical nightmare. "The Polish war does not resemble the war in Austria at all; without means of transport, everything becomes worthless," Napoleon wrote to Prince Eugène in December.[34]

The French army's supply system, such as it was, was in the hands of a transport corps, a military formation called *le train*, founded in 1807. In the course of 1811 and 1812 Napoleon gradually expanded the size of the train to twenty-six battalions, with 9,336 wagons drawn by some 32,500 horses, with six thousand spare horses. He put in hand the construction of heavy ox-drawn wagons which could be used to haul flour to the front line, where the oxen would be consumed along with the flour. He realised that these heavy wagons, capable of carrying one and a half tons, would have difficulty in negotiating all but the best roads. He therefore equipped eight of the battalions with lighter

This sketch of Napoleon, made by Anne-Louis Girodet-Trioson in March 1812, shortly before the Emperor's departure for the Russian campaign, shows the unhealthy plumpness which had recently set in, belying his attempt to appear aquiline.

The French army on campaign was a far cry from the glorious image suggested by most military prints, as this scene, drawn from life by Albrecht Adam, shows. Here, a party of cuirassiers are concerned with the vital business of herding food on the hoof.

Tsar Alexander I, painted by Gerhard von Kügelgen in 1804, the year he denounced Napoleon as an upstart and before he had been forced to flee from him on the battlefield of Austerlitz.

General Mikhail Barclay de Tolly, Russia's Minister of War and commander of the First Army, a brave but prudent soldier who knew the limitations of his force, by an unknown artist.

A group of Russian gunners, drawn from life by Johann Adam Klein. The Russian artillery was probably the best in the world at the time.

General Prince Piotr Ivanovich Bagration, commander of the Second Army, a firebrand who believed he should be in overall command, sketched at the start of the campaign by one of his staff officers, General Markov.

General Count Levin Gottlieb Bennigsen, who, despite having been soundly beaten by Napoleon at Friedland in 1807, believed he should be in command in 1812, by George Dawe.

Cossacks on the march, by Johann Adam Klein. A type of light cavalry recruited largely from the cossacks of the Don and the Kuban, they were used mostly to harry rather than to fight. Those depicted here belong to one of the regular cossack regiments.

Napoleon's stepson Prince Eugène de Beauharnais, Viceroy of Italy, loved throughout the army and esteemed by his peers for his bravery and chivalry, by Andrea Appiani.

Marshal Davout, a strict disciplinarian, probably the most capable and loyal of Napoleon's marshals, from a miniature on porcelain by Jean-Baptiste Isabey.

RIGHT Joachim Murat, King Joachim I of Naples, a legend of dash and flash who invented his own uniforms. Detail from a painting by Louis Lejeune, a colonel on the general staff.

BELOW Elements of the Army of Italy on the march on 29 June, shortly after crossing the Niemen, sketched by Albrecht Adam, an artist attached to Prince Eugène's staff. At the centre of the picture, an officer has drawn his sabre to try to get a cantinière's wagon out of the way of his marching men.

Portuguese infantry returning from a foraging expedition, by Faber du Faur.

A French infantryman returning from the maraude, on a small peasant pony typical of the type to be found in this part of the world, a drawing by Christian Wilhelm von Faber du Faur, a Swiss artist serving in the Württemberg contingent of the Grande Armée.

A Bavarian cavalryman haggling with Jewish traders, by Faber du Faur.

French cuirassiers returning from a foraging expedition, by Albrecht Adam.

ABOVE This scene, sketched by Albrecht Adam near Pilony on 29 June, shows how inexperienced many of the men were at the tasks of butchering and cooking.

ABOVE A detail of Württembergers on grave-digging duty, by Faber du Faur.

RIGHT Tents were not part of the French army's equipment, and the only shelter the men could hope for was one constructed with straw or foliage, like this one, depicted by Faber du Faur.

The fierce rainstorm that assailed the Grande Armée four days after it crossed the Niemen slowed progress. This engraving by Faber du Faur shows French artillery teams floundering through a sea of mud on 30 June.

The conditions took a heavy toll of the French army's horsepower, and it is thought that as many as 50,000 horses may have died in the space of twenty-four hours, often in harness, like these drawn by Albrecht Adam on 29 June.

The territory through which the French were moving was criss-crossed by a multitude of rivers, most of which had to be forded. Here, Albrecht Adam shows Prince Eugène's 4th Corps on 29 August crossing the river Vop, which was to prove a fatal obstacle on the way back.

In this print, made from drawings done on the spot by Faber du Faur, Russian light infantry can be seen fighting a rearguard action in the suburbs of Smolensk against German grenadiers of the Grande Armée.

French artillery facing Borodino on the afternoon of 5 September, by Albrecht Adam.

Russian prisoners being led away by French light infantry after the battle of Borodino, by Faber du Faur.

General, later Prince and Field Marshal, Mikhail Ilarionovich Kutuzov, named commander-in-chief of the Russian armies as their continuous retreat threatened disintegration. This etching by Hopwood clearly shows his sagging right eye, the result of a musketball passing through his head and severing the muscle behind it.

The field of Borodino, sketched on the morning after the battle by Albrecht Adam. Not until the first day of the Somme in 1916 were so many to be killed in a single day's fighting.

French soldiers looting as Moscow burns, by Albrecht Adam. The group at the centre provide a good illustration of the extraordinary range of objects taken from the houses and palaces of the gentry, most of which would be jettisoned along the road during the retreat.

Looters at work and surviving inhabitants among the ruins of Moscow, by Faber du Faur. Note the tin roof lying up-ended at the centre of the picture. These roofs were lifted in one piece and sent flying up into the air by the rush of hot air generated as the houses they covered went up in flames.

Count Fyodor Vassilievich Rostopchin, Governor and destroyer of Moscow, intelligent, cultivated, and possibly mad, by Orest Adamovich Kiprensky.

BELOW The troops passed their time as best they could in the ruins of the burnt-out city, waiting for Napoleon to decide on the next move, as this group, sketched by Faber du Faur on 8 October, shows. The weather was growing cold, which is why the sentry is wearing his greatcoat. Two of the card-players are wearing the bonnet de police, the forage cap worn off-duty.

A gloomy French sentry watches over the burnt-out suburbs of Moscow, a lithograph by Albrecht Adam after a sketch made on the spot which epitomises the hollowness of the French victory.

The roadside between Mozhaisk and Krimskoie, sketched on 18 September by Faber du Faur. Stragglers and wounded shelter in the corner of a stove, all that remains of a wooden dwelling, whose occupants' corpses were calcinated when it was burnt down. By all accounts, this was the picture all along the road travelled by the Grande Armée, and is characteristic of conditions in its rear.

As the French cavalry's horses perished, they were replaced by the little local ponies, with ludicrous results in the case of cuirassiers and carabiniers such as these, who were specially selected for their height and were supposed to be mounted on large horses. Pen, ink and wash by Faber du Faur.

Armand de Caulaincourt, drawn here by Jacques Louis David, was a wise and devoted friend to Napoleon, whom he had consistently discouraged from invading Russia, foreseeing the worst.

As this lithograph from a drawing by Faber du Faur shows, even the bridges over the Kolocha at Borodino in the Grande Armée's rear had not been cleared of the dead and debris of battle.

Three irregular cossacks enjoying a meal, by Jean-Pierre Norblin de la Gourdaine.

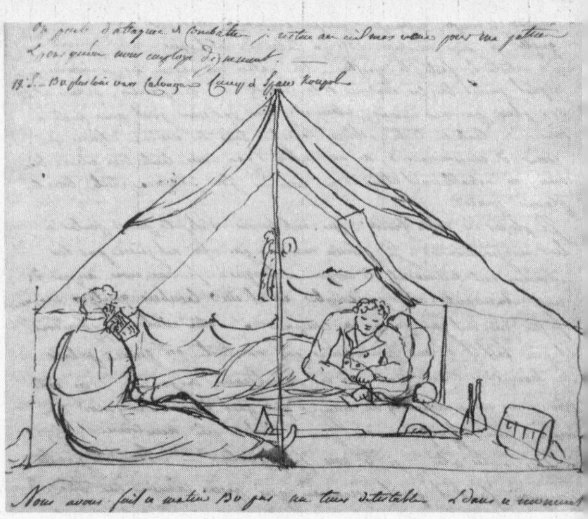

Lieutenant Aleksandr Chicherin of the Russian artillery made this sketch of himself writing his diary on 2 October while his comrade S.P. Trubetskoi looks on. Although they were in despair at the abandonment of Moscow, the Russians were beginning to realise that the French were in worse condition than themselves.

This forty-five-year-old retired officer who had returned to service in order to defend the fatherland was characteristic of a class of patriots, and was dubbed "Le nouveau Donquichotte" by Chicherin, who drew this picture of him in his diary. He would spend whole days dreaming of vanquishing the French, but when his favourite horse, "Mavr," was stolen, he went into a decline and fell behind.

wagons, but he was reluctant to increase the number of these, as that would only increase the number of horses needed: four horses could draw a heavy wagon laden with one and a half tons, but two horses could not manage a lighter wagon laden with three-quarters of a ton. And horses needed to be fed.

This detail was to be the crucial element in the forthcoming campaign. It even decided the timing of its start: as there could be no question of hauling fodder for the horses as well as supplies for the men along with the army, the horses would have to be fed on the new harvest of hay and oats (cavalrymen were issued with sickles for this purpose), and none of these crops would be ripe for harvesting before the end of June at the very earliest. This meant that while he had to begin moving up his troops in the inclement marching conditions of the winter and early spring, Napoleon could not open his campaign until the middle of the summer, which left him with very little time in which to achieve his victory.

On 1 January 1812 François Dumonceau, a Belgian officer in the Lancers of Berg, was on parade in the courtyard of the Tuileries. "The Imperial Guard, the Young and the Old, appeared more numerous and imposing than ever," he recalled. The band of the Polish Chevau-Légers played, led by a kettle drummer magnificently mounted, caparisoned, uniformed and plumed. There were also two Illyrian infantry regiments, "whose fine and robust bearing drew admiring looks." They had just marched up from the Balkans so that they could see their Emperor before they joined the Grande Armée in Germany, and they were hosted and shown around Paris by the Grenadiers of the Old Guard. A few days later Dumonceau and his men set off from their regimental depot in Versailles, marching through Brussels, Maastricht, Osnabrück, Hanover, Brunswick, Magdeburg and Stettin to Danzig.[35]

Soon the whole of Germany was covered in troops marching eastwards and northwards. There were files of cavalry: cuirassiers in helmets and breastplates, chasseurs in green uniforms and bearskin *kolpaks*, lancers in blue and crimson with four-cornered Polish caps, dragoons in helmets and uniforms of every hue. There were long convoys of artillery, in lighter blue with black shakos. And above all, endless columns of infantry.

The French footsoldier wore a standard cutaway blue coat, white breeches or trousers, white or black gaiters and a shako or a bearskin bonnet. It was not a uniform designed for convenience, let alone comfort. "I have never understood why under Napoleon, when we were constantly at war, the soldier should have been forced to wear the ghastly breeches, which, by pressing in on the hams at the back of the knee, prevented him from walking easily," wrote Lieutenant Blaze de Bury. "On top of that, the knee, which was covered by a long buttoned-up gaiter, was further strangled by another garter which pressed on the garter of the breeches. Underneath, the long undergarment, held in place with a cord, further restricted the movement of the knee. It was, all in all, a conspiracy by three thicknesses of cloth, two rows of buttons one on top of the other, and three garters to paralyse the efforts of the bravest of marchers." He wore shoes with unfashionably square toes – to prevent theft or the resale of military shoes to civilians. These shoes were supposed to last for a thousand kilometres of march, but usually fell apart far sooner.[36]

Every footsoldier carried a heavy musket 1.54 metres long without its bayonet – taller than Napoleon. On a bandolier slung across his shoulder he carried his *giberne*, a stiff leather case containing two packets of cartridges, a phial of oil, a screwdriver and other gun-cleaning implements. On his back he carried a pack made of stiffened cowhide, in which he carried a couple of shirts, collars, kerchiefs, canvas gaiters, cotton stockings, a spare pair of shoes, a sewing kit, clothes brush, pipeclay and bootwax, as well as a supply of hard tack, flour and bread. His rolled-up overcoat and other items of kit were strapped to the top of the backpack.

Most of the ornamental features of the uniform were absent on the march. Shakos and bearskins disappeared into oilcloth covers and strapped to the backpack or hung off it, plumes were put away in waxed canvas containers which were strapped to the swordbelt or bandolier. In their place the men donned the *bonnet de police*, a flat forage cap with a tasselled bob hanging down the side. Breeches and gaiters were replaced by baggy canvas trousers, and the Old Guard wore long blue tunics instead of their white-faced uniforms.

The men marched at the *pas ordinaire*, of seventy-six steps a minute, or the *pas accéléré*, of a hundred steps a minute. A usual day's march was between fifteen and thirty-five kilometres, but in a forced march they could cover anything up to fifty-five kilometres. Every unit, every horse, every man had a route prescribed for them, was instructed where to stop for the night and provided with food and accommodation. On arrival at a given stop, the unit's farrier or some other non-commissioned officer would go to the local military commander or *commissaire des guerres* and collect a visa from him. He would then take this to the town hall, where he would be given a full list of billets for the men, and chits for forage and victuals. These chits were then taken to the appointed provisioning merchants, and the unit would collect its supplies for that evening and the next day. The system worked like clockwork in France and Germany, with the required supplies ready and waiting for the tired men when they reached their prescribed halt.[37]

There were nevertheless jams of horses, cannon and wagons of every description, of straggling men and single platoons as well as large units snaking along over several miles of road, of officers in private carriages hurrying to join their units, and couriers trying to gallop in both directions, particularly at the crossings over the Rhine, at Wesel, Cologne, Bonn, Coblenz and Mainz.

Those still in Paris tried to make the most of what was a particularly glittering carnival, and the Comte de Lignières, Lieutenant in the Chasseurs of the Guard, found his order to march out when he returned to barracks at four in the morning after a ball. The notaries of Paris were kept busy writing wills, and some followed the departing troops, along with wives and lovers who wanted to see their dear ones for a few days longer.[38]

The Army of Italy came over the Brenner pass and down into the valley of the Danube, where it met up with the contingent from Bavaria. "Our march was like a brilliant and agreeable military promenade," noted Cesare de Laugier, a native of Elba. Lieutenant von Meerheimb found leavetaking from his native Saxony gloomy and tearful, but the mood changed as soon as they were on the march. "From the very first stage, every face reflected universal gaiety, and good humour reigned along the whole length of the snaking column," he wrote. They were warmly greeted as they trudged through southern Germany, and had many an amorous adventure along the way.[39]

"The ancients had a great advantage over us in that their armies were not tailed by a second army of penpushers," Napoleon frequently complained in conversation.[40] He was not referring to the vastly expanded commissariat he had organised for this campaign.

From the moment he became ruler of France, he had begun to take elements of government as well as a military staff off to war with him. And when he became Emperor, he began to take a skeleton court. For this campaign, whose scope and duration were both so imponderable, he decided to take his whole life-support system, the means to exercise government, and everything that was necessary to make a grand show wherever he went and whatever he might decide to do – be it enthroning a King of Poland or having himself crowned Emperor of India. Napoleon's equipage, under the command of the Master of the Horse Caulaincourt, consisted of some four hundred horses and forty mules carrying or drawing tents, camp beds, office, wardrobe, pharmacy, silver plate, kitchen, cellar and forges as well as an assortment of secretaries, officials, servants, cooks and grooms; and 130 saddle horses for the Emperor and his aides-de-camp. His baggage included a great many tents that would never be pitched and equipment that would never be unpacked. There was also a force of a hundred postillions attached to him. These would be posted along the road of the advance to supervise the rapid movement of mail between Paris and Napoleon's headquarters by passing over the locked boxes containing his state correspondence from one courier or *estafette* to another.[41]

The exigencies of administrating the army, its support services and the Emperor's entourage had exponentially inflated the numbers. Thousands of *commissaires* and lesser administrators, each with his servants, followed in the wake of the army. "The military administration was full of people who had never seen war and who said out loud that they had come on this campaign in order to make their fortune," complained Colonel de Saint-Chamans. Colonel Henri-Joseph Paixhans ranted against these people, "penetrated with the importance of their little persons," who together with their minions made up "a cloaca of ineptitude, baseness and rapacity."[42]

Every officer had at least one carriage, in which he kept spare uniforms, arms, maps, books and personal effects, driven by his own coachman and attended by at least one servant. General Compans, commanding the 5th Division in Davout's 1st Corps, was by no means a sybarite; if anything, he was one of the plainer-living officers. Yet his establishment at the outset of the campaign consisted of his *maitre d'hôtel* Louis; his *valet de chambre* Duval; his coachman Vaud; two valets, Simon and Louis; his gendarme Trouillet; three other servants, Pierre, Valentin and Janvier; five carriage horses, half a dozen saddle horses and some thirty draught horses; one carriage and several wagons.[43]

The unknown object of the campaign, the uncertainty as to where it might take them and the likelihood of great distances to be covered made many an officer stock up against all eventualities, and more than one had new uniforms made for himself and new liveries for his servants. Faced with the possibility of a long absence from home, many, particularly among the Italians, appear to have defied Napoleon's strict instructions and brought their wives along.[44]

As they marched across Germany and Poland, they had no clear vision of the aims of the campaign, and this dampened the ardour of some. "The future was vague, and its fortunes very distant; there was no inkling, nothing to exercise the imagination, nothing to awaken the enthusiasm," wrote Colonel Boulart of the artillery of the Guard.

This did not stop them from speculating wildly. Jakob Walter of Stuttgart thought they were being marched up to some Baltic port, from which they would be shipped to Spain. But most looked eastwards. "We thought that, together with the Russians, we would cross the deserts of that great empire in order to go and attack England in her possessions in India," wrote General Pouget. One soldier wrote home saying they were marching to England, overland through Russia.[45]

"Some said that Napoleon had made a secret alliance with Alexander, and that a combined Franco-Russian army was going to march against Turkey and take hold of its possessions in Europe and Asia; others said that the war would take us to the Great Indies, to chase out the English," remembered one volunteer.* "All of this concerned me very little: whether we were going to go to the right, to the left or straight ahead was a matter of indifference to me, as long as I could enter into the real world," wrote another. "My friends, my childhood companions were almost all in the army; they were already storing up glory. Was I, useless burden on this earth, to remain with my hands crossed to shamefully await their return? I was eighteen years old." Another, a fusilier in the 6th Regiment of the Guard, wrote to his parents telling them he was off to the "the Great Indies" or possibly "Egippe." "As for me, I don't care either way; I would like us to go the ends of the earth." He spoke for many.[46]

* Curiously enough, Alexander's secret information services had also reported to him that Napoleon's plan was to knock out Russia with a quick blow, force peace on her and then, using 100,000 Russian auxiliaries, march on Constantinople, thence to Egypt, and then to Bengal.[47]

6

Confrontation

As hundreds of thousands of men drawn from every corner of Europe tramped across Germany ready to fight and die for him, dreaming of an epic march to India or just of getting back home as quickly as possible, the Emperor of the French was setting the scene for the catastrophe that would engulf all but a handful of them.

Napoleon was about to pit himself against a huge empire while still engaged in a wasting war in Spain, with Germany in a state of ferment and Britain hovering on the sidelines ready to take advantage of any opportunity that might arise. It is customary before going to war to firm up as many allies as possible, and for one such as this it was an absolute necessity. As luck would have it, he had a number of them lining up to support him. Sweden was a natural ally, with a long history of francophilia and an interest in recovering Finland and her enclaves on the Baltic from Russia; Turkey, another traditional ally of France, was actually engaged in a bloody war with Russia; Austria, whose emperor was Napoleon's father-in-law, had many interests in common with the French; Prussia was begging to be allowed into an alliance with France; and the Poles were only waiting to be given the signal to rise up all over western Russia.

In the circumstances, Napoleon's behaviour is astonishing. On 27 January 1812, under the pretext that the Continental System was not being enforced rigorously enough there, he sent his armies into

Swedish Pomerania and took possession of it. He followed this up with a demand to Sweden for an alliance against Russia and a contingent of troops. When this was rejected by Bernadotte, he said he would allow the Swedes to recapture Finland, and offered some trading concessions. When this too was rejected, Napoleon offered to return Pomerania and threw in Mecklemburg as well as a large subsidy. But it was too late. His high-handed seizure of Pomerania had been taken as an insult in Sweden, and within two weeks of the news reaching Stockholm, Bernadotte's special envoy was in St Petersburg asking for a treaty with Russia, which was duly signed on 5 April.

As for France's other traditional ally, Turkey, Napoleon did nothing, assuming that she would go on fighting Russia unbidden. It is true that relations between France and Turkey had been strained by the treaty of Tilsit, which appeared to ally France with Turkey's enemy. It is also true that Napoleon had a low opinion of the three sultans who had followed each other in rapid and bloody succession. But at this stage any gesture of support for Turkey would have yielded real advantages: Alexander had just instructed his commander on the Turkish front, General Kutuzov, to start talks and make peace at almost any cost, as he needed all his troops to face the French.

Napoleon's treatment of Austria was hardly less offhand. The treaty he signed with her on 14 March stipulated that following a French victory Moldavia and Wallachia would be returned to Turkey, and that if Poland were to be restored, Austria could keep Galicia, or, if she preferred, receive compensation in Illyria. While the treaty suggested a common policy in central Europe and the Balkans, it kept everything vague, as Napoleon did not wish to tie his hands. For the same reason, he only asked for a small Austrian auxiliary force under Prince Schwarzenberg, which was to cover his right flank.

Frederick William of Prussia had begged Napoleon for an alliance which would restore some dignity to his country's enforced subjection to France. But Napoleon responded with a treaty, signed on 4 March, by which he graciously allowed Prussia to supply a small contingent of troops for the forthcoming campaign, on the most abject terms. This not only incensed the Prussian nationalists further, it also undermined the pro-French party in Berlin, paving the way for an explosion of anti-French feeling. It also meant that French troops had to be diverted to keeping an eye on the country, as Napoleon insisted that they march through Berlin every day and maintain strong garrisons in fortresses such as Spandau and Danzig.[1]

Finally, he refused to give the Poles an unequivocal signal, thereby strengthening the party in that country which mistrusted his intentions and believed that their best chance of survival lay with Alexander. The fact that Napoleon did not see fit to give such a signal speaks volumes both about his self-confidence and his unwillingness to damage Russia any more than was necessary. He wanted to frighten her, but he did not want to destroy her as a power. He wanted to co-opt her as an ally against Britain. There was no other reason for France to go to war with Russia: there was nothing Russia had that France could possibly have wanted. The only other conceivable motive for confronting Russia was to force her out of her newly dominant position in European affairs and neutralise her ability to threaten France.

In the first days of March, in a long conversation with one of his aides, Napoleon announced that he was determined to "throw back for two hundred years that inexorable threat of invasion from the north." He expounded a historical vision according to which the fertile and civilised south of Europe would always be threatened by uncivilised ravenous hordes from the north. "I am therefore propelled into this hazardous war by political reality," he affirmed. "Only the affability of Alexander, the admiration he professed for me, which I believe was real, and his eagerness to embrace all my schemes, were able to make me disregard for a while this unalterable fact ... Remember Suvorov and his Tartars in Italy: the only answer is to throw them back beyond Moscow; and when will Europe be in a position to do this, if not now, and by me?"[2]

He did not believe any of it. He had already shown that he was even prepared to add to Russia's power if that meant she would help him vanquish Britain. And, as ever when he thought of Russia and Britain, Napoleon's mind filled with a notion that lived uneasily beside the vision of himself as a latter-day Roman emperor throwing back the barbarian hordes, namely the Alexandrine dream of a joint march to India.[3]

To General Vandamme he gave a more perfunctory reason for going to war. "One way or another, I want to finish the thing," he said, "as we are both getting old, my dear Vandamme, and I don't want to find myself in old age in a position in which people can kick me in the backside, so I am determined to bring things to a finish one way or the other."[4] In effect, he had assembled the greatest army the world had ever seen, with no defined purpose. And, by definition, aimless wars cannot be won.

One cannot help wondering whether Napoleon did not realise this. In the weeks before setting off, he made more than his usual share of cryptically fatalistic comments. "And anyway, how can I help it if a surfeit of power draws me towards dictatorship of the world?" he said to one of his ministers who urged him to draw back from the war. "I feel myself propelled towards some unknown goal," he told another.[5] This fatalism would also explain the absence of the speed and determination which were his usual hallmarks. While the vast military machine was taking shape in northern and eastern Germany in March, the diplomatic niceties continued.

For all the talk of barbarian hordes being thrown out of Europe, the unfortunate Russian ambassador in Paris, Prince Kurakin, was finding it difficult to get away. He had never enjoyed his job, and had found it increasingly difficult to carry it out as tension mounted between Napoleon and Alexander. Things had not been made any easier when, in February, a spying scandal had broken over Paris involving Alexander's special envoy Colonel Chernyshev. He had for some time been paying a clerk at the French War Ministry to supply him with information on troop numbers and movements. The French police had got wind of this and informed Napoleon. On 25 February, just as Chernyshev was about to set off for St Petersburg with a personal letter from him to Alexander, the Emperor accorded him a long interview, in which he treated him with cordiality and respect. The following day the police broke into the apartments the departed Chernyshev had just vacated and brought the whole matter into the open.

Kurakin had to listen to torrents of outraged self-righteousness on the subject. As he watched troops leaving Paris bound for Germany, he found himself in a ridiculous position. He felt he should ask for his passports and leave, but every time he mentioned this to Maret or to Napoleon they evinced shocked surprise, affirming that there was no reason at all for him to go, and intimating that his departure would be interpreted as a declaration of war.[6]

On 24 April Kurakin called on Maret with a letter from Alexander stating that Russia would not negotiate until France withdrew all her troops behind the Rhine. This was rich, considering that only two weeks earlier Alexander himself had set off to join his armies on the frontier of the Grand Duchy of Warsaw. On 27 April Kurakin had an audience with Napoleon at the Tuileries to discuss this. The interview was not as stormy as might have been expected, and Napoleon handed him a letter for Alexander. It expressed regret that the Tsar should be ordering Napoleon where to station his troops while he himself stood at the head of an army on the frontiers of the Grand Duchy. "Your Majesty will however allow me to assure him that, were fate to conspire to make war between us inevitable, this would in no way alter the sentiments which Your Majesty has inspired in me, and which are beyond any vicissitude or possibility of change," he ended.[7]

But he could not delay any longer. He had to go and take command of his armies. Before doing so, he made arrangements for the defence and the administration of France. Although he had, as a long shot, made a peace offer to Britain, suggesting a withdrawal of all French and British troops from the Iberian peninsula, with Joseph remaining King of Spain and the Braganzas being allowed back into Portugal, he expected nothing to come of it. He therefore strengthened the coastal defences in order to discourage any British attempt at invasion, and organised a national guard of 100,000 men who could be called out to deal with any emergency.

He had considered leaving Prince Eugène in Paris as regent, but decided against it. In the event he left the Arch-Chancellor of the Empire Jean-Jacques Cambacérès in charge. The Arch-Chancellor would preside over the Council of State, which was a non-political executive composed of efficient and loyal experts.

At their last interview, on the eve of Napoleon's departure, Étienne Pasquier, Prefect of Police, voiced his fears that if the forces of opposition building up in various quarters were to try to seize power with the Emperor so far from Paris, there would be nobody on the spot with enough authority to put down the insurrection. "Napoleon seemed to be struck by these brief reflections," recalled the Prefect. "When I had finished, he remained silent, walking to and fro between the window and the fireplace, his arms crossed behind his back, like a man deep in thought. I was walking behind him, when, turning brusquely towards me, he uttered the following words: 'Yes, there is certainly some truth in what you say; this is but one more problem to be added to all those that I must confront in this, *the greatest, the most* difficult, that I have ever undertaken; but one must accomplish what has been undertaken. Goodbye, Monsieur le Préfet.' "[8]

Napoleon knew how to hide any anxiety he may have felt. "Never has a departure for the army looked more like a pleasure trip," noted Baron Fain as the Emperor left Saint Cloud on Saturday, 9 May with Marie-Louise and a sizeable proportion of his court.[9] It soon turned into more of an imperial progress.

At Mainz, Napoleon reviewed some troops and received the Grand Duke of Hesse-Darmstädt and the Prince of Anhalt Coethen, who had come to pay their respects. At Würzburg, where he stopped on the night of 13–14 May, he found the King of Württemberg and the Grand Duke of Baden waiting for him like two faithful vassals.

On 16 May he was met by the King and Queen of Saxony, who had driven out to meet him, and together they made a triumphal entry into Dresden that evening by torchlight as the cannon thundered salutes and the church bells pealed. His *lever* the next morning was graced by the ruling princes of Saxe-Weimar, Saxe-Coburg and Dessau. This was followed by a solemn Mass (it was Sunday), attended by the entire court and diplomatic corps. Napoleon went out of his way to greet the representative of Russia. The Queen of Westphalia and the Grand Duke of Würzburg arrived in Dresden later that day, and the Emperor Francis of Austria and his Empress the following day. A couple of days later Frederick William arrived in Dresden accompanied by his son the Crown Prince.

Napoleon had taken up residence in the royal palace, which Frederick Augustus had obligingly vacated, guarded by Saxon rather than French sentries. It was he who was the host, and he dictated etiquette, treating both the King of Saxony and the Emperor of Austria as his guests. At nine every morning he would hold his *lever*, which was the greatest display of power Europe had seen for centuries. It was attended by the Austrian Emperor and all the German kings and princes, "whose deference for Napoleon went far beyond anything one could imagine," in the words of Boniface de Castellane, a twenty-four-year-old aide-de-camp.[10] He would then lead them in to assist at the *toilette* of Marie-Louise. They would watch her pick her way through an astonishing assemblage of jewels and *parures*, trying on and discarding one after the other, and occasionally offering one to her barely older stepmother the Empress Maria Ludovica, who simmered with shame and fury. She loathed Napoleon for the upstart he was – and for having thrown her father off his throne of Modena many years before. Her distaste was magnified by the embarrassment and resentment she felt in the midst of this splendour, as the poor condition of the Austrian finances allowed her only a few jewels, which looked paltry next to those of Marie-Louise.

In the evening they would dine at Napoleon's table, off the silver-gilt dinner service Marie-Louise had been given as a wedding present by the city of Paris, and which she had thoughtfully brought along. The company would assemble and enter the drawing room in reverse order of seniority, each announced by a crier, beginning with mere excellencies, going on to the various ducal and royal highnesses, and culminating with their imperial highnesses the Emperor and Empress of Austria. A while later, the doors would swing open and Napoleon would stride in, with just one word of announcement: "The Emperor!" He was also the only one present who kept his hat on.

To some of the older people present, and particularly to the Emperor Francis, there must have been an element of the surreal about the proceedings. It was less than twenty years since his sister Marie-Antoinette had been shamefully dragged to the scaffold and guillotined to please the Parisian mob, yet here was this product of the French Revolution not only ordering them all about, but insinuating himself into the family, becoming his own son-in-law. At dinner one evening, the conversation having touched on the tragic fate of Louis XVI, Napoleon expressed sympathy, but also blamed his "poor uncle" for not having shown more firmness.[11]

His stay in Dresden was enlivened with balls, banquets, theatre performances and hunting parties. They were by no means just gratuitous show, but part of a carefully choreographed display of power. "Napoleon was indeed God at Dresden, the king amongst kings," was how one observer saw the proceedings. "It was, in all probability, the highest point of his glory: he could have held on to it, but to surpass it seemed impossible." Napoleon was flexing his muscles before the whole world, and he meant everyone to sit up and take note. On the one hand he wanted to remind all his German and Austrian allies of their subjection to him. More importantly, he was still hoping that Alexander's nerve would break; that when he saw himself isolated and faced with such an array of power he might agree to negotiate.

To many, this still seemed the most likely outcome. "Do you know that many people still do not believe there will be war?" Prince Eugène wrote to his pregnant wife from Płock on the Vistula on 18 May.

"They say it won't take place, as there is nothing to be gained from it by either party, and that it will all end in talk." Napoleon's secretary Claude-François Meneval noted "an extreme repugnance" to war on his master's part.[12]

Napoleon had convinced himself that Alexander was being manipulated by his entourage, and that if only he could talk to him directly or through some trusted third party, he would manage to strike a deal. He therefore sent a special envoy to the Tsar. For this delicate and, as he thought, crucial mission, he chose one of his aides-de-camp, the Comte Louis de Narbonne.

Narbonne was a fifty-seven-year-old general, who had, in turn, been Minister of War in the early stages of the revolution, an émigré and Napoleon's ambassador in Vienna. He was a man of vast education, with literary tastes and a special interest in the diplomacy of the Renaissance, on which he was something of an expert. He was generally believed to be the natural son of Louis XV, and exuded all the elegance and grace associated with the *ancien régime*. If anyone could inspire trust in Alexander, it must surely be him.

But Napoleon was deluding himself. Even had he wished to, Alexander could not afford to negotiate with him. "The defeat of Austerlitz, the defeat of Friedland, the Tilsit peace, the arrogance of the French ambassadors in Petersburg, the passive behaviour of the Emperor Alexander I with regard to Napoleon's policies – these were deep wounds in the heart of every Russian," recalled Prince Sergei Volkonsky. "Revenge and revenge were the only feelings burning inside each and every one." He may have exaggerated the strength and the universality of these feelings, but they were gaining ground, encouraged by popular literature, which scoured Russia's past for patriot heroes. "The upsurge of national spirit manifested itself in word and deed at every opportunity," wrote Volkonsky. "At every level of society there was only one topic of conversation, in the gilded drawing rooms of the higher circles, in the contrasting simplicity of barracks, in quiet conversations between friends, at festive dinners and evenings – one, and only one thing was expressed: the desire for war, the hope for victory, for the recovery of the nation's dignity and the renown of Russia's name."[13]

Reading the letters and memoirs of Russian nobles of the time, one is struck by the fact that nobody seems to have a good word to say of anyone in positions of authority, be it in the civil administration or the army. They reverberate with invective against "foreigners" running the country, and laments over "corruption," Freemasonry, "Jacobins," and any other bogey that came to mind. Much of this discontent settled on the figure of Speransky, who was heartily detested by Grand Duchess Catherine and her court, and by most of the nobility, who hated him for blocking their careers by introducing qualifying exams for senior posts in the civil service and who feared his alleged intention of emancipating the serfs. "Standing beside him I always felt I could smell the sulphurous breath and in his eyes the glimpse the bluish flames of the underworld," noted one contemporary.[14]

In February 1812 an intrigue was spun by Gustav Mauritz Armfeld, a Swede who was one of Alexander's military advisers, with the participation of the Minister of Police Aleksandr Dmitrievich Balashov, to show up Speransky as being in secret contact with the French (which indeed he was, with Talleyrand, on Alexander's orders). At the same time a rumour was spread to the effect that the police had uncovered a plot by Speransky to arm the peasants and call them out against their masters.

Alexander had Speransky put under surveillance by the police, but also had his Minister of Police followed and watched – this kind of paranoia was not a Soviet innovation. It is impossible to tell whether Alexander believed that Speransky had betrayed him or not, but it was certainly clear even to him that his State Secretary's unpopularity was not only tainting him, but even exposing him to danger.

On the evening of 29 March 1812 Speransky was summoned to an audience with the Tsar in the Winter Palace. There were no witnesses to the two-hour interview, but those waiting in the antechamber could see that something was wrong when the Minister emerged from the Tsar's study. Moments later the door opened again and Alexander himself appeared, with tears pouring down his cheeks, and embraced Speransky, bidding him a theatrical farewell. Speransky drove home, where he found Balashov waiting for him. He was bundled into a police *kibitka* and driven off through the night to exile in Nizhni Novgorod.[15]

His post as State Secretary was given to Aleksandr Semionovich Shishkov, a retired admiral and a particular hater of everything pertaining to France and her culture. He had denounced the Tilsit treaty and made frequent attacks on Speransky, and had of late achieved a certain notoriety through his *Dissertations on the Love of One's Fatherland*. He was astonished, and somewhat overwhelmed, to be hauled out of obscurity. But nobles up and down the empire rejoiced.

The fall of the hated Minister was an unequivocal signal to them that Alexander had understood he needed them during the uncertain times ahead. He was acutely aware that an invasion of Russia might well trigger a new Time of Troubles similar to that two hundred years before. It was probably for this reason that he agreed to give the post of Governor General of Moscow, which had fallen vacant in March, to his sister Catherine's protégé Count Fyodor Rostopchin. This erstwhile Foreign Minister to Tsar Paul was a lively, clever man, forthright in his opinions, but he was also something of a fantasist, and possibly mentally unbalanced. Alexander did not consider him up to the job, and had tried to resist his sister's request. "He's no soldier, and the Governor of Moscow must bear epaulettes on his shoulders," he argued. "That is a matter for his tailor," she riposted. Alexander gave way. It was, after all, a largely honorary post.[16]

He had removed the most obvious points of friction between himself and his people, and pacified the most vociferous centres of opposition. Now it was up to God. At 2 p.m. on 9 April, after attending a solemn service in the monumental new cathedral of Our Lady of Kazan, Alexander left St Petersburg for the army, accompanied part of the way by a crowd of wellwishers who ran alongside his carriage cheering and weeping. He had decided that his place was with his troops.

The Russian army was unlike any other in Europe, and could not have been more different from the French, particularly where the common soldier was concerned. He was drafted for a period of twenty-five years, which effectively meant for life. He was unlikely to serve that term, as no more than 10 per cent survived the dreadful conditions and the frequent beatings – including the practice of making them run the gauntlet of two rows of their comrades beating them as they went – let alone disease or death in battle.

When he was drafted, his family and often the entire village would turn out to see him off, treating the event as a funeral. His family and friends excised him from their lives, never expecting to see him again. As the children of men drafted into the army could not be looked after by working single mothers, they were sent to military orphanages to be brought up and trained to become non-commissioned officers when they grew up. But conditions in these institutions were so poor that only about two-thirds of them survived into adulthood.[17] If the conscript were to return after a quarter of a century (with no leave and no letters) he would be a stranger. And he would no longer be a serf, so there was no place for him in the rural economy any more. Those who did last out the twenty-five years would therefore either try to go on serving, or go off to towns looking for work.

When they entered their regiments the conscripts effectively joined a brotherhood, removed from the normal stream of Russian life and bound together by misery. Virtually the only respect in which they had an advantage over their French counterparts was that their uniforms, predominantly green in colour, were more practical and less constricting, as well as better made. In peacetime, their platoon functioned as a trading corporation, an *artel*, leasing their labour to local civilians, with the profits theoretically being shared out between them, though more often going into the pockets of their officers.

Desertion was difficult within the boundaries of the Russian empire, as an unattached peasant would stick out wherever he went. But when Russian armies were stationed along the western border it became frequent, and many would cross it and take service in the Polish or other forces. When they operated abroad, particularly when they were about to return home, desertion became common, and its scale testified to the misery of military life. In 1807, as they began their march back into Russia after Tilsit, Prince Sergei Volkonsky noted that his regiment, the élite Chevaliergardes, lost about a hundred men in four days, despite doubled sentries posted all around the camp perimeter.[18] The men nevertheless behaved with the greatest patriotism and loyalty in the face of the enemy.

Much of the training in the Russian army was directed at good performance on the parade ground rather than on the battlefield. The men were drilled mercilessly and marched about in formation until they learnt to operate as a mass, and taught to rely on the bayonet rather than musketry. In battle, obedience was considered to be a key factor. A special instruction addressed to infantry officers stipulated that on the eve of an engagement they must give their men a talk, reminding them of their duty and that they would be severely punished for any signs of cowardice. Even trying to dodge a cannonball while the unit was standing to was to be punished by caning. If a soldier or non-commissioned officer showed cowardice in the field, he should be executed on the spot. The same went for one who created confusion, by, for instance, shouting "We're cut off!," as he was to be considered a traitor.[19] All these factors conspired to generate solidarity, resilience and the ability to put up with almost any conditions. But they did not breed intelligence or initiative.

The chasm dividing officers from the other ranks was unbridgeable, and there was no possibility of promotion. The officers were drawn exclusively from the nobility. They were supposed to serve their apprenticeship in the ranks, but usually did this in cadet formations or officer schools, and kept contact with their troops down to a minimum. This was not a problem, since many could not sustain a conversation in Russian. But they did personally cane them for minor faults.

The pay of junior officers in the Russian army was lower than anywhere else in Europe. And as promotion to senior ranks was almost entirely dependent on influence at court, junior officers from the minor nobility were sentenced to a life of poverty and obscurity. As a result, such a career only attracted those of meagre talents. The operations of 1805–1807 had shown up grave faults in the command structure of the Russian army, lack of cooperation between units and arms, and other weaknesses, mainly to do with the low calibre and poor training of the officers. But all attempts at addressing these problems were vitiated by the rapid expansion of the armed forces over the next few years, which created a shortage of officers, with the result that in 1808 the length of training was actually cut.

Alexander did everything he could to prepare the army for its next showdown with Napoleon. He created a Ministry of the Armed Forces with the aim of making the army more effective, and lavished money on it. Military spending rose from twenty-six million roubles out of a total budget of eighty-two million at Alexander's accession to seventy million out of a budget of 114 million by 1814. He raised the draft, which took four men out of five hundred souls in 1805, to five out of five hundred, yielding 100 to 120,000 men each year, which meant that he conscripted more than 500,000 men between 1806 and 1811. In the course of that year 60,000 retired but capable soldiers were brought back into service. The total number under arms in Russia's land forces increased from 487,000 in 1807 to 590,000 in 1812, and in March of that year an extra draft of two men per five hundred souls yielded another 65 to 70,000 men. By September 1812 the total number of men under arms in the land forces would reach 904,000.[20]

In 1803 Alexander had charged General Arakcheev with modernising the artillery. His reforms did not yield fruit in time for the war of 1805, but by the end of the decade it was probably the most professional in Europe. Arakcheev got rid of small-calibre guns, and equipped it with six- and twelve-pounder field guns, and ten- and twenty-pounder "*licornes*," a kind of howitzer. He fitted these guns with the most sophisticated and accurate sights, and made sure that the gunners knew how to use them to best advantage.

The latest reform to see the light of day, in January 1812, was an ordinance for the command of large armies in the field. This laid out clearly who was responsible at every level of command, and gave the commander-in-chief almost unlimited powers in time of war. It also prescribed the channels through which all information should flow up from the furthest outpost to the commander-in-chief and how orders should be transmitted down from him to the company commander. Unfortunately, it was to be almost universally disregarded in the forthcoming campaign, with lamentable results.

The Russian forces in Lithuania were divided into two armies. The First, the stronger of the two, under General Barclay de Tolly, was deployed along the river Niemen in a slight arc almost a hundred kilometres long in advance of Vilna, a position from which it could either move out to attack or mount a defence. The Second Army, under General Bagration, was concentrated like a strike force, ready either to support the advance of the First Army by outflanking any enemy defence, or to ram the exposed flank of any force that attacked the First Army before Vilna. A Third Army, under General Tormasov, guarded the frontier to the south of the Pripet marshes. Exact figures for the strength of these forces are impossible to establish: the conflicting calculations of Russian historians put the total land forces as low as 356,000 and as high as 716,000, with as few as 180,000 and as many as 251,000 effectives in the front line. Most recent studies have been more consistent, but there is some confusion as a result of Russian historians' habit of breaking down effectives into "front-line" figures and totals, which include all the support services. The myth of the Russians being vastly outnumbered during this campaign has stemmed from the juxtaposition of Russian "front-line" figures with French totals. The strength of the First Army was 127,800 front-line and 159,800 in total, that of the Second 52,000 and 62,000 respectively,

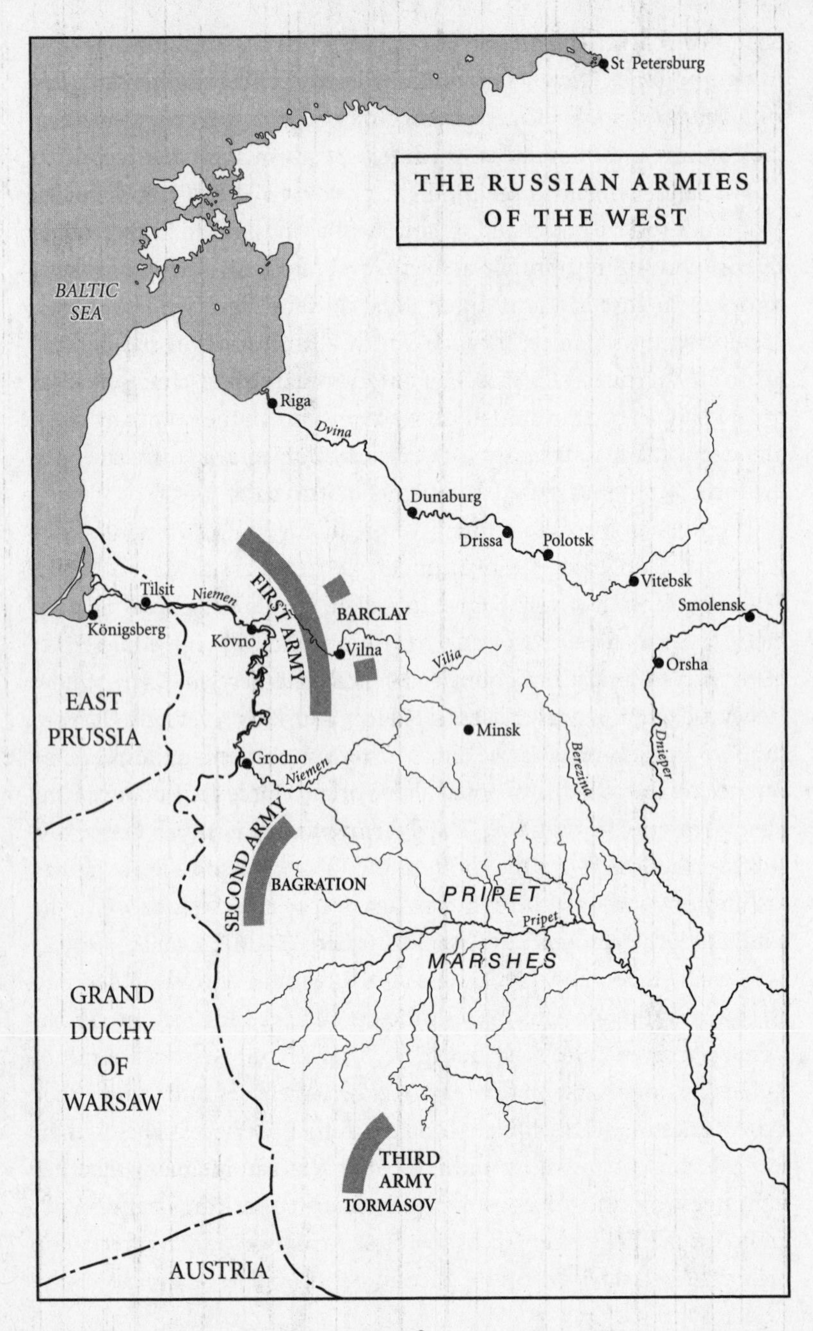
THE RUSSIAN ARMIES OF THE WEST
St Petersburg
BALTIC SEA
Riga
Dvina
Dunaburg
Drissa
Polotsk
Vitebsk
Smolensk
Orsha
Tilsit
Niemen
Königsberg
Kovno
FIRST ARMY
BARCLAY
Vilna
Vilia
Minsk
Berezina
Dnieper
EAST PRUSSIA
Grodno
Niemen
SECOND ARMY
BAGRATION
PRIPET
Pripet
MARSHES
GRAND DUCHY OF WARSAW
THIRD ARMY
TORMASOV
AUSTRIA

and of the Third 45,800 and 58,200, making up a force of 225,000 front-line and 280,000 in total positioned along the frontier, supported by just over nine hundred guns.

This force was backed up by two reserve corps, Ertel's of 55,000/ 65,000, and Meller Zakomelsky's of 31,000/47,000, bringing the total of the Russian forces facing Napoleon to 392,000. Behind them a second wave of units was being formed. As soon as Alexander's diplomatic demarches had secured peace with Sweden and Turkey, another 28,500/37,200 troops from Finland and 54,500/70,000 from Moldavia would be brought into play. The armies at the front were well supplied and supported for an offensive by a series of magazines at Vilna, Shvienchiany, Grodno and elsewhere, and a second line stretching behind that from Riga in the north to Kaluga in the south.[21]

It is difficult to gauge what Alexander intended by coming to Vilna, as he gave no indication that he meant to take command. His armies had been massing in the region for the past eighteen months, and were poised along the frontier in readiness to attack. The Tsar's arrival at advance headquarters would, in the circumstances, suggest that the decision had been taken to launch the attack, since there could be no conceivable point in His Imperial Majesty coming all that way just to review them. In the event, Alexander's presence helped to confuse what was already a highly confused situation.

The man ostensibly in command was the fifty-one-year-old Minister of War, General Mikhail Barclay de Tolly. He was an intelligent, sensible and competent man of strong character and independent judgement, who had shown his mettle in the field against Swedes, Poles, Turks and Frenchmen, being seriously wounded at Eylau. He was undoubtedly very brave and steady under fire. Alexander liked him and had promoted him over the heads of others, arousing their jealousy. He was reserved and stoical rather than affable, which did nothing to increase his popularity among his peers. As he could not count on their cordial cooperation, he was wont to look to everything himself. And although he had done more than any other to ease the lot of the common soldier, he was not the kind of commander who inspires devotion in the rank and file.[22]

While Barclay had been appointed commander of the First Army, he had not been nominated commander-in-chief. This may have been because Alexander felt this to be unnecessary since he was also Minister of War, or it may have been because he intended to command himself, or did not want to offend others vying for the post.

Ostensibly, Alexander took no part in military affairs beyond inspecting fortifications and attending parades. But he did interfere in day-to-day business. And his presence at headquarters ineluctably diminished Barclay's already fragile authority, as it gave huge scope for insubordination – and there was no lack of those who hated Barclay and resented the idea of serving under him.

Foremost amongst these was General Prince Piotr Ivanovich Bagration, commanding the Second Army. Bagration was a dashing battlefield general, recklessly brave, with a volcanic temper, a warm heart, and all the fiery bravado necessary to make him worshipped by his officers and men. Although younger than Barclay, he had been a general for longer and therefore felt he had a right to overall command. In the absence of any document stipulating that he must take his orders from Barclay, Bagration took the view that his command was an independent one. Alexander's arrival at headquarters provided him with an excuse to send all his reports to the Tsar, as supreme commander, bypassing Barclay entirely.

Bagration's position was a strong one. He was extremely popular with his fellow generals and had a great following at court. And, having been for a time Grand Duchess Catherine's lover, he also had a slight ascendancy over Alexander, who did not like him but could not be seen to put him down.

Matters were made no easier by the Tsar's younger brother Grand Duke Constantine Pavlovich, who commanded the Imperial Guard, an unbalanced blusterer whose only real pleasure in life derived from making troops parade before him. For him, the parade was a kind of ballet – perfect precision in the choreography and meticulous attention to dress were assured by the generous distribution of savage punishments for being a few inches out of line or missing a button. Between parades, Constantine could be counted on to support anyone who criticised Barclay or failed to carry out one of his orders.

This state of affairs opened the door for a comeback by General Levin Bennigsen. A Hanoverian by birth, this old professional soldier had taken service in Russia more than fifty years before. A competent commander, if a little fussy and ponderous, he had risen slowly but surely, and assured his survival in Alexander's good graces by being one of those who had murdered his father. He had ended his military career on a less than satisfactory note, leading the Russian army to defeat at Friedland. He was now sixty-seven years old. He had retired, and was living on his estate of Zakrent, which happened to be a few kilometres outside Vilna. Ever since Friedland he had longed for a chance to redeem his reputation, and felt it was he, not Barclay, who should be given overall command. Upon his arrival in Vilna, Alexander summoned Bennigsen and asked him to return to service in an unspecified capacity in his personal entourage.

This was already alarmingly large. Alexander was surrounded by a swarm of unofficial advisers, including his brother-in-law Prince George of Oldenburg, his uncle Prince Alexander of Württemberg, the Swedish adventurer Gustav Mauritz Armfeld, the French émigré Jean Protais Anstett, and a host of others. This was partly the result of Alexander's longer-term views and ambitions. "Napoleon means to complete the enslavement of Europe, and to do so he has to strike Russia down," he wrote to Baron vom Stein, inviting him to come and help plan the crusade for its liberation. "Every friend of virtue, every human being who is animated by the sentiment of independence and love of humanity is interested in the success of this struggle."[23]

Since Alexander also needed to carry on exercising political power, he had instructed his most important ministers to follow him to headquarters. He was soon joined in Vilna by Admiral Shishkov, who was slightly baffled to find himself, the effective Prime Minister and Minister of the Interior, stuck out at military headquarters. General

Arakcheev, head of the Council of State's Military Committee and Secretary to the Emperor for Military Affairs, was also in attendance. Chancellor Rumiantsev suffered a mild stroke on his way to Vilna, but this did not stop him joining the Tsar, though Alexander henceforth conducted his diplomatic business through his Secretary for Foreign Affairs Karl von Nesselrode.

The presence of so many different hierarchies of power had undesirable repercussions in the army at every level, and exacerbated a problem, created by the dearth of native officer material, which was to bring forth torrents of bad blood and bedevil the conduct of the campaign: the presence of large numbers of foreigners.

There were literally hundreds of French officers serving in the Russian army, most of them émigré aristocrats who had fled the revolution. They occupied a variety of posts, including some of the highest, with the Marquis de Traversay Admiral-in-Chief of the Russian Fleet, the Comte de Langeron and the Marquis Charles Lambert in command of army corps and General de Saint-Priest as chief of staff to the Second Army. There were also Italians, Swiss, Swedes, Poles and others: Barclay came of a family of Baltic barons who traced their ancestry to the Barclays of Towie in Scotland; Bagration's origins were Georgian. But those who caused the greatest problem were the Germans, and particularly the Prussians.

Hundreds of officers who had been cashiered as a result of the reduction of the Prussian army after Jena had taken service in Russia. More joined them over the next years, with a second great wave after Napoleon's defeat of Austria in 1809. And a final batch had just arrived, disgusted by the servile alliance Prussia had signed with France in February. Among these officers were the Prussian military reformers Major von Boyen and Colonel von Gneisenau, the future military theorist Major von Clausewitz, the staff officer Colonel Karl von Toll, and Baron Ludwig von Wolzogen, a native of Saxe-Meiningen and former aide-de-camp to the King of Württemberg.

All the Russian officers spoke French amongst themselves, and that was the language in which orders were normally given, but many of the Prussians lapsed into German when communicating with each other. As Barclay also spoke good German, they would address him in that language, creating an impression of a foreign club within the army, particularly at headquarters, since most of the Germans were employed as staff officers.

Alexander's presence in Vilna had a paralysing effect on the vital question of Russia's overall strategy. While he refrained from favouring any particular option, he lent a willing ear to anyone who expressed a view, and then asked others what they thought of it, thus seemingly opening to discussion something that should have been decided in small committee at the outset. There were a number of options to be considered.

There was the old plan, formulated by Barclay, Bennigsen, Bagration and others in the previous year, of a strike into Poland followed by an advance into Prussia to liberate it from French domination. Bagration repeatedly begged Alexander to implement this plan even at this late stage. "What have we to fear?" he wrote to his sovereign on 20 June. "You are with us, and Russia is behind us!"[24] According to some sources, Barclay still favoured this plan, although he was less sanguine about the chances of success than some of his colleagues. He was presumably also aware of his master's reluctance to be seen as the aggressor, and therefore prepared a second plan, of defending the frontier along the river Niemen. He stretched his forces along the frontier, ostensibly so they could contain and beat back any French attempt at crossing the river.[25]

Barclay had come up with another plan back in 1807, when he was lying in hospital recovering from a wound received at Eylau. The Russians had just been defeated by the French at Friedland, and he saw their only hope of avoiding total annihilation in a retreat deep into Russia. If the French were to follow them, they should avoid giving battle, but concentrate on consolidating their forces by marching back towards their bases. The further the French came after them, the more men they would have to leave behind, and the longer their lines of communication and supply would grow. In the end, the Russians would find themselves superior in numbers and resources and would be able to defeat the French.[26]

It was not a particularly original idea: the strategic asset provided by the vast country was something of a cliché, and Russian officers often brandished it as a threat in conversations with foreigners – Alexander himself had done so.[27] But Barclay had only considered it in 1807 as a last resort, a counsel of despair, at a moment when Russia did not really have an army left. Napoleon's willingness to treat with Alexander at Tilsit had saved the day, and no more was said of the plan.

While there was much interest in Russian and Prussian military circles in the concept of a long-drawn-out defensive war, inspired partly by Wellington's tactics in Spain, it was not one based on retreat. In a long memorandum written for Alexander at the end of July 1811, Barclay suggested moving out to attack the French, but not in a conventional battle – he advised loose manoeuvring by large numbers of light troops, which could harry and demoralise, dragging out the campaign and avoiding decisive engagements. This was to be carried out on enemy territory. Withdrawal into Russia was not something that could be seriously considered when there was a numerous and well-equipped army standing in defence of her borders, and neither Barclay nor Alexander, nor any of the Russian generals for a moment contemplated such a strategy.[28] It would have been politically inadmissible and militarily absurd. The troops were positioned for attack, not for retreat. Their stores and depots were as close behind them as they could be so that they could support an attack, and would be condemned to destruction or capture by the French if a retreat were ordered. And drawing an enemy into Russia raised all manner of terrifying possibilities, including peasant revolt – it was only four decades since a rebellion by the peasant leader Emelian Pugachov had brought the empire to the brink of collapse. The memory was fresh in people's minds, kept so by regular minor eruptions of discontent.

There was only one man at Russian headquarters who entertained a plan based on retreat, and it was nothing like the poetic vision of drawing the enemy in to be devoured by the expanses and forces of Russia. He was Karl Ludwig von Phüll. He had left the Prussian service after Auerstädt and joined the Russian army, in which he had been given the rank of Lieutenant General.

Phüll's plan was based on a tactic adopted by his hero Frederick the Great in 1761 when confronted by overwhelming forces. Frederick had fallen back into an entrenched camp and worn down the two enemy armies which besieged him there. Phüll suggested that in the event of a French invasion, the Russian First Army should fall back to a previously prepared position, drawing the French in behind them. The Second Army could then come up in the rear of the French and inflict great damage on them. For this purpose he, or rather his protégé Wolzogen, selected a site at Drissa, covering both the Moscow and the St Petersburg roads. Work had started in the last months of 1811 on the construction of massive earthworks that were to make the position impregnable.

The Drissa idea appealed to Alexander because it reminded him of Wellington's fallback to the lines of Torres Vedras in 1811. But he did not come down firmly in favour of this or any other plan, and he simultaneously entertained various other options. One was to launch a rising in the Balkans and Hungary in order to cause a diversion. The idea was the brainchild of Admiral Pavel Vasilievich Chichagov, an eccentric but competent sailor and erstwhile admirer of Napoleon, currently serving under Kutuzov on the Turkish front. He suggested that, having made peace with Turkey, Russia should use her army on that front to invade Bulgaria, whose population was Orthodox and therefore russophile, from there launch an attack on Napoleon's provinces along the Dalmatian coast, and thence into the heart of Napoleonic Europe via Italy and Switzerland. Alexander was entranced by the sheer scale of the scheme, and toyed with it for some time before Rumiantsev pointed out that it was unrealistic and diplomatically counter-productive, as it would rouse both Turkey and Austria against Russia and force them into Napoleon's camp.

Then there was the Polish card, which Alexander was still trying to find ways of playing. While in Vilna he devoted much effort to seducing the local Polish aristocracy, bestowing orders and honorific titles and making the odd allusion to the possibility of restoring Poland. He had a couple of trusted agents sounding out opinion, and wrote to Czartoryski asking whether now would not be a propitious moment to declare his intention of doing so. He was encouraged in this by Bernadotte, who wrote urging him to strike into Poland and to offer its crown to Poniatowski. Alexander sent Colonel Toll on a secret mission to Poniatowski to offer him high office (possibly even the crown) in a future Kingdom of Poland if he agreed to detach his corps from the French army and take it over to the Russian side. Poniatowski was astonished at this request, which would have been impossible to carry out, even had he wished to.

In his quest for ways of subverting the Poles, Alexander also instructed the notorious Catholic pugilist and Sardinian ambassador in St Petersburg, Joseph de Maistre, to employ the Jesuits (who had been disbanded by the Pope but had perversely been kept alive in Russia) to subvert Poland, using the argument that Alexander was the defender of the Papacy, while Napoleon was its enemy.[29]

Alexander's sister Catherine was urging him to leave the army. "If one of [the generals] commits a fault, he will be blamed and punished; if you make a mistake, everything falls on your shoulders, and the destruction of confidence in him on whom everything depends and who, being the only arbiter of the destiny of the Empire, must be the support to which everything bends, is a greater evil than the loss of a few provinces," she wrote.[30]

What she did not point out was that he had already done a great deal of damage by going to Vilna, and was compounding it by his irresolute behaviour. His refusal to commit himself to any of the options laid before him or to openly place his confidence in any one of his generals meant that nobody knew what to prepare for. His brother Grand Duke Constantine drilled his soldiers mercilessly, but nobody was preparing to meet the approaching Grande Armée. No serious attempts were being made to plot the enemy's movements, and the units had not even been issued with adequate maps of the areas they were to operate in.[31]

"In the meantime we held balls and parties, and our prolonged sojourn in Vilna resembled a pleasure trip rather than preparations for war," in the words of Colonel Benckendorff. Shishkov was astonished by the carefree atmosphere and the lack of any sense of imminent menace he found on his arrival in Vilna. "Our everyday life was so carefree that there was not even any news about the enemy, as though they had been several thousands of versts* away," he wrote. The troops had settled into their billets and savoured whatever pleasures the country life of Lithuania could provide. "The senior officers feared Napoleon, seeing him as a fearful conqueror, a new Attila," wrote Lieutenant Radozhitsky of the light artillery, "but we younger ones romped with the god of love, sighing and moaning from his wounds."[32]

People far away from the front could not understand why the Russian army, whose officers wrote home letters full of bravado, did not attack and drive the French out of Prussia and Poland. There was grumbling about the lack of action, reinforced by widespread fear of a French advance into Russia, not least because it might provoke social unrest.

In May the erroneous news reached St Petersburg that Badajoz and Madrid had fallen to the British and that a Spanish army had crossed the Pyrenees into southern France. Why, people asked themselves up and down the country, was Alexander not marching out to deal the final blow against Napoleon? He and his entourage appeared to be whiling away the time at balls and parties, and it was reported in the capital that the officers were indulging in "orgies."[33]

Russian estimates of the size of the Grande Armée were very low. Barclay and Phüll put the strength of the French forces at

---

* One verst = 1,060 metres, approximately five-eighths of a mile.

200–250,000; Bagration at 200,000; Toll at 225,000; Bennigsen at 169,000; and Bernadotte at 150,000. The highest estimate drawn up by anyone on the Russian side was 350,000, and that included all reserves and rear formations.[34] This meant that an attack on it would have been seen as perfectly feasible, and Alexander undoubtedly longed to launch one. His excitement about the Chichagov plan and his attempt to bribe Poniatowski can only be viewed in the context of an offensive. And there are other indications that he wanted to take command of it.[35]

But he was heavily influenced by Phüll's views, and Phüll was against any attack, believing as he did that the Russian army was not up to it.[36] Above all, Alexander wanted to be seen as the innocent victim rather than the aggressor, and his religious instincts told him to play the part of passive tool of the divine will.

In recent years he had made more and more references to the will of God in his letters and utterances, and he had been increasingly guided by the wish to make himself a worthy and righteous instrument of that will. "I have at least the consolation of having done everything that is compatible with honour to avoid this struggle," he had written to Catherine in February. "Now it is only a question of preparing for it with courage and faith in God; this faith is stronger than ever in me, and I submit with resignation to His will."[37]

Nesselrode was still advising Alexander to negotiate rather than provoke a war, but Alexander seems to have ruled out negotiations entirely as an option, and he was in no mood to talk when Narbonne arrived in Vilna on 18 May.[38] He received him and read the letter he had brought, but told him that as Napoleon had ranged the whole of Europe against Russia it was evident his intentions were hostile, and that there was therefore no point in negotiating. He reiterated that he would only consider doing so if Napoleon withdrew his troops beyond the Rhine.

"What does the Emperor want?" he asked Narbonne rhetorically. "To subject me to his interests, to force me to measures which ruin my people, and, because I refuse, he intends to make war on me, in the belief that after two or three battles and the occupation of a few provinces, perhaps even a capital, I will be obliged to ask for a peace whose conditions he will dictate. He is deluding himself!" Then, taking a large map of his dominions, he spread it on the table and continued: "My dear Count, I am convinced that Napoleon is the greatest general in Europe, that his armies are the most battle-hardened, his lieutenants the bravest and the most experienced; but space is a barrier. If, after a few defeats, I retreat, sweeping along the population, if I leave it to time, to the wilderness, to the climate to defend me, I may yet have the last word over the most formidable army of modern times."[39]

Although most people at Russian headquarters assumed that the only purpose of Narbonne's mission was to spy out their dispositions and rouse local patriots to stage an uprising, Alexander invited him to attend a parade on the following day, and to dine with him afterwards. But the next day Narbonne was informed by one of Alexander's aides-de-camp that a carriage generously provisioned for the journey back to Dresden would be waiting at his door that evening.[40]

7

The Rubicon

Narbonne's post-chaise, almost white from a thick coating of dust, rolled into the courtyard of the Royal Palace in Dresden on the afternoon of 26 May. He was shown upstairs and promptly summoned into the imperial presence. He gave a full account of his conversations with Alexander, laying stress on the Tsar's determination and on his parting words. "Tell the Emperor that I will not be the aggressor," Alexander had told him. "He can cross the Niemen; but never will I sign a peace dictated on Russian territory."[1]

It is difficult to know what Alexander expected Napoleon to make of this message. He had stipulated that he would not enter into any talks unless Napoleon evacuated all his troops from the Grand Duchy of Warsaw and Prussia, while he was himself poised with his army on the borders of those states. Napoleon had only two options: to disband his huge army and go home, exposing himself to attack as he did so, and leaving the whole of Poland and Germany open to invasion; or he could invade himself. He could only have taken Alexander's message as a taunt, "a sullen challenge" as the British historian William Hazlitt put it.[2] But he thought it was prompted by bravado rather than conviction. He therefore sent a courier to Lauriston in St Petersburg instructing him to go to the Tsar's headquarters at Vilna, as a last resort.

Napoleon was not afraid of war with Russia. "Never has an

expedition against them been more certain of success," he said to Fain, pointing out that all his former enemies were now allied to him. It was true that he had just received a somewhat disheartening reply to his last proposal for an alliance with Sweden. But it had only been a verbal one, and he assumed that in the event of his invading Russia Sweden would be unable to resist the opportunity of recovering Finland. "Never again will such a favourable concourse of circumstances present itself; I feel it drawing me in, and if the Emperor Alexander persists in refusing my proposals, I shall cross the Niemen!"[3]

He adopted a confident, even a blustering tone. "Before two months are out, Alexander will sue for peace," he declared, "the great landowners will force him to." He brushed aside Narbonne's warnings that this campaign would be difficult to win on account of the special nature of the nation and the terrain. "Barbarian peoples are simple-minded and superstitious," he asserted. "A shattering blow dealt at the heart of the empire on Moscow the great, Moscow the holy, will deliver to me in one instant that whole blind and helpless mass."[4]

But his plans were still dangerously confused, as he had come no closer to defining his goals. "My enterprise is one of those to which patience is the key," he explained to Metternich. "The more patient will triumph. I will open the campaign by crossing the Niemen, and it will end at Smolensk and Minsk. That is where I shall stop. I will fortify those two points, and at Vilna, where I shall make my headquarters during the coming winter, I shall apply myself to the organisation of Lithuania, which is burning to be delivered from the Russian yoke. I shall wait, and we shall see which of us will grow tired first – I of making my army live at the expense of Russia, or Alexander of nourishing my army at the expense of his country. I may well myself go and spend the harshest months of the winter in Paris." And if Alexander did not sue for peace that year, Napoleon would mount another campaign in 1813 into the heart of the empire. "It is, as I have already told you, only a question of time," he assured Metternich.[5]

He seemed to produce a different plan for every interlocutor. "If I invade Russia, I will perhaps go as far as Moscow," he wrote in his instructions to one of his diplomats. "One or two battles will open the road for me. Moscow is the real capital of the empire. Having seized that, I will find peace there." He added that if the war were to drag on, he would leave the job to the Poles, reinforced by 50,000 French and a large subsidy.[6]

He still refused to see Alexander as an enemy to be defeated, thinking of him rather as a wayward ally. Had it not been so, he would have declared the restoration of the Kingdom of Poland with its 1772 frontiers, thereby launching a national insurrection in the rear of the Russian armies. He could also have proclaimed the liberation of the serfs in Russia, which would have ignited unrest all over the country. This would have reduced the Russian empire to such a state of chaos that Alexander would have been in no position to mount a serious defence and Napoleon could have marched his troops about the country as he chose. But he wanted to bring Alexander back to heel with as little unpleasantness as possible and a minimum of damage. "I will make war on Alexander in all courtesy, with two thousand guns and 500,000 soldiers, without starting an insurrection," he had said to Narbonne back in March.[7]

Narbonne and Maret repeatedly put the case for creating a strong Polish state which would become a French satellite and a bulwark against Russian expansion. Napoleon did not rule this out. He did have to keep the Poles on his side, and he needed to prime, even if he did not need to fire it, the weapon of Polish national insurrection in Russia. In a word, he had to manipulate and deceive the Poles. And in order to do this, he must send a clever man to Warsaw as an unofficial personal ambassador.[8]

He had originally selected Talleyrand for this purpose, but for a number of diplomatic reasons his choice now fell on the Abbé de Pradt, Archbishop of Malines, "a priest more ambitious than cunning, and more vain than ambitious," as one contemporary described him. Pradt had made himself useful to Napoleon in the past, but he inspired neither confidence nor respect, and lacked the qualities necessary for the job in hand. He was described by one of the Poles with whom he would be working as "a nonentity, without a trace of dignity" who loved intrigue and gave the impression that he despised Napoleon. But whether anyone else could have done a better job in the circumstances is another question. Napoleon made it clear that Pradt was to encourage the Poles to announce their intention of resurrecting a Polish state and to start a national insurrection, without committing himself or his imperial master to backing it up.[9]

Napoleon even gave some thought to the question of whom to put on the Polish throne if he did decide to restore the kingdom. It would be too important a place for the volatile Murat or the inexperienced Prince Eugène, both of whom believed themselves to be in line for the job. He did consider Marshal Davout, who was a good soldier and administrator, and popular with the Poles, but the example of Bernadotte raised questions as to his future loyalty. One of his brothers might be a better bet in the circumstances. "I'll put Jérôme on it, I'll create a fine kingdom for him," he told Caulaincourt, "but he must achieve something, for the Poles like glory." He duly put Jérôme in command of an army corps and directed him to Warsaw, where he was supposed to win the love of the Poles. Napoleon could hardly have made a worse choice.

Jérôme made a regal entry into the Polish capital and announced that he had come to spill his blood for the Polish cause in the spirit of the crusaders of old. The Poles found him overbearing and ridiculous, and it was not long before all sorts of malicious stories were circulating about him, including one that he took a bath in rum every morning and one in milk every evening. His army corps, composed of German troops, behaved abominably, as did its French commander General Vandamme, who demonstrated his contempt for the locals by, among other things, putting his muddy-booted and spurred feet up on fine silk upholstery as he lounged in Warsaw drawing rooms. The Poles longed to be rid of Jérôme and his unruly soldiers.

"In truth, the king of Poland should have been Poniatowski," Napoleon admitted later, during his exile on St Helena. "He had every title to it and he had all the necessary talents." But at the time, the thought did not cross his mind, which was beset by more pressing considerations.[10]

His armies were now reaching their prescribed positions, and he needed to take charge. So, after thirteen days in Dresden, where he had settled nothing, he climbed into his travelling carriage, a yellow coupé drawn by six horses. His mameluke Roustam climbed onto the box next to the coachman, and Berthier installed himself inside with Napoleon.

The vehicle was fashioned to suit his every need and fitted out to allow him to make the best use of his time. It could be turned into a makeshift study, with a tabletop equipped with inkwells, paper and quills, drawers for storing papers and maps, shelves for books, and a light by which he could read at night. It could also be turned into a *couchette*, with a mattress on which he could stretch out, a wash-basin, mirrors and soap-holders so he could attend to his *toilette* and waste no time on arrival, and, naturally, a chamberpot.

One of the outriders from the Chasseurs à Cheval of the Guard noted that the Emperor took a long time over his farewells to Marie-Louise, and that there were tears in his eyes as he got into the carriage. But feelings of tenderness were quickly dispelled by unpleasant realities.[11]

Napoleon drove through Glogau in Silesia to the Polish city of Poznań, which he entered on horseback, riding under an arch inscribed with the words *Heroi Invincibili*. The whole town was illuminated and festooned with flags and garlands. He reviewed units of the Legion of the Vistula fresh from Spain, but was vexed at the sight of the recruits. "These people are too young," he complained to Marshal Mortier. "I want people capable of standing up to hardship; young people like this only fill up the hospitals." It was true. Teenagers made poor soldiers, not only because they were puny and prone to exhaustion and illness, but also because they could not stand up for themselves, and were easily bullied and pushed around, which led to demoralisation.[12]

After criticising the recruits he attended a ball in his honour at which he made a poor impression on the inhabitants, telling them he wanted to see them booted and spurred, not in dancing pumps. But it was not the attitude of the Poles that lay at the root of Napoleon's displeasure. On his arrival in Poznań he had sat down with the head of the commissariat, Pierre Daru, to review the provisioning of his troops, only to discover severe shortcomings. Matters only got worse as he progressed on his journey. By the time he reached Toruń, he was furious. He complained bitterly to General Mathieu Dumas, Intendant General in charge of supplies, that none of his orders had been carried out.[13]

The supply machine he had devoted so much time and thought to had never quite materialised. "The means of transport, whether supplied by the military teams belonging to the army or by auxiliary means, were almost everywhere insufficient," admitted Dumas. "This immense army, which crossed the Prussian lands like a torrent, consumed all the resources of the land, and supplies from the reserve could not follow it with enough speed." There was a shortage of draught horses from the outset, and the consequences grew serious as the army began massing in Poland.[14]

The troops had already been subjected to a rude awakening. For those who had not taken part in the 1807 campaign, there was an element of surprise at the exoticism and the backwardness of many of the areas east of the Oder. They marvelled at the emptiness of the landscape and the flocks of storks. Henri Pierre Everts, a native of Rotterdam and a major in the 33rd Light Infantry regiment in Davout's corps, could hardly believe his eyes when he beheld a Polish village for the first time. "I stopped in astonishment, and remained for some time sitting still on my horse observing those miserable wooden cottages of a type unknown to me, the small low church, also made of planks, and at the squalid appearance, the dirty beards and hair of the inhabitants, amongst whom the Jews seemed extraordinarily repulsive; all of this engendered some bitter reflections on the war which we were about to wage in such a country."[15]

The meat and potatoes washed down with beer or wine which they had got used to on the march through France and Germany were replaced by buckwheat gruel, and the best they could find to drink was bad vodka, mead or *kwas*, made of fermented bread. Even these had to be purchased, mostly at inflated prices, from the Jews who swarmed round them in every town and village. Communication took place in a variety of pidgin French, German and Latin. "Up to that point, our march had been no more than a pleasant promenade," wrote a rueful Julien Combe, a lieutenant in the 8th Chasseurs à Cheval.[16] From now on, it was to be an ordeal.

East Prussia and Poland were neither as rich nor as intensely cultivated as most of western Europe. The Continental System had diminished the amount of land under cultivation, since much of the produce had previously been exported, and had lost its markets as a result of the blockade. The traditional exports such as timber, potash, hemp and so on had also been cut off from their markets. To make matters worse, the previous year had seen a serious drought and the harvest had failed. This meant that landowners had been obliged to use up all their reserves of grain and fodder just to keep themselves and their peasants alive, so much so that there was a shortage of grain for sowing in the spring of 1812. The poorest peasants were eating bread made of acorns and birch bark, and pulling thatch off roofs in order to feed their horses and cattle.[17]

The need to raise an army almost twice as large as the territory and the population could realistically furnish or support put a terrible strain on the economy and the administration. The government of the Grand Duchy of Warsaw was insolvent, and no official had been paid for eight months. "The hardships we were suffering seemed so bad that things could not get much worse," wrote the wife of the Prefect of Warsaw to a friend at the end of March 1812, "but it turns out that things can get worse, and worse without limit."[18] Things did indeed get much worse as hundreds of thousands of hungry men and horses flooded into the area.

As there were no stores, military or otherwise, the troops took what they needed where they could find it. Giuseppe Venturini, a Piedmontese lieutenant in the 11th Light Infantry, bemoaned the fact that when he was ordered to go out and requisition supplies, he "reduced two or three hundred families to beggary." As the locals were unwilling to sell or give away the little that stood between them and hunger, the troops took it by force. The French system of provisioning effortlessly turned into looting.[19] And matters quickly degenerated from there.

"The French destroy more than they take or even want to take," noted an eighteen-year-old captain in the 5th Polish Mounted Rifles. "In the houses, they smash everything they can. They set fire to barns. Wherever there is a field of corn, they ride into the middle of it, trampling more than they feed on, without a thought for the fact that in a couple of hours their own army will come up looking for forage." The situation was aggravated by the multinational make-up of the army, as there was no sense of national pride or responsibility to restrain men who marched under a foreign flag. Everyone blamed another nationality, and even Polish troops looted their compatriots.[20]

A Polish officer travelling to join the army found himself moving though a scene of devastation: every window was smashed, every fence had been ripped up for firewood, many houses were half demolished; horse carcases as well as the heads and skins of slaughtered cattle lay by the roadside being gnawed by dogs and pecked at by carrion birds; people fled at the sight of a rider in uniform. "One felt that one was following a fleeing rather than an advancing army," wrote a Bavarian officer following in the wake of Prince Eugène's corps, astonished at the numbers of dead horses and abandoned wagons littering the road.[21]

The situation was no better in East Prussia, where violent national animosities also came into play. Even troops from other parts of Germany found the atmosphere hostile, and stragglers were attacked by locals. The soldiers responded in kind. The Dutchman Jef Abbeel and his comrades took full advantage of their position to show what they thought of the Prussians. "We would force them to slaughter all the livestock we judged we needed for our sustenance," he writes. "Cows, sheep, geese, chickens, all of it! We demanded spirits, beer, liqueurs. We were billeted in villages, and, since only the towns were provided with shops, we would sometimes demand the locals to drive three or four leagues to satisfy our needs. And they would be thanked on their return with blows if they failed to procure everything we demanded. They had to dance as we sang, or they would be beaten!"[22]

A cold start to the year meant that the harvest was late. "We were obliged to cut the grass of the meadows, and, when there was none, reap corn, barley and oats which were only just sprouting," wrote Colonel Boulart of the artillery of the Guard. "In doing so we both destroyed the harvest and prepared the death of our horses, by giving them the worst possible nourishment for the forced marches and labours to which we were subjecting them day after day."[23] Fed on unripe barley and oats, the horses blew up with colic and died in large numbers.

Without bread, meat or vegetables, the men, particularly the younger recruits, fell ill and perished in alarming numbers. Many sought salvation in desertion and a dash for home. Others, preferring quick release to the long-drawn-out pangs of hunger and the uncertainties that lay ahead, put their muskets to their heads and shot themselves. One major in the 85th Line Infantry of Davout's corps complained he had lost a fifth of his young recruits by the time he reached his position on the Russian border.[24]

Napoleon did not see the worst of this as he rushed ahead. Before leaving Poznań he had written to Marie-Louise that he would be back in three months; either the Tsar's nerve would break when he saw the Grande Armée come up to the border or he would be knocked out in a quick battle. Napoleon was now in a hurry to bring things to a head. He raced to Danzig, moving so fast that he left most of his household behind, arriving there on 8 June. He inspected troops and supplies, accompanied by the military governor, General Rapp. At Danzig he also met up with Marshal Davout, commander of the 1st

Corps, and with his brother-in-law Joachim Murat, King of Naples.

It would be hard to bring together two more different characters. Louis-Nicolas Davout was a year younger than Napoleon. He came from an old Burgundian family with roots in the Crusades, and was the most devoted as well as the ablest of Napoleon's marshals. He was strict and demanding, a hard taskmaster to those serving under him, feared and disliked by most of his peers, but loved by his soldiers because, in order to get the most out of them, he made sure they had everything they needed and were not tired out with unnecessary duties.

Joachim Murat, who was three years Davout's senior, was of a different cut in every way. The son of a Gascon innkeeper from Cahors, he had studied for the priesthood at a seminary in Toulouse before running away to join the army. Although not without a certain cunning, he was dimwitted, which allowed him to be absurdly and recklessly brave even though he lacked real courage. He was, in Napoleon's words, "an *imbecille* [sic] without judgement." But he was an instinctively brilliant cavalry commander in battle. He was also devoted to Napoleon. He had married the Emperor's sister Caroline, and in 1808 he was made King of Naples.[25]

In Danzig, Napoleon explained to Davout and Murat what part they were to play in his plans. Murat would command the huge body of cavalry, a great battering ram of four divisions, with a nominal strength of 40,000, which was to spearhead the attack. Napoleon wanted to fight and defeat the Russians as quickly as possible, so he decided to strike them at the point where they might feel strong enough to make a stand, which meant a frontal attack at Vilna. He would attack Barclay's First Army, using Davout's 1st Corps of 70,000 men, flanked by Ney's 3rd Corps of 40,000 to the north and backed up by the Guard, numbering some 40,000. Prince Eugène's 4th and St Cyr's 6th Corps, totalling 67,000 Italians, Bavarians and Croats as well as Frenchmen, would advance to the south of this thrust, driving a wedge between Barclay and Bagration. Further south, Jérôme was to advance against Bagration with three other army corps (5th Polish,

7th Saxon and 8th Westphalian), altogether some 60,000 men. In the north, Marshal Macdonald's 10th Corps, made up of Prussians as well as Frenchmen, would cross the Niemen at Tilsit and advance on Riga, while Oudinot's 2nd Corps supported both him and the main strike force by attacking Barclay's right wing. South of the Pripet, Schwarzenberg's Austrians were to mark Tormasov's Third Army.

It is impossible to be precise about the numbers involved. On paper, the overall strength of the forces poised for invasion was 590,687 men and 157,878 horses, while the total number of French and allied troops in the whole theatre of operations, including Poland and Germany, was 678,000. But these figures beg many questions.[26]

The strength of an army which has taken up positions, as the Russian had done over the months, can be established fairly accurately, as the units are concentrated in one place, and there is little reason or scope for anyone to absent themselves for more than the few hours it might take to report to headquarters or pick up some stores. But an army on the move is far more volatile.

Whatever the technical strength of any unit on campaign, it is never concentrated in a single place, or even area, at one time. It always leaves a skeleton force, sometimes a whole battalion, at its depot. It does not move, lock stock and barrel, from one place to another: its head races ahead, leaving its body and tail to catch up, which they occasionally do, only to be left behind once more, in the manner of a huge centipede. It is constantly leaving behind platoons or smaller clusters of men to hold, defend or police areas. Numbers vary, almost always downwards, with every day.

A company of 140 men marches out from town A on its way to town B. On the morning they are setting off, it turns out that three of the men are too ill to march, so they are left behind, in the care of a corporal and two orderlies. In addition, one of the captain's four horses is lame, and a second is out of condition, so they remain behind, in the care of an orderly. One of the company's ammunition *caissons* or luggage wagons has a broken axle, and remains in town A while it is being repaired, in the care of two men. One man failed

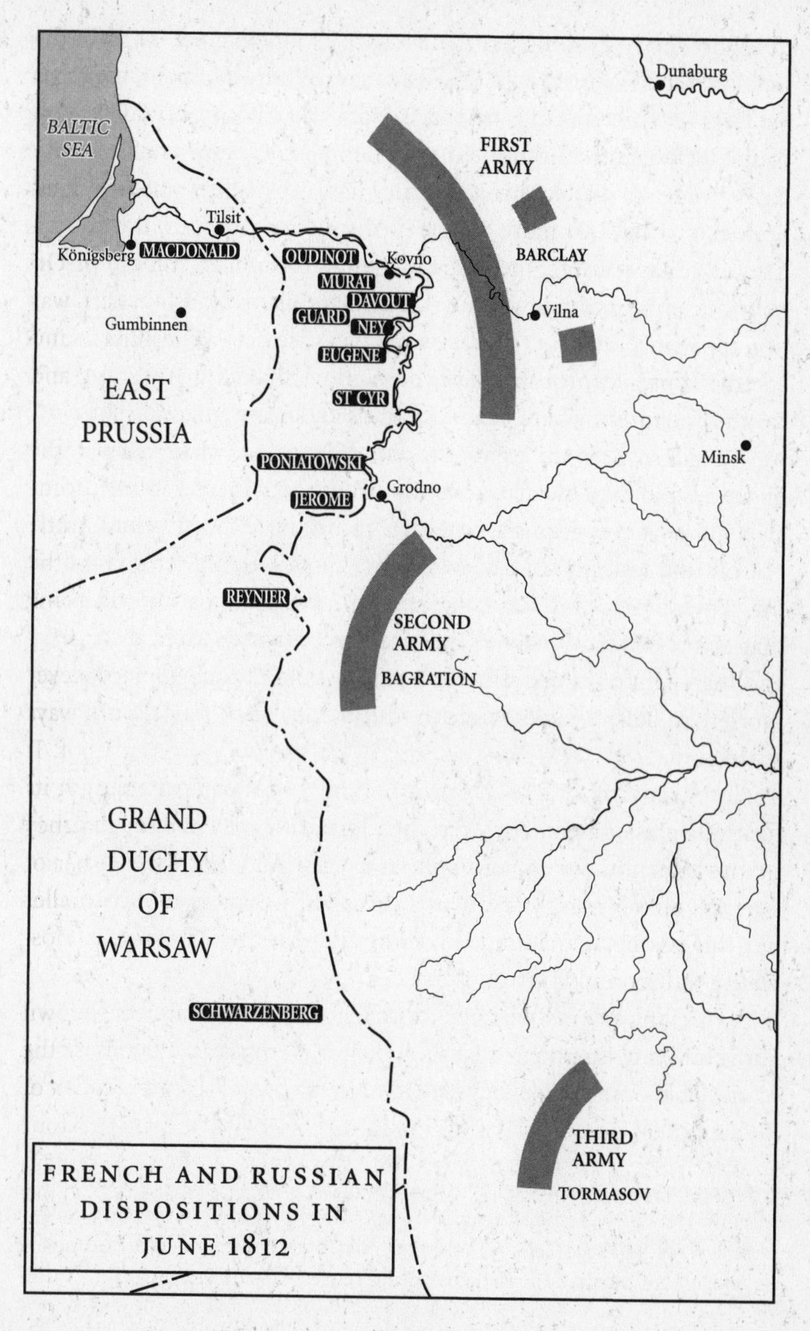

FRENCH AND RUSSIAN DISPOSITIONS IN JUNE 1812

to report for roll call before the company marched out. This means that only 130 men actually set off. Along the way, eight men are detailed to find supplies, and they set off into the countryside with a couple of wagons. Another ten men fall behind in the course of the day's twenty-five-kilometre trek, and, another of the wagons having broken a wheel, two more are detailed to look after it until it can be fixed. By that evening, the company with a technical strength of 140 men can only assemble 110 men in a single place. And that diminution took place without the intervention of disease, bad weather or the enemy. It would probably have been more drastic in the case of a cavalry squadron, where lameness and saddle sores played their part. And there has been no account taken of desertion, which is far easier on the march than in a fixed position, and which increases the further an army is from its home ground.* Some of the men left behind catch up, but the faster and further an army moves, the fewer do, and so the gap between those catching up and those falling away widens. If that same company had to make a forced march over three days and then fight on the fourth, its captain would be lucky to lead much more than half its paper strength into battle – less than a week after setting out.

Numbers arrived at by means of adding up the paper strength of the units present in an army can therefore serve only as a rough guide to the situation on the ground. It is generally accepted that the strength of the Grande Armée as it invaded Russia was about 450,000, but this has been arrived at by computing theoretical data, and the reality was certainly very different.

On 14 June Napoleon issued a circular to the commanders of every corps insisting that they must provide honest figures on the numbers of the able-bodied, the sick and deserters, as well as the dead and

---

* Desertion was easy and common, as J. Steininger's *Mémoires d'un vieux deserteur* show. He was the son of a deserter from Württemberg, and served, successively, under the following colours, deserting each time: Piedmont, Britain, Prussia, Holland, Austria, Bavaria, Piedmont again, Naples, Corsica, France.

the wounded. "It has to be made clear to the individual corps that they must regard it as a duty towards the Emperor to provide him with the simple truth," ran the order.[27]

This admonition was ignored. "He was led astray in the most outrageous way," wrote General Berthézène of the Young Guard. "From the marshal to the captain, it was as if everyone had come together to hide the truth from him, and, although it was tacit, this conspiracy really did exist; for it was bound together by self-interest." Napoleon was always angry when provided with dwindling figures, particularly if these could not be explained by battle casualties, so those responsible simply hid the losses from him. Berthézène went on to say that the Guard, which was usually written up as being nearly 50,000 strong, never exceeded 25,000 during the whole campaign; that the Bavarian contingent, given as 24,000, was never stronger than 11,000; and that the whole Grande Armée was no larger than 235,000 when it crossed the Niemen. One can quibble with his estimates, but not with his argument, which is supported by others.[28]

Russian estimates of the French forces at this stage were much lower than the generally accepted figures (and intriguingly close to Berthézène's), which has surprised historians and led them to believe that they must have had very poor intelligence. But it may simply be that while French figures were based on paper computations, the Russians based their estimates on reports from spies, and those reports may have been more accurate as to the numbers of troops actually present than the paper calculations.

It would be rash to try to be precise, but a sensible guess would be that no more than three-quarters and possibly as little as two-thirds of the 450,000 crossed the Niemen in the first wave, and that the remainder, if and when they caught up with the main body, were only plugging gaps left by men dropping away. At the same time, it would be difficult to *over*estimate the number of civilians following in the wake of the army, and a figure of 50,000 would certainly be on the conservative side.

* * *

Having fixed his plan, Napoleon applied himself to putting it into action. Speed was of the essence. He wanted to get at the Russian army before it had time to withdraw or concentrate. Speed was also essential for logistical reasons: with the shortage of supplies available, the ground was burning the feet of the Grande Armée. He was counting on being able to confront and defeat the Russian army inside three weeks, as he could not possibly take with him supplies for any longer.

From Danzig, he raced on to Marienburg, Elbing and Königsberg. At every point along his frantic journey he inspected troops, artillery parks and supply depots. During the four days he spent at Königsberg he inspected the stores and boatyards as, having seen for himself the state of the local roads and appreciated the shortage of draught animals, he had decided to despatch as many supplies as possible up the Niemen and its tributary the Vilia to reach him once he had occupied Vilna.

As there would be scant possibility of finding provisions along the way, he had given orders that every soldier should carry with him four days' ration of bread and biscuit in his knapsack, and every regiment a twenty-day ration of flour in its wagons. But his orders could only ever be as productive of results as the land was fertile in the necessary means, and they were meaningless where there was nothing to be had.

On 22 June General Deroy, a splendid warrior in his eighties who had more than sixty years' military service behind him and would soon die in battle commanding one of the Bavarian divisions, reported to his monarch that he did not see how they were going to survive at all. "I am looking forward to getting killed," one young soldier wrote home to his parents in France, "for I am dying as I march."[29]

When he saw for himself the poverty of the surroundings, Napoleon gave the order for the units in the principal strike force under his personal command to make a last-minute requisition and seize whatever they could in the way of provisions before marching out. Thus the unfortunate inhabitants of East Prussia suddenly found that their every cart was taken and filled with anything that came to hand. Napoleon brushed aside the complaints reaching him from all quarters about shortage of supplies and dwindling forces. There was nothing he could do about it anyway – except defeat the Russians as quickly as possible. And he trusted in his extraordinary ability to achieve what he wanted in the face of insuperable obstacles.

On 16 June he wrote to "Quiouquiou," as the King of Rome's governess the Comtesse de Montesquiou had been dubbed by her charge, thanking her for informing him that his son's teething was nearly over. Two days later he heard from Marie-Louise that she was not pregnant, as he had been led to believe by a hint from one of his courtiers. He registered his disappointment and his hope that they would have a chance to put that right in the autumn. He wrote to her daily, in short, scribbled, mis-spelt notes of remarkable banality. "I am often on horseback, and it is doing me good," he informed her on 19 June.[30]

The following day, at Gumbinnen, he was reached by a courier from the French embassy in St Petersburg who informed him that Lauriston had been refused an audience with the Tsar and forbidden to travel to Vilna. He and the diplomatic representatives of the various allied states had been instructed to call for their passports, which amounted to a declaration of hostilities.

Napoleon's propaganda machine, the *Bulletins de la Grande Armée*, which presented his troops and the world with his version of events, swung into action. The first Bulletin of the campaign detailed his long and painstaking efforts to keep the peace, and reminded the world of the generosity with which he had treated the defeated Russians in 1807, all to no avail. "The vanquished have adopted the tone of conquerors," the Bulletin announced, "they are tempting fate; let destiny then take its course." He announced to his soldiers that they would be required to fight soon. "I promise and give you my imperial word on it that it will be for the last time, and that you will then be able to return to the bosom of your families."[31]

The three corps – Davout's, Ney's and Oudinot's – which were to cross the Niemen first, along with Murat's cavalry, were massing in the low-lying ground between Wyłkowyszki and Skrawdzeń, concealed from view behind the high left bank of the river. The summer heat was intense, made more unbearable for the marching men by the clouds of dust kicked up by the hundreds of thousands of feet and hooves. On 22 June Napoleon set out from Wyłkowyszki, passing the marching columns, and reached Skrawdzeń at dusk. He had supper in the garden of the parish priest's house, and asked the priest whether he prayed for him or for Alexander, to which the man replied: "For Your Majesty." "And so you should, as a Pole and as a Catholic," replied Napoleon, delighted by the answer.[32]

At about eleven o'clock he climbed back into his carriage, which drove off in the direction of the Niemen, past large encampments of Davout's infantry and Murat's cavalry, which had been instructed to remain out of sight of the river. He did not want the Russians patrolling the other bank to see a single French uniform, and only Polish patrols, which were a familiar sight, were allowed to show themselves.

It was well past midnight when Napoleon's carriage rolled up to the bivouac of the 6th Polish Lancers. He got out, proceeded to swap his famous hat and overcoat for the cap and coat of a Polish lancer, and made General Haxo of the Engineers, Berthier and Caulaincourt do the same before they mounted horses and set off, escorted by a platoon of lancers. Napoleon rode into a village, from one of whose houses he and Haxo could, unnoticed, survey the city of Kovno on the other side of the river through their telescopes. He then rode up and down the bank, looking for the best place for a crossing. As he was riding along at full gallop a hare started just in front of his horse, causing it to shy abruptly, and Napoleon was thrown. He jumped up immediately and remounted without a word.

Caulaincourt and others of his entourage were astonished: normally Napoleon would have launched into a string of curses directed at his horse, the hare and the terrain, but this time he acted as though nothing had happened. "We would do well not to cross the Niemen,"

Berthier said to Caulaincourt. "This fall was a bad omen." Napoleon himself must have felt the same. "The Emperor, who was ordinarily so gay and so full of ardour at times when his troops were executing a major manoeuvre, remained very serious and preoccupied for the rest of the day," wrote Caulaincourt.[33]

Napoleon spent most of that day, 23 June, working in the tent that had been pitched for him. He seemed in sombre mood, and his entourage reflected this by maintaining a silence that many later interpreted as being full of foreboding. But this may have been hindsight. "Despite the uncertain future, there was enthusiasm, a great deal of it," recalled Colonel Jean Boulart of the artillery of the Guard. "The army's confidence in the genius of the Emperor was such that nobody even dreamed that the campaign could turn out badly."[34]

Amongst other things, Napoleon was working on a proclamation to be read out to his troops the following morning:

> *Soldiers! The Second Polish War has begun. The first ended at Friedland and Tilsit: at Tilsit Russia swore an eternal alliance with France and war on England. She is now violating her promises. She refuses to give an explanation of her strange behaviour unless the French eagles retire beyond the Rhine, thereby leaving our allies at her mercy. Russia is tempting fate! And she will meet her destiny. Does she think that we have become degenerate? Are we no longer the soldiers of Austerlitz? She has forced us into a choice between dishonour and war. There can be no question as to which we choose, so let us advance! Let us cross the Niemen! Let us take the war onto her territory. The Second Polish War will be glorious for French arms, as was the first; but the peace that we will conclude will be a lasting one, and will put an end to that arrogant influence which Russia has been exerting on the affairs of Europe over the past fifty years.*[35]

The proclamation would be greeted with enthusiastic shouts of "*Vive l'Empereur!*" when it was read out the following morning. Some were left cold, but according to Étienne Labaume, an officer on Prince

Eugène's staff who hated Napoleon, it "excited the ardour of our soldiers, always ready to listen to anything that flattered their courage." "His words," affirmed Boulart, "acted mightily on the imagination of all and awakened all the ambitions." "It was so fine, that I almost had it by heart," recalled an eighteen-year-old military surgeon.[36]

At six o'clock that evening, Napoleon mounted up and rode over to the riverbank once again. He spent the next six hours reconnoitring and then watching as, at ten o'clock, three companies of the 13th Light Infantry crossed the river silently in boats and fanned out on the other side, while General Jean-Baptiste Eblé and his men began putting in place three pontoon bridges. A patrol of Russian Hussars rode up to the infantrymen, and its officer challenged them with the regulation French "*Qui vive?*" It was not a particularly dark night, but uniforms were hard to make out. "*France!*" came the answer. "What are you doing here?" the Russian shouted, again in French. "F—k,* we'll show you!" they shouted back, letting off a volley of shots which scattered the Hussars.

Napoleon was annoyed by the sound of musketry, as he had hoped to keep the Russians in ignorance of his movements for as long as possible. He rode back to his tent to snatch a couple of hours' sleep, but at three o'clock in the morning he was back in the saddle, riding a horse named "Friedland" after his last victory over the Russians. By dawn the three bridges were in place, and General Morand's division, the first of Davout's corps, was on the other side, ready to cover the crossing of Murat's cavalry.

Napoleon took up his position on a knoll on which the sappers of the Guard had built him a small bower and a seat out of branches. From here he surveyed the scene, sometimes using the telescope which he held in his right hand, while his left was folded behind his back. There was no trace of the previous day's preoccupation, and he seemed happy, occasionally humming military marches to himself as

---

* Diarists of the period invariably display a reticence about spelling out certain words.

he looked down on what one witness termed "the most extraordinary, the most grandiose, the most imposing spectacle one could imagine, a sight capable of intoxicating a conqueror."[37]

"The army was in full dress, and from the top of the hill on which the Emperor stood, one could see it file across the three bridges on the Niemen in perfect order," recalled Nicolas Louis Planat de la Faye, an aide-de-camp standing close by. The units marched up, one after the other, converging from different directions, on the heights that dominated the left bank of the Niemen. They then descended to cross one of the three bridges over the river, and deployed on the flat right bank. "Each regiment marched behind its band, which played fanfares that mingled with cries of '*Vive l'Empereur.*' As there were no enemy troops to fight, it looked like an immense military parade."[38]

"All the finest men in full dress, all the most beautiful horses of Europe, were brought together there under our eyes, around the central point which we occupied," wrote Louis Lejeune, chief of staff to Berthier. "The sun shone on the bronze of twelve hundred cannon ready to destroy everything; it shone on the breasts of our superb carabiniers with their gilded helmets and scarlet manes; it shone on gold, on silver, on the tempered steel of helmets, of breastplates, of the weapons of men and officers, and on their rich uniforms." The men revelled in their shared sense of power. Some thought of Caesar, some of the crusades. "Nobody doubted the success of the enterprise; and at the sight of the means his will had brought together, our hearts beat with joy, pride and vainglory, making us dream of and smile at this forthcoming success," according to Lejeune.[39]

At midday, Napoleon remounted his horse, rode across the river and positioned himself on the Russian bank, where the men crossing the river could clearly see him. "*Vive l'Empereur!*" they shouted as they passed. Once he was satisfied that the essential elements of his strike force were across, he mounted a fresh horse, named "Moscow," and rode off to Kovno. That evening he took up quarters in a monastery on its outskirts.

"*Vive l'Empereur!*" Captain Fantin des Odoards wrote in his journal at a campfire just outside the city. "The Rubicon has been crossed. The shining sword which has been drawn from its scabbard will not be put away in it before some fine pages are added to the glorious annals of the great nation."[40]

8

Vilna

As Napoleon was taking up his quarters in Kovno at the end of an epic day, Alexander drove out to a ball. It was being given for him by his staff officers at the country house of General Bennigsen, at Zakrent, just outside Vilna. The house, which was built on the ruins of a former Jesuit convent, had extensive walled gardens, and as it was a balmy moonlit summer night, dinner was served around the fountains in the illuminated gardens, with only the dancing taking place indoors. During the dinner, Alexander wandered from table to table, pausing to talk to the guests. He was affability itself, captivating all with his usual charm. He then danced with his hostess, Countess Bennigsen, and with Barclay's wife, before turning his attention to the other ladies. He showed particular interest in some of the young beauties of Vilna society, whom he thrilled by choosing to dance with them. It was already quite late when Balashov sidled up to him and whispered something into his ear, so Alexander's departure soon afterwards did not cause any surprise, and the party went on.[1]

There had been a great many parties in the two months since Alexander's arrival in Vilna in April. And if the rumours of "orgies" were a little far-fetched, the time had indeed been whiled away in the pursuit of pleasure. "The fight is about to begin, and we are expecting to be attacked any day," Alexander had written to Aleksandr

Nikolaevich Galitzine only a week before. "We are all prepared and will do our best to carry out our duty. As for the rest, God will decide."[2]

In fact, nobody and nothing was prepared. Even the burning issues of command and strategy had not been addressed, and as a result Alexander found himself in a very uncomfortable position when he returned to his quarters in the former archbishop's palace in the early hours of 25 June. He had just learned that Napoleon had crossed the Niemen, and he had no idea what to do next.

He sent for Shishkov, who was fast asleep in his bed. When the State Secretary arrived, he found the Tsar writing at a table. Alexander informed him of the French invasion and instructed him to compose a proclamation to rally the nation in the face of foreign aggression, the gist of which he dictated himself. One of the few sensible things Alexander had done on his arrival in Vilna was to set up a printing press at headquarters, which would allow him to issue manifestos and leaflets with propaganda to combat the kind of fear and rumour that Napoleon's reputation generated in the lower ranks.[3]

A few hours later Alexander held a council of war in order to decide on a course of action. Attack was no longer an option. Bagration's plan, to strike up into what he believed to be the underbelly of the French armies, was hardly advisable in the absence of firm intelligence of French numbers and positions. A defence of Vilna was difficult to organise, as the Russian forces were scattered over too large an area. The French invasion had taken everyone unawares, and it would be days before they could be readied for battle. There was nothing for it but to withdraw. With Alexander's approval, Barclay gave orders for a fallback to Shvienchiany, where he hoped to be able to concentrate his forces. He wrote to Bagration, asking him to draw back with his Second Army. "Faced with the consideration that the defence of the fatherland is entrusted to us at this critical moment, we must put away all differences and anything that might at other times influence our mutual relations," he wrote, in an appeal for cooperation. "The voice of our fatherland is calling us to reach an understanding, which is the only guarantor of success. We must unite and defeat the enemy of Russia. The fatherland will bless us for acting in harmony."[4]

In an address to his troops, Barclay exhorted them to show bravery and to emulate their comrades who had defeated Charles XII and Frederick the Great, adding that if there were any cowards among them, these should be flushed out at once. In fact, the army was spoiling for a fight. "'War!' shouted all the officers, all delighted at the news [of the French invasion], while a warlike spark ran like electricity through the veins of all those present," according to Lieutenant Radozhitsky. "We thought that we would immediately go out to meet the French, fight them on the border, and chase them back." "We were all delighted and rubbed our hands," another recalled. "It never entered anybody's mind that we might retreat."[5]

Barclay also printed leaflets to be strewn in the path of the invaders calling on Frenchmen to go home, and on the German soldiers in Napoleon's ranks not to fight their Russian brothers, but to join the German Legion being organised in Russia to fight for the liberation of their own country, assuring them that Alexander would grant those who wanted it good land in southern Russia if they defected. Similar leaflets were printed for the Spanish and Portuguese, and for the Italians, who for their part produced an indignant and fiery response demanding that the Russians stand and fight so that they, the Italians, could redeem their honour.[6]

Alexander's proclamation was published later that day. It was addressed as much to the world as to the Russian people. It claimed that he had always sought peace, and that it was only the unfriendly actions and then the military preparations of the French that had forced him to mobilise his forces. "But even then, cherishing the hope of reconciliation, we remained within the boundaries of our Empire, not disturbing the peace, while being ready to defend ourselves." Now that he had been attacked, Alexander could do no more than call on God for assistance and on his soldiers to do their duty. "Soldiers! You are defending your faith, your fatherland and your freedom! I am with you, God is against the aggressor!" It unequivocally equated the eventual defeat of Napoleon with the liberation of Europe by Russia.[7]

At the same time, Alexander decided on something which some have interpreted as evidence of a sudden loss of nerve, but which was probably only a piece of window-dressing – namely, to accept Napoleon's proposal to negotiate. His earlier precondition to any talks had been that Napoleon's troops vacate Prussia and the Grand Duchy of Warsaw; now he merely demanded that Napoleon withdraw beyond the Niemen. That evening he summoned Balashov and handed him a letter he had penned to Napoleon, instructing him to take it to the French headquarters. He told Balashov that he did not expect his mission would stop the war, but at least it would demonstrate to the world that Alexander wanted peace. And as Balashov was taking his leave, the Tsar reminded him to make it clear to Napoleon that he would not negotiate while a single French soldier remained on Russian soil.[8]

Early next morning, after hearing a report that French cavalry had been spotted approaching the city, Alexander made a hasty and less than dignified exit from Vilna. When news of this spread, a general panic ensued, as staff officers and the many hangers-on struggled to find horses to make their own getaway. Some of the more prescient inhabitants led their horses up the stairs of their houses and concealed them on upper floors. Barclay kept his head and evacuated his troops and headquarters, after destroying the bridge over the river Vilia and setting fire to the military stores which had been so painstakingly accumulated in the city.

These were still burning when, in the early afternoon of Sunday, 28 June, Napoleon rode into Vilna. He had been greeted at the gate by a delegation, but the inhabitants had not managed to prepare a triumphal reception. The Emperor rode through the city to inspect the smouldering remains of the bridge over the Vilia, and then reconnoitred the high ground above the city, before taking up his quarters in the same archepiscopal palace which Alexander had vacated not forty-eight hours before.

"The Emperor's manoeuvres will prevent this from being a

particularly bloody campaign," Marshal Davout reported to his wife the following day. "We have taken Wilna without a battle and forced the Russians to evacuate the whole of Poland: such a beginning to the campaign is equivalent to a great victory."[9]

His enthusiasm was not shared by his imperial master. Napoleon was annoyed and somewhat mystified. He was annoyed that he had not managed to engage with the Russians, and mystified because he could not make out what they were up to. They had deployed a substantial army to face him, and appeared to be preparing to defend Vilna. But the moment he marched out to engage it they had decamped, abandoning an important city and all their stores. Their behaviour made no sense, and he could not be sure that they were not setting some kind of trap. He instructed all his corps commanders to proceed with the utmost caution and to expect a counterattack at any moment.

In the north, Macdonald was making for Riga more or less unopposed, and Oudinot was in hot pursuit of Wittgenstein, against whom he had scored a minor success. Ney and Murat were on the heels of Barclay. And to the south Jérôme was advancing with his Westphalians, Prince Eugène's Italians and Poniatowski's Poles. Napoleon himself, with the Guard and part of Davout's corps, remained in Vilna, partly in order to supervise the rebuilding of bridges, the erection of fortifications and the construction of bread ovens, and partly out of the necessity to deal with alarming new problems.

The troops had been on the move for days, struggling on the bad roads to reach their positions. Their provision wagons were left far behind, and when they stopped for the night they often did not have the time or the energy to cook the rice they had in their knapsacks, let alone turn their bags of flour into bread. The countryside between the Niemen and Vilna was sparsely populated, and there was no food to be had from the locals, who had already been bled dry by the Russian forces over the past months. It was swelteringly hot, as hot as Spain in summer according to some of those who had campaigned there. The marching columns kicked up clouds of fine dust, and the scarcity of habitations meant there were no wells at which the men could slake their parched throats. The horses also suffered from the heat and the lack of water, as well as from being fed on unripe barley and oats.

After four days of this, just as the exhausted men were about to bed down for the night – which meant lying down on the ground wherever they could find a place – a primeval storm burst upon the area to the south and west of Vilna. A torrent of freezing rain poured down on them all through the night, and they soon found themselves lying in pools of icy water. "By daybreak, the storm had passed, but it was still raining," recorded Colonel Boulart of the artillery of the Guard, who could hardly believe his eyes as he emerged from under the gun carriage where he had taken shelter. "What a sight offered itself to my eyes! A quarter of my horses were lying on the ground, some dead or dying, others shivering. I quickly ordered as many as possible to be harnessed, hoping to get my wagons away, and to set this sad crew in motion so that the poor creatures might generate some of the heat they so badly needed, and thereby prevent a good number of them dying."[10]

Conditions were indescribable as the various units struggled to get to somewhere they might find shelter and food for themselves and their beasts. The roads were littered with the carcases of horses and the corpses of men, with abandoned wagons and gun carriages. "One could see in front of every *caisson* two or three of these creatures in full harness, their traces on the swingletree, fighting against death or already lying lifeless," recalled Lieutenant Sauvage of the artillery. "The gunners and the soldiers of the train stood by in mournful silence, their eyes filled with tears, trying to avert their gaze from this afflicting scene." Adjutant Lecoq of the Grenadiers à Cheval, a veteran of the Italian and Prussian campaigns, watched in horror as an artillery unit struggled through the storm. "The horses had water up to their bellies, on a shifting sandy road," he recalled. "As they tried to pull they lost their footing, fell and drowned."[11]

The number of men who perished during the downpour was

probably quite small, although there were rumours of up to three grenadiers being struck by lightning. But losses among the horses were horrific. Most of the artillery units lost 25 per cent of theirs that night, and the situation in the cavalry was not much better. Abraham Rosselet, an officer in the 1st Swiss Regiment, estimated that Oudinot's corps, to which he belonged, lost upwards of 1500 cavalry and artillery horses. Colonel Griois, commanding the divisional artillery of Grouchy's cavalry corps, claimed that he lost 25 per cent of his. It is generally estimated that the combat units lost over 10,000 in a period of less than twenty-four hours. This is a conservative figure, and it does not take into account the losses of the supply columns, which according to one *commissaire* were probably as high as 40,000.[12]

The psychological damage left by the storm was hardly less significant. As the men trudged on through the quagmire that had replaced the sandy roads, they could see dead and dying men and beasts by the roadside, and rumours of grenadiers of the Old Guard having been struck by lightning passed from rank to rank. The young Anatole de Montesquiou-Fezensac joked with Baron Fain that if they had been Greeks or Romans in ancient times they would undoubtedly have turned about and gone home after such an omen. Others took it more seriously, the Italians in particular. "So many disasters were a sad augury for the future," wrote Eugène Labaume. "Everyone began to take fearful note of them, and if, following the example of the ancients, we had shown more respect for the warnings of heaven, the whole army would have been saved: but, the sun having reappeared over the horizon, our apprehensions were dissipated along with the clouds."[13]

Napoleon's apprehensions were not dissipated so easily. Soon after his arrival in Vilna he became aware of the appalling losses his army had sustained, and, what was worse, of the fact that his carefully-worked-out plans for supplying it had come to nothing. The heavy wagons drawn by oxen had sunk into the sand of the Lithuanian roads before being waterlogged by the storm. The supplies despatched from

Danzig through Königsberg by river and canal had had no trouble in reaching Kovno, but the river Vilia was too shallow in places for the barges, and they could not be brought up until some dredging was carried out or their loads were transferred to shallower boats.

The French found no supplies to speak of in Vilna, and now, thanks to the storm, there were no horses with which to bring any in. "We are losing so many horses in this country that it will take all the resources of France and Germany to keep the present effectives of the regiments mounted," Napoleon wrote to his Minister of War, General Clarke. And while no supplies were finding their way to Vilna, sick stragglers were limping or dragging themselves in at an alarming rate. There were soon 30,000 of them crowding the hospitals rapidly improvised to receive them.[14]

The only bright spot for Napoleon on an otherwise bleak landscape came when he learned that his rapid advance had driven a wedge between Barclay and Bagration's Second Army; if he exploited this breach, he would be able to annihilate the latter. He sent Davout with his two remaining divisions (two had been detached to support Murat) and Grouchy's cavalry corps in a south-westerly direction to cut off Bagration's retreat, and despatched orders to Jérôme, Eugène and the commanders of other units in the area instructing them to encircle Bagration. But getting orders through at all, let alone with the requisite speed, was another unexpected problem.

On the morning of 1 July one of the duty aides-de-camp, Boniface de Castellane, was summoned to the Emperor's quarters. He found Napoleon in his dressing gown, a red and yellow kerchief on his head. The Emperor showed him a point on the map where he would find General Nansouty with his cavalry corps. "I am manoeuvring to cut them off, I have 30,000 within my grasp, make haste," he said to Castellane as he handed him sealed orders to deliver to Nansouty.[15] Napoleon needed a victory to throw into the balance against the losses he had borne, and catching Bagration became a priority.

A few hours later Napoleon summoned Balashov, who had presented himself to the French advance posts and been brought to headquarters bearing Alexander's letter. He was now ushered into the very same room in the former archbishop's palace in which Alexander had delivered the letter to him six days earlier. Napoleon was in a filthy mood. "Alexander is making fun of me," he thundered in response to the letter. "Does he really think I have come all the way to Wilna to negotiate commercial treaties?" He had come to deal once and for all with the barbarians of the north. "They must be thrown back into their icy wastes, so that they do not come and meddle in the affairs of civilised Europe for the next twenty-five years at least."[16]

Balashov was hardly allowed to get a word in as Napoleon paced the room, giving voice to his thoughts and feelings, venting his disappointment and the anxiety that was beginning to eat away at him. The monologue veered from whining reproach to squalls of anger as disappointment and fury struggled against each other for expression. He complained that it was all Alexander's fault, that the war had been started by Russia, by her demands that he evacuate Prussia and by Kurakin requesting his passport. He professed his esteem and love for Alexander, and reproached him for surrounding himself with adventurers and turncoats such as Armfeld, Stein and the murderer Bennigsen. He could not understand why they were fighting instead of talking as they had at Tilsit and Erfurt. "I am already in Wilna, and I still don't know what we are fighting over," he said.[17]

The wheedling would turn to bluster, with Napoleon defying the Russian generals to fight him, accusing them of ineptitude and cowardice, and threatening to unleash the army of a restored Kingdom of Poland at them. He shouted, stamped his foot and, when a small window which he had just closed blew open again, tore it off its hinges and hurled it into the courtyard outside. The detailed accounts of the interview by Balashov and Caulaincourt make embarrassing reading.

He was no more polite or dignified at dinner that evening, to which he sat down with Balashov, Berthier, Bessières and Caulaincourt. He fumed and bullied, declaring that Alexander would regret his stubbornness and that Russia was finished as a great power. With his customary wishful thinking, he warned that the Swedes and the Turks would not resist the opportunity to take their revenge for past defeats, and would pounce on Russia as soon as he advanced further. But in the letter addressed to Alexander that he handed to Balashov, Napoleon professed continuing friendship, peaceful intentions and a desire to talk, without however acceding to Alexander's precondition of a withdrawal behind the Niemen.[18]

There can be no doubt that Napoleon still wanted to patch up the alliance with Alexander. "He has rushed into this war which will be his undoing, either because he has been badly advised, or because he is driven by his destiny," he declared after Balashov had gone. "But I am not angry with him over this war. One more war will be one more triumph for me." Much later, in exile on the island of St Helena, he stated that if he had felt that Alexander's letter was sincere, he would have fallen back behind the Niemen. "Wilna would have been neutralised, we would each have come there with two or three battalions of our guards, and we would have negotiated in person. How many proposals I could have suggested! . . . He would have been free to choose! . . . We would have parted good friends . . ." Another Tilsit.[19]

This continuing self-delusion could only hamper his conduct of the campaign, both at the military and the political level. Napoleon had hoped that he would be able to defeat the Russians and reach an agreement with Alexander before he had to confront the Polish question, since that would probably have been part of the agreement. But the time had come to make some sort of decision.

Many of the inhabitants of Vilna and its environs had come to terms with Russian rule over the past decade and a half, and a number of Polish aristocrats had left in the wake of the Russian army. Many of those that had stayed behind and longed for Lithuania to be reunited with an independent Poland were unimpressed by Napoleon's treatment of the Grand Duchy of Warsaw and wary of his intentions for the future. There was nevertheless much enthusiasm.

"Our entry into the city was triumphal," wrote Count Roman

Sołtyk, who was among the first to ride in with his squadron of Polish lancers. "The streets, the public squares, were full of people; all the windows were garnished with ladies who displayed the wildest enthusiasm; some of the houses were decorated with precious carpets, kerchiefs were waved by every hand, and repeated shouts of joy were echoed far and wide." Another who was deliriously greeted as his troop of Chasseurs à Cheval trotted into town was Lieutenant Victor Dupuy, who was showered with sweets and flowers by excited ladies.[20]

On the very day Napoleon had his interview with Balashov, the Polish patriots of Vilna had held a solemn *Te Deum* in the cathedral, followed by a ceremonial act of reunification of Lithuania and Poland (a confederation of patriots had already proclaimed the resurrection of Poland in Warsaw). They expected him to make a public announcement endorsing this and proclaiming the restoration of the Kingdom of Poland. In an attempt to duck the issue, on 3 July he set up a government for Lithuania, which was to administer the country and, most important, gather supplies and raise troops. But this evasion was recognised by many locals for what it was, and the ardour of the patriots began to cool. "Even though we were Poles, the population received us quite coldly," an infantry lieutenant in the Legion of the Vistula noted in his diary. "The troops which had preceded us had exhausted most of the enthusiasm – and most of the victuals as well."[21] The question of victuals was of the essence.

The moment they crossed the Niemen, the men of the Grande Armée began to suffer from a shortage of food. The first wave might find people in the villages willing to give or sell them some, but the units coming up behind saw only deserted towns and villages with a handful of Jews who had stayed behind to sell whatever they had managed to preserve. Not surprisingly, as they ran out of food the men began pillaging.

In his proclamation launching his "Second Polish War," Napoleon had given his troops the impression that from the moment they crossed the Niemen they were in enemy territory, so they felt licensed to behave as they liked. And they behaved atrociously. "All around the city and in the countryside there were extraordinary excesses," noted a young noblewoman of Vilna. "Churches were plundered, sacred chalices were sullied; even cemeteries were not respected, and women were violated." Józef Eysmont, a local squire, had greeted a French cavalry unit with traditional bread and salt when they arrived at his small estate outside Vilna, but within an hour they had emptied his barns and stables, scythed the entire crop in the fields, thoroughly looted his house, smashing every window and anything they could not carry, and left him and all the peasants in his village destitute.[22]

There were instances of local gentry rising up against the retreating Russian troops as the French approached, capturing arms and supplies which they handed over to the incoming liberators, only to be subjected to looting and rapine themselves. Some peasants took the opportunity of rising against hated landlords, but most, particularly in northern Lithuania, were with their masters in greeting the French. But as soon as they saw how the French behaved, they took themselves and their livestock off to the forests, as they had done since time immemorial every time the Tatars raided. "The Frenchman came to remove our fetters," the peasants quipped, "but he took our boots too."[23]

The situation in the south was even worse. "To begin with we greeted the armies of Napoleon with open arms as liberators of the motherland and as benefactors, as everyone, in the manor and the village, felt that they were going into battle in the Polish cause," wrote Tadeusz Chamski, the son of a landowner. People prevented the retreating Russians from burning bridges and stores, and greeted the French warmly. In Grodno, the French forces were met by a procession with religious images, candles, incense and choirs. In Minsk, Davout's troops were welcomed and a *Te Deum* was held to thank God for the liberation. Resplendent in full dress uniform, General Grouchy personally handed around the plate at Mass, but at the other end of town his cuirassiers were breaking into shops and warehouses, and ill-treating the locals.[24]

The peasants lost interest once they realised that Napoleon was not going to emancipate them, while the gentry were quickly forced to regret the arrival of the French by the depredations of the troops. "The path of Attila in the age of barbarism cannot have been strewn with more horrible testimonies," in the words of one Polish officer who was accosted by a beggar pleading for bread only to discover that it was a friend of his, a local prince.[25]

Napoleon fumed when reports of these depredations reached him, and sent out patrols of gendarmes with orders to shoot all those caught looting. But the firing squads had little effect on the marauders. "They went to their execution with an incredible insouciance, with their pipes stuck in their mouths," wrote young Countess Tiesenhausen. "What did they care whether they died sooner or later?" Even the most radical methods of restoring discipline were bound to fail in the circumstances, as one lieutenant in the Polish Chevau-Légers pointed out: "Our generals tried a new method of punishment: the guilty man would be completely stripped and tied by his hands and feet in a square or street, whereupon two troopers were ordered out of the ranks to lash and cut him with whips until the skin peeled off him and he looked like a skeleton," he wrote. "After this punishment the whole regiment would file past so that they should see this horrific sight: but even this did not help much."[26]

Many of the culprits had left the ranks altogether, and therefore evaded punishment. At least 30,000 deserters, and possibly three times as many, were roaming the countryside in gangs, attacking manor houses and villages, raping and killing, sometimes in collusion with mutinous peasants. They drove about on stolen carts filled with booty, avoiding organised French units. There was no way of enforcing the law in view of their numbers, and those rounded up simply deserted again at the first opportunity. Officials were not safe: imperial *estafettes* carrying the Emperor's mail were attacked, and the man appointed by Napoleon as governor of Troki was robbed and beaten up.[27]

The miseries visited on the population were such that, in the words of one Polish officer, "the inhabitants who were at first overjoyed to see their reputed liberators, soon began to regret that the Russians had abandoned them." Another Polish officer found himself greeted with ill-concealed resentment where he should have been embraced. "The Muscovites were far more polite than you gentlemen," a young noblewoman explained to him.[28]

On 11 July, eight deputies from the Confederation in Warsaw arrived in Vilna, led by Józef Wybicki, one of the most pro-Napoleonic men in Poland. But the Emperor kept them hanging about for three days before giving them an audience, and listened impatiently to their request that he announce the restoration of the Kingdom of Poland. "In my position, I have many different interests to reconcile," he told them, but added that if the Polish nation arose and fought valiantly, Providence might reward it with independence. The delegates were crestfallen. "They left full of fire," recalled Pradt in Warsaw, "they returned with ice in their souls. The chill communicated itself to the whole of Poland, and it was not possible to warm her after that."[29]

The investigation set up by the Russians after the war to identify who had collaborated with the enemy and who had remained loyal reported that there had been no single recorded manifestation of loyalty to the Tsar in the entire province of Vilna during the whole period of French occupation. And many of the Russian commanders had first-hand experience of open hostility. The province was entirely for Napoleon in political terms. Yet he did not take advantage of it. A group of students from the University of Vilna volunteered to create a kind of political guerrilla group which would fan out over the country behind the Russian lines, raising the villagers against the Russians, but Napoleon replied that he did not want social unrest or revolution. He did not want anything that might stand in the way of reconciliation with Alexander. "He's an excellent emperor!" he told Jan Sniadecki, rector of the university, when the Tsar was mentioned.[30]

If Napoleon did not want revolution, he did want troops. He had hoped that many Polish officers and men in Russian service would desert to join up with their brothers on the French side, and although some did, the numbers were disappointing. In Lithuania he intended to raise five regiments of foot and five of horse, a total of 17,000 men, as well as a large national guard to keep order in his rear. He could probably have raised twice that number, but he would have had to produce some funds as well as a firm promise of the country's future independence. In the event, no more than 12 to 15,000 joined up.[31]

Napoleon not only mishandled the locals, he failed to inspire confidence in them. "This man will not achieve any more great things, he is assailed by some malingering illness," noted the rector's brother, the distinguished naturalist and physician Jędrzej Sniadecki, after observing Napoleon at length. And his attempt to charm the aristocracy went down poorly. "It was as impossible for Napoleon to emulate Alexander's perfect amiability in a drawing room as it would have been for Alexander to defy the talent and genius of Napoleon on the battlefield," as Countess Tiesenhausen put it.[32]

But Franco–Polish relations were not all bad. Captain Fantin des Odoards of the Grenadiers of the Guard confessed to being "enraptured" at a ball given in Vilna by the patriotic Count Pac as, "in the sweet embrace of the waltz," he made out "how white and rounded were the things that swayed this way and that under the national colours."[33]

Castellane, the duty officer who had been sent off on 1 July, had immense difficulty finding Nansouty. The roads were atrocious, there were no signposts of any sort, and there was nobody of whom to ask the way, as the already sparse population had fled to avoid the marauders, whom he himself had to dodge as well. His case was by no means exceptional. Napoleon was operating in huge army corps at distances that would have presented a problem in well-mapped areas with good roads. In the present circumstances, the problem assumed vast proportions as couriers and staff officers struggled to find their way down sandy roads, through boggy wilderness and interminable forests. It was extremely difficult for them to locate the commanders they were looking for, as they were themselves on the move, and many of the troops encountered along the way were not familiar enough with the marshals and generals to recognise them, not to mention the fact that some of them could not speak French. This meant that Napoleon could not act or react as fast as he was wont to, which frustrated his designs.

Jérôme had got off to a slow start, which allowed Bagration time to begin an orderly withdrawal. The Second Army left its positions on the same day Napoleon entered Vilna, moving in a north-north-easterly direction in order to join up with Barclay. But on 4 July Bagration discovered that Davout had cut across his line of retreat, so he swerved to the south and made for Minsk. On 4 July a courier arrived at Napoleon's headquarters with a letter from Jérôme which contained no useful information, only complaints. Napoleon was annoyed. "You tell me neither the number of Bagration's divisions, nor their names, nor the places where they were stationed, nor any of the information you will have found at Grodno, nor what you are doing," he fumed. "It is impossible to make war in this manner." The following day he told Berthier to send Jérôme further instructions. "You will tell him that it would be impossible to manoeuvre more ineptly than he has," he added.[34]

Davout reached Minsk before Bagration and paused there for three days to tidy up his forces. Bagration had also paused, at Nieshviezh. He was in a trap, with Jérôme's three army corps behind him and Davout barring his way. Had Jérôme pressed forward with energy, Bagration would have been done for. But Jérôme found the difficulties of the terrain beyond him, and could not bring his main forces up fast enough. When the Polish lancers of his advance guard finally did make contact with the Russians, at Mir, they were given a drubbing by Platov's cossacks.* Bagration then turned south and slipped out of the trap.

---

* The cossacks were a type of light cavalry with a characteristic style of fighting based on harrying the enemy rather than engaging him. They were mostly recruited from among the indigenous "*kazaks*" of the Ukraine, the Don and the Kuban.

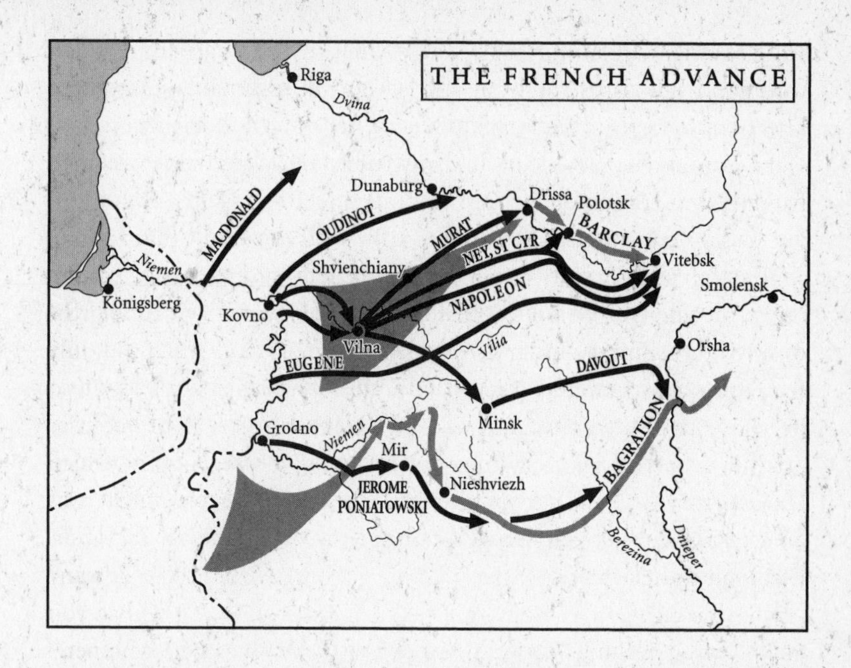

Napoleon could hardly contain his frustration. "If you had the most elementary grasp of soldiering, you would have been on the 3rd where you were on the 6th, and several events which would have resulted from my calculations would have given me a fine campaign," he wrote to Jérôme. "But you know nothing, and not only do you consult nobody, you allow yourself to be guided by selfish motives." He also reprimanded Prince Eugène for his sluggishness and failure to press the Russians. To Poniatowski, who had explained his inability to pursue the enemy by lack of food and fodder, he had Berthier write that "the Emperor can only be pained to discover that the Poles are such poor soldiers and have so little spirit as to bring up such paltry privations."[35]

The failure to cut off and destroy Bagration was entirely Napoleon's own fault. It had been his idea, politically motivated, to place his brother, who had never been to war, in command of three army corps, one of which was commanded by the only slightly more experienced Eugène. There had been problems between Jérôme and General Vandamme, who as his chief of staff was supposed to command the corps for him, and whom he summarily dismissed. There had also been friction between Poniatowski and Jérôme.

To cap it all, Napoleon had instructed Davout to oversee the combined operation of the various units operating against Bagration, but had failed to apprise Jérôme of this, with the result that Jérôme at first refused to carry out Davout's orders and then, in a fit of pique, decided to go home, taking his royal guards with him. On 16 July he began his march back to Cassel. "You have made me miss the fruit of my cleverest calculations, and the best opportunity that will have presented itself in this war," Napoleon wrote to him. He was far better off without him. But for good measure, he reproved Davout for his handling of the situation.[36] In the circumstances, he might have derived some comfort from what was going on at Russian headquarters.

Shortly after leaving Vilna, Alexander wrote to Barclay saying that he would give no more orders and implicitly handing over command to him. But in breathtaking contradiction to this, he went on issuing instructions. To Barclay himself he penned letters going into matters as minute as how much forage each cart should carry. To others, who were under Barclay's command, he sent orders and demanded reports without bothering to notify the unfortunate commander.[37]

The large number of advisers and officers without specific duties who surrounded the Tsar felt they had to justify their positions. "Project followed project, plans and dispositions, each one contradicting the other, each entailing envy and slander, disturbing the peace of mind of the commander-in-chief," noted A.N. Muraviov, an officer on Barclay's staff. "In the Tsar's entourage everyone was bent on getting himself noticed and showing off his importance," concurred Friedrich von Schubert, a young staff officer. "From every side advice and campaign plans were proposed and it took all of Barclay's steadiness not to lose his head while warding off all the fresh projects and the budding intrigues against himself." Barclay himself described headquarters as "a real wasps' nest of intrigue" in a letter to his wife. Grand Duke Constantine held his own court at his divisional headquarters, badmouthing Barclay and supporting all those who attacked him. Bennigsen and a dozen other ill-wishers were constantly slipping poison into the Tsar's ear and imputing every conceivable mistake or neglect to Barclay. Alexander would then tax the commander with these.[38]

Barclay had taken something of a risk by appointing a new chief of staff, Colonel Aleksei Petrovich Yermolov. He was a tall, striking man with small, deepset grey eyes and a Roman profile, a firebrand popular with the lower ranks, and sometimes referred to as "the hero of the subalterns." Barclay knew that Yermolov was an old friend of Bagration and a hater of all the "Germans" in the army. But Yermolov was intelligent and highly competent, and would not let this affect his ability to collaborate with them. Barclay had also just appointed as his quartermaster the thirty-five-year-old Colonel Karl Friedrich von Toll, whom Yermolov learned to value, even if he considered him opinionated and a little too pleased with himself. And it is more than likely that Yermolov's appointment was intended to capture some good will from Bagration.[39]

Before leaving Vilna, Barclay had written to Bagration asking him to move back fast in order to bring the two armies together. Indignant at being ordered to retreat, the fiery Bagration vented his fury to anyone who would listen. "We were brought to the frontier, scattered along it like pawns, then, after they had all sat there, mouths wide open, shitting along the whole length of the border, off they fled," he wrote to Yermolov. "It all disgusts me so much it's driving me crazy." To Arakcheev, he wrote that they should have attacked rather than retreating, and gave voice to a suspicion he would do much to propagate. "You won't convince anyone in the army or in Russia that we have not been *betrayed*," he wrote.[40]

Although the Tsar complained that Bagration was not moving fast enough, or in the right direction, he would not order him to obey Barclay's instructions, with the result that Bagration felt himself entirely at liberty to ignore them. In the event, this probably saved the Second Army, as Bagration selected his own priority, of getting out of Davout's trap, over Barclay's of bringing the two forces together at all costs. But the commander could hardly be expected to appreciate this.

His own retreat was being carried out with orderliness and deliberation – the one thing Barclay did have control over was the First Army itself, and he quite reasonably decided that he must preserve its integrity and battleworthiness at all costs. A hurried retreat would threaten both – stragglers and materiel would get left behind and fall into enemy hands if the columns marched any faster. The French were impressed by the orderliness of his withdrawal, and surprised at how few abandoned wagons they encountered. There were far fewer stragglers than might be expected, and his only significant losses were 10,000 or so soldiers recruited in the Lithuanian provinces who deserted, either to go home or to join the advancing Polish units.[41]

Alexander himself felt no great sense of urgency as the First Army fell back, and his remarkable complacency was reflected in his entourage. "Generally, everything is going well," Nesselrode wrote to his wife shortly after quitting Vilna. "Do not be frightened by our retreat, *on n'a reculé que pour mieux sauter*." The Tsar was cheered by the arrival of Leo von Lützow, a Prussian guards officer who had been cashiered in 1806. He had entered Austrian service in 1809 in order to carry on fighting the French, and after the defeat of Austria he had gone to Spain. He was taken prisoner by the French at Valencia in 1811, but managed to escape from prison in the south of France and made his way on foot, via Switzerland, Germany and Poland, to Russia, where he now offered his sword to the Tsar. It was eloquent evidence of what Alexander had come to represent in the minds of many Europeans. "Now that the war has begun," he wrote to Bernadotte on 4 July, "it is my firm resolution to make it last for years, even if I have to fight on the Volga." In the interim, he was hoping to make a stand at Drissa.[42]

Major Clausewitz was sent to Drissa to inspect the camp and report back on its state of readiness, but as he spoke no Russian and the only piece of paper he had with him was an order signed by Phüll written in French, he was arrested as a spy when he got there.* When he did extricate himself and report back to Alexander, he minced his words out of consideration for his friend Phüll, but clearly gave Alexander to understand that the camp was of no military value. His assessment was backed up by others present, but it did not bring about a change of plan.

Alexander reached Drissa on 8 July, and Phüll took him on a conducted tour of the fortifications, explaining their finer points. The Tsar's entourage maintained a stony silence and averted their eyes whenever he looked round for confirmation of Phüll's assertions. Finally, it was Colonel Alexandre Michaud, a capable officer formerly in the Sardinian service, who summoned up the courage to declare out loud that the camp was of no military value. One Russian general claimed he saw Alexander start to weep with despair, but he quickly recovered himself.[44]

The First Army trudged into Drissa on 11 July. Alexander issued a resounding proclamation to the troops promising action and a victory of the magnitude of Poltava. This cheered everyone, as the troops could not understand why they had been forced into a two-week retreat before they had been given a chance to fight. But the following day Alexander caved in to reason and decided to abandon the Drissa camp and fall back on Vitebsk, a more suitable position in which to give battle.

Alexander had undoubtedly taken the right decision, as the Drissa camp would have proved a trap in which the Russian army would have been surrounded and destroyed. He nevertheless now found himself in an untenable position. His undignified flight from Vilna and the hasty withdrawal of the army posited the questions of why it

---

* The language problem could have far more serious consequences: during a skirmish at Nieshviezh a few days later, a Colonel Mukhanov was run through and killed by cossacks when he shouted an order in French to one of his officers in a mêlée.[43]

had been stationed along the border if it did not mean to defend it, and what Alexander was doing there if he was going to run at the first news of a French advance.

On hearing of the retreat, people in Moscow and St Petersburg were bewildered. Alexander's only justification had been that he was carrying out the preordained plan of drawing the French on to the defences of Drissa. But now that plan had been abandoned, he was left without a figleaf. All he had to show for two weeks' campaigning was the abandonment of vast tracts of his empire to the enemy, including its largest city after Moscow and St Petersburg, an army tired and reduced by about a sixth, and the loss of a large quantity of stores.

It was clearly time to heed his sister's advice and leave the army, but he was afraid that he would be accused of deserting his troops at a critical moment. And while the memory of Austerlitz haunted him, he longed to emulate his namesake Alexander Nevsky, who had thrown the foreign invader out of Holy Russia at the head of his troops.

Fortunately for Holy Russia, Shishkov, who had come to the conclusion that the army was in peril of disintegrating, now took the matter into his own hands. He decided that whatever happened, Alexander must be persuaded to leave headquarters. He discussed the matter with Balashov, who agreed, and together they persuaded Arakcheev to back them up. Shishkov penned a memorandum in which he explained that the Tsar's place was in his capital, not with his army, and that at this juncture his duty was to rally his people and raise more troops. The three of them signed the document, which was left on Alexander's desk among his papers.

The Tsar made no mention of it as they all prepared to leave Drissa on the morning of 16 July. But on arrival at Polotsk that evening he turned to Arakcheev and said: "I have read your paper." A couple of hours later he mounted his horse and rode out to see Barclay, whom he found having a frugal supper in a stable. They spent an hour together, and when they emerged from the stable Alexander embraced Barclay, saying: "Farewell, General, once more farewell, *au revoir*. I commend my army to your keeping. Do not forget that it is the only one I have." He mounted his horse and rode back to Polotsk, where he gave orders for his departure for Moscow on the next day.[45]

9

Courteous War

While Alexander's presence at headquarters had been detrimental to the operations of the Russian armies, Napoleon's absence from the front line was disastrous for the French. He had allowed himself to get bogged down with political and administrative concerns in Vilna, where he spent two precious weeks during which he effectively lost the initiative.

Having failed to trap and defeat Bagration, he now ordered Davout to keep alongside him and diverted Prince Eugène northwards in order to prevent the Second Army from joining up with Barclay's First. This was being hotly pursued by Murat with his cavalry corps. Such a pursuit would in normal circumstances either have forced the Russians to face the French or turned their retreat into flight, but forced marches were something the Russian army was good at. In the event, Murat's hot pursuit only contributed to the destruction of the Grande Armée's cavalry, with catastrophic consequences later in the campaign.

This was not entirely Murat's fault. In putting together a corps of 40,000 cavalry, Napoleon had created the greatest forage problem in the history of warfare. By giving the corps the role of a mobile spearhead, he condemned its stock to attrition. After a long day's march, often involving skirmishes with cossacks or other units of the Russian rearguard, the men and horses would have to bivouac out in the open, with no shelter and often no food for the men or any kind of feed for the horses, which were lucky to get some old thatch pulled off a peasant hovel. In the morning they would be saddling and tacking up early for "*la Diane*," a long stand-to while all the pickets came in and reported back, a rigmarole that could take up to a couple of hours. The long marches and the lack of rest meant that the horses suffered from saddle sores and other ailments as well as exhaustion, but these factors were aggravated by Murat himself.

Murat was one of the most colourful characters of his day. "He always wore grandiose or bizarre outfits deriving from the Polish and the Mussulman, combining rich cloth, striking colours, furs, embroidery, pearls and diamonds," wrote a contemporary. "His hair fell in long curls on his broad shoulders, his thick black sideburns and his sparkling eyes contributed to an ensemble which aroused astonishment and made one think of him as a charlatan." Determined to look his best on this campaign, Murat brought with him not only a variety of his operatic costumes but also, according to one of his officers, a whole wagonload of scents and cosmetics.[1]

Murat's great virtues were his reckless bravery and his ability to inspire valour in his men. He thought nothing of standing under fire or leading charges, often disdaining to draw his bejewelled sabre and brandishing no more than a riding crop. But he had absolutely no tactical, let alone strategic, sense. He was the master of unnecessary and suicidal cavalry charges.

While he was a fine horseman, Murat showed a total lack of concern for the well-being of the beasts, which communicated itself to the generals serving under him, as a captain in the 16th Chasseurs à Cheval explained. "I will quote only one example among a host of similar ones," he wrote. "Sent off on picket duty with a hundred horsemen in the evening after the combat at Viasma, I was left at my post without being relieved until noon on the following day, and with strict orders not to unbridle the horses. The horses had been bridled since before six o'clock in the morning on the previous day. Having nothing for my pickets, not even water nearby, in the course of the night I sent an officer to explain my situation to the General, asking him for some bread and above all for some oats. He replied that he was there to make us fight, not to feed us. And so our horses went for thirty hours without being watered or fed. When I came in with my pickets it was time to move on; I was given one hour to refresh my detachment, after which I had to rejoin the column at a trot. I had to leave behind a dozen men, whose horses could no longer walk."[2]

When he came across Murat's cavalry on the march, barely three weeks into the campaign, Prince Eustachy Sanguszko, one of Napoleon's aides-de-camp, noted that "the horses swayed in the wind." Caulaincourt watched them skirmishing with the enemy rear-guard at about the same time, and was shocked to see that after some of the charges the troopers were obliged to dismount and walk their horses back, and if there was a counterattack they had to abandon their mounts and save themselves by running, as they could do so faster than the exhausted creatures could carry them.[3]

When Murat reported that Barclay had taken up position at Drissa, a new plan began to form in Napoleon's mind. He would make for Polotsk, thereby turning Barclay's left wing and cutting him off not only from Bagration but also his supply lines to the east. He would then attack and push him westwards towards the sea and into the path of the advancing corps of Oudinot and Macdonald. Leaving Maret in Vilna to manage his relations with the outside world, he set off in pursuit.

The Russians had indeed exposed themselves by wasting time at Drissa, and Napoleon's plan would undoubtedly have culminated in the destruction of the First Army had it not been for his uncharacteristic lack of decision and speed. While the Russians were discussing the merits of Phüll's position, he himself had been procrastinating in Vilna. On the very day he finally marched out, 16 July, the last units of the Russian army were evacuating the Drissa camp.

When Napoleon learned that the Russians had abandoned Drissa he amended his plan and made for Vitebsk rather than Polotsk, hoping to be able to outflank them there. But when he reached Beshenkoviche on 25 July he discovered that Barclay had slipped through ahead of him. Murat, leading the advance with his cavalry corps, had stumbled on his rearguard near Ostrovno. This time, the Russians stood and fought. Napoleon was delighted when he heard of this. "We are on the eve of great things," he wrote to Maret in Vilna, adding that he would soon be announcing a victory.[4]

Barclay appeared to have decided to give battle before Vitebsk. He had left Count Ostermann-Tolstoy with his 4th Corps of some 12,000 men barring the road at Ostrovno, with instructions to delay the French advance and win him time to deploy his forces. With his habitual impetuousness, Murat had launched his cavalry at the Russians. General Piré's Hussars executed a brilliant charge in which they captured a Russian battery, and Murat himself led a couple of charges which pushed the Russians back in disorder. But he could not make any real headway against them, particularly when they took up positions in a wood, as he had no infantry or artillery to support him. That evening he was reinforced by the arrival of General Delzons' infantry division, and the Russians were pushed back once more despite putting up a determined resistance.

This first serious engagement of the campaign had been the baptism of fire for many on both sides, and Lieutenant Radozhitsky of the Russian light artillery was horrified by the carnage. "My heart shuddered at such a sight," he wrote. "An unpleasant feeling took hold of me. My eyes grew dim, my knees gave way." Some of his old soldiers had told him that the fear would go once the waiting was over and they went into action, and he did indeed find that once he was firing off his guns at the French he felt only fury.[5]

The French felt an eerie sense of disappointment when they occupied Ostrovno that evening. "There was nobody who could pay to the courage of these soldiers the tribute of admiration which they had so richly deserved," wrote Raymond Faure, a doctor in the 1st Cavalry Corps. "There was not a table at which they could sit down and recount the exploits of the day."

On the following morning Faure rode out onto the battlefield. "The sward was ploughed up and strewn with men lying in every position and mutilated in various ways. Some, all blackened, had been scorched by the explosion of a *caisson*; others, who appeared to be dead, were still breathing; as one came up to them one could hear their moans; they lay, some with their heads on the body of one of their comrades who had died a few hours before; they were in a sort of apathy, a kind of sleep of pain, from which they did not appear to wish to awake, paying no attention to the people walking around them; they asked nothing of them, probably because they knew that there was nothing to hope for."[6]

The battle resumed that morning some eight kilometres to the east, as the Russians, who had been reinforced with Konovnitsin's division and General Uvarov's cavalry corps, fought to hold the French advance. They mounted a counterattack, sowing confusion in the French ranks, but the situation was redressed by a spectacular charge led by Murat in person. The charge was backed up by Delzons' infantry, and the Russians were defeated. Only the timely arrival of reinforcements in the shape of General Stroganov's division of grenadiers steadied their ranks and allowed them to retreat in good order.

Napoleon spent much of that night in the saddle, hurrying his forces forward, sensing the possibility of a battle. By midday on the following day, 27 July, he could see Barclay's troops drawn up outside Vitebsk, behind the river Luchesa. The approaches to the river were defended by Count Pahlen's cavalry corps, supported by infantry and cossacks, and Napoleon himself directed the operation of clearing them away, taking the opportunity to reconnoitre the Russian positions. The fighting was heavy. Many of Napoleon's units were still coming up along the road, so he decided to put the battle off till the next morning. It was uncharacteristic of him, and it was also a grave mistake, for, as Yermolov's aide-de-camp Lieutenant Grabbe pointed out, if he had pushed home his attack that evening, the Russians, who were also unprepared for battle, would have been routed.[7]

Napoleon was in the saddle until ten that night, directing the deployment of units as they arrived. Those that had already made camp were busily preparing for the next day. Days of battle were regarded as holidays, and as such demanded the finest turnout possible. The men unpacked their dress uniforms, polished the brass buckles and plaques, and pipeclayed the belts with an enthusiasm born of frustration. Everyone was looking forward to the battle, and none more so than Napoleon himself. There was no hiding his excitement as he bade Murat goodnight. "Tomorrow, at five o'clock. The sun of Austerlitz!" he said, alluding to the magic moment, fixed in the memory of all those who had been there, when the sun broke through the morning mist on the day of the legendary battle.[8]

"The men had spent the night preparing themselves, and on [28] July, the rising of a magnificent sun revealed us in the colourful dress of a parade," wrote Henri Ducor of the marine infantry. "Weapons glinted, plumes fluttered; joy and delight were visible on every face; the mood of gaiety was universal." But the gaiety turned to disappointment, then anger, and even a degree of despair when, as they marched out to take up their positions for battle, they realised that the enemy had vanished overnight.[9] Napoleon was stupefied.

It is difficult to be sure whether Barclay had really intended to give battle. In his account of the campaign he maintains that he did. He had no more than about 80,000 men at his disposal, but he was only confronting the corps of Ney and Eugène, Murat's cavalry and the Guard, rather than the whole of the Grande Armée, so although he would have been outnumbered, the disproportion would not have been that great. And he was under enormous pressure to fight. The murmuring against him and all the "Germans" at headquarters had turned into open accusations of incompetence, cowardice and even treachery, and this was beginning to seep down to the ranks. "I beg Your Majesty to rest assured that I will allow to pass no opportunity to harm the enemy," he wrote to Alexander on 25 July as he began taking up position at Vitebsk, "nevertheless, the most thorough regard for the preservation and security of the army will remain an inseparable element of my efforts against the enemy forces."[10]

This last phrase suggests that Barclay was open to any excuse not to fight, and on the afternoon of 27 July, while the two armies were skirmishing and just before Napoleon decided to adjourn the battle to the morrow, he was handed one. A courier arrived from Bagration informing him that the Second Army had failed to break through to Orsha, from where it could have come to his aid. It had found its path blocked at Saltanovka by Davout who, although greatly outnumbered, had beaten it back on 23 July, forcing it to wheel south. This meant not only that Bagration would not be able to come to the support of Barclay, which the latter must have known anyway, but, as Barclay did not hesitate to point out, that even if he scored a victory over Napoleon at Vitebsk he would have Davout's corps (whose strength the Russians consistently overestimated) in his rear, and when he turned about to defeat this he would have Napoleon, at best beaten but not routed, in his rear. He therefore took the only sensible course. He let Napoleon believe he would fight, and slipped away quietly as soon as darkness fell, leaving only a skeleton force of cossacks to keep the campfires burning.[11]

This was not the result of some new plan to draw the enemy into Russia, but of dire necessity. "I cannot deny that although many reasons and circumstances at the beginning of military operations made it necessary to abandon the frontiers of our land, it was only with the greatest reluctance that I was obliged to watch as these backward movements continued all the way to Smolensk itself," Alexander wrote to Barclay a couple of weeks later. And Smolensk was the very furthest fallback position contemplated.[12]

Barclay did the right thing. If he had stood and fought, he would almost certainly have been defeated, and that would have entailed the loss of, in Alexander's words, the only army he possessed. In the highly unlikely event that Barclay had won, the victory would not have been a decisive one, as it would only have defeated part of Napoleon's forces, and it would not have expelled the French from Russia. Finally, robbing Napoleon and his army of the chance of a

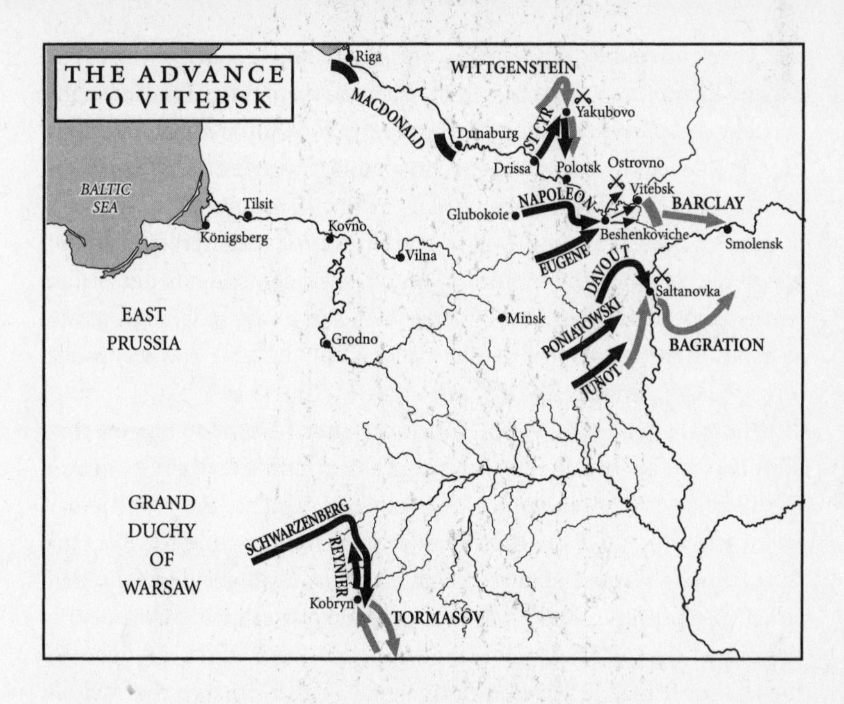

fight had a negative effect on their morale, and it was in very low spirits that they took possession of Vitebsk on 28 July.

The men were at the end of their endurance. They had been on the march for up to three months, and some, like one regiment of flankers of the Young Guard made up of youths barely out of school, had come all the way from Paris with only a single day's rest at Mainz and another at Marienwerder. Some units were marched for thirty-two hours at a stretch, with only a couple of hours' rest in brief intervals, covering as much as 170 kilometres. Carl Johann Grüber, a Bavarian cuirassier officer, recalled that his men were so tired after some of the marches that "hardly had the cry of 'Halt!' rung out, than they fell to the ground and into a deep sleep without even bothering to cook up the slightest meal." And even if they were not moving out early in the morning, they had to go through the tedious ritual of "*la Diane*."[13]

As they moved from Germany into Poland the quality of the roads declined, and when they crossed the Niemen these turned into troughs of sand or mud, according to the weather. Wading through either required twice the effort of marching along a firm road. The terrain between the Niemen and the borders of Russia proper is wooded and boggy in places. It is cut across by a multitude of small rivers, many of which trickle along the bottom of deep ravines. Existing bridges often consisted of no more than a few tree trunks, so if the sappers were not at hand, infantry, cavalry, artillery and supply wagons had to tumble down into the ravines and scramble up the other side.

Where they ran into woods, the tracks often became so narrow that infantry had to break ranks, while *caissons* and guns got jammed. Cavalry had to contend with low branches, which were a particular menace to those, like the Italian Guardia d'Onore, whose tall Roman-style helmets seemed made to catch on them. "Many riders who had fallen asleep in the saddle out of sheer exhaustion hit their heads on the branches," recorded Albrecht Adam, an artist who accompanied the staff of Prince Eugène on campaign. "Their helmets either fell off or hung lopsidedly by the chinstraps, and several of the men tumbled to the ground."[14]

The sheer volume of traffic on these country roads and tracks ground the dust finer and churned the mud to its most viscous. It also caused jams, delays, arguments over precedence, and sometimes fights. A gun team which had to stop in order to repair a piece of harness or tend to a lame horse would lose its place and then have to literally fight against infantry units unwilling to let it back into the stream of traffic, with the result that it would get cut off from its battery and might not be able to rejoin it for a day or more. And the multinational make-up of the army meant that arguments over precedence could turn ugly.

Seasoned soldiers were used to hardship, and needed to be, as marching for hundreds of miles burdened with a heavy backpack, a full cartridge case, a sword and a musket was no picnic. But none of them could remember as difficult and painful a march as this. By the middle of July most of the footsoldiers were barefoot, as their boots had fallen to pieces, and even the proverbially jolly French infantrymen had stopped singing their songs as they marched.

In France, Germany, Italy and even Spain, troops would be found billets in towns and villages on the march, and only had to bivouac out in the open on the eve of battle. But there could be no question of lodging anyone in the miserable villages of eastern Poland and Russia. The troops had not slept under any kind of cover since crossing the Niemen five weeks before, as they hardly ever had the time to construct shelters out of branches and whatever else came to hand. Prescient officers had had canvas sleeping bags run up for themselves, but their men had only their greatcoats to sleep under. The young recruits, away from home for the first time, adjusted particularly badly to this.

On warm summer evenings a bivouac could be a pleasant experience, with either some of the regimental bandsmen or odd strummers playing music and the men smoking their pipes as they chatted by the fireside. That is certainly how it struck the nineteen-year-old Baron Uxküll, an officer in the Russian Imperial Guard. "What a sight tonight!" he wrote in his diary on 30 July. "Imagine a dense forest that had taken in two cavalry divisions beneath its majestic, bushy roof! The campfires, now gleaming brightly, now dying down, could be seen burning through the foliage their heat kept agitating; at every moment they revealed to astounded eyes groups of men sitting, standing, and lying down around them. The confused noise of the horses and the axe strokes that were cutting away to feed the fires, all this added to the blackest night I've ever seen in my life – created, in fact, an effect that was as novel as it was bizarre; it resembled a magic tableau. I thought involuntarily of Schiller's *The Robbers* and of mankind's primitive life in the forest." It is worth noting that this particular young man had set off to war with Madame de Staël's *Delphine* and the poems of Ossian in his saddlebag.[15]

He omits to mention the swarms of mosquitoes that tormented the men in the predominantly marshy areas they were marching through, or the fact that, as often as not, they had nothing to eat as they sat around their campfires. The sweltering heat of the July days was frequently followed by bitterly cold nights, and they sometimes had to lie down to sleep under torrential rain. "The rain and the cold of the night meant that we had to keep campfires burning all night in front of the shelters," recalled another Russian officer. "The smoke from the damp brushwood, mixed with that of tobacco from the men who smoked, stung our eyes and tickled our throats, making us weep and cough."[16] Even in summer there is dew, so their clothes were invariably damp when they rose; some never did, having died of hypothermia in the night.

Because of the forced marches they usually stopped for the night quite late, and instead of being able to rest had to busy themselves making fires, gathering supplies and cooking something to eat. "It was, to my mind, the hardest part of this campaign," recalled the Comte de Mailly, an officer in the Carabiniers à Cheval. "Imagine what it was like for us, after having marched ten leagues, in atrocious heat, with a helmet and cuirass, and often without having had enough to eat, to start killing a sheep, skinning it, plucking geese, making the soup, and keeping the fire going so as to roast what we would eat or take with us for the next day."[17]

Often they had nothing with them, and had to send out men to forage in the locality. By the time these returned with victuals, or the regimental supply wagons rolled up, the men had fallen asleep, so they would take it in turns to cook the food while most of them slept, and the men would eat their dinner before setting off in the morning. This meant that they had to wolf it down in a hurry; sometimes, if there was an alarm or an order to move out urgently, they had to leave it behind uneaten.

The sheer misery comes through the pages of the journal kept by Giuseppe Venturini, a Piedmontese conscript. "Horrible day!" reads the entry under 20 July. "Bivouacked in the mud, thanks to our two cretinous generals. Ditto on the 21st, 22nd, 23rd. On the 24th, in a nice meadow. I felt as though I was in a palace. I was on sentry duty at

General Verdier's. I was lucky that day; I ate a good soup. On the 26th six men died of hunger in our regiment."[18]

Hunger was the worst affliction of all. "You who have never felt the pangs of hunger, you whose palate has never been parched by thirst, you do not know what real need is, a need which never lets up for an instant, and which, only partly satisfied, becomes all the more urgent and more acute," wrote one of the paymasters in Davout's 1st Corps. "In the midst of the great events which unfolded before my eyes, one dominant thought preoccupied me: to eat and to drink, that was my only goal, the nub around which my whole mind was concentrated."[19]

The regimental supplies carried on wagons and the herds of cattle driven along in the wake of the advancing troops inevitably fell behind if the tempo of the march quickened, and in many cases they were three or four days behind. As a result, regular distributions of rations rarely, if ever, took place. Major Everts of Rotterdam, serving in Davout's corps, noted that his men had not had a crust of bread for thirty days when they reached Minsk. Captain François of the 30th of the Line claims that outside Vilna they received two rations of mouldy bread, the only such distribution during the entire campaign. These were not isolated experiences. "As soon as we had crossed the Vistula all regular supplies and normal distribution of food ceased, and from there as far as Moscow we did not receive a pound of meat or bread or a glass of brandy through orderly distribution or normal requisitioning," reported General von Scheler to the King of Bavaria.[20]

Ultimately, it was down to the men to find something to eat and to prepare it themselves. The most resourceful were, by all accounts, the French and the Poles, who procured – by purchase or looting – cooking pots and utensils, and knew how to make out of the most unpromising raw materials a dish that would at least lull the stomach even if it did not nourish. They were also, according to General von Scheler, better at bringing in supplies. They would find what they needed quickly and bring it straight back to the unit for all to share; if everyone did this there would be plenty to go round. Scheler's

Bavarians were slow off the mark, and if they found something to eat would set about filling their own bellies. They would only then try to rejoin their unit with the rest of their booty, by which time it would have moved off and they would either have to jettison their load or fall behind for good.[21]

But necessity is a great instructor, and Jakob Walter of the Württemberg division in Ney's corps soon learned how to find the jars of pickles, the barrels of stewed cabbage, the honey, potatoes and sausages hidden under floors and piles of wood or buried in the orchards of deserted and apparently devastated villages. "Here and there a hog ran around and then was beaten with clubs, chopped with sabres and stabbed with bayonets," he wrote, "and, often still living, it would be cut and torn to pieces. Several times I succeeded in cutting off something; but I had to chew it and eat it uncooked, since my hunger could not wait for a chance to boil the meat."[22] Eating undercooked pork is notoriously dangerous, but even without that, the diet they were subjected to, rich in meat but poor in bread, rice and vegetables, upset their stomachs and made them prone to diarrhoea and dysentery.

Near Korytnia, Murat's corps moved into a camp recently vacated by the Russians. "The shelters constructed out of branches were still standing, the fires were only just extinguished," noted an officer in the Lancers of Berg. "Behind this camp was a ditch in which the soldiers had gone to satisfy their natural needs, and I noted the considerable volume of the piles of excrement with which this area was covered, and came to the conclusion that the enemy army had been abundantly fed." Heinrich von Roos noted that on coming upon a recently vacated encampment, the surest way of telling whether its last occupants had been friend or foe was to seek out the latrines, explaining that "the excreta left behind by men and animals on the Russian side testified to a good state of health, while ours showed in the clearest possible way that the entire army, horses as well as men, was suffering from diarrhoea."[23]

Thirst also tortured them all from the moment they crossed the

Niemen. Daytime temperatures reached 36°C (97°F), and many of those who had campaigned in Egypt claimed they had never marched in such heat.* On 9 July the 11th Light Infantry lost one officer and two men to heatstroke. "The air along the wide sandy tracks running through endless dark pinewoods was really like an oven, so oppressively hot was it and so unrelieved by the slightest puff of wind," recorded a Russian cavalryman retreating before them. The occasional downpour would drench them without refreshing them or the country they marched through, and steam would rise from their uniforms while the water seeped into the sandy soil. Due to the sparseness of the population there were few wells, and the ponds and ditches contained only brackish water. The men would dig holes in the ground and wait for them to fill with water, but it was so full of worms that they had to filter it through kerchiefs before they could drink it. One of Berthier's staff officers had equipped himself with his own cow and a range of essences, and was able to treat his colleagues to ice cream, but the rank and file on the march had to make do with what was to hand. "How many times did I not throw myself down on my belly in the road to drink out of the horsetracks a liquid whose yellowish tinge makes my stomach heave today," recalled Henri Ducor, and he was certainly not the only one to drink horse's urine out of the ruts in the road.[24]

Not surprisingly, many died of dehydration or malnutrition. Others got dysentery. The German contingents seem to have been most vulnerable. Their men were less resourceful than the French and others at building shelters, harvesting crops, grinding corn, baking bread or making up a pottage of whatever was going. The Württembergers suffered particularly badly from dysentery – Carl von Suckow's company had dwindled from 150 to just thirty-eight, without having so much as seen the enemy – and the Crown Prince himself became so ill that he had to leave the army. The Bavarians also

* All temperatures recorded during this campaign were in Réaumur. For the reader's sake I have converted them into Celsius and Fahrenheit throughout.

suffered badly, and by the time their contingent of 25,000 men reached Polotsk it was down to 12,000. The Westphalians succumbed to the heat in droves. At the end of a forced march in a temperature of over 32°C (90°F) one regiment was down to 210 men out of a complement of 1,980.[25]

The more resilient simply suffered from diarrhoea and soldiered on, clutching their bellies and making sudden dashes to the side of the road to drop their pants. This was more than a nuisance to Aubin Dutheillet de la Mothe, a twenty-one-year-old officer in General Teste's brigade. At one point he felt such an urgent need to defecate that he rode off the road, dismounted and dropped his breeches without pausing to tether his horse. As he squatted helplessly a squadron of cuirassiers clattered by and his own mount trotted off with them, bearing his sabre and much of his kit, never to be seen again.[26]

The roadside was littered not only with excrement, but with the carcases of horses and the bodies of men who died on the march. "On some stretches of the road I had to hold my breath in order not to bring up liver and lungs, and even to lie down until the need to vomit had died down," wrote Franz Roeder, a Hessian Life Guard officer.[27]

The horses too were having a terrible time of it. Unused to the kind of diet they were being subjected to, they suffered from colic and diarrhoea or constipation. One artillery officer recorded that he and his men would have to plunge their arms up the poor creatures' anuses up to the elbow in order to pull out rock-hard lumps of dung. Without such attentions, their stomachs would blow up and explode.[28]

As they were continually in action, the horses had also developed saddle sores. "Entire columns consisting of hundreds of these poor beasts had to be led along in the most pitiful condition, with sores on their withers and backs stopped up with bits of hemp and dripping with pus," noted a witness. "They had lost so much weight their ribs stood out and they presented a picture of the most abject misery." As the fine-bred creatures brought from France and Germany died they were replaced with whatever the country could offer, which for the most part were shaggy little peasant horses known throughout the army as "*cognats*," from "*koń*," the Polish for horse.[29]

To the discomforts of the march have to be added the swarms of vicious wasps, horseflies and mosquitoes that are a feature of summer in that part of the world, and the terrible clouds of dust churned up by the men and horses as they marched. "The dust on the roads was so thick that whether a horse be a bay or a grey it was the same colour, and there was no difference in the colour of the uniforms or of the faces," recorded a lieutenant in the 5th Polish Mounted Rifles. The dust was so thick in places that infantry had to have the drummers beating constantly at the head of every company, so they did not lose their way.[30]

When they could see around them, they were confronted with an empty, desolate landscape stretching away into the unknown. Towns and villages were few and far between, and most were devastated by the passage of troops. "What I have seen in the way of distress in the past two weeks is beyond description," the artist Albrecht Adam wrote to his wife. "Most of the houses are deserted and without roofs. In the areas we are marching through most of them have thatched roofs, and the old straw from these has been used as fodder for the horses. The houses have been destroyed or ransacked, and the inhabitants have fled, unless they are so poor that they have died of hunger, having had all their food taken away by the soldiers. The streets are strewn with dead horses which give off an awful stink in the hot weather we are having, and every moment more horses collapse. It is a horrible war. The campaign of 1809 seems like a pleasant promenade by comparison."[31]

The population, such as it was, shocked the French and their allies by its abject poverty and backwardness. The Jews, who were such a dominant feature of every town and village in these former Polish lands, revolted the troops and brought out all their innate prejudice as they surrounded them to buy, sell or barter. "If you knew the countryside through which we are wandering, my dear Louise," General

Compans wrote to his young bride, "you would know that nothing in it is beautiful, not even the stars, and that if one were to try and form a harem of four wives for a sultan here, one would have to tax the entire country."[32]

Another factor, enervating as well as striking by its novelty, was the shortness of the summer nights this far north. "Many times when we went into bivouac for the night, the great glow of the sun was still in the sky so that there was only a brief interval between the setting and the rising sun," wrote Jakob Walter from Stuttgart. "The redness remained very bright until sunrise. On waking, one believed it was just getting dark, but instead it became bright daylight. The night-time lasted three hours at most, with the glow of the sun continuing."[33]

Whether they came from Germany or Portugal, the troops felt very far away from home, and their anxiety grew with every step further they took. Such feelings were particularly strong among the young recruits, but even the old soldiers later reflected that the advance into Russia had been in some respects worse than the notorious retreat. Some deserted and headed back homeward. Hundreds committed suicide. "Every day one heard single shots coming from the woods lining the road," recalled Carl von Suckow. A patrol would be sent out to reconnoitre and would return with the report that a man had shot himself. And it was not just unhappy recruits who took their lives. On 14 July Major von Lindner of the 4th Bavarian infantry cut his throat with a razor from despair.[34]

According to the *commissaire des guerres* Bellot de Kergorre, the whole army had been reduced by a third by the time it reached Vitebsk, without fighting a single battle. The Legion of the Vistula had lost between fifteen and twenty men in every company. "In a normal campaign," explained one of its officers, "two proper battles would not have managed to reduce our effectives to such an extent." The Army of Italy was down by one-third overall, though some units were even more depleted – one had lost 3,400 out of a total of 5,900 men. Ney's 3rd Corps was down from 38,000 to 25,000.[35]

The German contingents had suffered more than most. "The food is bad, and the shoes, shirts, pants, and gaiters are now so torn that most of the men are marching in rags and barefoot. Consequently, they are useless for service," General Erasmus Deroy reported to the King of Bavaria. "Furthermore I regret to have to tell Your Majesty that this state of affairs has produced a serious relaxation of discipline, and there is such a widespread spirit of depression, discouragement, discontent, disobedience, and insubordination that one cannot forecast what will happen."[36]

But the worst affected were Poniatowski's Poles and the Westphalians, now under the command of General Junot, whose march had taken them through the most desolate areas and who, because they had been set in motion after the main force of the Grande Armée, had been obliged to make up for lost time when Napoleon changed his plans and gave them the job of chasing Bagration.

Ironically, the losses in men were in fact beneficial to the Grande Armée. The ranks had been cleared of the weakest, who should never have been sent off to war in the first place. But before they died they had helped to slow down the operations of the army, to ravage the country through which they passed and to overload the supply machine to an extent from which neither recovered. And the sight of them dying in their thousands had an unsettling effect on those who remained.

Napoleon was usually very active when on campaign. He was always up by two or three o'clock in the morning and never retired before eleven at night. He ranged all over the theatre of operations, travelling in his carriage so he could work, and mounting one of his saddle horses which were led along behind in order to go and reconnoitre a position or inspect some troops. His aides-de-camp, who had to ride behind his carriage and often did not have a spare horse to hand, would have trouble in keeping up with him as he galloped off.

But the sheer size of the Grande Armée at the outset of this campaign and its wide dispersion in so many corps meant that Napoleon never saw most of his troops, on the march or at bivouac, as he usually did. As a result he did not see the condition or catch the mood of the army, and the men did not feel his presence and commitment in the way they had grown used to. He studied lists and figures, many of them wildly inaccurate, and optimistic reports from unit commanders eager to please. General Dedem de Gelder noted that when reporting their numbers, all the commanders inflated the figures in order to make themselves look good, and according to his calculations Napoleon must have thought he had about 35,000 troops more than he did at this stage.[37]

Napoleon only saw his troops when they were in action or on parade before him, when, buffed up and motivated, they looked and felt their best. He therefore tended to disregard the occasional honest reports on their condition that did reach him, and as he did not like to hear them, dismissed them as exaggerated scaremongering. Had he taken a closer look, he would have seen that the situation called for some urgent reorganisation.

Murat's huge cavalry formation had outlasted its usefulness, since it performed no real military purpose and was incapable of feeding itself. The cavalry would have found it much easier to feed its horses and would have been of far more use if it had been broken down into brigades and divisions and dispersed among the various corps.

It had also become obvious to most of the commanders on the ground that the artillery could have done with some weeding. The Russian artillery was not only of a very high standard, it also used guns of greater calibre, which meant that the dozens of light field pieces being dragged along by the French were largely useless. Napoleon had given every infantry regiment its own battery of four-pounders, primarily for psychological effect. They could hold their own in skirmishes between small bodies of troops, but not in a set-piece battle in which the Russian artillery was deployed. But their presence required a vast amount of effort, thousands of horses and tons of fodder. Many had been left behind along the way, at Vilna and elsewhere, for lack of draught animals, but they were all laboriously dragged up as soon as possible instead of being abandoned.

According to some sources, on entering the rooms which had been prepared for him in Vitebsk, Napoleon took off his sword and, throwing it on the table covered with maps, exclaimed: "I am stopping here; I want to take stock, rally and rest my army, and to organise Poland; the campaign of 1812 is over! The campaign of 1813 will accomplish the rest!" Whether the scene was quite as theatrical as this is questionable, but the gist of what he said is not. He told Narbonne that he would not repeat the mistake of Charles XII by venturing deeper into Russian lands. "We must settle down here this year, so as to finish the war next spring." And to Murat he is reported to have said that "the war against Russia will be a three-year war."[38]

Napoleon had taken his quarters in the residence of the Governor of Vitebsk, the Tsar's uncle Prince Alexander of Württemberg, and there he installed his travelling office, with his *portefeuille* and boxes containing papers as well as a couple of long mahogany cases with his travelling library. Presumably alarmed at the thought of the *longueurs* that lay ahead, he wrote to his librarian in Paris asking him to send "a selection of amusing books" and novels or memoirs that made for easy reading. The weather was extremely hot, with temperatures of 35°C (95°F) at night recorded by his secretary Baron Fain, and while his troops cooled themselves by bathing in the river Dvina, he sweated as he worked at tidying up the army. He gave orders for a new route of communication and supply to be opened via Orsha to Minsk, instructed General Dumas to start building up a sizable magazine and to construct bread ovens, and he set up a local administration. He bought and demolished a block of houses in front of his quarters in order to create a square on which he could review his troops, and held regular parades, knowing this to be good for the soldiers' morale. He was aware that they were grumbling about the supply situation, so he used one of these parades, on 6 August, to vent his anger publicly at the *commissaires* and those in charge of the medical service. Shouting loud enough for the soldiers to hear, he reproved them for failing to comprehend "the sanctity of their mission," sacked the chief apothecary and threatened the doctors that he would send them all back to treat the whores of the Palais-Royal. It was pure theatre, but the soldiers cheered their Emperor who cared so much for them.[39]

Napoleon issued confident-sounding and mendacious Bulletins, wrote to Maret instructing him to publicise non-existent successes, and blustered in front of the men; but in the privacy of his own quarters he was irritable and often in a bad mood, shouting at people and insulting them in a way he rarely did. He made contradictory statements and appeared at a loss as to what to do next.

His instinct was to pursue the Russians and force them to fight. He had nearly managed it at Vitebsk, and felt sure they would make a stand in defence of Smolensk, a larger city of some moral significance to them. There was no logic whatever in stopping at Vitebsk while there was still an undefeated Russian army in the field. For one thing, he would quickly starve. He had reached a more fertile area, but the country would still not be capable of feeding his army over a long period of time, while supplying it from Germany was not realistic. He could hardly go into winter quarters in July, and he was the first to see that this was not a good position, since the rivers that provided some kind of defence in summer would freeze over in winter, making it highly vulnerable.

The wider strategic situation was problematic. On Napoleon's northern flank, Macdonald was besieging Riga and Dunaburg. Oudinot, after an inconclusive engagement against Wittgenstein at Yakubovo which both sides claimed as a victory, had fallen back to cover Polotsk, where he was reinforced by St Cyr's 6th corps. In the south, General Reynier's Saxons had suffered a minor defeat at the hands of Tormasov near Kobryn, but he and Schwarzenberg had then pushed Tormasov back, clearing the Russians out of Volhynia entirely.

It was high time he made a decision on whether to play the Polish card or not. He had been greeted in Vitebsk by delegations of local

Polish patriots, and had evaded their expectant questions as to his intentions by heaping abuse on Poniatowski and the alleged cowardice of the Polish troops, which, he claimed, was largely responsible for the failure to catch Bagration. "Your prince is nothing but a c—," he snapped at one Polish officer.[40]

From Glubokoie he had written to Maret telling him to instruct the Polish Confederation in Warsaw to send an embassy to Turkey with the request for an alliance. "You realise how important this *démarche* is," he wrote. "I have always had it in mind, and cannot imagine how I forgot to instruct you accordingly." Poland and Turkey were united in enmity to Russia, and Turkey had never reconciled herself to the removal of her ally from the map. A firm declaration of intent by Napoleon to restore Poland might well have persuaded Turkey to resume her war with Russia.[41]

Many argued that this was the moment to send Poniatowski south into Volhynia. This would have raised an insurrection in the whole of the old Polish Ukraine, which would have yielded men and horses in plenty as well as abundant supplies. More important, it would have tied down all the Russian forces in the south, under Chichagov and Tormasov, neutralising any threat they might otherwise pose to Napoleon's flank. But a few days later, from Beshenkoviche, he wrote to Reynier leaving it up to him whether to encourage the local Poles to rise against the Russians.[42]

In the event, Reynier's Saxons behaved so badly that they aroused hostility among even the most patriotic Poles in the area. Those who had hoped that Napoleon might restore Poland were disenchanted. "The mask of his good intentions towards us was beginning to slip," in the words of Eustachy Sanguszko, one of his aides-de-camp. For his part, Napoleon told Caulaincourt that he was disappointed by the Poles, and that he was more interested in using Poland as a pawn than in restoring her independence.[43]

While at Vitebsk he received news of the ratification of the treaty of Bucharest between Russia and Turkey, and details of that between Russia and Sweden signed in March. What he did not know was that

Russia had also signed a treaty of alliance with Britain on 18 July. From Berlin he was receiving intelligence that the British were planning a joint landing in Prussia with the Swedes. But he was cheered by the news of the outbreak of war between Britain and the United States of America.

"While the Emperor meditated on new and more decisive blows, a great cooling off was taking place around him," according to Baron Fain. "Two weeks' rest gave people time to reflect on the enormous distance at which they found themselves, and on the singular character which this war was assuming." He added that there was "anxiety and discouragement" in the various staffs. Napoleon sensed this, and for the first time he drew a wider group of generals into his confidence, asking them what they thought should be done. Berthier, Caulaincourt, Duroc and others felt it was time to call a halt. They cited losses, provisioning difficulties and the length of the lines of communication, and expressed the fear that even a victory would cost them dear, on account of the lack of hospitals and medical resources.

But Napoleon clung to his original assessment. The Russians were now stronger than they had been, and were on the borders of Russia proper, so they were more likely to accept battle – and that was what he staked everything on. "If the enemy holds at Smolensk, as I have reason to believe he will, we shall have a decisive battle," he wrote to Davout. Once he had defeated them, Alexander would sue for peace, and the Russians would furnish him with all the supplies he needed. "He believed in a battle because he wanted one, and he believed that he would win it because that was what he needed to do," wrote Caulaincourt. "He did not for a moment doubt that Alexander would be forced by his nobility to sue for peace, because that was the whole basis of his calculations."

But Napoleon remained agitated, as Narbonne's diary shows, for he was far too intelligent not to see the truth of all the arguments against proceeding. This proverbially decisive man seemed panicked by the very fact that he could not reach a decision, and, leaping out of his bath at two o'clock one morning, suddenly announced that they must advance at once, only to spend the next two days poring over maps and papers. "The very danger of our situation impels us towards Moscow," he said to Narbonne finally. "I have exhausted all the objections of the wise. The die is cast."[44]

10

The Heart of Russia

"The greatest regret in my life," Alexander's sister Catherine later said, "is not having been a man in 1812!"[1] Had she been one, she might well have ended up in his place. Alexander knew that as the Russian armies fell back, allowing the French invaders to strike into the very heartlands of the empire, he would be blamed. And he could not forget what had happened to his father and grandfather. Arakcheev, Balashov and Shishkov had convinced him that he must galvanise Russia and rally it to his cause. But he could not be sure which way his subjects would swing, particularly as many of them had only become his subjects as a result of conquest.

The territory occupied by the French thus far had only been part of the Russian empire for between seventeen and fifty years, and Alexander could not expect much devotion to the cause of the Tsar and fatherland from the preponderantly Polish gentry or the mass of peasants, who had no defined sense of nationhood. Nor could he play the religious card there: in the areas acquired by Russia in 1772, the population consisted of 1,500,000 Uniates and 1,300,000 Catholics, 100,000 Jews, 60,000 Old Believers, 30,000 Tatar Muslims and three thousand Karaim Jews, as against only 80,000 Russian Orthodox. And even some of these were not reliable. Varlaam, Orthodox Archbishop of Mogilev, went so far as to swear an oath of allegiance to Napoleon, encouraging his flock to do likewise and attend services in the

Emperor's honour. The Russian troops considered the locals to be foreign and ill-disposed to Russia, and while they maintained amicable relations with them during the eighteen months or so they had been posted there, they began plundering manors and villages as they withdrew.[2]

In the event, most of the Polish nobility opted for Napoleon in principle, though the majority did so without much enthusiasm, and many adopted a wait-and-see attitude. The peasants seem to have reacted in more pragmatic ways. Some took the opportunity to rebel, or at least to refuse to carry out their labour obligations. Others apparently made a run to freedom: in 1811 the administrative regions of Mogilev, Chernigov, Babinovitse, Kopys and Mstislav contained 359,946 serfs belonging to landowners and the Church; in 1816 there were only 287,149 of them.[3]

The Russian authorities were nervous of how the Jews would react, since they represented the only group in this area, apart from the Polish nobility, who could have been of use to the French in administering it. Napoleon had emancipated the Jews in every country he had marched through, and in 1807 he had called a Grand Sanhedrin or gathering to which he had invited Jews from all over the world. In the event, many Jews did make themselves useful to the French as traders, as guides, and sometimes as informers. But most remained indifferent, while a number proved surprisingly loyal to the Tsar.[4]

In the Russian heartlands, which the French were now entering, there were none of these alien elements (the Jews were banned from them), but this did not make the authorities any less apprehensive. In what seems a chillingly modern kind of operation, the police in the province of Kaluga rounded up its foreign residents – Frenchmen, Germans, Swiss, Danes, Englishmen, a Dutchman, a Pole, a Spaniard, a Portuguese, a Swede and an Italian, who variously plied every trade from doctor, tailor, hatter, pastrycook, dancing master and gardener to governess and hairdresser – and packed them off away from the war zone.[5]

Another potential threat were the Old Believers, a sect which had split from the Orthodox Church a century and a half earlier in protest at reforms brought in by the Tsar. They regarded Alexander, not Napoleon, as the Antichrist. But although they were ubiquitous, they were few in number and passive by nature.

Amongst Russians, first reactions to the invasion had been positive, even if some exaggerated the patriotic surge. In an unfinished novel set in 1812, Pushkin depicts Moscow society as francophile and dismissive of the rather "simple" defenders of things Russian, whose patriotism "was limited to passionate condemnations of the use of French," until the French invaded. From that moment, "social circles and drawing rooms filled with patriots: some threw the French snuff out of their snuffboxes and began to use the Russian variety; others burned French pamphlets by the dozen; others turned away from the Lafitte and took to sour cabbage soup instead." Pushkin was certainly being a little ironical. Others were less so. "News that the enemy had invaded made people of every age and condition forget their private joys and sorrows," noted the nationalist Filip Vigel in Penza. "When news of the intrusion of Napoleon's countless hordes spread through Russia, one can truly say that one feeling inspired every heart, a feeling of devotion to the Tsar and the fatherland," wrote Prince N.V. Galitzine, a serving officer. In Moscow, "old ladies would cross themselves, spit and curse Napoleon," while young society girls imagined themselves in the roles of Amazons or nurses.[6] But such enthusiasm was by no means shared by the population as a whole.

Over 90 per cent of this consisted of serfs, about half of them owned by the nobility, the rest by the Church or the state. Serfs were chattels. They could be bought and sold, included in a gambling debt, marriage contract or loan. They could be flogged by their masters, and killed in the process. There was no identifiable revolutionary urge among the serfs, mainly because they had no leaders, but there was always the potential for bloody mutiny, and there had been a terrible one, led by Pugachov, in living memory.

The intrusion of the French army could not fail to affect the peasants' attitude towards their rulers, or at least bring to the surface latent grievances. The authorities had anticipated this, and the police intensified their snooping in taverns and alehouses all over the country. There were confused reports of agitators travelling around the countryside encouraging serfs to rise, and rumours circulated among peasants and house serfs to the effect that Napoleon had told Alexander to emancipate them, otherwise he would. Alexander had stationed half-battalions of three hundred men in every province as a precaution, which was not misplaced. There would be sixty-seven minor peasant revolts in thirty-two different provinces in the course of 1812, more than twice the annual average.[7]

The serfs were nevertheless Russians, deeply attached to their land and their faith, and it was hoped that they would rally to the defence of these. But first signs were not promising. D.I. Sverbeev, the nineteen-year-old son of a landowner to the south of Moscow, recalled how, on hearing news of the invasion, his father called his serfs together after church on Sunday. The seventy-two-year-old master announced that he and his son were going to go to Moscow to enlist, and called on his serfs to volunteer. There was much shuffling of feet, and finally one old peasant volunteered, while some of the others offered a few pennies. Other squires fared even less well. There were instances of rebellion and sacking of manors as the French armies drew near.[8]

It was up to Alexander to ensure that patriotic feeling, or rather the determination to stand by the whole political, social, religious and cultural edifice which he embodied, spread through every class of the nation. A huge propaganda exercise was required, and in this Shishkov was to be invaluable. Alexander would hand him a draft proclamation, written in French, and Shishkov would turn it into stirring Russian. The manifesto issued in Polotsk on 6 July announced that Napoleon had come to destroy their "great nation." "With guile in his heart and flattery on his lips he is bringing for it eternal flails and shackles," it warned. Alexander would chase Napoleon from the land, but as Napoleon disposed of enormous strength, the Tsar would need to gather new forces. The manifesto represented the whole nation as being engaged in a life and death struggle in defence of their wives, children and homes, and called on nobles, clergy and peasants to emulate heroes of the past. Shishkov knew how to combine love of family and home with love of fatherland and Tsar. His proclamations were couched in biblical vein, equating Russia with the chosen people which must one day tower above others, and brimmed with pious confidence in divine providence.[9]

Alexander also brought the propaganda machine of the Orthodox Church into play. He wrote to bishops urging them to mobilise their clergy into action against the common threat of the alien and godless "army of twenty tongues," as the multinational Grande Armée was sometimes referred to. The Synod issued its own proclamation, calling on everyone to take up arms in defence of faith and fatherland against the godless intruders who had offended the Almighty by overthrowing the throne and the altar in France. It called on priests up and down the country to arm simple souls with the correct sentiments. At Alexander's request, Augustine, Vicar of Moscow, wrote a special prayer in which the faithful could beg God to defend Russia, inspire devotion to the Tsar and imbue him with all the wisdom and courage required to give him victory, quoting the examples of Moses and Gideon, David and Goliath.[10]

Issuing ringing manifestos was one thing. Keeping people calm was another. The febrile mood of St Petersburg veered from absurd optimism to darkest despair with astonishing ease. "Everyone is expecting a courier to appear at any moment with news of a victory, rumours of which are circulating in the city," wrote an inhabitant of the capital to a friend in the county on 21 July. "They say that Bagration has beaten the King of Westphalia. They give the number of prisoners taken as 15,000." There was talk of resounding victories: Wittgenstein had thrashed Oudinot and Macdonald, Ney and Murat had been beaten outside Vitebsk, and so on.[11]

But there was also much anger at the retreats, and at the optimistic bulletins being issued by the authorities. "In these they write about our successes, about the slow pace of Napoleon's advance, about his lack of confidence in his forces, but the facts themselves show us something quite different; we have been successful only in retreating, and while the enemy has not conquered, he has simply helped himself to entire provinces," complained Varvara Ivanovna Bakunina in a letter to a friend, adding that "despair and fear grow by the hour, while they try to deceive us, assuring us . . . that all this is happening according to some very clever plan." Hundreds of refugees from Riga had turned up in St Petersburg, sowing panic. "Grief, fear and despair has taken hold of everyone," she reported on 6 July.

The ratification of peace with Turkey, announced a week later, steadied nerves. But fast on this came news that Alexander had abandoned Drissa, which threw St Petersburg into a panic. Tidings of Wittgenstein's stand at Yakubovo restored a semblance of calm. On 25 July there was a service of thanksgiving for this in the Tauride Palace, followed three days later by one for Tormasov's success at Kobryn. But when they heard that Napoleon was in Vitebsk, many people began to pack their bags and some actually left St Petersburg, expecting it to be the next goal of his advance.[12]

The mood was only a little steadier in Moscow. On 27 June, three days after Napoleon crossed the Niemen but before she had heard of it, Maria Apolonovna Volkova wrote to her friend Varvara Ivanovna Lanskaia that she "had always been of the opinion that one should not concern oneself too much with the future." But her philosophy did not stand up to the test when the fatal news broke. "Peace has abandoned our lovely city," she wrote on 3 July. A couple of days earlier, a German inhabitant of Moscow was nearly stoned to death by the mob, who thought he was French. A week later, Maria Apolonovna took up the pen once more. "Five days ago they were saying that Ostermann had won a great victory. This turned out to be a fabrication," she wrote. "This morning news reached us of a brilliant victory won by Wittgenstein. This news comes from a reliable source, and Count Rostopchin is confirming it, but nobody dares believe it."[13]

What Alexander had to do was to convert these fears and this anger into action. In the first place, he needed more men – and these belonged, physically, to the landed nobility. There had already been much grumbling in March when he had squeezed another extra draft out of them, and he would need all their good will to get them to give him more now. He also needed large quantities of cash, and he must therefore appeal for donations, over and above taxation, which had itself been increased dramatically in the run-up to war. It would be the greatest test his legendary charm was ever put to.

In Smolensk, where he went first, many nobles came and offered themselves and their wealth to the cause. Typical of them was Nikolai Mikhailovich Kaliachitsky, who offered his three sons for "either a determined defence or a glorious death," as well as wagonloads of supplies for the army, which virtually spelt ruin for his small estate.[14] Emboldened by the effusion of patriotism and devotion to his person he had witnessed here, Alexander set off for one of the most important meetings of his life – that with the inhabitants of Moscow.

He did not want to make a triumphal entry into the city, as he was not at all sure of the reception he would get in this stronghold of the "*starodumy*," the defenders of tradition whom he had done his utmost to placate by ditching Speransky and appointing Rostopchin. He asked the Governor to drive out and meet him at the last posting station before Moscow on the afternoon of 23 July. They had a long talk, during which Rostopchin reassured him that he had the city under control and that Alexander had nothing to fear. The Tsar nevertheless resolved to drive into Moscow at midnight, hoping that by then everyone would be in bed. But he was to be disappointed.

The nobility had gathered to meet him in the Kremlin. The great rooms were filled with aristocrats and high officials, while the open spaces inside the precincts were crowded with the populace. Suddenly a rumour spread through the throng that Orsha had fallen to the French, and the people surged around the Kremlin howling about treason. Then someone suggested that the Tsar had not shown up yet because he was dead. "A tremor ran through the crowd," recalled one of the young noblewomen in the upstairs rooms. "They were ready to believe anything and to fear everything." Someone standing next to her, hearing the roar of the populace outside, whispered "Rebellion!" The word flew through the room, and bedecked aristocrats began to panic. Happily for all concerned, a courier then appeared announcing the Tsar's imminent arrival.[15]

Crowds had also gathered on the Poklonnaia, the hill of Salutation, to greet him, but instead of waiting, they moved further and further out along the road. It was a warm, starry night, and when his carriage was still fifteen versts from the city, Alexander found the road lined with peasants clutching candles and priests holding aloft icons and blessing him. When he reached the city, the crowd unharnessed his horses and hauled his carriage through the streets, with people kneeling as he passed.

The next morning he attended a solemn service in the Kremlin's Uspensky cathedral to celebrate the ratification of the peace with Turkey. Afterwards Metropolitan Platon blessed him with an icon of St Sergei which had accompanied Tsar Alexei Mikhailovich in his wars against the Poles and Peter the Great against the Swedes. "Lead us where you will, lead us, father, we will die or conquer!" people shouted all around him. But he was nevertheless nervous when, on the following day, he set off to the Sloboda Palace, where the nobility and the merchants had gathered to meet him. Alexander looked "pale and thoughtful," and unlike himself without his usual warm smile. In order to reassure him, Rostopchin had parked a row of police *kibitkas* outside the palace, and was prepared to bundle any troublemakers into them and send them off to Kazan. But they would not be needed. Alexander harangued the nobles and the merchants about the need to make sacrifices in order to defend the fatherland, and then left them to debate, separately, on the matter.[16]

The nobles began to discuss the viability of raising one man in twenty-five, when someone suggested one in ten. It later transpired that the proposal had been put forward by a man who owned no land in the province and only wanted to gain influence at court. His suggestion hit the mood and was endorsed in a rush of enthusiasm. The merchants' meeting was held in a no less exalted atmosphere. "They struck themselves on the head, they tore out their hair, they raised their hands to heaven; tears of fury flowed down their faces, which resembled those of ancient heroes. I saw one man grinding his teeth," recorded a witness. "It was impossible to discern exact words in the general uproar; all one could hear were wails and shouts of indignation. It was a spectacle quite unique." The provost of the merchants gave a huge sum, saying: "*Je tiens ma fortune de Dieu, je la donne à ma patrie.*" As well as offering to give one man in ten to the militia, the nobility came up with a pledge of three million roubles, and the merchants contributed eight million.[17]

Alexander had achieved something far greater than merely obtaining the men and the cash he needed. According to Prince Piotr Andreievich Viazemsky, who was living in Moscow at the time, the war had been regularly discussed in the English Club and in drawing rooms, but the tone of the discussions had always been a touch academic, as though it did not really concern those present. But everything changed with Alexander's arrival on the scene. "All vacillation, all perplexity vanished; everything seemed to harden, to become tempered and came together in one conviction, in one holy feeling that it was necessary to defend Russia and save her from the invading enemy."[18]

His visit to Moscow had a profound effect on Alexander himself, and he left the old capital on the night of 30 July a stronger man. He was filled with a new determination and strength by the effusion of devotion to his person. "I have only one regret, and that is not to be able to respond in the way I should wish to the love of this admirable nation," he said to one of the Tsarina's ladies-in-waiting, Countess Edling. "How so, Sire? I do not understand." she replied. "Yes, it needs a leader capable of leading it to victory, and unfortunately I have neither the experience nor the talents required at such a moment." Despite himself, he began to think once more of assuming command of the army. But a timely letter from his sister Catherine, which pointed out that he had let down Barclay by his indecision, virtually ordered him not even to think of it.[19]

The Russian army was no happier than the French at its failure to stand and fight at Vitebsk, and the troops were in a state of dejection as they trudged back towards Smolensk. They reached the city on 1 August and made camp on the north bank of the Dnieper. Barclay issued a proclamation to the effect that they were about to be joined by Bagration's Second Army, and would then be ready to take on the French, which lifted spirits.

On the following day, Bagration himself rode into Barclay's camp wearing all his decorations and accompanied by a suite of generals and staff officers. Barclay looked sober and underdressed as he came out to meet him. "They greeted each other with all possible marks of courtesy and the appearance of friendship, yet with coldness and distance in their hearts," according to Barclay's chief of staff Yermolov. Bagration's Second Army was only a day's march behind him, and he graciously placed himself and it under Barclay's command.[20]

"This news filled everyone with extraordinary joy," recalled Nikolai Mitarevsky, a fledgling artillery officer in General Dokhturov's corps. "We thought there would be no more retreating and the war would take on a different character." The very look of the Second Army buoyed up the spirits of the men of the First, as Yermolov explained: "The First Army, exhausted by the retreat, had begun to mutter and had given rise to disorders, a sure sign of the collapse of discipline. The unit commanders had cooled towards the commander, the lower ranks felt their confidence in him shaken. The Second Army was fired by a completely different spirit! The constant sound of music, the ubiquitous strains of singing coming from them, revived the spirits of the soldiers."[21]

The anti-Germans and russophiles expected the brave spirit of Bagration to dominate, and the consciousness that they were now standing in defence of old Russian lands was expected to have an effect as well. "The spirit of the nation is awakening after a two-hundred-year slumber, feeling the threat of war," wrote Fyodor Glinka, a passionately russophile officer, alluding to the Polish wars of 1612. Songs and odes were composed in honour of what was to be a heroic and successful stand at Smolensk.[22] Others talked of taking the offensive and chasing the French out of Russia.

The Russians were now in a stronger position than they had been at any time since the start of the war. They had, it is true, given up a vast part of their territory, lost up to 20,000 men, a couple of dozen guns and huge stores of supplies. But they now had some 120,000 men grouped in the centre, with two forces, of 30,000 in the north and 45,000 in the south, threatening Napoleon's flanks. From their own intelligence and the questioning of prisoners, they knew about the difficult conditions in the French army, whose main attacking force they now estimated, somewhat conservatively, at about 150,000.

Yet, as Clausewitz pointed out, their new strength was a strategic rather than a tactical one, and the French would still be bound to win a pitched battle. But the ability of the French to operate effectively was reducing with every day. There was, as a result, no point whatsoever in the Russians taking the offensive at this stage.[23] Yet that is exactly what they were determined to do. The whole army, from the top down to the last ranker, was fed up with the continuous retreat. They had been told of brilliant victories won by Tormasov, Wittgenstein, Platov and others, and they could not conceive why they were themselves giving up territory without a battle. Now that Bagration had joined forces with Barclay, there seemed to be no further excuse for retreating, and there was a universal desire to stand and fight.

Barclay, who realised the pointlessness of giving battle at this stage, was being put under great pressure to do just that, from above by Alexander himself, and from below by everyone down to the ranks, and he was in no position to oppose. On 3 August he wrote to Alexander that he was going to attack the exposed corps of Ney and Murat. But everything points to the fact that he was still hoping to be able to avoid giving battle. On 6 August he held a council of war at which he argued his case, but he was powerfully outnumbered by the hawks, and reluctantly agreed to the attack, urging everyone to proceed with extreme caution.

They set out on the morning of the following day, 7 August, in three columns, which were to attack and overwhelm Murat's cavalry and Ney's corps, encamped around Rudnia, ahead of the rest of the French forces. The enterprise could have been successful had it been carried out with speed and determination. It would have made no difference to the strategic situation, yet it would have raised the morale of the Russian troops and made it more difficult to retreat again afterwards (which they would have to do anyway), so in this case success would have had mixed benefits. In the event, success was not going to be an issue.

On the night of 7 August, at the end of the first day of the offensive, Barclay received intelligence, inaccurate as it turned out, that a large French force had occupied Poriechie, to the north of his line of attack. This could only be either an attempt at outflanking him or an exposed enemy unit which could be easily cut off. He therefore ordered his three columns to wheel round to face in a northerly direction. Bagration did not understand the thinking behind this order, and obeyed only with extreme reluctance. But the order never reached the cavalry, under Platov and Pahlen, and while Barclay and Bagration marched northwards on the following day, the cavalry continued to move westwards. At Inkovo it stumbled on General Sebastiani's cavalry division, which it surprised and routed, taking a couple of hundred prisoners, but it was subsequently beaten back by a French counterattack.

When Napoleon learned of the Russian attack the next day, he deduced that Barclay had decided to fight in defence of Smolensk. But he was by now wary of the Russians. Determined to make sure that they would not escape this time, he put into action a plan to encircle them and strike them in the back. He instructed Prince Eugène to move south and join Ney in keeping an eye on the area where he assumed the Russian army to be, and ordered all other units

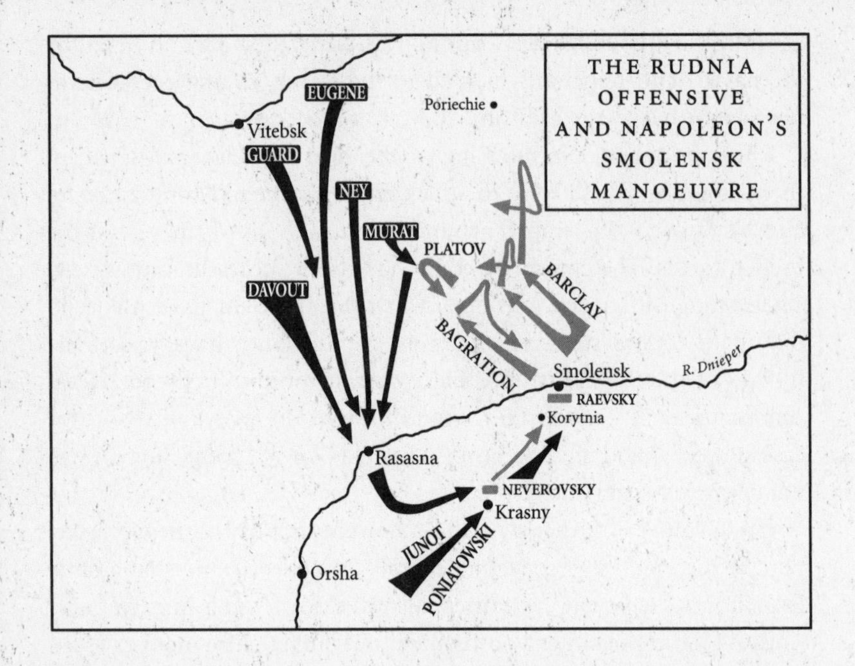

in the Vitebsk area to make for Rassasna on the Dnieper. They were to cross the river and join Junot's and Poniatowski's corps, then sweep into a presumably unoccupied Smolensk, recross the river and appear behind Barclay's back.

But Barclay's forces were by now in a state of such confusion that Napoleon need not have bothered. After assuring himself that there were no French at Poriechie, Barclay had marched back to his starting position and ordered Bagration to do the same, intending to implement the original plan of a frontal attack on Rudnia. But when Bagration, who was sullenly trudging back towards his starting positions, got what was now his third order to about turn and march in a different direction, he was beside himself with irritation. "For the love of God," he wrote to Arakcheev, "give me a posting anywhere away from here – I'll even accept command of a regiment, in Moldavia or the Caucasus if necessary – I just cannot stand it here any longer; headquarters is so full of Germans a Russian cannot survive there."[24]

He was so angry that he resolved to ignore Barclay's order. He was moving in the direction of Smolensk, and he decided to keep going. So while two of the three Russian columns were now wheeling back to march on the French, the third was resolutely moving away in the opposite direction. This act of wilful insubordination was to save the Russian army.

All the changes of order had added to the usual degree of confusion surrounding the movements of an army, with the result that the area in front of the French forces was covered in Russian units marching and countermarching in different directions, some of them lost, most of them confused, and all of them increasingly fed up. "As this was the first time we had advanced after so many retreats, the joy felt by the whole corps, which was longing to be able to attack the enemy at last, would be hard to convey," wrote Lieutenant Simansky of the Izmailovsky Life Guards.[25] The change of plan, which they assumed to herald a new retreat, was therefore greeted with fury.

Junior officers like Simansky wondered whether their commanders knew what they were doing. "Our lack of experience in the art of war reveals itself at every step," Captain Pavel Sergeevich Pushchin of the Semeonovsky Life Guards noted in his diary on 13 August. The feeling that the commander-in-chief was out of his depth was gaining ground. Staff officers were railing against "Germans," and the word "treason" was being muttered more and more frequently. The iron discipline gripping the Russian soldier was beginning to relax, and instances of desertion and looting multiplied.[26] Had Napoleon delivered an energetic frontal attack at that moment, both Russian armies would have been annihilated.

But Napoleon was busy implementing his own plan. At dawn on 14 August the divisions of Davout's, Murat's, then Ney's and Prince Eugène's corps began to cross the Dnieper at Rassasna on three bridges constructed during the night. The troops then took the great Minsk–Smolensk highway, a wide straight road running between rows of silver birches, laid out by Catherine the Great for the rapid movement of mail and troops. They were joined along this road by

Junot's Westphalians and Poniatowski's Poles, coming up from Mogilev. In the early afternoon they stumbled on General Neverovsky's 27th Division, which Bagration had left covering the southern approaches to Smolensk at a place called Krasny.*

Neverovsky had no more than about 7,500 men, many of them raw recruits, with which to face Murat's entire cavalry corps. But he did not lose his head. He sent his cavalry and his guns back to cover his retreat, and formed his men into an extended square formation. Luckily for him, Murat did not deign to stop and wait for his artillery to come up, but threw his cavalry at the Russians, meaning to sweep them out of the way. Where any other army would have laid down its arms or scattered, Neverovsky's peasant soldiers retreated in a solid mass. "The very lack of experience of those Russian peasants making up this unit gives it a force of inertia which amounts to resistance," Baron Fain mused. "The dash of the cavalrymen was cushioned by this crowd which clung together, pressing and filling every gap. The most brilliant valour was exhausted by striking this compact mass which they could only hack at without breaking up."[27]

Neverovsky retreated nearly twenty kilometres under constant pressure from Murat's cavalry, which delivered some thirty charges. He lost about two thousand men and seven guns, but he reached Korytnia, where he was reinforced on the following morning by troops sent out of Smolensk, and escorted into the city. He also lost one of his regimental bands, which some French grenadiers found cowering in the ruins of a burnt-out church, brandishing their instruments as a mark of their peaceful intentions and begging for mercy in the broken French of one of their number, a native of Tuscany.[28]

That evening, 15 August, Napoleon reached Korytnia, and was greeted by a hundred-gun salute – it was his birthday. But he had

* The town is referred to by diarists and historians by a variety of names, most commonly "Krasnoie." I give the spelling used by the contemporary local historian Voronovsky.

nothing to celebrate. His manoeuvre had failed. He had hoped to find Smolensk undefended, which would have permitted him to occupy the city and use the bridges across the Dnieper to penetrate into Barclay's rear. As it was, the city was garrisoned – thanks to the insubordination of Bagration.

Napoleon vented his irritation on Poniatowski, whose 5th Corps had just rejoined the main force of the Grande Armée. When the Prince came to Napoleon's bivouac to pay his respects, the Emperor unleashed a string of gross insults at him, shouting for all to hear, accusing him and his Poles of cowardice and laziness, and saying that the only thing they were good at was playing with Warsaw whores. At the same time, he was furious when he heard that the Poles were down to 15,000 men – some units had lost half of their effectives through forced marches, sickness and fighting. He taunted Poniatowski, and said he and his Poles would have a chance to show their mettle at Smolensk on the following day.[29]

Smolensk was a city of 12,600 inhabitants, and had no particular economic or strategic importance. It did have a certain spiritual significance, as one of its churches housed a renowned miraculous icon of the Virgin, and the city had been the scene of several desperate struggles for dominion over the area between the Poles and the Russians, who had finally wrenched it back 150 years before. It was surrounded by massive brick walls twenty-five feet high and fifteen feet thick, with a deep dry ditch in front, and strengthened by thirty massive bastion towers.[30] There was no advantage for Napoleon in taking Smolensk, as his real aim was to defeat the Russian army, and since the chance of crossing the Dnieper and penetrating into their rear had been denied him here, he should have immediately gone in search of a crossing point further east. Had he done so, he would have forced the Russians to fight and would certainly have won Smolensk without a blow, probably with its magazines intact. He did send Junot off up the Dnieper in search of a crossing. But he had willed himself to believe that the Russians would come out and face him in defence of their holy city, so he decided to attack it. In doing so he committed, according to Clausewitz, his greatest error of the whole campaign.[31]

Murat and Ney reached Smolensk early on the morning of 16 August and launched a first attack. This was successfully beaten off by General Nikolai Raevsky with his 7th Corps, and at midday the defenders were reinforced with more of Bagration's troops. In the afternoon the French could see large columns of Russian troops coming into view on the opposite bank of the Dnieper, their bayonets glinting. It was Barclay, who had been obliged to abandon his attack on Rudnia by Bagration's refusal to cooperate with him. He had heard of Napoleon's crossing of the river on the night of 14 August and had hurried back to Smolensk. On seeing this, Napoleon, according to some, rubbed his hands, saying: "At last! Now I've got them!"[32]

Barclay was under tremendous pressure to defend the city. He realised that it was a hopeless position, which could have been held at most for a week or two, at immense cost and to no advantage. But his authority was by now so shaky that even though his instinct told him to fall back, he had to make a show of defending Smolensk. As he was afraid that Napoleon might try to cross the Dnieper further east, and no doubt also to be rid of him, he sent Bagration down the Moscow road towards Dorogobuzh, with orders to prevent any French crossing the river and to keep the line of retreat open. On the evening of 16 August he relieved Raevsky with 30,000 men under Dokhturov, who were to hold the city. He stationed the rest of his forces and set up his batteries on the north bank.[33]

On the morning of 17 August the French attacked the outlying suburbs outside the city walls. In hand-to-hand fighting they managed to push back Dokhturov's men, but were counterattacked by a large force and fell back. Napoleon hoped that this Russian sortie meant the whole Russian army was going to come out of the city to face him, but he was disappointed.

The action had taken up most of the morning, and was accompanied by heavy shelling of the Russian positions by French artillery. At midday there was a lull in the fighting. Lieutenant Hubert Lyautey

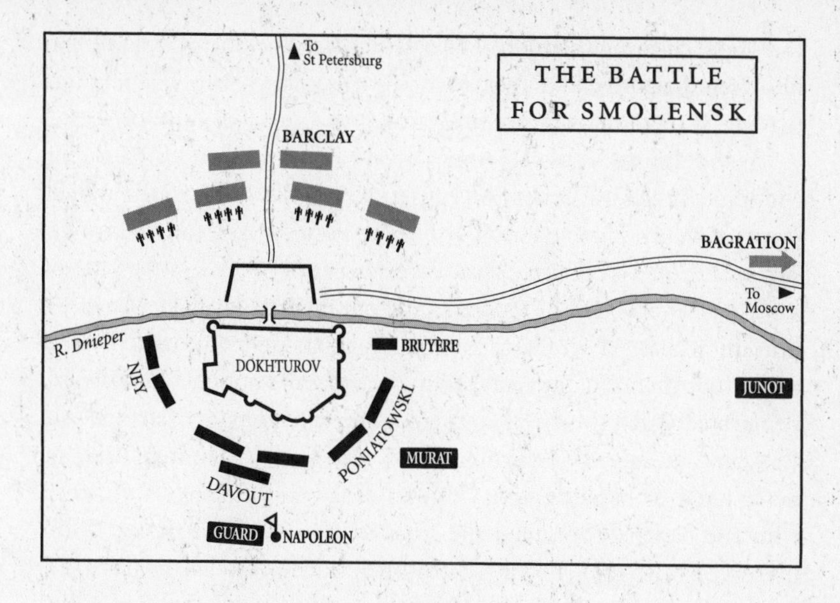

of the artillery of the Guard had been in action all morning, and he took the opportunity to lead his horses down to the river for watering. The Russian batteries on the opposite shore had come to do the same. "The Russians drank on one side, we on the other, we communicated with each other with words and gestures, we exchanged drinks, tobacco, in which we were richer and more generous," Lyautey wrote in a letter home. "Soon afterwards, these good friends were exchanging cannon shots."[34]

At 2 p.m., seeing that the Russians were not going to come out and give battle, Napoleon gave the order for a general assault on the city. Over two hundred guns opened up, and three corps of the Grande Armée went into action. It was an unforgettable sight for those present. The city of Smolensk lies on a slope descending to the river Dnieper, on the side of a great amphitheatre, the other side of which is made up by the slope rising from the other side of the river. This was occupied by Barclay's army, which could see the whole city across the river and the French attacking it from three sides. As the French attackers went into action, their comrades watched from the top of the slope, cheering them on. The weather was fine, the troops were in full dress uniform, and they went into the attack with their bands playing. It was a magnificent spectacle.

In the centre were Davout's three divisions, under Generals Morand, Gudin and Friant; on the left Ney with two divisions, one of them of Württembergers; on the right Poniatowski with two divisions, and on his right General Bruyère's cavalry division; a total of some 50,000 men. The French columns lumbered forward. Bruyère's horsemen charged a body of Russian dragoons under General Skallon and swept them off the field, killing their commander. The infantry penetrated the suburbs and forced the defenders to retreat. The Russians attempted a counterattack, but this was beaten back in fierce hand-to-hand fighting. "On both days at Smolensk, I attacked with the bayonet," recalled the Russian General Neverovsky. "God preserved me, and I only had three bulletholes in my coat."[35] Eventually the French reached the city walls, which they then tried, vainly, to escalade. They had no ladders, so all they could do was attempt to climb them.

Auguste Thirion of the 2nd Cuirassiers went to get a better view from the emplacement of one of the French batteries, which was being shelled by Russian guns from the heights on the other side of the river. "I cannot conceive how a single man or a single horse could escape that mass of cannonballs coming from two sides and crossing in the midst of those batteries," he wrote, but it did not prevent him enjoying the spectacle. "We also saw at our feet infantry laboriously descending into the ditches, or rather the ravines which made up the moat of the fortress. It was a Polish division which was trying to storm those rocks with a courage, a desperation worthy of greater success; these brave men tried to scale them by climbing on each other's shoulders. But the nature of the terrain would not permit of success, and it was a unique and curious spectacle to see that ant-like mass of men crawling over the rocks in such a picturesque manner while above their heads the cannon which were the object of their efforts thundered against their brothers in arms. Opposite them, the French batteries fired projectiles which occasionally fell short, showering the rocks with a mass of fragments of the walls."[36]

Charles Faré, a lieutenant in the 1st Grenadiers of the Old Guard, told his mother in a letter home that he had never seen French troops fight with greater dash. Ney himself said that the attack by one battalion of the 16th Regiment of the Line was the bravest feat of arms he had ever seen. But not everyone was cheering. General Eblé and his colleague the cartographer of the Grande Armée, General Armand Guilleminot, could see no point to the grand frontal assault on the city walls. They knew that the twelve-pounder field guns Napoleon had brought up were of no use, as their cannonballs were simply absorbed by the soft brick fortifications, and that they could not make a breach in the massive walls. "He always wants to take the bull by the horns!" exclaimed Eblé, shaking his head. "Why doesn't he send the Poles off to cross the Dnieper two leagues upstream of the city?"[37]

Others who were not enjoying the spectacle were the citizens of Smolensk. In the morning the inhabitants of the suburbs had taken weapons from the bodies of dead soldiers and made up gaps in the ranks of the defenders; priests holding crucifixes aloft placed themselves at the head of the militiamen and died with them. When the troops were beaten back, the civilians tried to follow. "The inhabitants fled in horror, dragging their valuables; here I saw a good son, bearing on his shoulders an infirm father, there a mother making her way along a safe path towards our positions clutching her little ones in her arms, having sacrificed everything else to the enemy and to the fire," recalled one artillery officer.[38]

Their fate was not much more to be envied when they did manage to get into the old city within the walls, according to a fifteen-year-old officer in the Simbirsk Infantry Regiment. "What an awful confusion I witnessed within the walls: the inhabitants, believing that the enemy would be repulsed, had remained in the city, but that day's strong and violent attack had convinced them that it would not be in our hands by the morrow. Crying out in despair, they rushed to the sanctuary of the Mother of God, where they prayed on their knees, then they hurried home, gathered up their weeping families and left their houses, crossing the bridge in the utmost confusion. How many tears! How much wailing and misery, and, in the end, how many victims and blood!"[39]

The grandiose spectacle of the afternoon turned into a scene from hell as the evening drew in. The mortar shells that the French had been lobbing into the city had set many of the predominantly wooden houses on fire, and this spread rapidly. Baron Uxküll of the Russian Chevaliergardes speaks for many who watched impotently as one of their old cities and its inhabitants were engulfed in the flames. "I was standing on the mountain; the carnage was taking place at my very feet. Shadows heightened the brilliant sheen from the fire and the shooting," he wrote. "The bombs, which displayed their luminous traces, destroyed everything in their path. The cries of the wounded, the Hurrahs! of the men still fighting, the dull confused sound of the rocks that were falling and breaking up – it all made my hair stand on end. I shall never forget this night!"[40]

The French onlookers were equally gripped by the "sublime horror" of the spectacle; thoughts turned to the fall of Troy. "Dante himself would have found here inspiration for the hell he set out to depict," according to Captain Fantin des Odoards. As the French still tried to storm the walls, much of the city was on fire, and the defenders showed up as black silhouettes against the flames behind them, looking for all the world "like devils in hell," as Colonel Boulart of the artillery put it. Similar thoughts were going through the head of Caulaincourt as he stood in front of Napoleon's tent, watching. Suddenly he felt a slap on his shoulder. It was the Emperor, who had come out to watch, and who compared the sight to an eruption of Vesuvius. "Don't you think, *Monsieur le Grand Écuyer*, that this is a fine spectacle?" he added. "Horrible, Sire," was Caulaincourt's only answer.[41]

Grand spectacle or not, Napoleon had nothing to be pleased about. As the fighting died down that night it became apparent that the

French had gained nothing and lost at least seven thousand men in dead and wounded. Barclay too had little to rejoice over. Aside from the satisfaction of denying the French an easy victory, he had achieved nothing, beyond the loss of over 11,000 men and two generals.

Barclay realised that he could not stay where he was much longer, as it was only a matter of time before Napoleon crossed the Dnieper upstream and cut him off. He had made a symbolic gesture in defending Smolensk for two days, and it was now time to think of saving the army. He therefore ordered Dokhturov to evacuate the city after setting fire to all remaining stores and anything else that could be of use to the enemy, and to destroy the bridges after him. The holy icon of the Virgin of Smolensk had already been removed from its shrine, placed on a gun carriage and escorted over the bridge to the northern bank of the river.

Barclay's orders for the city to be abandoned provoked a general outcry. "I cannot express the indignation that prevailed," wrote General Sir Robert Wilson, who had just arrived to take up his post as British "commissioner" at Russian headquarters. A succession of senior officers came to beg Barclay to reconsider his decision, or, if he were determined to retreat, to allow them to fight on to the last drop of blood. Bagration wrote him a note demanding that Smolensk be defended regardless of cost. Bennigsen, in stark contradiction to his earlier assertion that there was no point in a battle at this point in the retreat, came out in favour of a last-ditch stand. He stormed into headquarters, accompanied by Grand Duke Constantine and a bevy of generals, demanding that Barclay change his plans. The Grand Duke virtually commanded him to rescind his "cowardly" order and launch a general attack on the French. "You German, you sausage-maker, you traitor, you scoundrel; you are selling Russia," he shouted at Barclay for all to hear. "I refuse to remain under your orders," he added, saying he would move the Guards corps under Bagration's command. He continued to heap invective on Barclay, who eyed him in silence. "Let everyone do their duty, and let me do mine," he finally interjected, cutting short the argument. That evening Constantine received an order from Barclay to take an important letter to the Tsar and hand over command of the Guards to General Lavrov.[42]

Two hours before dawn the last of Dokhturov's men trudged back across the bridges and set fire to them. A short while earlier, a *voltigeur* company of the 2nd Polish Infantry managed to make a breach in the walls and entered the blazing city. In the morning, once one of the gates had been opened and its approaches cleared of the dead and dying bodies heaped around it, the French made their entry into the city.

It was a veritable charnelhouse, its streets strewn with corpses, mostly blackened by fire. In the ruins of houses that had been engulfed by fire lay the remains of inhabitants or wounded soldiers who had taken refuge inside. "One had to walk over debris, dead bodies and skeletons which had been burned and charred by fire," recalled one French officer. "Everywhere unfortunate inhabitants, on their knees, weeping over the ruins of their homes, cats and dogs wandering about and howling in the most heart-rending way, everywhere only death and destruction!" The Russian wounded had been laid out in makeshift hospitals, which had then been swept by fire as their comrades evacuated the city. "These unfortunates, abandoned in this way to a hideous death, lay in heaps, calcinated, shrunken, conserving only just a human form, amidst the smoking ruins and burning beams," in the words of Lieutenant Julien Combe of the 8th Chasseurs à Cheval. He was not the only one to notice that the bodies of the burnt soldiers had shrunk; some thought they were those of children. "Soldiers who had wanted to flee had fallen in the streets, asphyxiated by the fire, and had been burned there," observed Dr Raymond Faure. "Many no longer resembled human beings; they were formless masses of grilled and carbonised matter, which the metal of a musket, a sabre, or some shreds of accoutrement lying beside them made recognisable as corpses."[43]

One German soldier could hardly believe what he saw as he tramped up streets carpeted with human remains. "Like thousands of others, I was marching along when, between two burnt-out houses, I saw a small orchard whose fruit had been carbonised, underneath the trees of which were five or six men who had been literally grilled," he wrote. "They must have been wounded men who had been laid out in the shade before the fire started. The flames had not touched them, but the heat had contracted their nerves and pulled up their legs. Their white teeth jutted from between their shrivelled lips and two large bloody holes marked the place where their eyes had been."[44]

All this had a profound effect on the army. "It marched through these smouldering and bloodied ruins in good order with martial music and its habitual pomp, triumphant on these deserted ruins, and having nobody but itself as a witness to its glory!" in the words of Ségur. "Spectacle without spectators, victory practically without fruit, bloody glory, of which the smoke which surrounded us, and which seemed to be our only conquest, was the only too faithful emblem!"[45]

The city was not in fact entirely deserted. A considerable number of the inhabitants had not managed to get away, and they cowered among the ruins or thronged the churches, which, being of brick and stone, were refuges from the fire. There were also thousands of wounded Russian soldiers, and as the French prepared to march into Smolensk a delegation from the city authorities came out to ask Napoleon to help take care of them. He detailed sixty physicians and medical staff to go into the city and organise hospitals.[46]

This was easier said than done. Several large buildings such as monasteries and warehouses were designated for the purpose and the wounded were brought in, but there were no beds or mattresses to lay them on, and it was days before straw could be found to place on the floor under them. The lightly wounded were laid side by side with the sick, and infection spread quickly as they lay sweating in what even Napoleon termed "dreadfully hot" weather. The sheer numbers of the wounded meant that many were not dressed for a day or two, by which time supplies had run out. Surgeons were sewing up wounds with tow in lieu of thread and dressing them with strips torn from their own uniforms and paper taken from the city archives. "Without medicines, without broth, without bread, without linen, without lint and even without straw, they had no other consolation as they lay dying other than the sympathy of their comrades," in the words of General Berthézène. Napoleon sent Duroc around the hospitals to give the wounded money, but while in Austria or Italy this would have meant they would have been able to procure food and other necessities for themselves, here they would be able to buy nothing, while the possession of the coins made them vulnerable to robbery and murder.[47]

"That is the hideous side of war, the one to which I shall never grow accustomed," noted Captain Fantin des Odoards of the Grenadiers of the Guard as he walked through the city on the following day. "To see so much misery and not to be able to provide assistance is torture." Seasoned soldiers could not afford to dwell on such thoughts if they were to survive. General Dedem de Gelder, whose division bivouacked on the main square that night, did his best. "I spent the night on a very luxurious settee which the men had found in one of the neighbouring town houses," he recalled, and he dined on jam, stewed fruit, two fresh pineapples and some peaches. "I would have preferred a good soup, but in war, one eats what one finds."[48]

Barclay had remained on the north bank of the Dnieper throughout the day of 18 August, holding the suburb on that side of the river and preventing the French attempts at rebuilding the burnt bridges. But that night he withdrew. As the road to Moscow ran for several miles along the north bank within range of French guns, he set a course that started in a northerly direction, gradually swinging round to rejoin the Moscow road at Lubino. So as to avoid encumbrance along the small country roads they would have to use, he divided his force in two. But this only complicated matters, as during the first stage of the withdrawal, on the night of 18 August, several units lost their way. Progress was slower than expected, with guns and supply wagons getting stuck at the crossing of the many streams dissecting the roads. The inclines were so steep that in several places guns and heavy wagons rolled down, dragging their teams of horses and men to a nasty death at the bottom of a ravine, and in turn obstructing progress further.

In the meantime, Ney had repaired the bridge at Smolensk, crossed the river and started to advance down the Moscow road, while Junot had begun to cross further upstream, at Prudichevo. Fearing that they might reach it before his retreating men did, Barclay had sent General Pavel Alekseievich Tuchkov with a small force to Lubino to cover the point at which the wheeling Russian columns were to rejoin the Moscow road.

Ney, who began to move along the Moscow road in the morning, was checked by what he thought was a counterattack developing on his left flank. In fact it was Ostermann-Tolstoy's division, which had got lost in the night, and after marching in a circle for ten hours reappeared outside Smolensk. Ney deployed against it, which gave Tuchkov some time, but soon the French were pushing the Russians back along the Moscow road.

On hearing of the fighting, Napoleon rode out to the scene. Assuming that this was no more than a rearguard action, he ordered Davout to back up Ney with one of his divisions. Together they pushed Tuchkov back, but he too was reinforced by other units and the timely arrival of Barclay himself, who rallied the troops and steadied the situation. Wilson was impressed by Barclay, who "seeing the extent of the danger to his column, galloped forward, sword in hand, at the head of his staff, orderlies, and rallying fugitives, and crying out, 'Victory or death! we must preserve this post or perish!' by his energy and example reanimating all, recovered possession of the height, and thus under God's favour the army was preserved!"[49]

The Russians took up strong positions at Valutina Gora. Junot with his Westphalians was actually behind their left wing, and could have taken them in the back, which indeed Napoleon ordered him to do. But the usually fearless Junot, who had been acting strangely and complaining of heat stroke, made a number of incoherent replies

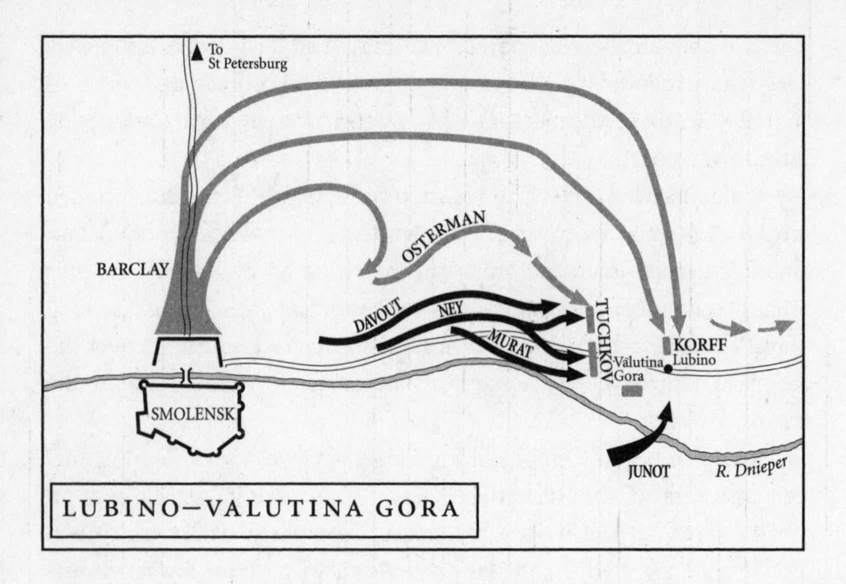

and would not move, even when Murat galloped up in person to tell him to attack.

"If we had attacked, the Russians would have been routed, so all of us, soldiers and officers, were eagerly awaiting the order to attack," wrote Lieutenant Colonel von Conrady, a Hessian in Junot's corps. "Our ardour to go into battle was expressed vociferously, with whole battalions shouting that they wanted to advance, but Junot would not listen, and threatened those who were shouting with the firing squad . . . Grinding our teeth, we were reduced to the role of spectators, while honour and duty beckoned. Never was an opportunity to distinguish oneself more shamefully lost! Several officers and soldiers in my battalion wept with despair and shame."[50]

There were by now some 20 to 30,000 Russians facing, and outflanked by, as many as 50,000 French. According to Barclay's aide-de-camp Woldemar von Löwenstern, Tuchkov rode up and asked for permission to fall back, to which Barclay allegedly replied: "Return to your post and get yourself killed if you must, for if you fall back I shall have you shot!" Aware that the fate of the Russian army was in his hands, he held on, but it was touch and go. At one point Yermolov, who was watching, seized his aide-de-camp by the elbow. "Austerlitz!" he whispered in horror.[51]

If the French had been able to defeat Tuchkov, they could have sliced through the middle of the Russian forces on the march, and these would have stood no chance. "Never had our army been in greater danger," Löwenstern later wrote. "The fate of the campaign and of the army should have been sealed on that day."[52]

It was unlike Napoleon not to sense the reason behind the Russian stand, but at about five o'clock in the afternoon he left Ney to get on with it and rode back to Smolensk. "He seemed to be very annoyed, and broke into a gallop when he came up with us, whose acclamations appeared to importune him," noted an officer of the Legion of the Vistula who watched him ride by.[53]

Tuchkov stood his ground, and his men fought like lions. Ney's divisions, supported by Davout's Gudin division, also fought with dash and determination, and the battle developed into a massacre which only ceased when darkness fell. The field was strewn with seven to nine thousand French and nine thousand Russian dead and wounded, but the living lay down to sleep among them, too exhausted to build a camp.[54]

The following morning, Napoleon rode out to the scene. "The sight of the battlefield was one of the bloodiest that the veterans could remember," according to one of the Polish Chevau-Légers who escorted him.[55] He took the salute of the troops drawn up on this field of death and proceeded to enact one of the rituals that made him such a brilliant leader of men. He had decreed that he would award the coveted eagle that topped the standards of regiments which had proved their valour to the 127th of the Line, made up largely of Italians, which had distinguished itself on the previous day. "This ceremony, imposing in itself, took on a truly epic character in this place," in the words of one witness. The whole regiment was drawn up as if on parade, the men's faces still smeared with blood and blackened by smoke. Napoleon took the eagle from the hands of Berthier and, holding it aloft, told the men that it was to be their rallying point, and that they must swear never to abandon it. When they had sworn the oath, he handed the eagle to the Colonel, who passed it to the Ensign, who in turn took it to the centre of the élite company, while the drummers delivered a deafening roll.

Napoleon then dismounted and walked over to the front rank. In a loud voice, he asked the men to give him the names of those who had particularly distinguished themselves in the fighting. He then promoted those named to the rank of lieutenant, and bestowed the Légion d'Honneur to others, giving the accolade with his sword and giving them the ritual embrace. "Like a good father surrounded by his children, he personally bestowed the recompense on those who had been deemed worthy, while their comrades acclaimed them," in the words of one officer. "Watching this scene," wrote another, "I understood and experienced that irresistible fascination which Napoleon exerted when he wanted to, and wherever he was."[56]

By this extraordinary ceremony, Napoleon managed to turn the bloody battlefield into a field of triumph, sending those who had died to immortality and caressing those who had survived with kind words and glorious rewards. But many asked why he had not been there himself to direct the battle. And his entourage wondered what, if anything, had been achieved by the past four days of bloodletting.[57]

11

Total War

According to his secretary Baron Fain, Napoleon felt disheartened and disgusted at the turn events had taken. He had beaten the Russians and taken a major city. But while he had inflicted heavy casualties on them, he had lost as many as 18,000 seasoned troops himself in the two engagements, and had failed to force the Russian army to accept defeat. As is abundantly clear from his contradictory utterances, he did not know what to do next.

"In abandoning Smolensk, one of their holy cities, the Russian generals have dishonoured their arms in the sight of their own people," he said to Caulaincourt. "This gives me a good position. We will push them away a little distance in order to be at our ease, and I will consolidate. We will rest and use this strongpoint to organise the country, and we shall see how Alexander likes that. I will take command of the corps on the Dvina, which are doing nothing, and my army will be more formidable, my position more threatening to Russia than if I had won two battles. I will take up quarters at Vitebsk, arm Poland and later I will choose between Petersburg and Moscow."[1]

But he knew he was talking nonsense. All the arguments militating against stopping in Vitebsk went, in magnified form, for Smolensk. The burnt-out city represented neither an effective bastion nor a resource for his army. But to retreat now was even more unthinkable than it had been at Vitebsk. He had led himself into a trap.

He vented his frustration on anything that came to hand. He wrote to Maret complaining that the Poles of Lithuania had failed to raise enough troops and supplies. He complained that the army was losing men in needless foraging expeditions, and reprimanded the corps commanders. He raved about abuses. He went into a rage when he discovered that a Parisian wine merchant had been using wagons supposedly bringing medical supplies to ship wine for sale to the troops. When he came across some soldiers looting one day, he attacked them with his riding crop, yelling obscenities at them. And he was uncharacteristically ill-tempered and rude with his entourage.

In his desperation to find a way out, he clutched at every straw. General Pavel Alekseyevich Tuchkov, who had been taken prisoner at Valutina Gora, was treated with the greatest consideration by Berthier, who supplied him with shirts from his own wardrobe and offered him the choice of any city in Napoleonic Europe as a place of captivity. He was then granted an audience by Napoleon, who treated him with the utmost consideration. The Emperor poured out a torrent of self-justification and professions of friendship for Alexander, and asked Tuchkov to write to his sovereign telling him that all he wanted was peace. The embarrassed Russian wriggled out of this by saying that he was only a brigade general and that protocol forbade him to write to his Emperor, but he finally agreed to write to his elder brother, who was senior in rank.

"Alexander can see that his generals are making a mess of things and that he is losing territory, but he has fallen into the grip of the English, and the London cabinet is whipping up the nobility and preventing him from coming to terms," Napoleon said to Caulaincourt. "They have convinced him that I want to take away all his Polish provinces, and that he will only get peace at that price, which he could not accept, as within a year all the Russians who have lands in Poland would strangle him as they did his father. It is wrong of him not to turn to me in confidence, for I wish him no ill: I would even be prepared to make some sacrifices in order to help him out of his difficulty." He would probably have given Alexander the whole of Poland,

and Constantinople as well, in order to get out of the present impasse with a semblance of honour.[2]

But as he could not stop where he was, and as he would not retreat, he could only advance, in the hope of "snatching" a victory from the Russians. Moscow was only about four hundred kilometres, or eight days' forced march away, and the Russians would surely make a stand in defence of their old capital. There were still two months of decent campaigning weather ahead. "It was therefore reasonable to think that one would be able to bring the enemy to fight before the bad weather set in," argued General Berthézène. "The strength of our army, its morale, the confidence it had in its leader, the ascendancy the Emperor exerted on the Russians themselves, all this gave us a sense of the certainty of success, and none of us questioned that."[3]

A large number of senior officers, however, believed they had gone far enough. "Everyone felt they had endured enough fatigues and had enough glorious encounters for one campaign, and nobody wanted to go any further; the need and the wish to stop were felt and frankly expressed by all," wrote Colonel Boulart. Many in Napoleon's entourage, led by Berthier, Duroc, Caulaincourt and Narbonne, begged him to call a halt. But he was adamant. "The wine has been poured, it has to be drunk," Napoleon retorted to Rapp, who questioned the advisability of further advance. When Berthier nagged him once too often about the inadvisability of proceeding, Napoleon turned on him. "Go, then, I do not need you; you're nothing but a —. Go back to France; I do not force anyone," he snapped, adding a few lewd remarks about what Berthier was longing to get up to with his mistress in Paris. The horrified Berthier swore that he would not dream of abandoning his Emperor in any circumstances, but the atmosphere between them remained frosty for several days, and Berthier was not invited to the imperial table.[4]

"We are now committed too far to draw back," Napoleon finally declared. "Peace lies before us; we are only eight days' march from it; so close to the goal there can be no discussion. Let us march on Moscow!" While older men and senior officers shook their heads and grumbled, the younger ones were excited by the prospect. "If we had been ordered to march to conquer the moon, we would have answered: 'Forward!,' " recalled Heinrich Brandt of the Legion of the Vistula. "Our older colleagues could deride our enthusiasm, call us fanatics or madmen as much as they liked, but we could think only of battles and victories. We only feared one thing – that the Russians might be in too much of a hurry to make peace."[5] There was little danger of that.

Colonel Boulart had been filled with sadness by the fire of Smolensk, "not so much on account of the moral effect which a great disaster always produces and of the resources of every kind the flames had devoured, but rather because it announced, on the part of the enemy, an exasperation which left no more hope of negotiation and because it shed light, so to speak, on our future." His unease was shared by many others in the Grande Armée as they began to appreciate that they were entering alien territory in more senses than one. "This kind of warfare is horrible and does not resemble in any way that which we have been used to until now," noted Jean-Michel Chevalier.[6]

The French soldier of 1812, even if he was conscripted against his will (and sometimes brought to the colours in fetters if he had tried to avoid the draft), knew that he was in essence a free citizen who had another life outside the army, to which he would return if he survived. His behaviour while he was in the army was to a large extent dictated by his hopes for that moment. He would do all he could to survive, and to profit from his time in the ranks by gaining reputation, promotion and booty. He could be roused to acts of selfless courage by a mixture of patriotism, *esprit de corps* and love of his Emperor, but he did not believe in unnecessary butchery. Unless he and his comrades had been whipped into some exceptional frenzy, he was always calculating chances and options, and if he was surrounded without hope of relief, he saw nothing wrong in surrender. A free citizen under arms would decide, privately or collectively, at what point his or his unit's welfare demanded this. The same was true to a greater or lesser extent for every soldier of the Grande Armée, whatever his nationality.

The same had also been true of every enemy Napoleon's soldiers had faced: a certain basic human solidarity meant that the men of both sides, however desperate they may have been to destroy them as a force, respected the others' desire to survive. "Soldiers kill without hating each other," explained Lieutenant Blaze de Bury, who had taken part in campaigns all over Europe. "During a ceasefire, we would often visit the enemy's encampment, and while we were ready to murder each other at the first signal, we were nonetheless prepared to help each other if the occasion presented itself."[7] This had even held in Spain, where the *guerrilla*, or little war, had introduced a hitherto unknown level of national and religious fanaticism into the proceedings. But it was not true in Russia.

Frederick the Great is alleged to have said that one first had to kill the Russian soldier and then push him over. Napoleon's troops were reaching the same conclusion after the fighting at Krasny, Smolensk and Valutina Gora. Russian soldiers did not lay down their arms. They had to be hacked to pieces. Clausewitz, who had the advantage of observing the phenomenon from within the Russian army, put it down to "motionless obstinacy." The French were nonplussed, and ascribed the phenomenon to more or less poetic stereotypical atavisms. "I could never have imagined that kind of passive courage which I have since seen a hundred times in the soldiers of that nation, which stems, I believe, from their ignorance and credulous superstition," wrote Lubin Griois, who had watched them stand impassively as his batteries pounded them at Krasny, "for they die kissing the image of St Nicholas which they always carry with them, they believe they will go straight to heaven, and almost give thanks for the bullet which sends them there."[8]

Belief in an afterlife was certainly a factor. The unfree Russian soldier, drafted for twenty-five years, did not think in terms of a return to another, normal life on earth. The army was his life. And death, which held out the prospect of heaven, was in many ways preferable to that life. The iron discipline of the army, supplemented by his experiences in the fighting against the Turks or tribesmen in Georgia or the Caucasus, vicious and genocidal, with quarter neither expected nor given, meant that the concept of surrender was not part of his military consciousness. The decision to surrender is essentially an assertion of human rights against the army and its master the state, and there was no such subversive concept in Russia.

The French were dismayed by all this. This was not how war was supposed to be. What was alarming for the simple soldier was that his opponents' uncompromising approach to warfare bound him to the actions of his commander and implicated him in his commander's crimes. He could not say, as soldiers down the centuries have said, that he was an innocent pawn of kings and generals. The whole army was answerable, and it was now looking as though this would be a fight to the death. This became increasingly apparent as the Grande Armée marched out of Smolensk in that last week of August.

They were now moving through fertile country, down a fine road, straight as an arrow and broad enough for columns of infantry and cavalry to march abreast under a double avenue of birch trees on either side of the central causeway, which was reserved for the artillery and the army's wheeled vehicles. But it was not an easy march. "We trotted along from two or three o'clock in the morning until about eleven at night, without dismounting, except to answer an urgent call of nature," wrote the Dutch Carabinier Jef Abbeel. "The rare pauses we passed in trying to rid ourselves of the vermin that infested us."[9]

"The heat in this part of the world at this time of year is nothing like the heat of southern Europe," explained Julien Combe. "It was not just the heat of the sun we had to bear, but the vapours emanating from the baking earth. Our horses kicked up a cloud of burning sand as fine as dust, with which we were so covered that it would have been difficult to distinguish the colour of our uniforms. This sand, which got into our eyes, subjected us to excruciating pain." The men wrapped scarves round their noses and mouths, and some even made protective masks out of foliage in an attempt to keep the dust out. To no avail. When Napoleon appeared before the 6th Bavarian

Infantry, they could not shout "*Vive l'Empereur!*" because, as Christian Septimus von Martens, one of its officers, pointed out, "our tongues were stuck to our gums."[10]

These discomforts were added to by the fact that the Russians had adopted a new tactic now that the invaders were in Russia proper. They evacuated the entire population and took the civil administration with them as they retreated, leaving towns and villages deserted. The French began to regret the Jews who had been so useful to them in the former Polish provinces. Lieutenant Charles Faré of the Grenadiers complained in a letter to his mother that food was in short supply and the *cantinières* were charging extortionate prices. Normally, he expected to make money on campaign, but this time they were all being ruined, and were in a hurry to get to Moscow where they might find pots of gold or at least some fine furs they could bring back and sell in Paris.[11]

The Russians had also taken to encumbering the road with overturned carts, felled trees and other obstacles. They were now leaving behind quantities of dead men and horses, which decomposed rapidly in the scorching heat. More importantly, they had begun demolishing and burning farms and villages in the path of the French, and setting fire to haystacks, wheatfields and anything else that might burn. The smoke of the burning mixed with the fine dust to make the march one of the hardest the veterans of the Grande Armée could remember. "At night, the whole horizon was on fire," in the words of the artilleryman Antoine Augustin Pion des Loches.[12]

The scorched-earth policy now being applied by the Russians tested the resourcefulness of even the most accomplished practitioners of "*la maraude.*" "The very existence of the army was a miracle, renewed every day by the active, industrious and cunning minds of the French and Polish soldiers, their habit of overcoming every difficulty, and their taste for the dangers and fortunes of this terrible game of adventure," in the words of Ségur. Dr René Bourgeois of the medical staff could not help but marvel at the men. "By their activity and their industry they ensured themselves against excessive privations, and managed to make the means of existence and succour appear one might almost say out of nowhere," he wrote.[13]

Every regiment was followed by a multitude of wagons and carts, carrying not only regulation supplies, but a whole range of items picked up along the way which constituted its life-support system, as well as flocks of sheep and cows, driven along by those soldiers who in normal life had been shepherds or stockmen. Every man brought the skills of his trade to the support of the unit. "The necessisties of this way of life had turned us into millers, bakers, butchers or artisans," as Jef Abbeel put it.[14] But the men were uneasy about the turn events had taken, and began to murmur about "Scythian tactics" and some diabolical trick.

"It has to be said that we were beginning to grow anxious as we followed a powerful enemy without being able to reach him," confessed Colonel de Pelleport, commanding the 18th of the Line in Ney's corps. They were growing acutely aware that every step was a step further from home, and it was noted that even the Italians had lost much of their "*brio*." The only thing that kept the troops going was their faith in Napoleon. "Fortunately we have unbounded confidence in the vast genius of the one who is leading us, for Napoleon is for the army its father, hero, demi-god," noted Jean-Michel Chevalier.[15]

Napoleon was made uneasy by the sight of the burning villages, but attempted to dispel his fears by heaping ridicule on the Russians and calling them cowards. "He sought to avoid the serious reflections which this terrible measure raised as to the consequences and duration of a war in which the enemy was prepared to make, from the very outset, sacrifices of this magnitude," explained Caulaincourt. On the evening of 28 August Napoleon was walking in the garden of a country house he had stopped at just outside Viazma. Murat was arguing with Davout, trying to convince him that they should go no further, and the argument grew heated, but the Emperor merely listened pensively and then went into the house without saying a word.[16]

Uncertainty would be succeeded by bluster. Two days later, when he and his entourage stopped for lunch by the roadside, Napoleon walked up and down in front of them, holding forth about the nature of greatness. "Real greatness has nothing to do with wearing the purple or a grey coat, it consists in being able to rise above one's condition," he declaimed. "I, for instance, have a good position in life. I am Emperor, I could live surrounded by the delights of the great capital, and give myself over to the pleasures of life and to idleness. Instead of which I am making war, for the glory of France, for the future happiness of humanity; I am here with you, at a bivouac, in battle, where I can be struck, like any other, by a cannonball . . . I have risen above my condition . . ."[17]

The following day he entered the pretty town of Viazma, which delighted the French with its low, brightly painted houses. The retreating Russians had set fire to it, but the fires were soon put out. The even prettier town of Gzhatsk, with its blue-painted wooden houses, was intact when they entered it on 1 September hot on the heels of the Russians, but by that evening it was on fire through the carelessness of the soldiers, who would light campfires in inappropriate places. At Gzhatsk they also found large stores of wheat and spirits, which helped the supply situation.

Davout wrote to his wife telling her they would be in Moscow in a matter of days. "This campaign will have been one of the Emperor's most extraordinary, and not the least useful to our children, for it will protect them from the invasions of the hordes of the north." But on the following day an *estafette* from Paris brought Napoleon the unwelcome news that Marshal Marmont had been defeated by Wellington outside Salamanca on 22 July. "Anxiety was clearly visible on his usually serene brow," according to General Roguet of the Young Guard, who lunched with him that day.[18]

The mood at Russian headquarters was hardly better, even though the general situation was changing rapidly in their favour. Clausewitz saw the fighting at Smolensk as a strategic victory for the Russians: they had lost a great many men, but French losses had been heavy, and while the Russians would be able to make good their losses as they moved back to meet their reinforcements, the French would not.

But that was of little comfort to the Russian soldiers as they trudged through the heat and dust, with only slightly better provisioning than the French – General Konovnitsin's rearguard sometimes went for two days without any food. And even when they did have food, they had to decamp before they could prepare or eat it, according to Lieutenant Uxküll, who had waxed so poetic about bivouac life. "We're running away like hares," he noted in his diary outside Dorogobuzh on the night of 21 August. "Panic has seized everyone."[19]

The rearguard never managed to shake off the French pursuers, so the pace had quickened, and they were covering up to sixty-five kilometres a day. The retreat was a good deal less orderly than before, and they were now leaving behind them a trail of abandoned wagons and dead or dying men and horses. "We continue to retreat, without knowing why," Prince Vassilchikov wrote to a friend. "We lose men in rearguard actions, and we are losing our cavalry, which can hardly move any more . . . I believe that within a couple of weeks we shall have no cavalry left at all."[20]

"All this retreating is incomprehensible for me and for the army, which has to leave its positions and flee, in the heat and at night," Bagration wrote to Rostopchin. "We are tiring our men and leading the enemy on behind us. I am afraid that Moscow may suffer the same fate as Smolensk." The junior officers and the other ranks were bewildered. "Gathering in small groups, officers talked of the impending destruction of the fatherland and wondered what fate awaited them," writes Lieutenant Radozhitsky of the artillery. "The arms which they had borne so bravely in the defence of their fatherland now seemed useless and cumbersome." Ensign Konshin felt "a heavy bleakness" oppressing his soul. "Our courage is crushed," wrote Uxküll in his diary. "Our march looks like a funeral procession. My heart is heavy."[21]

Like the French, the Russians were disturbed at the turn the campaign had taken. "The war had gone beyond the bounds of humanity, becoming desperate, implacable, exterminatory; its conclusion could only lie in the destruction of one or other of the warring sides," noted Radozhitsky. For the Russians, a new and unfamiliar factor had come into play.[22]

"The destruction of Smolensk acquainted me with a feeling I had never experienced before, which wars carried on outside one's own frontiers do not contain," wrote Yermolov. "I had not seen my own land laid waste, I had not seen the cities of my fatherland in flames." Lieutenant Luka Simansky of the Izmailovsky Life Guards had also experienced strong sensations as he watched Smolensk burn and the civilians streaming out of it. "I was vividly reminded of my own family, which I had left behind," he wrote, adding that while he was still prepared to die for his country, he had now understood what this would mean to them, and he began praying to his guardian angel for protection. He had suddenly become aware of the real cost of war. The fifteen-year-old D.V. Dushenkievich, who had fought so heroically with his Simbirsk regiment in defence of Smolensk, was overwhelmed by sorrow, but also felt a growing anger.[23]

This anger was echoed in the feelings of many officers, particularly at staff level, where the customary restraints of deference and even discipline were fast breaking down. The muttering about "foreigners" had grown louder, and everyone was on the lookout for traitors. "[Napoleon] knows our movements better than we do," Bagration wrote to Rostopchin after the Rudnia fiasco, "and it looks to me as though we advance and retreat at his orders."[24] The conspiracy theorists were soon crowing over what seemed like a piece of real evidence.

Among the papers that fell into Russian hands when Platov's cavalry had overrun Sebastiani's camp at Rudnia was a letter from Murat informing Sebastiani that he had received intelligence of an impending Russian attack. When this came to be known at Russian headquarters, there was a general outcry and demands that the "spies" should be rooted out. Suspicion hovered over all foreign officers, but fell most heavily on Ludwig von Wolzogen and Waldemar von Löwenstern, who were known to have spent time in France. What was significant about the fact that these two were picked on was that both always spoke German with Barclay – in other words, by naming them, their accusers were pointing the finger at him.

Yermolov was demanding that Löwenstern be sent to Siberia, but Platov suggested a more reliable expedient. "This is how to deal with the matter, brother," he said to Yermolov. "Get him ordered to go and make a reconnaissance of the French positions and send him off in my direction. I'll make it my business to separate the German from everyone else. I'll give him guides who will show him the French in such a way that he'll never see them again."[25]

There had in fact been no treachery, and it was later discovered that Murat had gleaned his intelligence from an intercepted letter written by a Polish staff officer to his mother, whose estate lay in the path of the offensive, warning her to remove herself. But the anti-German party was in full hue and cry, and Barclay did not have enough authority to oppose it. He had Löwenstern sent to Moscow under guard, and the staffs were purged of other foreigners, such as officers of Polish descent. Löwenstern's brother Eduard, himself serving in Pahlen's corps, was outraged. "The army and the nation wanted to believe that Russia had been sold and betrayed from the start," he wrote. "One had to let these people clutch at this idea, as one leaves a naughty child with its toy, to stop it crying any more."[26]

This did not reduce the pressure on Barclay, who, far from running away, was now desperately looking for a favourable position in which to give battle. Toll located one at Usviate which was suitable, according to Clausewitz, but Bagration criticised it. When Toll attempted to defend his choice and point out its virtues, Bagration flew into "a violent passion," accused him of insolence and insubordination, and threatened to have him demoted to the ranks. Rather than stick up for his quartermaster, Barclay agreed to fall back further. He found another favourable position before Dorogobuzh, but Bagration objected to this one too, and another damaging quarrel ensued.[27]

Yermolov urged Bagration to write to the Tsar demanding

Barclay's removal from command, and if Bagration did not quite dare to do that, he was writing to Arakcheev, Rostopchin, Chichagov and others. He accused Barclay of being a "fool" and a "coward," he petulantly declared that he was ashamed to wear the same uniform as him, and repeatedly bragged that if he had been in command, he would have "pulverised" Napoleon. He even threatened to take his army off and do the deed on his own. He was not the only one making trouble. Generals and influential officers throughout the army were writing to friends in high places demanding the removal of Barclay, and in some cases even his execution as a traitor.[28]

All this was having a detrimental effect on the army and Barclay's authority. "Senior officers accused [Barclay] of indecision, junior officers of cowardice, while the soldiers murmured that he was a German who had been bribed by Bonaparte and was selling Russia," recorded one of Yermolov's aides-de-camp. "The army began to complain that the commander-in-chief, a German, does not attend religious services, does not give battle, and there were those who called the conscientious and brave Barclay a bogeyman," according to Nikolai Sukhanin. The rank and file started referring to the commander by the disparaging wordplay on his name "*boltai da i tolko*," which roughly translates as "all bark and no bite." As he rode past marching columns of troops, the unfortunate commander could hear shouts of: "Look, look, there goes the traitor!"[29]

Had it not been for the essentially passive attitude of the Russian conscript and the framework of iron discipline within which he functioned, the army would have been in trouble. If he felt let down by authority, the "foreign traitor" excuse allowed him to continue to trust in the regimental structure and in his immediate superiors. There was thus no threat of mutiny. But desertion was becoming rife. More importantly, things had come to such a pass that, according to Yermolov's aide-de-camp Grabbe, if they had given battle now, everyone would have suspected treason at the slightest setback and consequently not obeyed orders they did not clearly see the point of, resulting in a free-for-all.[30]

Yet a battle was what Barclay was hoping for. He identified a strong position outside Viazma, and on 26 August, as his men began to dig in, he wrote to the Tsar that "the moment has come for our advance to begin." He needed two full days to prepare the positions and tidy up his army, but he was not to get them, as Konovnitsin's rearguard failed to hold back the advancing French, and he was obliged to fall back once again, "like one who has lost his balance and cannot stop himself," in the words of Clausewitz. For once, Bagration had approved of the position chosen by Barclay, so now he could wax indignant about the other's order to continue the retreat. "I say forward, he says back!" he wrote to Chichagov the next day. "In this manner we shall soon find ourselves in Moscow!" But Barclay was now determined to make a stand, whatever the consequences, and began to dig in at Tsarevo-Zaimishche, just 160 kilometres – three to four days' march – short of Moscow.[31]

This was too close for comfort. News of the fall of Smolensk had had a devastating effect around the country, spreading panic in its wake. Many thought that all was lost. People began to pack up and flee, even when they were nowhere near the front. Kursk filled up with refugees from Kaluga. In Kharkov, a merchant found that none of his regular clients would sell on credit. Even in faraway cities, people were calling in debts, selling at discounts, and going liquid.[32]

Moscow had hitherto been calm, and it was still basking in the patriotic glow produced by Alexander's visit. Its military governor, Count Rostopchin, was determined that it should remain so. He was a fine-looking man of fifty with polished manners, a broad education and a jaunty wit. In his privy he had installed a fine bronze bust of Napoleon, suitably adapted to serve the lowest function. He was a prized raconteur, and made an impression on Madame de Staël, who passed through Moscow accompanied by the poet August Wilhelm Schlegel just before the middle of August, and whom he entertained to dinner and showed around the city. For all his liberal French education, Rostopchin was a xenophobe and a reactionary. Over the years he had talked himself into believing in a vast conspiracy of Freemasons, Jacobins, democrats, Martinists and other freethinkers aimed at destroying Russia, and was convinced that the French invasion could be the catalyst for this, by sparking off popular rebellion.

He was determined to control the mood of the people through propaganda and news-management. He composed proclamations, full of demagogic patriotism and braggadocio, which were posted at street corners for all to read. They painted Napoleon and the French in the blackest of colours, and plucked every xenophobic chord in order to forestall any French appeal to the lower orders; at the same time they gave those lower orders an object of hatred that would distract them from any hostile feelings they might have entertained towards the nobility. Rostopchin also enjoyed sowing shameless lies. "I gave instructions that a rumour should be spread to the effect that the Turks are now going to be supporting us, and this morning I received reports that the peasants are saying: 'The Turks have submitted and have promised our Tsar to pay him a tribute of 20,000 Frenchmen's heads a year,'" he wrote to Alexander with satisfaction on 23 July.[33]

Rostopchin inflated every skirmish into a victory, and organised grandiose religious thanksgiving services. On 17 August everyone in Moscow was rejoicing in rumours of a victory over the French at Smolensk. General Tuchkov had apparently beaten Napoleon, and it was said that he had killed 17,000 Frenchmen and taken 13,000 prisoners. Two days later, Rostopchin was reporting to Balashov that the city was calm; sixty people swore to having seen a vision of God blessing Moscow appear over the Danilovsky monastery; a French resident who had been extolling the wonders of French liberty was flogged and sent into exile; in Bogorodsk, a Russian worker who had been saying that Napoleon would bring freedom to Russia, causing the whole of his factory to down tools, was flogged and imprisoned, while his comrades were driven back to work.[34]

The shock produced by the truth about Smolensk, when it finally reached Moscow, was predictable. "Moscow was shaken with horror;

all thoughts turned to flight and to either removing or burying valuables, or walling them up," remembered a young noblewoman. "Houses were cluttered with trunks, streets filled up with wagons, heavy carriages and light breaks crammed with whole families and their entire wealth." Churches remained open night and day, crowded with praying multitudes. Most of the nobles in Moscow had estates, on which they would normally spend the summer, and many of those who had stayed behind in town or come up for Alexander's visit now made for the country. "Every day one could see hundreds of carriages driving across the city, mostly occupied by women and children," recalled Rostopchin's daughter. Men of military age who were spotted leaving were jeered and sometimes even threatened by the rabble. "In order to avoid the taunts and the insults of the populace, men of all ages adopted the costumes of their wives and mothers, hoping to save themselves from any disagreeable comments with the aid of this disguise."[35]

They were being replaced now by refugees from Smolensk, who told tales of horror, and wounded officers evacuated from the front, whose complaints about Barclay and the German "traitors" began to circulate in the city. Soon even Moscow's coachmen were cursing Barclay for a German traitor.[36]

Rostopchin himself began evacuating the treasuries of churches, libraries and the Kremlin. But he continued to pen his proclamations, which grew increasingly warlike, and walked the city streets accosting people and telling them that they need not fear, that the French would soon be beaten, and that he would burn the city sooner than let them into it. "The people here, who are faithfully devoted to their sovereign and filled with love for their country, are resolved to die under the walls of Moscow and, if God refuses us His succour in our noble enterprise, then, in accordance with the old Russian saying 'You will not fall into the hands of the wicked,' the city will be reduced to ash, and instead of a rich prize, Napoleon will find only a heap of dust where the ancient capital of Russia stood," he wrote to Bagration on 25 May.[37]

On Rostopchin's orders, "spies" and "agitators" were being arrested every day. They were flogged and either incarcerated or, if they were foreign, sent to some far-off town under surveillance. French inhabitants of the city who appeared to be too pleased at Napoleon's successes were exiled to Nizhni Novgorod. Rostopchin proudly reported to Bagration on 24 August that he had the situation under control, that they had nothing to fear from the lower orders as the only people who had been heard extolling Napoleon and French liberty were a few drunks. "The mood of the people is such that every day I find myself shedding tears of joy," he wrote to Balashov.[38]

On 18 August hundreds of peasant draftees were brought into the city, accompanied by keening wives, mothers and children who had come to see them off. They were the first batch of the 24,835 men raised in the province to date. They were given their militia uniforms – grey knee-length peasant kaftans and baggy pants tucked into Russian boots, cloth forage caps with earflaps that could be tied under the chin, adorned with a brass cross and the motto: "For Faith and Tsar." They were then paraded before the military governor, exhorted by the historian Nikolai Karamzin and blessed by the metropolitan Bishop of Moscow, who sprinkled them with holy water and gave them sacred banners to carry into battle, before being marched out to the front.[39]

Rostopchin was nothing if not conscientious, and he was prepared to consider almost anything that might contribute to the destruction of the French. He was taken in by a German charlatan by the name of Leppich, who claimed that he could build a huge aerostat which would sail over the French army and destroy it at a stroke by pouring fire down on it. Having relieved the city treasury of a great deal of money, Leppich set to work, in secret.

But Rostopchin's activities were sowing confusion and generating an increasingly febrile atmosphere. People who used any language other than Russian in the street were set upon by angry mobs. The proclamation issued by Rostopchin on 30 August, which announced that he would lead the people of the city out to face the enemy, armed with hatchets and pitchforks if necessary, actually caused rioting, with shops being broken into and innocent citizens being roughed up on the streets on suspicion of being French spies.[40] All this boded ill for the ancient capital of the tsars.

12

Kutuzov

After what he had experienced in Moscow, Alexander found the mood of St Petersburg depressingly defeatist when he returned there at the beginning of August. There were some at court who were calling for peace, and even those who were against treating with Napoleon showed little of the exalted courage of the Muscovites. Many, including Alexander's mother the Dowager Empress, had been packing up and sending away valuables, others had theirs crated up in readiness for a quick evacuation, and most people had horses and carriages waiting. They were all living, as one wit put it, on axle grease.

The only manifestations of patriotism were the beards and Russian clothes sported by some nationalists, and the public's boycott of the French theatre, where the celebrated Mademoiselle Georges played to empty houses. Performances of *Dmitry Donskoi*, by contrast, were packed out, but according to one witness took place in an atmosphere more redolent of a church than a theatre, with half of the audience in tears.[1]

Alexander had retired to his summer residence on Kamenny Island and buried himself in work. He saw less of his mistress and more of the Tsarina, and did not show himself much in public. But he could not ignore the stream of letters from his brother Constantine telling him Barclay was an incompetent coward and a traitor, those from Bagration to Arakcheev, which he saw, saying much the same thing, and the gossip at court – everyone was getting letters from someone in the army, all of them complaining and accusing. There was, in the words of the American ambassador John Quincy Adams, "an extraordinary clamour" against Barclay.[2]

Alexander stood by his commander as long as he could. He was desperately hoping for some good news, and after hearing that Barclay had abandoned Vitebsk he hoped he would make a stand before Smolensk, and would later express his profound disappointment that he did not. "The ardour of the soldiers would have been extreme, for it would have been the entry into the first really Russian city that they would have been defending," he wrote to him.[3] With the fall of Smolensk, he could no longer go on supporting the despised general without exposing himself to similar feelings. Particularly as public opinion had already selected Barclay's successor – Kutuzov.

"Everyone is of the same mind; everyone says the same thing; indignant women, old men, children, in a word all conditions and all ages see in him the saviour of the fatherland," wrote Varvara Ivanovna Bakunina to a friend.[4] Alexander hated Kutuzov for his immorality, his slovenly manner and his attitude, as well as for the memories of Austerlitz and his father's murder. He also had a low opinion of his competence. He had rewarded him for making a rapid peace with Turkey by making him a prince, but had then given him an insignificant post. A delegation of St Petersburg nobles came to Kutuzov begging him to accept command of the militia the city was raising, which he did, having obtained Alexander's approval. This brought him to the capital and back into the limelight.

On the evening of 17 August, just as the battle for Smolensk was beginning, Alexander convened a meeting of senior generals, presided over by Arakcheev, to advise him on the choice of a successor to Barclay. After three and a half hours' deliberation, they settled on Kutuzov. But Alexander did not act on their advice, prevaricating for a full three days, during which he considered nominating Bennigsen, and even of inviting Bernadotte over from Sweden. His sister urged him to bow to the inevitable. "If things go on like this, the enemy will be in Moscow in ten days' time," she wrote, adding that he must under no circumstances even think of assuming command himself. Rostopchin wrote declaring that Moscow was clamouring for Kutuzov. There was nothing for it but to go along with the general mood. "In bowing to their opinion, I had to impose silence on my feelings," Alexander later wrote to Barclay. "The public wanted him, so I appointed him, but as far as I am concerned, I wash my hands of it," he said to one of his aides-de-camp.[5]

Immediately after nominating Kutuzov, Alexander set off for Finland, where he had arranged to meet Bernadotte. According to the agreement their foreign ministers had negotiated in April, Russia was to indemnify Sweden for the loss of Finland (seized by Russia) by permitting her to conquer Norway from Denmark and annex that. Russia was also to support a Swedish invasion of Pomerania to reclaim it from the French. But with Napoleon now advancing into the heart of Russia, there was a distinct possibility that Sweden might seize the opportunity of taking back Finland.

Bernadotte and Alexander met on the island of Åbo, and took an immediate liking to each other. Alexander was relieved to discover that Bernadotte hated Napoleon as much as he did. Bernadotte, it appeared, was taking a longer view, and for good measure Alexander encouraged him in this and at the same time wove him into his own plans for the future by suggesting that when Napoleon was finally defeated and the French throne became vacant, the Crown Prince of Sweden might become the King of France. They parted as friends, and Alexander could safely withdraw the three divisions protecting Finland and redeploy them against Napoleon.[6]

On his return to St Petersburg, Alexander found that even more people and valuables had left, and those who remained were in despondent mood. They had received news of the fall of Smolensk a few days before, and this confirmed many in the conviction that it was time to open negotiations with Napoleon. Grand Duke Constantine, sent away from headquarters by Barclay after Smolensk, was persuading people that the situation was hopeless. Another who had arrived from Smolensk was the British "commissioner" General Robert Wilson. He brought news of dissension and strife at headquarters which confirmed that Alexander had been right to replace Barclay. He had also appointed himself spokesman for an indeterminate group of patriotic senior officers who, according to him, demanded that the Tsar sack his Foreign Minister Rumiantsev and others tainted with sympathy for France. In the politest terms, Alexander brushed off this preposterous meddling.[7]

When Kutuzov had walked out of Alexander's palace on Kamenny Island on the evening of 20 August after receiving his nomination, he had himself driven straight to the cathedral of Our Lady of Kazan. There, taking off his uniform coat and all his decorations, he lowered his great bulk to his knees and began to pray, with tears pouring down his face. On the next day he went, this time accompanied by his wife, to pray in the church of St Vladimir. Two days later, on 23 August, he set off for the front. His carriage could hardly move, such was the throng of people cheering and wishing him well. He stopped at the cathedral once again, to attend a religious service, and knelt throughout. Afterwards, he was presented with a small medallion of Our Lady of Kazan, which was blessed with holy water. "Pray for me, as I am being sent forth to achieve great things," he is alleged to have said as he left the church.

At a posting station along the way, he encountered Bennigsen travelling in the opposite direction. Alexander had insisted that Kutuzov appoint Bennigsen as his chief of staff, as a safety measure against the new commander-in-chief's possible treachery as well as his imputed incompetence. Bennigsen was far from pleased when he heard of this, as he had been on his way to St Petersburg hoping to prevail upon Alexander to give him overall command. "It was not pleasant for me to serve under another general, after I had commanded armies against Napoleon and the very best of his marshals," he later wrote. But when Kutuzov handed him Alexander's letter begging him to accept, he could not do otherwise.

Another who was far from happy was Barclay. He wrote to Alexander, saying that he was prepared to serve under the new commander, but begging to be relieved of the post of Minister of War, as this would place him in an anomalous and difficult situation. Within a few days of Kutuzov's arrival at headquarters he wrote again, this time asking to be relieved of command altogether, without success.[8]

Kutuzov's arrival in the Russian camp at Tsarevo-Zaimishche was greeted with an explosion of joy in the ranks. "The day was cloudy, but our hearts filled with light," noted A.A. Shcherbinin. "Everyone who could rushed to meet the venerable commander, looking to him for the salvation of Russia," according to Radozhitsky. Kutuzov was idolised by junior officers, on whom he lavished a garrulous charm and an avuncular concern. The rank and file trusted the old man, whom they called their "*batiushka*," their little father. They perked up instantly, carrying out even the most humdrum tasks at the double, as though they were about to go into action, and that evening there was singing around the campfires for the first time in weeks. Old soldiers reminisced about Kutuzov's feats in the Turkish wars, and assured their younger comrades that henceforth everything would be different.[9]

Their confidence was not shared by many of the senior officers, who had serious reservations about Kutuzov's competence and his fitness for the task in hand at his advanced age of sixty-five. This certainly aggravated a natural laziness. "For Kutuzov to write ten words was more difficult than for another to cover a hundred pages in writing," recorded his duty officer Captain Maievsky. "A pronounced gout, age and lack of habit were all enemies of his pen." Bagration actually wrote to Rostopchin that replacing Barclay with Kutuzov was like exchanging the frying pan for the fire, or, as he put it, "a deacon for a priest," and referred to the new commander-in-chief as a "goose." But the situation had got so bad that any change was welcome. "Many Russians who did not attribute treachery to foreigners, yet believed that the household gods of their country might be indignant at their employment, and that it was therefore unlucky," wrote Clausewitz, adding that "the evil genius of the foreigners was exorcised by a true Russian."[10]

Kutuzov's distinguishing characteristic was slyness. He had built up an extraordinary reputation, and had managed to convince many that his quirks and eccentricities were marks of genius. One of these quirks was a total disregard for form. He dressed sloppily, preferring a voluminous green frock-coat and round white cap to the uniform of his rank. He addressed generals and subalterns by familiar nicknames, and occasionally struck a populist note by lapsing into foul language. But he always wore all his decorations, and was prone to displays of arrogance. And while writers and film-makers have relentlessly portrayed him as some kind of son of the Russian soil, he was cultivated and refined in his tastes, and gave all his orders in impeccable French.

Unfortunately, his disregard for form extended to the way he gave these. He disdained the proper channels, issuing orders through whoever came to hand. He was secretive and mistrustful, and avoided writing them down if possible. He was not beyond instructing a given unit to carry out some operation without informing the commander of the division or corps of which it formed part, so that generals not infrequently found a section of their command moving off in a different direction as they prepared to go into action. He had an unfortunate habit of assenting to some suggestion without considering its implications for other measures he had taken. He changed his mind often, and did not always inform all the relevant people of the change of plan.

Some of this may have been the result of senility – Clausewitz certainly thought so. But some of the confusion was the result of Bennigsen's tendency to overstep his prerogatives and of Kutuzov's entourage, which included his impetuous son-in-law Prince Nikolai Kudashev and the busybody Colonel Kaisarov, who, in Barclay's words, "thought that, being both confidant and pimp, he had no less right to command the army." All three issued orders in Kutuzov's name, sometimes failing to notify him.[11]

"In the army, people were attempting to work out who was the person who was really in command," wrote Barclay. "For it was evident to all that Prince Kutuzov was only the cypher under which all his associates acted. This state of affairs gave birth to parties, and parties to intrigues."[12] This was particularly galling to Barclay, who had put so much work into laying down correct procedure. But he continued to serve as commander of the First Army, although he found the situation intolerable.

"The spirit of the army is magnificent, there are plenty of good generals, and I am full of hope," Kutuzov wrote to his wife just after his arrival in camp. But to Alexander he wrote that he had found the army disorganised and tired, and complained about the large numbers of deserters. He liked the position Barclay had chosen at Tsarevo-Zaimishche and his first intention was to give battle there. "My present objective is to save Moscow itself," he wrote to Chichagov.[13]

But after assessing the state of his army he felt he could not face Napoleon, whose strength he gauged at 165,000. There were 17,000 reinforcements on their way under General Miloradovich, and Kutuzov decided to fall back and incorporate these into his army, hoping to find a good position to give battle somewhere near Mozhaisk. "If the Most High blesses our arms with success, then it will be necessary to pursue the enemy," he wrote to Rostopchin, asking him to send supplies and to prepare hundreds of wagons to take away the wounded. He also asked Rostopchin to send him the death-wreaking aerostat he had heard about.[14]

Barclay suggested a position outside Gzhatsk, but Bennigsen did not like it, and Kutuzov decided to gain more time for reinforcements to be brought up by falling back further. He sent Toll ahead to find another position. He wanted a strong defensive one, as he knew that the Russian soldier was at his most firm when given a rampart or a ditch to defend, and because he believed himself to be seriously outnumbered by Napoleon.

On 3 September Kutuzov rode out to inspect the positions Toll had selected near the village of Borodino. They met with his approval, and soon militiamen were clearing woods, digging earthworks and dismantling entire villages which impeded the field of fire. That evening he established his headquarters at Tatarinovo, some two kilometres east of Borodino, and stayed up late into the night writing. His exhausted men had been arriving all day long, and as they reached their prescribed positions they stacked their muskets in the regulation pyramids and fell asleep.

Kutuzov spent much of the following day supervising the preparations. His positions lay along a ridge which ran at an angle to the main road, behind a small tributary of the Moskva river, the Kolocha. In the north, his right wing was anchored in the angle made by the Moskva and the Kolocha, and ran along the high ground behind the Kolocha. He fortified this high ground with four batteries protected by earthworks, and ordered the construction of two redans astride the old Moscow road, in which he positioned twelve heavy guns. The centre of the line was also atop high ground behind the Kolocha, and it was dominated by a hillock which commanded a good field of fire. Kutuzov fortified this with a strong earthwork which was to go down in history as the Raevsky redoubt, about two hundred metres long with embrasures for eighteen cannon. Further along the line, the high ground was cut by the course of a stream, the Semeonovka, and fell away, leaving an expanse of flat ground between the centre and the right wing, which ended on a knoll by the village of Shevardino. Kutuzov closed this gap by constructing three V-shaped redans, or *flèches*, and had a pentagonal redoubt built on the Shevardino knoll. As the slight depression formed by the bed of the Kamionka stream did not permit the *flèches* to be built further forward, the Russian positions actually curved backwards at this point, leaving the Shevardino redoubt hanging in a rather exposed position. As a result, nobody was quite sure whether this redoubt was really the tip of the left wing or just an outpost. According to Toll, Kutuzov did not mean it to be part of the line, but only an outpost from which he

could "observe the enemy's strength and dispositions," but he did not bother to inform anyone of this, except for Toll. "Kutuzov kept his real intentions on the left wing secret from General Bennigsen," according to Toll.[15]

The lack of clarity probably stemmed from the fact that Kutuzov's dispositions had to remain fluid. He had taken up a diagonal position athwart the new Smolensk road, which presupposed that the main French thrust would be delivered along that road. "I hope that the enemy will attack us in this position, and if he does I have great hopes of victory," he wrote to Alexander that evening. "But if, finding my position too strong, he starts manoeuvring along other roads leading towards Moscow, I cannot vouch for what might happen." The other road which particularly worried him was the old Smolensk road, running past the southern tip of his positions. If Napoleon turned this, he would be forced to fall back on Mozhaisk, "but whatever happens, it is essential that Moscow be defended," he added.[16]

Early on the morning of 5 September, Murat's advance guard reached the little monastery at Kolotskoie, from where he could see the Russian army preparing for battle. He immediately notified Napoleon, who appeared at midday. He marched into the monks' refectory and wished them "*Bon appétit!*" in bad Polish, then rode out to look at the positions the Russian army had taken up.

He could see that to the north of the new Smolensk road they had occupied high ground whose approaches were made difficult by the river running in front of them, while to the south of the road the flatter ground provided an easier approach. As his own right wing, consisting of Poniatowski's 5th Corps, was anyway advancing along the old Smolensk road, Napoleon was naturally drawn to aim his principal thrust in this area. Approached from this angle, the tip of the Russian position, the Shevardino redoubt, was a salient that impeded the deployment of his troops. He therefore ordered Davout to liquidate it as a preliminary.

At five o'clock in the afternoon, General Compans' division duly attacked the redoubt, which was held by Neverovsky's 27th, and took it. But a spirited Russian counterattack, supported by two fresh divisions, threw the French out. The fighting spilled over to the area around the redoubt, which passed from one side to another several times before, at about eleven o'clock that night, the Russians finally gave up trying to recapture it, and fell back, having lost some five thousand men and five guns. But they took with them eight French guns. This allowed Kutuzov to write to Alexander, announcing a first victory over the French.[17]

The next day was spent in preparations. "There was something sad and imposing in the spectacle of these two armies preparing to murder each other," recorded Raymond de Fezensac, one of Berthier's aides-de-camp. "All the regiments had received orders to wear full dress as on a holiday. The Imperial Guard in particular seemed to be preparing for a parade rather than for combat. Nothing could be more striking than the *sang-froid* of these old soldiers; there was no trace of either enthusiasm or anxiety on their faces. A new battle was for them no more than another victory, and in order to share in that noble confidence one only had to gaze on them."[18]

A similar confidence could be felt in the Russian camp. That morning Kutuzov rode out on his white charger to inspect his positions, cheered by the troops, and at one point someone noticed an eagle soaring overhead. News of the assumed omen spread through the camp, filling the men with hope for the morrow. But in stark contrast to the quiet confidence felt by the rank and file, the Russian command was engulfed by argument and recrimination provoked by the fall of the Shevardino redoubt and the pointless loss of so many men. The event also raised questions about the Russian dispositions as a whole.

Whatever his original intentions had been, the fall of the redoubt compelled Kutuzov to bend back his left wing so that his now truncated front could not be outflanked. What seems extraordinary is that as he rode about that day surveying the terrain and the enemy positions, he did not notice that Napoleon was not politely positioning himself symmetrically opposite his front. While Kutuzov had stretched his forces out along a front of some six kilometres diagonally on either side of the Smolensk–Moscow road, Napoleon was coming up in a more compact mass south of that road, and at a different angle, in such a way that he threatened the Russian left wing, which was occupying the weakest positions.

Kutuzov's northern or right wing, up to and including the Raevsky redoubt, was manned by Barclay's First Army. The entrenched batteries were protected by screens of *jaeger* light infantry, skirmisher sharpshooters like the French *tirailleurs*, occupying forward positions in the village of Borodino and dispersed throughout the bushes and brushwood along the banks of the Kolocha, then by massed ranks of infantry drawn up in columns and by units of cavalry, deployed in front of the earthworks. Behind, Barclay had a strong cavalry force in the shape of General Fyodor Uvarov's corps and Platov's cossacks, as well as a number of infantry reserves to call on. This sector was thus more than adequately defended.

The same could not be said of the centre and left, occupied by Bagration with his Second Army. Bagration, with no more than about 25,000 men at his immediate disposal, was seriously overstretched, and the positions he was occupying were poor in natural defences. The only obstacles his attackers would face were the marshy ground in the area at the confluence of the Kamionka and Semeonovka streams, the waist-high walls of the dismantled village of Semeonovskoie, and the three hurriedly built *flèches*. Kutuzov reinforced the southern wing with General Nikolai Alekseievich Tuchkov's 3rd Corps, consisting of eight thousand regulars, seven thousand Moscow militia and 1,500 cossacks, which he positioned in the woods behind the village of Utitsa. They were concealed from the enemy, and were to remain so until the last moment, as their purpose was to suddenly rise up and launch a flank attack at any French force that tried to turn Bagration's flank.

Both Bennigsen and Barclay, neither of whom had been informed of the Tuchkov ruse, saw the weakness of the left wing and badgered Kutuzov to reinforce it. Characteristically, he listened to them but said and did nothing. In this case, his disdain for his subordinates worked against him. While riding around the positions later that day, Bennigsen stumbled on Tuchkov's corps and, as its purpose had not been explained to him, proceeded to reposition it, bringing it out into the open further forward and making it close up with Bagration's left flank. "I have resorted to artifice to strengthen the one weak sector in my line, on the left wing," Kutuzov wrote to the Tsar that evening, unaware that his ambush had been dismantled. "I only hope that the enemy will attack our frontal positions: if he does, then I am confident that we shall win." To Rostopchin he wrote assuring him that if he were beaten he would fall back on Moscow and stake everything on a defence of the ancient capital.[19]

When the temperature dropped on 6 November, Napoleon exchanged his characteristic Chasseur uniform, greatcoat and tricorn hat for a Polish-style fur-lined coat and hat. He was sketched by Faber du Faur as he warmed himself by the roadside at a fire made of the limbers and wheels of an abandoned gun near Pneva on 8 November. Behind him stand Berthier and, in plumed hat, Murat.

The French were now assailed by hordes of irregular cossacks, more feral and vicious than their brothers in regular regiments, who preyed mercilessly on the stragglers. This drawing, by Aleksander Orlowski, captures the character of these fearsome, if not fearless, horsemen.

When Faber du Faur caught up with his fellow Württembergers on the morning of 7 November, after the first heavy snowfall and frost, he was surprised to find them still apparently asleep in their makeshift shelters. As he tried to rouse them he discovered that they had frozen to death. (See p.391.)

As can be seen from this watercolour by an unknown participant, the retreating column was in places little more than a procession of stragglers.

This lithograph by Faber du Faur depicts a group of men cooking up at Krasny on 16 November. The soldier on the left has donned a lady's pelisse, originally booty destined for sale or for his beloved in Paris. The pair on the right have acquired a small pan, a life-saving implement which permitted men to make an improvised pancake out of whatever was available.

Marshal Ney, known as "the bravest of the brave," whose exploits at Borodino and during the retreat earned him the title of Prince de la Moskowa, from a miniature on porcelain by Jean-Baptiste Isabey.

Artillerymen of the 25th Württemberg Division, part of Ney's 3rd Corps, spiking the guns they no longer have the horses to draw and throwing them into the river Dnieper before moving out of Smolensk. The spike was driven into the firing-breach, making the guns unusable even if recovered. A lithograph by Faber du Faur.

The river Berezina near Studzienka on the night of 25 November. The French pontoneers have just begun building the struts for the bridge. In the bottom right-hand corner their commander, General Eblé, can be seen directing the work. Across the river, a cossack picket is keeping an eye on the proceedings, and, beyond the woods in the top left-hand corner, the sky reflects the campfires of the Russian force which was supposed to prevent the French from crossing. The scene was drawn from life by François Pils, a grenadier in Oudinot's corps. (See p.464.)

In the early hours of 26 November, Napoleon reached Studzienka. He is pictured here by Pils talking to Oudinot in front of the bridge-struts, which are ready to be hauled down to the river. The tallest struts, at about three metres, were nearly twice as tall as Napoleon.

LEFT This sketch by Pils shows Oudinot's corps crossing the river on the first bridge, as Napoleon watches, flanked by Berthier and a plumed Murat. On the left of the picture, sappers are at work on struts for the second bridge.

RIGHT General Eblé sketched by Pils on 28 November as he exhorts stragglers to cross the river before he must burn the bridges. Note the soldier in the foreground cutting open a horse's stomach with his sabre to get at its heart and liver, the most prized nourishment.

The Grande Armée bivouacking on the right bank of the Berezina on 27 November, by Faber du Faur. In the background, some generals who have taken over a hut are trying to prevent their shelter being dismantled from outside by men desperate for firewood, a frequent occurrence.

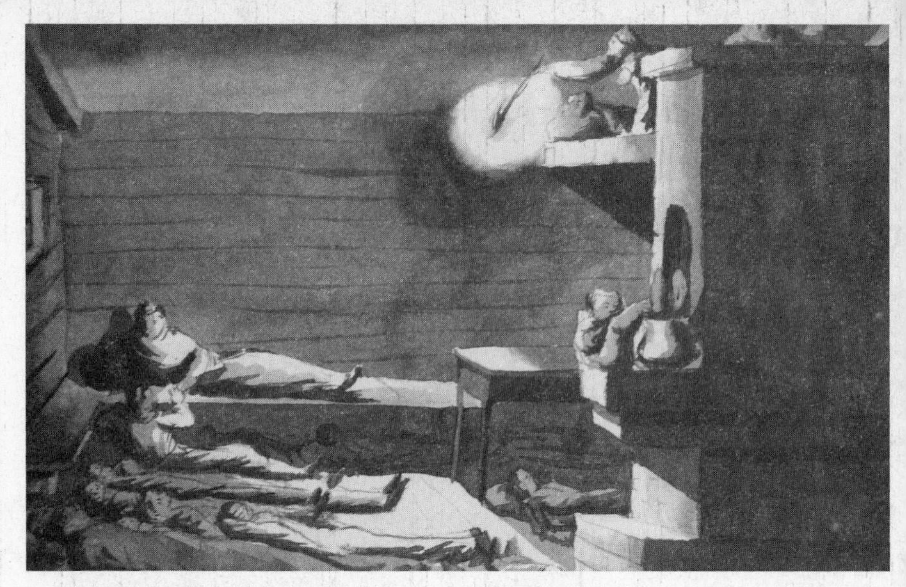

"Billets – understand? Billets! Not a bivouac, not a camp, but real billets – a palace, paradise!" noted Chicherin in his diary on 11 November, the first time they had not camped in the snow for two weeks. In his drawing of the scene one can see him and his brother officers pressed like sardines on the floor of a peasant cottage, with the luckiest ones on the shelf above the stove – the warmest place.

The pursuing Russians were under severe strain as well, both physical and psychological, and iron discipline had to be applied. This grim watercolour in Chicherin's diary records the execution of a Russian officer.

A dying man being stripped by his fellows, a watercolour by an unidentified participant.

This detail from a watercolour by an anonymous French soldier shows men cutting up a horse for meat.

"The sight of these people so numbs the heart that in the end one ceases to feel anything at all," Chicherin noted over this drawing of a group of stragglers seeking a remnant of warmth among the corpses and embers of a burnt-out cottage.

Many witnesses noted that as they froze to death, men looked haggard and even appeared to be drunk. Detail of a lithograph by Faber du Faur.

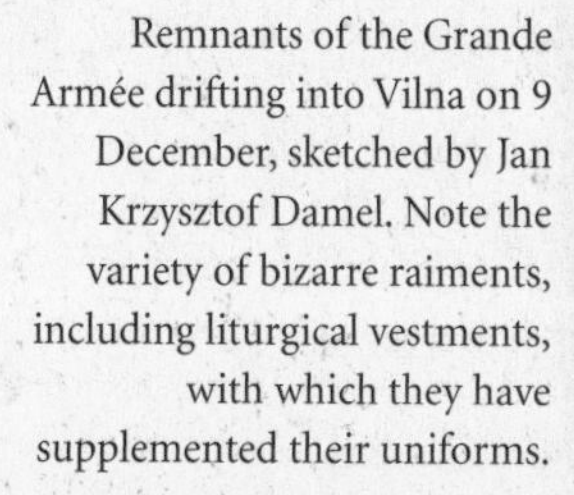

Remnants of the Grande Armée drifting into Vilna on 9 December, sketched by Jan Krzysztof Damel. Note the variety of bizarre raiments, including liturgical vestments, with which they have supplemented their uniforms.

Charles de Flahault, one of Napoleon's aides-de-camp, carrying his servant David up the slope of Ponary outside Vilna, assisted by his secretary Boileau. Alerted to the fact that David had collapsed at the foot of the slope, Flahault had gone back to fetch him. By the time he arrived, the man's boots had been stolen by a needy soldier. David could only wail: "Mon Dieu! Mon Dieu!" as he was carried up, and died before they reached the top. A painting by Horace Vernet, executed under Flahault's direction.

"I cannot render the horror that overpowered me this morning," Chicherin wrote in his diary on 20 December, after visiting one of the improvised prisons in Vilna. Returning to his billet, he painted this watercolour, in which one can see corpses being thrown out of the "hospital" windows, and others being carted away on a sledge, while emaciated French soldiers fight amongst themselves or wander aimlessly, wearing, in the case of one of them, little more than a horse-blanket.

Had Napoleon been in anything approaching his usual form, Kutuzov would undoubtedly have been routed and the Russian army destroyed. Kutuzov had taken up entirely passive positions which did not give him much scope for manoeuvre and were flawed by the weak underbelly to the south of the Raevsky redoubt. He had compounded the problem by overmanning his right wing, which Napoleon was evidently ignoring, and leaving his vulnerable left wing seriously denuded. Luckily for him, Napoleon was to deliver probably the most lacklustre performance of his military career.

The Emperor had also been busy reconnoitring the field. He was in the saddle at two o'clock in the morning. Accompanied by a suite of staff officers, he had gone to see the redoubt captured the night before, and then rode along the whole line, dismounting several times to observe various points along the Russian line through his telescope. He could not get a clear enough view, and, as it later turned out, made some mistaken assumptions about the terrain. He did not get back to his tent until nine o'clock that morning, and spent the next few hours poring over maps and figures.

He was feeling ill. He had caught a cold, and this had precipitated an attack of dysuria, a condition affecting the bladder from which he suffered periodically. On the previous night he had summoned his personal physician, Dr Mestivier, who had noted that Napoleon had a bad cough and was breathing with difficulty. He could only pass water with great pain, and the urine came in drops, thick with sediment. His legs had swollen and his pulse was feverish. The Emperor's valet, Constant, recorded that his master had shivering fits and complained of feeling sick. Others noted similar complaints, and all those around him could see that he was unwell over the three vital days of 5, 6 and 7 September.[20]

At two o'clock that afternoon Napoleon rode out once more for a final survey of the enemy's positions, during which he explained to his marshals his plan for the morrow. He had spotted the weakness of the Russian left wing, and meant to exploit it. Davout and Ney were to attack the *flèches* (Napoleon had only spotted two of them through his telescope) while Poniatowski turned the Russian left wing, and then all three, supported by Junot, were to roll up the whole Russian army in a northerly direction, pinning it against the Moskva river and annihilating it completely. Davout suggested an even deeper flanking move, to be delivered by himself and Poniatowski while Ney tied down the Russians, which would have achieved even more spectacular results with greater economy.

But Napoleon was uncharacteristically cautious. He feared that a force sent into the Russian rear might get lost in the unfamiliar terrain. Also, the Russians might retreat the moment they saw their flank threatened, cheating him once more of a chance to annihilate them. He would deliver a strong frontal attack that would suck in Kutuzov's main forces and defeat them, and in order to deliver this, he actually weakened Davout's 1st Corps by transferring two of his best divisions, Morand's and Gérard's, to the command of Prince Eugène on the left wing, who was to launch the main attack on the Russian centre.

Napoleon's caution was dictated in part by the number and condition of troops under his command. On 2 September, at Gzhatsk, he had ordered a roll call, and this had yielded a figure of 128,000 men, with another six thousand capable of rejoining the ranks within a couple of days, giving a total of 134,000. It is questionable whether this was an accurate figure, given the well-known tendency of unit commanders to inflate their numbers. Anecdotal evidence supplied by officers who had no interest in magnifying or diminishing the figures would suggest that the official ones are on the high side, particularly where the cavalry was concerned. One officer noted that his squadron of Chasseurs à Cheval was down from an initial strength of 108 men at the outset to no more than thirty-four; several regiments, with an original strength of 1,600, were down to no more than 250; one cavalry division had dwindled from 7,500 to one thousand. Russian historians currently estimate the French forces at no more than 126,000.[21]

Either way, the French were outnumbered. While earlier accounts by Russian historians rated Kutuzov's forces at no more than 112,000, current estimates vary between 154,800 and 157,000. It is true that these include about 10,000 cossacks and 30,000 militia, whose role in the battle would be limited. But if one discounts these, one Russian historian has recently argued, then one must exclude from the French tally the 25,000 or so men of the Imperial Guard, who never fired a shot all day.[22]

In the event, apart from taking an active part in the fighting, the militia performed the vital task of carrying away the wounded, which meant that regular soldiers could not use this as an excuse to leave the front line – often never to return. The militia also made a cordon behind the front line, which prevented anyone, even senior officers, who were not obviously wounded from leaving it. "This measure too was highly salutary, and many soldiers, and, I am sorry to have to say it, even officers, were thus forced to rejoin their colours," wrote Löwenstern.[23]

More significant than the disparity in numbers was the condition of the two armies facing each other. The French units were depleted and disorganised by the need to keep themselves fed, and although they put on their fine parade uniforms and pipeclayed their crossbelts, a great many went into battle virtually barefoot, as their shoes had long since fallen apart. They had not been properly fed for days, and, as one officer in the Grenadiers of the Old Guard pointed out, "If General Kutuzov had been able to put off the battle for several days, there is no doubt that he would have vanquished us without a fight, for an enemy mightier than all the arms of the world had laid siege to our camp: that enemy was a vicious hunger which was destroying us."[24] The Russians, on the other hand, were relatively well fed and supplied by a flow of carts coming out of Moscow.

The horses of the French cavalry were in particularly bad condition, and many a charge on the next day would never break into anything more urgent than a trot. The less numerous Russian cavalry were better mounted, and were able to deliver some fierce attacks on the following day.

The greatest discrepancy between the two armies was in the quality of their artillery. The Russians, with 640 pieces, enjoyed outright superiority over the French, who had 584, and a far greater proportion of their guns were heavy-calibre battery pieces, many of them *licornes* with a longer range than any French gun. Over three-quarters of the 584 French guns were light battalion pieces, useful only in close support of infantry attacks.

Napoleon had a pleasant surprise when he returned from his afternoon reconnaissance. The Prefect of the imperial palace, Louis Jean François de Bausset, had just arrived from Paris with administrative papers. Before his departure, Bausset had gone to François Gérard's studio to see the painter finish his latest portrait of the King of Rome, lying in his cradle, toying with a miniature orb and sceptre. He had taken the picture in his carriage as he set off on the thirty-seven-day journey to Napoleon's headquarters. "I had thought that being on the eve of fighting the great battle he had so looked forward to, he would put off for a few days opening the case in which the portrait was packed," Bausset wrote. "I was mistaken: eager to experience the joy of a sight so dear to his heart, he ordered me to have it brought into his tent immediately. I cannot express the pleasure which that sight gave him. Regret at not being able to press his son to his heart was the only thought which troubled such a sweet joy."[25]

Like any doting father, Napoleon called in his entourage to have a look, and then had the portrait placed on a folding stool outside his tent, so that everyone could share in his private joy. The troops began to queue up. "The soldiers, and above all the veterans, seemed to be deeply moved by this exhibition," wrote one staff officer, "the officers on the other hand seemed more preoccupied with the fate of the campaign, and one could see anxiety on their faces."[26]

"*Ma bonne amie,*" Napoleon scrawled to Marie-Louise that evening, "I am very tired. Bausset delivered the portrait of the king. It is a masterpiece. I thank you warmly for thinking of it. It is as beautiful as you are. I will write to you in more detail tomorrow. I am tir[ed]. *Adio, mio bene.* Nap."[27]

Napoleon was not only ill and tired. He was deeply preoccupied. Another who had arrived along with Bausset was Colonel Fabvier, bearing despatches from Spain with details of Wellington's victory over Marmont at Salamanca. The reverse itself was of no great consequence, but its propaganda value was tremendous, as the Emperor well knew. All his enemies would take heart, as they had after Essling, and this meant that the next day's battle must be decisive.

He was not the only one who realised this. "Many a mind was anxious, many eyes remained open, many reflections were made on the importance of the drama which had been announced for the morrow and whose stage, so far from our motherland, allowed us the choice of either winning or perishing," in the words of Julien Combe. Colonel Boulart, of the artillery of the Guard, felt similar forebodings. "If we are beaten, what terrible risks will we not run! Can a single one of us expect to return to his native country?" Captain von Linsingen, who had walked over to where his Westphalians were sleeping or sitting around campfires, wondered how many would be alive the following evening. "And suddenly I found myself hoping that this time too the Russians would decamp during the night," he noted in his diary. "But no, the sufferings of the past days had been too great, it was better to end it all. Let the battle begin, and our success will assure our salvation!" Raymond de Fezensac put it more succinctly: "Both sides realised they had to win or perish: for us, a defeat meant total destruction, for them, it meant the loss of Moscow and the destruction of their main army, the only hope of Russia."[28]

Such reflections were not lightened by the circumstances. There was no bolstering spirits with a shot of drink and a good pipe. They were in an arid place which had been trampled by two armies. "Not a blade of grass or of straw, not a tree; not a village that has not been looted inside out," noted Cesare de Laugier. "Impossible to find the slightest nourishment for the horses, to find anything for oneself to eat, or even to light a fire." The men settled down to a cheerless and cold night. "A miserable plateful of bread soup oiled with the stump of a tallow candle was all I had to eat on the eve of the big battle," recalled Lieutenant Heinrich Vossler of the Württemberg Chasseurs. "But in my famished condition even this revolting dish seemed quite appetising."[29]

The Russians were more fortunate, as they had plenty of food, and looked forward to the next day with greater enthusiasm. "We all knew that the battle would be a terrible one, but we did not despair," recalled Lieutenant Nikolai Mitarevsky of the artillery. "My head was full of things remembered from books about war – the '*Trojan War*' in particular would not leave my thoughts. I was eager to take part in a great battle, to experience all the feelings of being in one, and to be able to say afterwards that I had been in such a battle." As they lay beside their campfires gazing up at the stars, they wondered about what it felt like to be dead.[30]

Several Russian officers were struck by the calm that descended on their camp that day, which was a Sunday, and by the almost spiritual manner in which the soldiers readied themselves for the battle. While the grenadiers of the Old Guard were taking heart from Gérard's picture of the King of Rome, the Russians were seeking solace from a different image. Kutuzov had ordered the miraculous icon of the Virgin of Smolensk, which had followed the army on a gun carriage, to be taken around the Russian positions in procession. The procession, made up of Kutuzov and his staff and a body of monks with candles and incense, would pause at every regiment's position, in every battery and every redoubt, and prayers would be said and hymns sung. "Placing myself next to the icon, I observed the soldiers who passed by piously," wrote one artillery officer. "O faith! How vital and wondrous is your force! I saw how soldiers, coming up to the picture of the Most Holy Virgin, unbuttoned their uniforms and taking from their crucifix or icon their last coin, handed it over as an offering for candles. I felt, as I looked at them, that we would not give way to the enemy on the field of battle; it seemed as though after praying for a while each of us gained new strength; the live fire in the eyes of all the men showed the conviction that with God's help we would vanquish the enemy; each one went away as though inspired and ready for battle, ready to die for his motherland."[31]

"As for us," noted General Rapp as he watched from outside Napoleon's tent while the procession wound its way round the Russian camp, "we had neither holy men nor preachers, nor indeed supplies; but we carried the heritage of years of glory; we were going to decide whether it would be the Tatars or us who gave laws to the world; we were at the confines of Asia, further than any European army had ever ventured. Success was not in question: that was why Napoleon watched Kutuzov's processions with delight. 'Good,' he said to me, 'they are at their mummeries, they won't escape me now.'" He had already decided on an appropriately resonant name for the victory – "La Moskowa," after the name of the river flowing nearby.[32]

"Night fell," continued Rapp. "I was on duty, so I slept in Napoleon's tent. The place in which he sleeps is usually separated by a canvas wall from that in which the duty aide-de-camp sleeps. This prince slept very little. I woke him several times to hand him reports that came in from the outposts, all of which confirmed that the Russians were preparing to receive an attack. At three o'clock in the morning he called for a valet and had some punch brought, which I had the honour of drinking with him." They chatted as they drank, and Napoleon mused that if he had been Alexander, he would have chosen Bennigsen rather than Kutuzov, who was a very passive commander. He asked Rapp what he thought of their chances for the coming day, and, when Rapp replied optimistically, went back to studying his papers. "Fortune is a fickle courtesan," Napoleon suddenly said. "I have always said so, and now I am beginning to experience it."

Rapp did not like the tone of resignation in his master's voice. "Your Majesty will remember that at Smolensk you did me the honour of telling me that the wine had been poured and it must be drunk. That is now the case more than ever; the moment for retreat has passed. The army is well aware of its position: it knows that it will only find supplies in Moscow, and that there are only thirty more leagues to go."

"Poor army, it is much reduced," interjected Napoleon, "but what is left is good; and my Guard is intact."[33]

13

The Battle for Moscow

Napoleon was in the saddle by three o'clock in the morning, and rode over to the Shevardino redoubt. The troops were already moving up to their positions, cheering as they passed their Emperor. "It's the enthusiasm of Austerlitz!" Napoleon observed to Rapp. By half past five, all the units were in their designated positions, drawn up as if on parade. "Never has there been a finer force than the French army on that day," recalled Colonel Seruzier of the artillery of Montbrun's Cavalry Corps, "and, despite all the privations it had suffered since Vilna, its turnout on that day was as smart as it ever was in Paris when it paraded for the Emperor at the Tuileries."[1] The commanding officers of every unit then read out a proclamation penned by Napoleon the night before:

> *Soldiers! This is the battle that you have looked forward to so much! Now victory depends on you: we need it. It will give us abundance, good winter quarters and a prompt return to our motherland! Conduct yourselves as you did at Austerlitz, at Friedland, at Vitebsk, at Smolensk, and may the most distant generations cite your conduct on this day with pride; let it be said of you: "He was at that great battle under the walls of Moscow!"*[2]

"This short and bold proclamation galvanised the army," according to Auguste Thirion. "In a few words it touched on all its concerns, all its passions, all its needs, it said it all." By alluding to the legendary battle it also reminded them who was in now command of the Russian army – the man they knew as "*le fuyard d'Austerlitz*," the runner of Austerlitz. Napoleon never missed an opportunity to allude to his most famous victory, and when the sun broke through the morning mist, as it had on that glorious day, he turned to those around him and exclaimed: "*Voila le soleil d'Austerlitz!*"[3]

He had taken up position on the rise at the back of the Shevardino redoubt, from where he could see the entire battlefield. The Imperial Guard was drawn up alongside and behind him. He was brought a folding camp chair, which he turned back to front and sat astride, leaning his arms on its back. Behind him stood Berthier and Bessières, and behind them a swarm of aides-de-camp and duty officers. Before him he could see a formidable sight.

The reeds and bushes along the Kolocha were alive with Russian *jaegers*. Behind them, on the rising ground, were the Russian infantry and cavalry drawn up in massed ranks in front of the redoubts, on the parapets of which gleamed brightly polished bronze cannon. Behind the redoubts could be seen more massed bodies of men. Kutuzov had put all his cards on the table, probably in order to provoke Napoleon into concentrating on a frontal assault.

He had set up his command post in front of the village of Gorki. A cossack brought him his folding chair, and Kutuzov sat down on it heavily, dressed in his usual frock-coat and flat round white cap. He could not see the battlefield from where he was, but his mere presence was enough. "It was as though some kind of power emanated from the venerable commander, inspiring all those around him," in the words of Lieutenant Nikolai Mitarevsky. The fifteen-year-old Lieutenant Dushenkevich had been thrilled when the commander had driven past the bivouac of his Simbirsk Infantry regiment. "Boys, today it will fall on you to defend your native land; you must serve faithfully and truly to the last drop of blood," he had exhorted them. "I am counting on you. God will help us! Say your prayers!"[4]

At six o'clock, the French guns opened up, the Russians answered, and as nearly a thousand cannon spewed out their charges, to those present, even those who had been in battle before, it seemed as though all hell had been let loose. The sound brought a sense of relief to many. "There was great joy throughout the army when we heard the sound of the cannon," according to Sergeant Bourgogne of the Vélites of the Guard.[5] He was lucky, as the Guard were out of range, and from where he stood he had a fine view of the bombardment. The battlefield was so compact that it was possible for most of the men to observe the action. The French guns pounded the Russian positions, particularly the earthworks, throwing up clouds of dust which mixed with the smoke from the defending guns to create the impression of a vast swirling sea. The Raevsky redoubt, whose eighteen guns were firing as fast as they could, looked like an erupting volcano to some of the spectators, and poetic comparisons were made with Vesuvius.

There was nothing poetic about the bombardment to those who had to endure it. Most of the troops, Russian and French, were positioned within range of the enemy's guns, and they found themselves under fire as they waited to go into battle. There were three types of projectile raining down on them: ball, shell and canister. Ball shot was a solid iron ball weighing anything between three and twenty pounds. Shells consisted of a thick steel casing filled with explosive which was detonated by a fuse. The shell might explode after it had landed among the enemy ranks, or above their heads, in both cases scattering jagged fragments of the casing in all directions. Canister, also known as grape or case shot, was like a giant shotgun cartridge, which shot out of the cannon's mouth in a hail of iron balls a couple of centimetres in diameter.

The old soldiers stood impassively as they watched the cannonballs fly through the air or bounce along the ground towards them. To lift their spirits, the Russians laughed at the militiamen, who tried to dodge the projectiles – old soldiers' wisdom had it that there was no point, as each one had somebody's name on it anyway. Veterans also had to remind recruits not to put out their foot to try to stop what seemed to be spent cannonballs as they rolled by, for the harmless-looking objects could still tear off a leg. The tension and the fear could be terrible as they watched men standing next to them cut in two. When the order was given to move off, the sense of relief was such that many soldiers experienced an urgent need to defecate, producing a comic rush to squat beside the columns as they lumbered forward.[6]

The Delzons division from Prince Eugène's corps, made up of French and Croat infantry, opened the action by sweeping away the advanced Life Guard *jaegers*, who lost half of their men in the short action, and occupying the village of Borodino. Two more of Eugène's divisions crossed the Kolocha, pushing back the Russian infantry with shouts of "*Viva Italia!*" coming from the Tuscans and the Piedmontese as well as "*Vive l'Empereur!*," but they got so carried away that they were caught in a counterattack and thrown back across the river. They then began to prepare a fresh attack.

Meanwhile, Davout had launched two divisions on the southernmost of Bagration's *flèches*, which had been subjected to half an hour's softening up by artillery bombardment. General Compans was wounded as he led his division towards the earthwork, but his men nevertheless occupied it and General Desaix led his division up in support. Further south, Poniatowski pushed back Tuchkov's division and occupied the village of Utitsa.

A Russian counterattack soon expelled the French from the southern *flèche*. But a second French assault was mounted, with General Rapp now leading the Compans division, supported by Desaix and Junot, while General Ledru's division from Ney's corps attacked the next one along. Both earthworks were taken after fierce hand-to-hand fighting, in which both Rapp and Desaix fell, and in which Prince M.S. Vorontsov, commanding the Grenadier division which held them, was himself wounded. "Resistance could not be long," he explained, "but it only came to an end, so to speak, with the existence

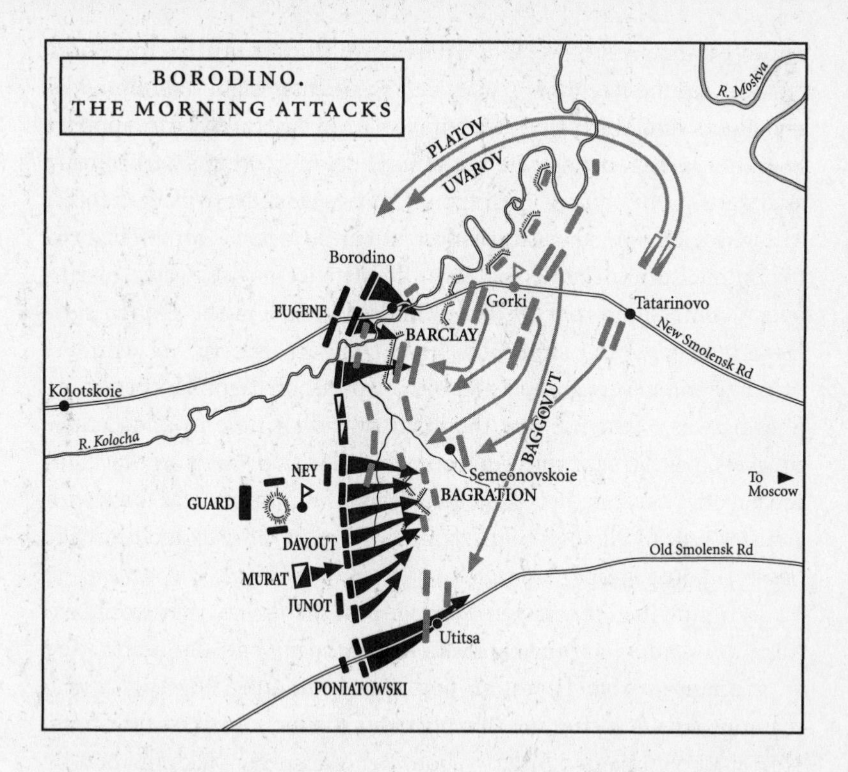

of my division."[7] By eight o'clock he and his division were out of action, having lost 3,700 of its four thousand men and all but three of its officers in the space of two hours.

The *flèches* were in fact traps for the victors. They were no more than V-shaped earthworks, open at the back, so once the French had taken them, they found themselves having to face the next line of Russians while stuck in a funnel with a wall at their backs. And it was only after they had taken the second *flèche* that they realised there was a third. While the Russian guns poured a murderous fire into the confused ranks of the French, General Neverovsky mounted a counterattack that expelled them from the earthworks once again. Undeterred, the French rallied for a new assault.

Over the next three hours, these earthworks were to be stormed, captured and retaken no fewer than seven times as both sides poured reinforcements into the fray. At seven o'clock Kutuzov had three Guards regiments, three cuirassier regiments, eight battalions of grenadiers and thirty-six guns moved from his reserves to support Bagration. An hour later he sent another hundred guns and, shortly afterwards, a brigade of infantry. At nine o'clock he sent in General Miloradovich with the 4th Infantry and 2nd Cavalry corps. Thus the 18,000 men originally dedicated to the defence of this sector rose to over 30,000, supported by three hundred guns. On the French side, Davout's corps was joined by Ney's, Junot's and part of Murat's, bringing some 40,000 men and over two hundred guns to bear. The fighting was so intense that the infantry had no time to reload their muskets, which had by now become fouled with powder anyway, and the bayonet became their principal weapon. But the air was thick with canister shot flying in both directions. "I had never seen such carnage before," noted Rapp.[8]

Each time the French were evicted from the *flèches*, they would re-form and launch another assault. Their bearing and discipline were so magnificent that Bagration applauded, shouting "Bravo!" as the columns advanced towards his positions for the fourth or fifth time. Ney, who complained bitterly about being made to "take the bull by the horns," was in the forefront, clearly visible on his white horse. Davout, who had been wounded and carried off the field during the first attack, was back in the saddle, encouraging his men. Murat was everywhere, drawing eyes and bullets by his theatrical costume. The attacks and counterattacks succeeded each other like a tidal ebb and flow which left thousands strewn across the field at every turn. This dogged slogging-match over a line of earthworks and the carnage it involved were entirely novel elements in European warfare, in which outnumbered or outmanoeuvred units had hitherto tended to fall back rather than fight to the last drop of blood.

At about ten o'clock the French had once again captured all three *flèches*, but Bagration rallied his troops for one final effort and led them into the attack once more. It was successful, but at the moment of triumph Bagration was hit in the leg. He tried to carry on as though nothing had happened, but he had no strength in his shattered leg and after a moment slid from his horse. He wanted to remain on the scene, but was carried away, still protesting. Barclay's aide-de-camp Löwenstern spotted him and came over. "Tell the General that the fate of the army and its preservation is in his hands," Bagration said, in a belated, grudging admission of respect for Barclay's competence. "So far all is going well, but let him look after my army, and may God help us all."[9]

News quickly spread among the Russian troops that their beloved commander had been killed, and although Konovnitsin tried to steady them, they could not resist the next French assault, which finally cleared them from the *flèches* and pushed them back across the ravine of the Semeonovka all the way to the ruins of the village of Semeonovskoie, whose houses were collapsing "like theatrical stage-sets" under the French artillery's bombardment. "There are no words to describe the bitter despair with which our soldiers threw themselves into the fray," wrote Captain Lubenkov. "It was a fight between ferocious tigers, not men, and once both sides had determined to win or die where they stood, they did not stop fighting when their muskets broke, but carried on, using butts and swords in terrible hand-to-hand combat, and the killing went on for about half an hour."[10]

Semeonovskoie was in French hands. Ney and Murat, who could see into the rear of the entire Russian army through the gap they had created, glimpsed victory lying within their grasp, but they could not forge ahead and seize it with the tattered units at their disposal. They sent urgent requests to Napoleon for reinforcements.[11]

But Napoleon did not respond. Although he had a good view of the whole battlefield, from where he sat he could not make out clearly what was really happening on the ground, and he did not, as usual, mount his horse to take a look. He sat very still most of the time, showing little emotion, even when listening to the reports of panting officers who, without dismounting, retailed news from the front line. He would dismiss them without a word, and then go back to surveying the battlefield through his telescope. He had a glass of punch at ten o'clock, but brusquely refused all offers of food. He seemed very absorbed, but his concentration did not yield any results.

"Previously it was above all on the battlefield that his talents had shone with the greatest *éclat*; it was there that he seemed to master fortune herself," wrote Georges de Chambray, who could not recognise in the tired old man of that day the god of war who had galloped about the field of so many battles, spotting the right moment and the weak point at which to launch the decisive attack.

Many observed that Napoleon was not his usual active self that day. "We did not have the pleasure of seeing him, as in the old days, go to electrify with his presence those points where a too vigorous resistance was prolonging the fighting and called success into question," Louis Lejeune, an officer on Berthier's staff, noted in his diary that evening. "We were all surprised not to see the active man of Marengo, Austerlitz, etc. We did not know that Napoleon was ill, and that this state of discomfort rendered it impossible for him to take an active part in the great events taking place before him, exclusively for the sake of his glory." People from all over Europe and half of Asia were fighting under his gaze, the blood of 80,000 French and Russian troops was flowing in a struggle to affirm or destroy his power, and all he did was sit and watch calmly, Lejeune reflected. "We did not feel satisfied; our judgements were severe."[12]

There was, as Davout complained to one staff officer, no unified superior direction to the operations. While Ney and Davout were engaged in their battle over the *flèches*, Napoleon had launched further attacks against the Russian centre in the Raevsky redoubt. The first, by two of Eugène's infantry divisions, was repulsed, but a second, by Morand's division, would prove more successful. It was a fine display of French military prowess.

Captain François of the 30th of the Line was leading his company straight at the redoubt while salvoes from the Russian guns raked the attackers. "Nothing could stop us," he remembered. "We hopped over the roundshot as it bounded through the grass. Whole files and half-platoons fell, leaving great gaps. General Bonamy, who was at the head of the 30th, made us halt in a hail of canister shot in order to rally us, and we then went forward at the *pas de charge*. A line of Russian troops tried to halt us, but we delivered a regimental volley at thirty paces and walked over them. We then hurled ourselves at the redoubt and climbed in by the embrasures; I myself got in through an embrasure just after its cannon had fired. The Russian gunners tried to beat us back with ramrods and levering spikes. We fought hand-to-hand with them, and they were formidable adversaries."[13]

Raevsky, who had been wounded in the leg in an accident a few days before, just managed to hobble away, but the remainder of the redoubt's defenders were cut down, including General Kutaisov, the able and popular commander of the whole Russian artillery. The French infantry then spilled out into the area behind the redoubt. As Raevsky himself pointed out, if Morand had been properly supported, that would have been the end of the Russian centre, and the battle would have been over by ten o'clock in the morning.[14] That it was not owed nothing to the Russian command.

Kutuzov had realised as early as seven o'clock that he must reinforce his left wing, and had begun a progressive transfer of units from his reserves and his idle right wing to the southern sector. A couple of divisions under Baggovut had been sent to reinforce Tuchkov, who was the only thing standing between Poniatowski and the Russian rear. Baggovut took over from Tuchkov, who had been mortally wounded, and stabilised the situation by forcing Poniatowski to fall back a small distance. A number of reinforcements had also been despatched to plug the holes in Bagration's defences as the French knocked more and more of his units out at the *flèches*.

None of this was part of a coherent strategy – Kutuzov was simply reacting to appeals for help and alarming reports. A staff officer would gallop up with some unit commander's request or suggestion, and Kutuzov would wave his hand and say: "*C'est bon, faites-le!*" Sometimes he would turn to Toll and ask him what he thought, adding: "Karl, whatever you say I will do." According to Clausewitz, the old General contributed nothing to the proceedings. "He appeared destitute of inward activity, of any clear view of surrounding occurrences, of any liveliness of perception, or independence of action," he wrote. It never occurred to Kutuzov that, Kutaisov having been killed, someone should take over directing the artillery, and as a result the reserve park stood idle all day and the Russian superiority in this arm was never brought into play.[15]

On hearing of Bagration's wound, Kutuzov sent Prince Eugene of Württemberg to the *flèches*. The Prince tried to stabilise the situation around Semeonovskoie by pulling back a short distance, but Kutuzov would not accept this and heaped insults on him. Yet when Dokhturov, whom he sent to take over, asked for reinforcements, the request was denied, only to be granted later. Kutuzov did at one stage mount his white horse and ride out to have a look at what was going on, but soon returned to Gorki. Later, he seems to have taken up a position even further back, where according to one staff officer he did justice to a fine picnic, assisted by his entourage of elegant officers from the best families. Luckily for him, Bennigsen and Toll kept visiting the battlefield, and a number of his subordinates showed remarkable initiative. At the same time everyone was acting on his own, and there was a great deal of mistrust. When Bagration sent an officer to Konovnitsin with an order, the latter kept the officer as a hostage, suspecting a trick on Bagration's part.[16] In the final analysis what saved Kutuzov's reputation that day was the stoicism of the Russian soldier, who fought and died – often pointlessly – where he had been ordered to.

When the French occupied the Raevsky redoubt, the situation was saved by a combination of lucky flukes. Barclay had gone to Gorki, but General Löwenstern took the situation in at once, galloped up to a battalion of infantry standing by, and swept it up into a counterattack. At the same time Yermolov, who happened to be riding by with reinforcements for the southern sector, also saw the peril, and on his own initiative deployed them against the French in the redoubt. General Bonamy and his 30th Regiment were thus caught in a counterattack from two quarters, while the troops holding the line on either side of the breach also began to close in. The French infantry retreated into the redoubt and brought a few guns to bear, but without support from their side they could not hold it, and the Russians swarmed in, taking Bonamy prisoner. Only eleven officers and 257 other ranks of the 30th of the Line, which had numbered 4,100 men that morning, managed to scurry down the hill to the safety of their own lines.[17]

It was shortly after the unfortunate Bonamy, weakened by his fifteen wounds, had been brought to Kutuzov that Toll came up to the commander with a request from General Platov. Platov with his 5,500 cossacks and Uvarov with 2,500 regular horse had been sitting idly on the right wing, and they requested to be allowed to cross the Kolocha and make a raid into the French rear. Kutuzov gave his assent to the suggestion without seemingly giving it much thought, and soon the eight thousand riders with their thirty-six guns were fording the Kolocha. They wreaked predictable havoc in the rear of Prince Eugène's corps, panicking the Delzons division into flight. But they were soon stopped in their tracks when the French infantry formed squares and fired off a few salvoes of grapeshot in their direction. The cossacks darted out of range, while the regular cavalry fled in disorder, pursued by French dragoons. Platov and Uvarov were given a cool greeting by Kutuzov on their return. As several Russian officers pointed out, the raid had been devoid of any tactical sense, as it could have yielded little but exposed the cavalry to serious loss.[18] But it did have some unexpected consequences.

Between eleven and twelve o'clock the French attacks had ground to a standstill, in spite of a number of local victories. In the centre, the Raevsky redoubt was back in Russian hands; further along, the Bagration *flèches* and then the village of Semeonovskoie had been taken; but new Russian lines of defence sprang up every time, and on the extreme right wing Poniatowski's attack had been halted. None of the forces engaged in these actions was strong enough to push the advantage home, and that was why Ney, Murat and Davout repeatedly called for reinforcements.

This would have meant sending in the Imperial Guard. Napoleon was loath to commit this, as it constituted his last and surest reserve, not something he wanted to gamble with so far from home. But he was, apparently, prepared to commit part of it. He sent the artillery of the Guard forward to bombard the Russian positions around Semeonovskoie, and ordered the Young Guard, under General Roguet, to move forward.

But at the very moment when he was preparing to throw this force into the scales, Platov and Uvarov appeared on his left flank, and he halted everything while he reassessed the situation.[19] Thus at the most critical moment, when the Russian defences had been breached, Napoleon virtually called a halt. Over the next two hours the French armies did not move, and this gave the Russians valuable time to patch their defences and bring up reserves.

But while the fighting subsided, the cannonade did not, and since most of the troops were massed within range of the enemy's guns the carnage continued. The most vulnerable were the massed ranks of Murat's cavalry, which had been positioned in the centre, under the guns of the Raevsky redoubt.

Roth von Schreckenstein of the Saxon cavalry pointed out that to be made to stand still under fire "must be one of the most unpleasant things cavalry can be called on to do . . . There can have been scarcely a man in those ranks and files whose neighbour did not crash to earth with his horse, or die from terrible wounds while screaming for help." Jean Bréaut des Marlots, a captain of cuirassiers, rode down the line of his squadron to keep his men steady under the withering fire. He congratulated one of his subalterns, by the name of Grammont, on his bearing. "Just as he was telling me that he lacked for nothing, except perhaps a glass of water, a cannonball tore him in two," wrote the Captain. "I turned to another officer to tell him how sorry I was to have lost this M. de Grammont. But before he could answer me, his horse was hit by a cannonball which killed him. And a hundred other incidents of this sort. I gave my horse to a trooper to hold for half a minute, and the man was promptly killed." The inaction was unbearable, and the young Captain could only stop himself from running away, as he explained in a letter to his sister Manette, by telling himself: "It's a lottery, and if you do survive it you will still have to die; and would it be better to live dishonoured or to die with honour?"[20]

Auguste Thirion, another cuirassier, experienced similar emotions. "In a charge, which in any case never lasts very long, everyone is excited, everyone slashes and parries if he can; there is action, movement, man-to-man combat; but here our position was quite different. Standing still opposite the Russian cannons, we could see them being loaded with the projectiles which they would direct at us, we could distinguish the eye of the gunner who was aiming at us, and it required a certain dose of *sang-froid* to remain still." One of his men lost his nerve and was about to run, so Thirion comforted him and offered to share a small crust of bread he had been saving. But at the moment he took it out of his pocket the man was hit in the head by a ball. Thirion brushed the brains from his crust, and ate it himself.[21]

One cannot be surprised at such apparent callousness. All the officers and men were undernourished and ravenously hungry. They suffered cruelly from thirst, as the tension and the smoke dried their throats. They were surrounded by death. Many of them were unwell, and the length of the battle magnified every kind of discomfort. "At Dorogobuzh I again fell victim to the terrible diarrhoea which had afflicted me so cruelly at Smolensk, and in the course of this day I endured the most awful agony imaginable, as I was unable to quit my post or dismount," wrote Lieutenant Louis Planat de la Faye, aide-de-camp to General Lariboisière. "I will not describe exactly how I managed to dispose of that which was tormenting me, suffice it to say that in the process I lost two kerchiefs which I disposed of as discreetly as I could by throwing them into the trench of the earthworks we passed."[22]

It was not until shortly after two o'clock in the afternoon that the French began massing for a general assault on the Raevsky redoubt. As some two hundred cannon pounded the earthwork and the gunners inside, Prince Eugène drew up three divisions of infantry – Gérard's, Broussier's, and Morand's from Ney's corps. Shortly before three o'clock the dense columns of French infantry, a great sea of blue but for the white uniforms of a battalion of Spaniards from the Joseph Napoleon regiment, lumbered forward up the incline. They were joined by two masses of heavy cavalry moving up on either side, Grouchy's 3rd Corps on the left, Latour-Maubourg's 4th and Montbrun's 2nd (now under the command of General Auguste de Caulaincourt, younger brother of the diplomat) on the right. These two, moving at a trot, overtook the advancing columns of infantry and made for the left flank of the redoubt and the area in its rear.

To Colonel Franz von Meerheimb of the Saxon Zastrow regiment, the handsome Latour-Maubourg looked absurdly boyish in his resplendent uniform, and far too young to lead the mass of horsemen. But as they approached the redoubt he drew them into a charge with great aplomb and, sweeping round the earthworks, they poured into it, some through the openings at the rear, some over the ditches already filled with French and Russian corpses and the pulverised earthen parapets strewn with debris. The first in were the Saxons and Poles of General Lorge's cuirassier division, followed by Caulaincourt's cuirassiers, whose gallant young leader fell dead as he rode up to the redoubt. When they came over the breastworks, the horsemen were met by a volley of musketry and plunged into a mass of bayonets. As they fell dead or wounded, their comrades followed, trampling over a writhing mass of wounded men and horses, and the defenders struggled in vain to keep them out.[23]

Colonel Griois of Grouchy's artillery, watching the scene from the rear, could hardly contain himself when he saw the glinting helmets of the cuirassiers inside the redoubt. "It would be difficult to convey our feelings as we watched this brilliant feat of arms, perhaps without equal in the military annals of nations. Every one of us accompanied with his wishes and would have liked to give a helping hand to that cavalry which we saw leaping over ditches and scrambling up

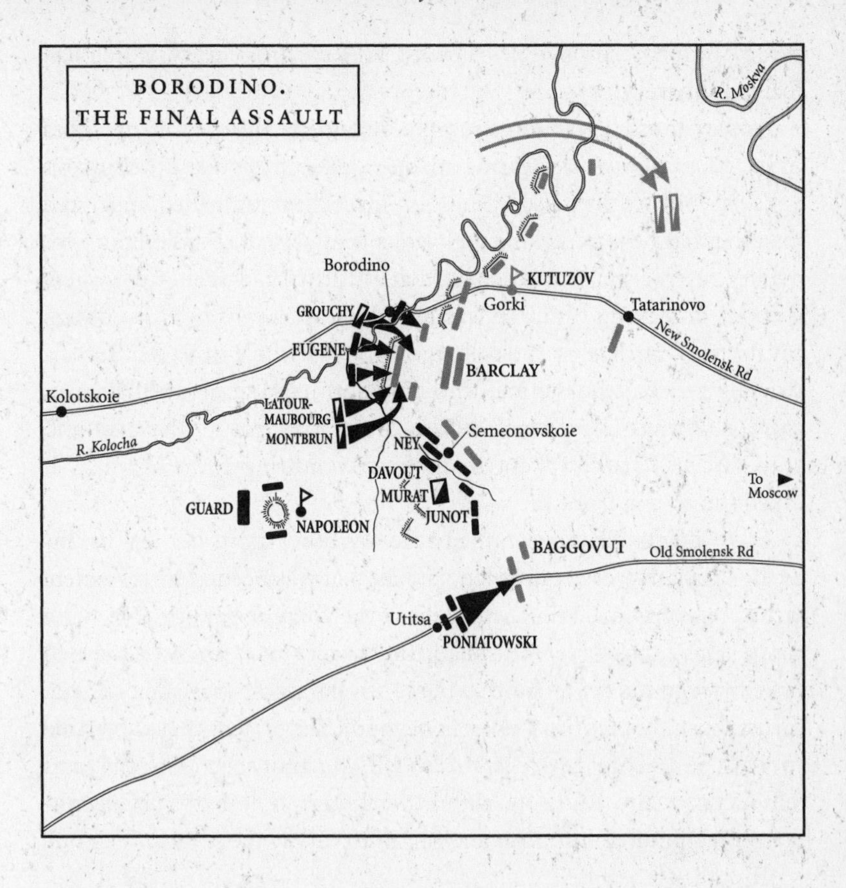

ramparts under a hail of canister shot, and a roar of joy resounded on all sides as they became masters of the redoubt."[24]

"Inside the redoubt, horsemen and footsoldiers, gripped by a frenzy of slaughter, were butchering each other without any semblance of order," wrote Meerheimb.[25] As the horsemen hacked away at the infantry and gunners defending the redoubt, the French infantry poured over the breastworks, and all resistance was quickly extinguished. It was half past three. Grouchy's cavalry had swept into the area behind the redoubt, where they were followed by other French units, only to find that Barclay had formed up a second line of defence, about eight hundred metres behind the redoubt. The cavalry found itself powerless against the Russian infantry, which had formed squares.

Barclay himself was directing the defence of this sector, cool and collected as always, appearing to one staff officer like a beacon in a storm. But he was also displaying a recklessness which suggested to some that he was seeking a glorious death. He had ordered up his cavalry reserve, only to discover that Kutuzov had sent it elsewhere without informing him. He was nevertheless able to muster enough cavalry to launch a counter-charge, and the whole area was soon a swirling *mêlée* of horsemen, thrusting and hacking at each other at close quarters. The French fell back to the line of the redoubt, and Barclay's artillery kept them from venturing forth again as it raked the area in front.[26]

According to Clausewitz, the battle was now "on its last legs" as far as the Russians were concerned, and all that was needed for complete victory was for the French to deliver the *coup de grâce*.[27] But none came. The cannonade continued, the cavalry of both sides clashed once more in the centre, and in the south Poniatowski made one final thrust, pushing the Russians back beyond Utitsa. The sky had become overcast and a cold drizzle started to fall. At about six o'clock the guns fell silent as the Russians withdrew about a kilometre. Napoleon painfully mounted his horse and set off to survey the results.

He rode down the slope from which he had been watching the action all day. At the bottom he found the ground covered in spent musket-balls and grapeshot lying as thick as hailstones after a storm. As his horse picked its way between the debris of men, horses and equipment, he saw what one general called "the most disgusting sight" he had ever seen. Since most of the carnage had been performed by artillery fire, the ground was covered in mangled corpses with exposed entrails and severed limbs. Wounded men struggled to free themselves from under dead men and horses, or dragged themselves in the direction of some perceived succour. Wounded horses crushed them as they themselves attempted to get to their feet. "One could see some which, horribly disembowelled, nevertheless kept standing, their heads hung low, drenching the soil with their blood, or, hobbling painfully in search of some pasture, dragged beneath them shreds of harness, sagging intestines or a fractured member, or else, lying flat on their sides, lifted their heads from time to time to gaze on their gaping wounds," recalled an appalled Belgian lancer.[28]

The Raevsky redoubt presented a gruesome sight. "The redoubt and the area around it offered an aspect which exceeded the worst horrors one could ever dream of," according to an officer of the Legion of the Vistula, which had come up in support of the attacking force. "The approaches, the ditches and the earthwork itself had disappeared under a mound of dead and dying, of an average depth of six to eight men, heaped one upon the other." Inside, the Russian defenders, who had died in their ranks, looked as though they had been scythed down.[29]

The Russian wounded lay stoically waiting for death or tried to drag themselves free, the French called for help or implored to be put out of their agony with a bullet. "They lay one on top of the other, swimming in pools of their own blood, moaning and cursing as they begged for death," according to Captain von Kurz. Some were able to drag themselves along the ground, hoping to find help or at least a drink of water. "Others just sought to get away, hoping to escape death by fleeing from the place where she reigned in all her horror," in the words of Raymond Faure, the doctor attached to the 1st Cavalry Corps.[30]

Single soldiers wandered about rifling through the haversacks and pouches of the dead in search of a crust of bread or a drop of liquor. Others stood or sat, grouped in their units, dazed and uncertain of what to do next. "Around the eagles one could see the remaining officers and non-commissioned officers along with a few soldiers, hardly enough to guard the flag," recalled the Comte de Ségur. "Their uniforms were torn by the ferocity of the struggle, blackened by powder and sullied with blood; and yet, in the midst of these tatters, of this misery, of this disaster, they maintained a proud look and even managed, at the sight of the Emperor, a few cheers; but they were rare and contrived, for in that army, which was capable of clear-sightedness as well as enthusiasm, each one was assessing the overall position." Officers and men were struck by how few prisoners had been taken, and they knew that it was by the number of prisoners, guns and standards captured that one could gauge the scale of a victory. "The number of dead testified to the courage of the vanquished rather than to the scale of the victory."[31]

Napoleon rode back to his tent, which had been brought forward and pitched on the battlefield near the spot from which he had commanded. He wrote to Marie-Louise telling her he had beaten the Russians and sent instructions to the bishops of France to sing *Te Deum*s in thanks for the victory. He was joined for dinner by Berthier and Davout, but he ate little and looked ill. They all agreed that they had won a decisive victory, but there was none of the usual sense of elation. Napoleon spent a sleepless night, according to his valet Constant, who heard him sigh, "*Quelle journée! Quelle journée!*"[32]

There were no songs around the bivouacs that night, no enthusiastic exchange of experiences and tales of glory. The men settled down where the end of the action had found them, huddling round fires made out of broken musket stocks and gunlimbers, piling corpses one on top of the other to sit on. It was the third day they had received no food, and whatever private supplies they had, as well as the *cantinières*, were behind in the bivouacs of the previous night. They had to make do with whatever they could scavenge, making gruel with the buckwheat found in the knapsacks of the Russian dead, with water from one of the streams criss-crossing the battlefield, already thick with blood. One who ate better than most was a *voltigeur* who had managed to shoot a hare which found itself in the path of his advance at the beginning of the battle, and which he now skinned and cooked. As the men sat around their mean fires, wounded comrades came crawling or dragging themselves towards them, begging to be allowed to share their meagre rations. The Russian wounded had to content themselves with chewing on the carcase of some dead horse. "The night of [7 September] was terrible," in the words of an officer of the Grenadiers of the Old Guard. "We spent it in the mud, without fires, surrounded by dead and wounded, whose plaintive cries broke one's heart."[33]

The wounded had been evacuated from the field of battle during the fighting by special details of soldiers with stretchers – and also by malingerers who would take the opportunity of carrying back a wounded comrade in the hope of then hanging about the dressing station and avoiding a return to the front line. But when night fell the evacuation of the wounded ceased, because of the dark and also because the dressing stations were swamped.

As most of them had been inflicted by cannon or musket at extremely close range, there were unusually few light wounds of the sort that could be dressed – a straightforward procedure which involved washing the wound, strapping on a piece of lint, binding it up and letting nature take its course. There was a severe shortage of surgeons, particularly on the French side, as so many had been left in hospitals along the way. Those remaining had been busy all day, carrying out operations and amputations in improvised conditions, washing their hands and instruments, as Dr Heinrich Roos remembered, in a nearby stream. Due to the shortage of draught animals, the ambulances and much of the medical equipment had been left behind in Vilna. When they ran out of bandages they had to tear up the shirts of the wounded.

As they had so little time to spend on each man, the simplest treatment for any wound to an arm or a leg was to amputate. The men were tied or held down on a table, given a lead bullet or a piece of wood or leather to bite on, and, if they were lucky, a shot of spirits to drink. Some struggled and screamed, cursing fate or calling their mothers, but many showed unimaginable stoicism. After the operation they would be set down on the ground, where they lay untended while the severed limbs piled up.

The surgeons continued to work through the night by the light of flickering candles. It was extremely hard work, and emotionally draining, even for such an experienced medic as Dr La Flise. "It is impossible to imagine what a wounded man feels when the surgeon has to inform him that he will certainly die unless one or two limbs are removed," he wrote. "He has to come to terms with his lot and prepare himself for terrible suffering. I cannot describe the howls and the grinding of teeth produced by a man whose limb has been shattered by a cannonball, the screams of pain as the surgeon cuts through the skin, slices through the muscles, then the nerves, saws right through the bone, severing arteries from which blood spatters the surgeon himself."[34]

As the surgeons could not attend to them immediately, many of the wounded were taken straight to "hospitals" improvised in the Kolotskoie monastery and whatever houses had remained intact in the village of Borodino. But since the cavalry had consumed all the straw for miles around, they lay on the bare ground. Some of them did not have their wounds dressed for days. "Eight or ten days after the battle three-quarters of these unfortunates were dead, from lack of attention and food," wrote Captain François, who was dumped along with 10,000 others in the Kolotskoie monastery, where he only survived because his servant cleaned his wound and brought him food and water.[35]

The more fortunate Russian wounded were evacuated to Moscow, but most never made it further than Mozhaisk, where they were packed into any available building and abandoned. When the French marched in two days later they found half of them dead from hunger or lack of water. The streets of the little town were full of corpses and heaps of amputated limbs, some of them still clad with gloves or boots.[36]

Ironically, spirits on the evening after the battle were higher on the Russian side than on the French. The fact that they had actually stood up to Napoleon and not fled before him gave the soldiers a sense of triumph. "Everyone was still in such a rapturous state of mind, they were all such recent witnesses of the bravery of our troops, that the thought of failure, or even only partial failure, would not enter our minds," recalled Prince Piotr Viazemsky, adding that nobody felt they had been vanquished. They knew the French had won, but they did not feel they had been beaten.[37]

Although the Russian front line had been withdrawn that evening some two kilometres back from its positions of the morning, the French did not follow it, and as soon as night fell cossacks, singly or in groups, ranged over the battlefield in search of booty (one party managed to kill two Russian officers who were having a conversation in French). The French did not post forward pickets or fortify their line, as, having beaten and pushed back the Russians, they felt no need to do so. They just camped where they were. For obvious reasons, nobody bedded down in the charnel house of the Raevsky redoubt, and this permitted a small party of Russian troops to "reoccupy" it briefly.[38]

Ever aware of the power of propaganda, Kutuzov determined to claim victory for himself. When Colonel Ludwig von Wolzogen, one of Barclay's staff officers, delivered his report on the situation on the front line at the cease of fighting, from which it was evident that the Russians had been forced to abandon all their positions and had suffered crippling losses, Kutuzov rounded on him. "Where did you invent such nonsense?" he spluttered. "You must have spent all day getting drunk with some filthy bitch of a suttler-woman! I know better than you do how the battle went! The enemy attacks have been repulsed at every point, and tomorrow I shall place myself at the head of the army and we shall chase him off the holy soil of Russia."[39]

He ordered preparations to be made for a general attack in the morning. "From all the movements of the enemy I can see that he has been weakened no less than us during this battle, and that is why, having started with him, I have decided this night to draw up the army in order, to supply the artillery with fresh ammunition, and in the morning to renew the battle with the enemy," he wrote to Barclay. The news that the battle was to continue on the morrow was received with joy by the rank and file, and the men settled down to rest in an orderly fashion.[40]

Meanwhile, Toll and Kutuzov's aide-de-camp Aleksandr Borisovich Galitzine carried out a tour of inspection of the whole army, and came back to inform the commander that he would only be able to muster a maximum of 45,000 men for battle in the morning. "Kutuzov knew all this, but he waited for this report, and it was only after listening to it that he gave the order to retreat," according to Galitzine, who was convinced that the commander had never intended to give battle on the following day, and only said he would "for political reasons."[41]

This is highly likely. On the one hand, it would help present the battle as a victory to the outside world, which Kutuzov immediately set about doing. He wrote a long letter to Alexander describing the course of the action, stressing the bravery and tenacity of the Russian troops, declaring that he had inflicted heavy losses on the French and not given an inch of ground. But, he continued, as the position at Borodino was too extended for the number of men at his disposal, and since he was more interested in destroying the French army than merely winning battles, he was going to pull back six versts to Mozhaisk. He wrote a similar letter to Rostopchin, assuring him that he would make a stand there. He also issued a public bulletin describing the battle, saying all French attacks had been unsuccessful. "Repulsed at all points, the enemy fell back at nightfall, and we remained masters of the battlefield. On the following day General Platov was sent out in pursuit and chased his rearguard to a distance of eleven versts from Borodino." "I won a battle against Bonaparte," he wrote, more succinctly, to his wife.[42]

The announcement about fighting the following day might also have been a ploy designed to forestall a rout. It would prevent anyone from anticipating a retreat, and therefore keep the units together. Also, as none of the officers and men knew the overall position, such news would lead them to assume that the army was in better shape than it actually was. Kutuzov's bluster and confidence certainly had their effect, and, as Clausewitz pointed out, "this mountebankism of the old fox was more useful at this moment than Barclay's honesty."[43] Had the Russians known the full extent of their losses, they might well have given way to despair.

It had been the greatest massacre in recorded history, not to be surpassed until the first day of the Somme in 1916. One does not have to look far to see why. Two huge armies had been massed in a very small area. According to one source, the French artillery fired off 91,000 rounds. According to another they fired only 60,000, while the infantry and cavalry discharged 1,400,000 musket-shots, but even that averages out at a hundred cannon shots and 2,300 rounds of musketry per minute.[44]

The Russians had drawn up their troops in depth, so that a hundred paces or so behind each line there was another line. This undoubtedly saved the day for them, as every time the French broke through they would come up against a fresh wall of troops and therefore fail to achieve a decisive breakthrough. But it did mean that the entire Russian army, including the units being held in reserve, was within range of the French guns throughout the day. Prince Eugene of Württemberg, for instance, recorded that one of his brigades lost 289 men, or nearly 10 per cent of its effectives, in half an hour while standing at ease in reserve.[45]

For no such good reason, Napoleon also stationed many of his reserves, and almost all of his cavalry, within range of the enemy guns. Captain Hubert Biot's 11th Chasseurs à Cheval stood under fire for hours, and lost one-third of its men and horses without taking part in any fighting. One regiment of Württemberg cavalry lost twenty-eight officers and 290 men out of 762, over 40 per cent of its complement.[46]

Calculations of Russian casualties vary from 38,500 to 58,000, but most recent estimates put the figure at around 45,000, including twenty-nine generals, six of them killed, among them Bagration, who died of his leg wound, Tuchkov and Kutaisov. But if Kutuzov's assertion that he would only be able to muster 45,000 for battle the next day is true, then the losses must have been much higher. French casualties came to 28,000, including forty-eight generals, eleven of whom died. In the spring of 1813, the Russian authorities clearing the field would bury 35,478 horse carcases.[47]

The Russian losses were crippling. The army had not only lost a vast number of men, it had actually lost half of its fighting effectives – the dead and wounded were overwhelmingly from the front-line regiments, not militia or cossacks. A very high proportion of the dead were senior officers. The result was that entire units had been rendered inoperational. The 1,300-strong Shirvansk regiment was down to ninety-six men and three junior officers by three o'clock. Of Vorontsov's division of four thousand men and eighteen senior officers, only three hundred men and three officers answered roll call that evening. Neverovsky's entire division could muster no more than seven hundred men. Its 50th Jaeger Regiment was down to forty men, in the Odessa Regiment the senior officer left alive was a lieutenant, in the Tarnopol Regiment a sergeant major. "My division has all but ceased to exist," Neverovsky wrote to his wife the next day.[48]

The French losses were lighter by comparison, with most units losing no more than 10 to 20 per cent of their strength, and while a number of fine generals and senior officers had perished, the system of promotion meant that there would be no difficulty in replacing them. There were fresh troops on their way from Paris, so there would be no major problem in filling the gaps in the ranks. Yet the French army's losses were far more important strategically than those of the Russians, for Napoleon had all but destroyed his cavalry.

Kutuzov's army was in no condition to give battle on any positions, however strong. Unless it was given several weeks' rest and massive reinforcement it would cease to exist altogether. As it could not possibly hope to defend Moscow, whatever Kutuzov may have written, the logical thing for him to do would have been to veer southward and withdraw towards Kaluga. Such a move would have brought him close to his supply base and forced the French to follow him, thus luring them away from Moscow. But if he did this, he would have to fight again or keep retreating, and in either case the remains of his army would disintegrate. More than anything else, he needed to get Napoleon off his tail, and he could only do that by distracting him with something else. In the only brilliant decision he made during the whole campaign, Kutuzov resolved to sacrifice Moscow in order to save his army. "Napoleon is like a torrent which we are still too weak to stem," he explained to Toll. "Moscow is the sponge which will suck him in."[49]

He therefore fell back on Moscow, announcing that he would fight at Mozhaisk, and then at a point closer to the city. Although the retreat was disorderly, the numb resignation which had settled on the troops after the exaltation of battle kept them from dispersing or deserting in large numbers. One of the horses in Nikolai Mitarevsky's gun team had had its lower jaw shot off by a shell fragment, so the gunners took it out of the traces and let it go, but the animal, which had been with the battery for ten years, merely followed it. Similar instincts inspired many a soldier. On the afternoon of 9 September the retreating units began taking up defensive positions on some heights just outside Moscow. They spent the next day digging in and preparing for battle, and Kutuzov wrote to Rostopchin, affirming that his army was in good shape and ready to repulse the French.[50]

Rostopchin had prepared Moscow for any eventuality, as he explained in a letter to Balashov on 10 September. He and tens of thousands of the inhabitants were ready to reinforce the army in a stand outside the city walls. "The people of Moscow and its environs will fight with desperation in the event of our army drawing near," he wrote to Balashov. Following assurances from Kutuzov, he had just declared that the city would be defended "to the last drop of blood." In one of his proclamations to the citizens he assured them that they could use pitchforks on the French, who were so puny that they weighed less than a bale of hay. To all who doubted his determination, he would declare his readiness to defeat the French with a hail of stones if necessary.[51]

On 13 September, Kutuzov set up headquarters in the village of Fili and gave all the appearances of intending to defend the position he had selected. He asked Yermolov, in front of several generals, what he thought of the position, and when Yermolov replied that he felt it was not very good, enacted a charade of taking his pulse and asking if he was quite well. But Rostopchin, who had driven out from Moscow at Kutuzov's request, was struck by the atmosphere of uncertainty he found at headquarters. He reported that he had evacuated the city, which could, if necessary, be abandoned to the enemy and then fired, but everyone seemed outraged at the idea of not defending Moscow. It was as though nobody dared admit the unpalatable truth. The only exception was Barclay, who was convinced that a battle fought at the gates of the city would destroy what was left of the army. "If they carry out the foolishness of fighting on this spot," he told Rostopchin, "all I can hope for is that I shall be killed."[52]

Rostopchin was by now convinced that there would be no attempt to defend Moscow. But Kutuzov assured him of the contrary, in the course of what Rostopchin termed "a curious conversation which reveals all the baseness, the incompetence and the poltroonery of the commander of our armies." Kutuzov asked him to return on the following morning, with the Metropolitan Archbishop, two miraculous icons of the Virgin, and enough monks and deacons to stage a procession through the Russian camp before the battle.[53]

After Rostopchin's departure, Barclay and Yermolov came to see Kutuzov and pointed out that, following a full reconnaissance, they had come to the conclusion that the position was indefensible. "Having listened attentively, Prince Kutuzov could not conceal his delight that it would not be him who would bring up the idea of retreat," wrote Yermolov, "and, wishing to divert from his person any possibility of reproach, ordered that all senior generals be called to a council of war at eight o'clock."

When this convened, Kutuzov began by stating that in the circumstances it would be impossible to hold the position they had chosen, as it could easily be broken and turned. If they were obliged to disengage and fall back, they would find themselves fleeing through Moscow, which would probably result in confusion and the loss of most of the artillery. "As long as the army exists and is in a condition to oppose the enemy we can preserve the possibility of bringing the war to a favourable conclusion," he explained, "but if the army is destroyed, Moscow and Russia will perish."

Kutuzov then asked the others to give their opinion, but seemed uninterested in what they had to say. Uvarov and Osterman-Tolstoy agreed with him, but others were incensed. Bennigsen suggested going onto the offensive and delivering a vigorous attack on one of the French corps while they were on the march, in which he was enthusiastically supported by Dokhturov, Yermolov and Konovnitsin. But Barclay pointed out that even if they had the men, they simply lacked enough experienced officers to carry out an offensive manoeuvre. Raevsky agreed that since the army was so weakened, and since the Russian soldier was unsuited to offensive tactics, the only thing to do was to abandon Moscow.

Bennigsen pointed out that nobody would believe they had won the day at Borodino if the only consequence of the victory was retreat and the surrender of Moscow. "And would we not then be obliged to admit to ourselves that we had in truth lost it?" he questioned. After about half an hour of this discussion, Kutuzov broke in, declared that they would be leaving Moscow, and ordered a general retreat.[54]

Konovnitsin claimed the hair on the back of his neck bristled at the thought of abandoning Moscow, and Dokhturov was unsparing in his indictment of "those small-minded people" who had made the decision. "What shame for the Russian people; to abandon one's cradle without a single shot and without a fight! I am in a fury, but what can I do?" he wrote to his wife that evening. "I am now convinced that all is lost, and since that is so nobody will convince me to remain in service; after all the unpleasantness, the hardships, the insults and the disorders permitted by the weakness of the commanders, after all of that nothing will induce me to serve – I am outraged by all these goings-on! . . ."[55]

Another who was outraged was Rostopchin, who at seven o'clock that evening received a note from Kutuzov informing him that he would not be making a stand in defence of Moscow after all. Rostopchin flew into such a fit of despair that it was all his son could do to calm him down. "The blood is boiling in my veins," he wrote to his wife, who had left the city some time before. "I think that I shall die of the pain."[56] He then busied himself with countermanding all the preparations he had made for the defence and hastened the final evacuation of the city.

Ever mindful of his reputation, Kutuzov sent Yermolov to Miloradovich, who was in command of the rearguard, with orders to "honour the venerable capital with the semblance of a battle under its walls" once the main army had moved through. Miloradovich was indignant. He saw through the ploy: if he were to score some success, Kutuzov could say he had mounted a defence of Moscow; if he was defeated, Kutuzov could blame him. Later that night Kutuzov wrote to the Tsar, explaining that he was abandoning the city to the enemy, declaring that it was an ineluctable fact that the fall of Smolensk had entailed the fall of Moscow, thus deftly shifting the blame from his shoulders to those of Barclay.[57]

The retreat began at once, and by eleven o'clock that night the artillery was rolling through the streets of the old capital. Staff officers with platoons of cossacks were posted at strategic points to direct the columns and keep order. "The march of the army, while being executed with admirable order considering the circumstances, resembled a funeral procession more than a military progress," according to Dmitry Petrovich Buturlin. "Officers and men wept with rage."[58]

But order soon began to break down. As news of the retreat swept through the city, people spilled out into the streets. Over the past weeks Rostopchin had managed to inflame the inhabitants to such a degree with his rabid proclamations that they had reached fever pitch. There had been fights and brawls during the past few days. Some hurled insults at the retreating soldiers, others opened their shops and houses and began handing out their goods so the French would not get them. Rostopchin was busily evacuating everything he could, and setting fire to stores of wheat and other victuals. Tens of thousands more civilians now left the city, taking the same direction as the troops and encumbering their march. Stones as well as abuse were hurled at the carriages of departing nobles, and in some instances wounded men and officers were dragged off wagons and out of carriages to make way for fugitives and their chattels. "There was shouting and lamenting everywhere," recalled N.M. Muraviov, "the streets were full of dead and wounded soldiers."[59]

In the midst of this violence and confusion people tried to locate relatives or friends, and officers who were from Moscow went home to at least provide themselves with some necessities and a change of clothes. Captain Sukhanin decided to call on an acquaintance, Count Razumovsky. The Count was not at home, but the servants welcomed him as though nothing was happening. "The Count's cook prepared breakfast for us, they brought wine, the musicians laid out their notes and began to play," Sukhanin wrote.[60]

As the streets became choked with horses and vehicles, the waiting soldiers wandered off to find food and particularly drink. They were soon breaking into wine shops and cellars, helped by the city's criminals; Rostopchin had opened the gaols and asylums. He had ordered a number of buildings to be set on fire, and the looting added to the fires blazing throughout the city.[61]

Rostopchin himself narrowly escaped being lynched when a baying mob surrounded his palace, only saving his skin by handing over to them a young man accused of being a French spy, who was promptly butchered as the Governor drove off. As he was leaving the city, he came face to face with Kutuzov. The commander had asked his aide-de-camp Galitzine, a Muscovite, to lead him through back streets so that he should not encounter anyone, so he was particularly annoyed. Galitzine recorded that Rostopchin tried to say something but was cut by Kutuzov, while Rostopchin, more credibly, claims that Kutuzov told him he would be going into action against Napoleon soon, to which he replied nothing, "as the reply to a *bêtise* can only be a *sottise.*" Either way, the meeting was enjoyable to neither.[62]

Another altercation took place between the commander of the small Moscow garrison and Miloradovich. The elderly general marched his men out of the Kremlin with their band playing. But as he proceeded through the streets he came across Miloradovich, whose rearguard was now moving through the city. "What swine ordered you to play?" Miloradovich roared at the garrison commander. The latter pointed out that according to regulations laid down by Peter the Great himself, a garrison must always march out of its fortress to the sound of a military band. "And what do the regulations of Peter the Great have to say about surrendering Moscow?" snarled Miloradovich.[63]

Miloradovich had reason to be tense. As he approached Moscow with the rearguard in order to march through it and out the other side, he discovered that Polish Hussars from Murat's corps were already riding into the city, and that he and many other small units or groups of soldiers were in effect swamped by the rest of the King of Naples' corps. Had Murat pressed on, he could have scooped up not only Miloradovich's rearguard, but also a large portion of the Russian army, which was still straggling through the streets and in no position to defend itself, including a great many officers.

Miloradovich sent an officer to Murat to say that he was willing to hand over Moscow without a fight if Murat would only halt his troops for a few hours and give him time to pass through it. If Murat refused, he threatened to set fire to it as he retreated. Murat, who knew Napoleon wanted Moscow whole, and who, like most of the French, thought the war was to all intents and purposes over, agreed to this. He noticed that the cossacks escorting the officer who had brought him the message looked at him with awe, and positively drooled when he pulled out his watch, so he gave it to one of them and ordered all his aides-de-camp to give theirs as well.[64]

The truce merely formalised a situation which was already developing naturally. A quartermaster of one of the cossack regiments in the

Russian Second Army watched in astonishment as a French cavalry division allowed a Russian brigade to trot through their ranks and out of encirclement. Heinrich Roos, a medical officer with the Württemberg Chasseurs, noted that there were hundreds of Russian stragglers in the streets, and that nobody on the French side bothered to pick them up, as everyone considered the war to be over. He came across some lightly wounded Russian officers, so he dressed their wounds and directed them to their units, which they were able to rejoin. When his Chasseurs met a regiment of Russian dragoons beyond the city, there was amiable fraternisation rather than fighting.[65]

If the French believed the war was over, the Russians felt that the world had come to an end. "The entire army is as though undone," wrote Uxküll as he watched the flames shooting up from the city they had evacuated that morning. "There is much talk of treachery and traitors. Courage has been undermined, and the soldiers are beginning to revolt." They were also now deserting in large numbers. Lieutenant Aleksandr Chicherin felt that "the last day of Russia" had dawned as he marched through Moscow, and he could not accept any of the practical strategic arguments put forward for abandoning the city. "Wrapping myself in my greatcoat, I spent the whole day in an unthinking torpor, doing nothing, unsuccessfully trying to repress the waves of indignation which surged over me again and again," he wrote in his little diary.

Many now accepted that Alexander would have to make peace with Napoleon, and even the diehards thought this to be inevitable. Some were already talking of going off to fight against the French alongside the English in Spain.[66]

14

Hollow Triumph

As the Grande Armée's marching columns reached the top of the Poklonnaia hill on the afternoon of 14 September, the men saw Moscow laid out at their feet. "Those who had reached the highest point were making signs to those who were still behind, shouting: 'Moscow! Moscow!'" remembered Sergeant Adrien Bourgogne of the Vélites of the Guard, and the columns quickened their pace, the men jostling each other to catch a glimpse of the goal of their seemingly endless trek. "At that moment," he recalled, "all the suffering, the dangers, the hardships, the privations, everything was forgotten and swept from our minds by thoughts of the pleasure of entering Moscow, of taking up comfortable winter quarters in it and of making conquests of another kind, for that is the character of the French soldier: from the fight to lovemaking, and from lovemaking to battle."[1]

Before them lay one of the most beautiful cities in the world, and one which immediately struck them by its exoticism. "This capital looked to us like some fantastical creation, a vision from the thousand and one nights," remembered Captain Fantin des Odoards. According to statistics drawn up in January that year, it covered 34,337,304 square metres with its 2,567 stone houses and 6,584 wooden ones, 464 factories and workshops, its gardens, churches and monasteries, and had a population of 270,184. "This magnificent spectacle surpassed by far everything that our imagination had been able to conjure in terms of

Asiatic splendour," wrote Lieutenant Julien Combe. "An incredible quantity of bell towers and domes painted in bright colours, topped with gilded crosses and linked to each other with chains which were also gilded, stood out even at a distance in the reddish tinge of the declining sun. The vast Kremlin, and its bell tower ending in a great cross which everyone claimed was of solid gold, but which was certainly of sparkling silver-gilt, dominated this magnificent picture."[2]

Napoleon was surprised to find no delegation waiting to greet him. "It is customary, at the approach of a victorious general, for the civil authorities to present themselves at the gates of the city with the keys, in the interests of safeguarding the inhabitants and their property," wrote a French officer in Russian service. "The conqueror can then make known his intentions concerning the governance of the city, and order the authorities to continue to police it and exercise their pacific functions."[3] As nobody came out to meet him, Napoleon sent some of his aides into the city to seek out some officials with whom he could make arrangements for its occupation.

French troops had already entered the city. The first in, at about two o'clock in the afternoon, were a squadron of the 1st Polish Hussars, followed by other units of the 2nd Cavalry Corps. They picked their way through the streets, which were still full of Russian soldiers, some armed, some not, and reached the Kremlin, which they found occupied by a rabble which had raided the city arsenal. The defenders fired a few shots at the French, but were soon dispersed by a salvo from Colonel Seruzier's artillery, and the French rode into the Kremlin.

While he waited, Napoleon surveyed the city through his telescope, asking Caulaincourt about various buildings. At length, one of Berthier's aides appeared accompanied by a French merchant established in Moscow, and they conversed for a while. Other officers returned with whomever they had been able to find, but none of this satisfied Napoleon, who wanted someone official. In the end, it became plain that the Russians were simply leaving the city to him unconditionally. "The barbarians, they really mean to abandon all this?" he exclaimed. "It is not possible. Caulaincourt, what do you think?" "Your Majesty knows very well what I think," replied the Master of the Horse.[4]

Napoleon did not make his entry into Moscow that day, and spent the night in a wooden house just inside the city limits. At six o'clock on the following morning he rode into the Kremlin and took up his quarters there, while his Imperial Guard in full parade dress made a triumphant entry behind its regimental bands.

About two-thirds of the inhabitants had left, and the remainder, including many foreign tradesmen, servants and artisans, were cowering in their homes. Even members of the several-hundred-strong French colony kept out of the way. The shops were closed and there was little traffic in the streets, although there were still numbers of Russian soldiers wandering about.[5]

Sergeant Bourgogne, whose regiment marched in behind its band, was disappointed. "We were surprised not to see anyone, not even one lady, come to listen to our band, which was playing *La Victoire est à Nous!*," he recalled. "The solitude and the silence which greeted us there calmed down in a disagreeable way the frenzy of happiness which had made our blood race a few moments before, and caused it to be succeeded by a vague sense of anxiety," according to Lieutenant Fantin des Odoards. There was certainly something sinister about the inhabitants' apparent refusal to acknowledge, let alone greet, the arrival of the French, and it made many of them uneasy. "This means we will soon be defending Paris," General Haxo remarked gloomily to Colonel Louis Lejeune as they rode through the silent streets.[6]

In a normal surrender, the city authorities would have been obliged to find all the men billets and make arrangements for feeding them, but in the present circumstances there was a free-for-all to find lodgings and obtain the necessities of life. Generals and groups of officers selected aristocrats' palaces and noblemen's town houses, while their men settled in as best they could in the surrounding houses, stables and gardens. Some did well. Roman Sołtyk and a group of officers on

Berthier's staff found a fine-looking town house which turned out to be the property of Countess Musin-Pushkin, whose servants met them at the door. "At their head was a butler or intendant, dressed with elegance in silk stockings, who asked me in quite good French what I desired, adding that the Countess had before leaving given instructions that we should be suitably received, and had left behind a sufficient number of servants to wait on us," he recalled. She had also left behind her French *dame de compagnie* and a French governess, who entertained the officers at dinner.[7]

Napoleon had appointed Marshal Mortier Governor of Moscow with the stern injunction that there was to be no looting, and according to most sources, the French occupation began in a relatively civilised manner. As all the shops were closed and shuttered, the famished soldiers went from house to house looking for people from whom they could buy or beg victuals and clothing. Some were polite, and most were willing to pay. But as many of the owners had left, the men began to break into shops and private houses and help themselves. While not averse to taking money if they found it, they were at this stage preoccupied almost exclusively with filling their bellies and acquiring shirts, socks, boots and other essentials. "When soldiers enter a city that has been abandoned by its inhabitants, where everything is at their disposal, and take for themselves victuals and items of clothing, can one say that this is looting?" wrote the Saxon Sub-Lieutenant Leissnig. "There was nobody there to give to the men that to which they had a right, so what could the French soldiers do?"[8]

But there were many instances of bad behaviour. I.S. Bozhanov, a priest attached to the Uspensky cathedral, was set upon by a group of soldiers and forced to take them to his house, where he had to feed them before they set about ransacking it. The monks of the Donskoi monastery were visited by a couple of hundred soldiers who rifled through the whole place, stealing anything of value they could find and beating up the monks. This kind of thing was soon to become the norm, thanks in large measure to the exalted nature of the city's Russian Governor.

Rostopchin had several times let slip that if he did have to abandon Moscow, he would make sure the French found nothing but a pile of ashes. Even before he had been informed by Kutuzov that the army would not defend the city, he had made preparations for anything that might be useful to the French – food stores, granaries, warehouses containing cloth and leather – to be torched. He had also ordered all the fire pumps to be evacuated along with the men who manned them. Before leaving the city himself, he gave orders to Police Superintendent Voronenko to set fire not only to the supplies, but to everything he could.

Voronenko and his men went to work, and fires flared up at various points around the city as the last units of the Russian army were leaving. Voronenko seems to have ceased his work that night, but it was carried on the following day by others, probably from the city's criminal elements as they went about looting, by careless French soldiers engaged in the same activity, and abetted by a strong wind which got up that day. As night fell on 15 September, large parts of the city were on fire, an alarming development as more than two-thirds of the houses were made of wood. Napoleon ordered Mortier to send out firefighting details and to arrest the incendiaries. There was not much the soldiers could do against the fires without pumps or other equipment, but they were more successful in arresting incendiaries, real or imagined, who were promptly shot.[9]

The fire raged out of control and spread to several districts of the city. By four o'clock on the morning on 16 September, as the sea of flame began lapping around the walls of the Kremlin, Napoleon was woken up by his fearful entourage. There were large stores of powder in the Kremlin arsenal, and it was feared that one of the sparks and embers swirling through the air might ignite it. He was finally prevailed upon to quit the city, and rode out along flaming streets, followed by most of his guard. He took up residence in the imperial country palace at Petrovskoie, a few kilometres outside Moscow.

From there, the spectacle was beautiful, if terrifying. It was possible to read at night and to feel the heat, and even at this distance the fire made a roar as of a distant hurricane. "It was the most grand, the most sublime, and the most terrific sight the world ever beheld!" Napoleon reminisced on St Helena. The Dutch General Dedem de Gelder thought it "the most beautiful horror one could ever witness . . . I gazed in wonder all night at this unique spectacle, horrible, yet majestic and imposing." But inside the city, it was infernal. "The whole city was on fire, thick sheaves of flame of various colours rose up on all sides to the heavens, blotting out the horizon, sending in all directions a blinding light and a burning heat," in the words of Dr Larrey. "These sheaves of fire, swirling in every direction through the violence of the wind, were accompanied in their upward rise and onward progress by a dreadful whistling and by thunderous explosions resulting from the combustion of powders, saltpetre, resinous oils and alcohol contained in the houses and shops." Tin tiles from the roofs flew through the air, borne upward by the rush of hot air, while whole tin roofs and domes buckled with a bang and flew upwards in one piece.[10]

The fire had displaced not only Napoleon. Those inhabitants who had stayed put in their houses were flushed out by the onward march of the flames, and gathered in herds in the larger squares, trying to find a way out of the inferno. There were scenes of horror as the fire reached the hospitals where the Russian wounded from Borodino had been laid. "When the flames took hold on the buildings in which they were crammed," wrote Chambray, "they could be seen dragging themselves along corridors or throwing themselves out of windows, yelling with pain."[11]

With Napoleon and most of the military authorities out of the way, there was nothing to stop the soldiers from looting. A city ablaze is not conducive to the niceties of respect for other people's property, particularly if it has been abandoned by the owners. Even to those who might have had qualms, it seemed wrong to allow precious supplies and indeed precious objects to be destroyed. The instinctive desire to salvage such things from the flames turned everyone into looters. Officers and even generals joined in the frenzy – and frenzy it was, because there was no time to lose, and reticence was out of place in the face of the rapidly advancing flames.

And once they had rescued things from a burning house or shop, people felt little compunction in rescuing them from houses that were still intact, but which were condemned to burn as well. The German painter Albrecht Adam, attached to Prince Eugène's staff, was shocked when a senior general took him into a palace which contained a fine art collection and exclaimed: "Come, Monsieur Adam, now we must become picture thieves!" But it did not stop him from taking an Italian Madonna for himself. The situation degenerated rapidly as the soldiers struggled to get hold of as much as they could before the roaring fire destroyed it. As they often passed through cellars they were also for the most part drunk. "The men of Marshal Davout's 1st army corps, which was stationed in Moscow and its environs, flooded into the city, penetrating into every accessible place, and particularly into the cellars, looting everything they could find and indulging in all the excesses of drink," in the words of Colonel Boulart. "One could see a continuous procession of soldiers carrying off to their camp wine, sugar, tea, furniture, furs, and so on."[12]

In this, they were ably seconded and even incited by some of the inhabitants and Russian soldiers who had stayed behind. The convicts who had been released were amongst the first to start looting, and in many cases servants left behind helped themselves to their masters' possessions as soon as they realised their actions could be blamed on others. In some cases they showed the soldiers where the masters had walled up or buried their most valued possessions. The Russian looters were soon accosted by French or allied soldiers, who not only robbed them but forced them to help by carrying their booty.

"The army had dissolved completely; everywhere one could see drunken soldiers and officers loaded with booty and provisions seized from houses which had fallen prey to the flames," wrote Major Pion des Loches. They would stumble on better pickings, and dump what they had already looted in order to make room for more valuable booty. "The streets were strewn with books, porcelain, furniture, and clothing of every kind."[13]

The roar of the fire was pierced by the screams of people being beaten up and women being raped, and by the howls of chained-up dogs being burnt alive. "All these excesses of avarice were joined by the worst depravations of debauchery," according to Eugène Labaume. "Neither the nobility of rank nor the candour of youth nor the tears of beauty were respected in a rush of cruel licentiousness which was inevitable in this monstrous war in which sixteen united nations differing in language and customs felt at liberty to give full rein to their lusts safe in the knowledge that their depredations would only be attributed to one of them."[14]

The *cantinières* were in the forefront, determined to stock up for the next few months, and they were among the most determined and pitiless looters, ripping the clothes off women in their search for precious items. Anyone who put up any resistance or tried to safeguard their valuables was likely to be bludgeoned to death, irrespective of age or sex. And any French looter who became isolated from his companions and wandered into a cellar where a larger number of the inhabitants were hiding was likely to meet the same fate.

Captain Fantin des Odoards remembered seeing three drunken soldiers being drawn in a gilded chariot by half-starved nags, people carrying precious supplies of flour wrapped in costly silks and vodka in gilt chamberpots, the only receptacle that had come to hand, and raddled old *cantinières* flouncing around in looted ballgowns. "The saturnalia of the carnival back home never came close to these hideous and grotesque sights," he recalled.[15]

Those of the remaining inhabitants who ventured out were beaten up, stripped even of their shirts and often forced to carry the very things stolen from them back to the looters' camp. As they were dispossessed, so they too were forced to join in the scavenging in order to survive, and old men, women and children were soon all busy, mainly at night so as to avoid the French.

One minor official stranded in Moscow with his family was robbed and turned out of his house by a gang of soldiers. They were then set upon in the street by another gang, who took everything the first lot had left them. As the family huddled in a courtyard they were accosted by a third gang who, finding nothing to steal, simply beat them up. They were then picked on to carry things by various groups of looters.[16]

After three days the fire began to abate, and on 18 September Napoleon rode back into Moscow. The fire died out the following day, order was restored and normality of a sort returned. Some of the inhabitants who had fled actually began to drift back into the city. But nothing seemed normal about this campaign any more to the men of the Grande Armée, who were horrified at the Russian burning of the city. "How can one make war on barbarians like these?" complained Lieutenant Henckens, echoing a widely-held view.[17] And Napoleon himself was baffled.

According to the parameters by which he, and most European states and statesmen, operated, he had won the war. The fact that the Russian army slunk away rather than surrendering did not alter this seemingly obvious fact; that is why he made no attempt at vigorous pursuit or at rounding up the odd stray unit and the thousands of stragglers and walking wounded.

The absence of a delegation formally surrendering Moscow to him was a blow, but that did not change the fact that he was in possession of the ancient capital. The fire, which had destroyed about two-thirds of the city, had robbed him of a wealth of material resources, but it did not affect the supply situation in a critical way.[18] It did have a psychological effect on him and on his troops, but it had no strategic significance.

The real problem was that Napoleon was losing the initiative. He had calculated that if Alexander continued in his refusal to negotiate, he would play on the natural divisions within Russian society to produce a political crisis which would oblige the Tsar to treat with him or risk being replaced by a man who would. Napoleon was a master of propaganda, and he was usually – with the notable exception of Spain

– able to persuade local populations that their armies were beaten and their governments or rulers politically bankrupt. He had been confident that he would find enough discontented merchants and liberal aristocrats, not to mention rebellious servants, through whom to foment some kind of revolution if necessary. But as he sat in an empty Moscow, he found himself in a propaganda black hole. He could not even find spies – "neither for silver nor for gold could one find a single person ready to go to St Petersburg or to infiltrate the army," Caulaincourt noted.[19] The burnt-out city no longer represented a political asset or even a forum. Napoleon could find nobody to talk to and no way of getting his message out. He was at a loss as to what to do next.

He had never meant to linger in Moscow, and the fire only confirmed him in this intention. When he returned to the Kremlin from Petrovskoie he began to make plans for a withdrawal. But there was no logical place to draw back to, short of Vilna, and that would mean losing face as well as the initiative. He therefore considered leaving the main body of his army in Moscow and setting off in the direction of St Petersburg with Prince Eugène's corps and a few other units. He could defeat Wintzingerode and perhaps Wittgenstein as well, which might frighten the capital and force Alexander to treat. And if he needed to, he could veer back towards Vitebsk, while the forces he had left in Moscow could march back to Smolensk.

Prince Eugène was apparently keen on the plan, but others in Napoleon's entourage raised endless objections – so much so that, according to Baron Fain, "they managed for the first time to make him doubt the superiority of his own assessment." Some of them wanted to fall back and take winter quarters in Smolensk; others suggested a march on the industrial cities of Tula and Kaluga followed by a foray through the rich lands of the south. But Napoleon would be leaving behind all his supplies and his lines of communication, both of which were tied to Minsk and Vilna. Also, in the Ukraine he would have been at the mercy of Austria.[20]

In the absence of any obvious military course to follow, he fell back on the idea of negotiation, assuming that if he could somehow get across to Alexander that he was prepared to be generous, the Tsar would come to realise that a settlement would be the best way out of the impasse for both of them.[21] The problem was how to open up a channel of communication.

The only Russian of any standing left in Moscow when the French entered was General Ivan Akinfevich Tutolmin, who had taken over the directorship of the city's great orphanage on his retirement from the army. He had remained with his charges, and when the French entered the city he asked for and obtained from them a regular guard of gendarmes to protect his institution. On the day of his return to Moscow, Napoleon sent for the General and gave him money for his orphanage. He also asked him to write to its patroness, the Dowager Empress, with a view to opening negotiations.[22]

Another potential intermediary was Ivan Alekseevich Yakovlev, a man of substance who had been unable to get away from Moscow on time. On 20 September he summoned Yakovlev to the Kremlin, where the unfortunate Russian was subjected to the usual self-justificatory harangue, part bombast, part pleading, delivered in a tone that veered from the cajoling to the bullying. There had never been any reason for war, Napoleon declared, and if there had been, then the battlefield should have been in Lithuania, not in the heart of Russia. The retreat into the heartland and the refusal to negotiate were not dictated by patriotism but by barbarism. Peter the Great would call them barbarians for having burnt Moscow. "I have no reason to be in Russia," he complained. "I do not want anything from her, as long as the treaty of Tilsit is respected. I want to leave here, as my only quarrel is with England. Ah, if only I could take London! I would not leave that. Yes, I wish to go home. If the Emperor Alexander wants peace, he only has to let me know."[23]

Napoleon gave Yakovlev and his family safe conduct out of Moscow on condition he delivered a letter to Alexander for him. The letter, dated 20 September, informed the Tsar that Moscow had been burnt on the orders of Rostopchin, which Napoleon condemned as an act of barbarism and for which he expressed heartfelt regret. He reminded Alexander that in Vienna, Berlin, Madrid and every other city he had occupied the civil administration had been left in place, and this had guaranteed life and property. He expressed the conviction that Rostopchin's conduct had not been in accord with Alexander's wishes or orders. "I have made war on Your Majesty without animosity," he assured Alexander, saying that a single note from him would put an end to hostilities.[24]

Napoleon also sent a minor civil servant, the commissar Rukhin, to St Petersburg with a proposal for peace, but the poor man was set upon at the first Russian outpost he reached and tortured as a suspected French spy. It was only after a couple of weeks that he was able to hand on Napoleon's letter.[25]

On 3 October, Napoleon asked Caulaincourt to go to St Petersburg in person to open negotiations, but Caulaincourt excused himself, saying that Alexander would not receive him. Napoleon then decided to send Lauriston, who had reached headquarters just before Gzhatsk. "I want peace, I need peace, I must have peace!" Napoleon told him as he set off two days later. "Just save my honour!"[26]

"Like everyone else, the Emperor realised that his repeated messages would, by showing up the difficulty of his position, only confirm the enemy in his hostile dispositions," argued Caulaincourt. "Yet he kept sending him new ones! For a man who was so politic, such a good calculator, this reveals an extraordinary blind faith in his own star, and one might almost say in the blindness or the weakness of his adversaries! How, with his eagle's eye and his superior judgement could he delude himself to such a degree?"[27]

Nor did Napoleon draw the right conclusions from the fire. He dismissed Rostopchin's firing of the city as the irresponsible act of a deranged Asiatic, and did not believe that it had been in any way an expression of popular feeling. He was right in a sense, but he failed to grasp that the blame for the destruction of Moscow would fall on him, while the symbol of the burning city would unite the Tsar with his nation and turn the war into a fight to the death.

Napoleon's reaction to the fire was to demonstrate that if it had been meant to deny him the supplies he needed, it had failed in its purpose. He backed this up by giving the impression that, fire or no fire, he was prepared to sit it out in Moscow, spending the winter there if necessary. He ordered fresh troops to come and reinforce him, and talked of raising levies of "Polish cossacks" who would sweep the countryside and keep lines of communication secure. He also spoke of bringing the actors of the Comédie Française to Moscow to entertain them through the winter months. He imagined that all this would put Alexander under increased pressure to negotiate. He was, to some extent, bluffing.[28]

On the face of it, there was nothing to stop Napoleon from taking his winter quarters in Moscow. Although much had been destroyed, enough had survived in cellars and buildings that had escaped the flames to feed and clothe his army for some months. There were even quantities of cannons, muskets, cartridges, shot and powder left in the city's arsenal.[29] The only area in which supplies were deficient was fodder for the horses, but this was crucial, for without horses he would be able neither to keep his lines of communication open nor to open a fresh campaign in the spring.

Another crucial factor was the situation in his rear and on his wings. From the moment that St Cyr had pushed Wittgenstein back from Polotsk, there was not much activity on that wing. Conditions in the 2nd and 6th Corps were not bad, although there was a persistent shortage of victuals. The state of the troops varied a great deal, with the French, Swiss, Portuguese and Croat infantry and the French and Polish cavalry in good shape, but the Bavarians in very poor condition. They had been prone to disease, and after General Wrede took over command following the death of General Deroy, they went to pieces. "The Bavarian soldiers left the colours in their hundreds and came to Wilna, pretending to be ill, in order to get into the hospitals," according to General van Hogendorp, Governor General of Lithuania. He rounded up 1,100 of them, and found that only about a hundred were actually ill, so he sent the rest back to their units, only to find them deserting once more.[30]

The troops positioned on Napoleon's extreme right wing, in the south, under the command of Prince Schwarzenberg, were in far better shape, principally because their commander studiously avoided fighting and had an unspoken agreement with his Russian opposite number not to engage in unnecessary hostilities.

In order to strengthen his position, Napoleon had ordered Marshal Victor, who had been stationed in East Prussia with his 9th Corps of some 40,000 men, to move forward into Russia and take up position in the Smolensk area, from which he could come to the assistance of the main army or either of the wings if necessary. Theoretically, Napoleon's position was quite strong. "He knew to the last man how many men he had stationed between the Rhine and Moscow," according to Rapp, and the numbers told him that he was still strong enough to deal with any eventuality.[31] What he could not see was the condition of the troops.

General Pouget had been lightly wounded, so in September he was given the post of Governor of Vitebsk. The garrison consisted of nine hundred men and two four-pounder cannon, which does not sound insignificant. In actual fact, apart from sixteen gendarmes and two dozen soldiers from the Young Guard, the rest were a motley crew of stragglers, rounded-up deserters and men who had come out of hospital, with every nationality in the Grande Armée represented. Most of them had fallen behind shortly after crossing the Niemen and had never seen the enemy. They were poorly trained, with no idea of how to look after their weapons, carry out regular patrols or perform picket duty. They were unmotivated, lazy and dirty. When the time came to retreat, they would break ranks at the first sight of a cossack and defy Pouget's attempt to make a stand, with the result that he was captured along with the rest of them. He later claimed that had he been alone with the gendarmes and the two dozen Young Guard, he would have got through.[32]

Lieutenant Jean-Roch Coignet of the Grenadiers of the Guard was on his way to join the Grande Armée in July, and as he came through

Vilna he was given the job of taking a column of some seven hundred stragglers forward to rejoin their units. Since he had started life as a shepherd boy, this should have presented no problem. But the 133 Spaniards in the column promptly deserted, and when he went after them they fired at him. He had to find a cavalry unit to help him round them up, and then it was only by making them draw straws and shooting half of them that he could get the column to stay together at all.[33]

The whole area behind the Grande Armée was awash with soldiery which was of no military use and only served to ravage the country, arousing the fury of the inhabitants. Bands of deserters from various units and of every nationality, usually under the chieftainship of a Frenchman, established themselves in manor houses a small way off the main road and, buying the good will of the locals in exchange for protection, preyed on the traffic travelling along it.

General Rapp, who had travelled from Danzig to Napoleon's headquarters at Smolensk, was horrified at the state of the army's rear. According to him, the Grande Armée left behind more debris than a beaten one, with the result that echelons of recruits marching up to join it were demoralised by what they saw. Many actually died of hunger along the road, as did the fresh remounts being walked from France and Germany, and the cattle being driven up from Austria and Italy. "From the moment of our leaving Vilna, in every village, in every farm, we found isolated soldiers who were abandoning the army under various pretexts," wrote Prince Wilhelm of Baden, who marched in with Victor's corps in September.[34]

Very little could be done about this state of affairs, since the dispositions made by Napoleon in these areas had been confined to a minimum. Wishing to keep political options open, he had not set up proper organs of local government. As a result, the administration of the occupied areas was chaotic and venal. Beginning with the devious Pradt in Warsaw and the despotic Hogendorp in Vilna, and ending with the venal *commissaires* in the various towns along the way, there was no real sense of purpose, no dedication, and no single authority capable of restoring order. At the end of September, more than six weeks after the battle, the streets of Smolensk were still strewn with corpses, on which stray dogs from the surrounding countryside were happily feeding. "Worse organisation, grosser negligence I have never seen, never dreamed of," wrote Captain Franz Roeder of the Hessian Life Guards who marched through on his way to Moscow.[35]

Those who suffered most as a result of this state of affairs were the sick and the wounded lying in the hospitals at Vilna, Minsk, Vitebsk, Polotsk, Smolensk and, perhaps most of all, the survivors of Borodino, cooped up in the monastery of Kolotskoie and at Mozhaisk. There were thousands of French wounded, including twenty-eight generals, scattered in various buildings at Mozhaisk. The *commissaire des guerres* Bellot de Kergorre, who was in charge, claimed that no provision had been made for them. The more mobile would drag themselves into the street and beg from passers-by, while he pilfered food from passing supply trains in order to feed the rest. They died of hunger and of dehydration, as there was little water nearby and he had been given no buckets or vessels of any sort. He had no dressings, no lint, no bandages, no stretchers, no beds, no candles and no nurses. He appealed to Junot, whose corps was stationed at Mozhaisk, for help, but the Westphalian soldiers were more trouble than they were worth. When his charges died, all he could do was dump them in the street outside. He also had hundreds of Russian wounded, who subsisted on the stalks of cabbages dug up in neighbouring gardens and the occasional dead horse.

However many men and horses Napoleon may have had, and whatever the quantities of food and fodder at his disposal, the manner in which these resources were being husbanded meant that he could not possibly remain in Moscow for more than a few weeks without his forces beginning to disintegrate. But rather than order a gradual evacuation of all sick and wounded westwards, he was ordering the call-up of another 140,000 men in France, 30,000 in Italy, 10,000 in Bavaria, and smaller contingents from Poland, Prussia and Lithuania, and he begged Marie-Louise to write to her father asking him to reinforce Schwarzenberg. "Not only do I want to have reinforcements sent from all quarters," he wrote to Maret in Vilna, "I also want those reinforcements to be exaggerated, I want the various sovereigns sending me reinforcements to publish the fact in the papers, doubling the number they are sending."[36]

What he did not take into account was that as St Petersburg was the administrative capital, housing all the institutions of state, the loss of Moscow did not in any way weaken the Russian state's ability to function or affect the interests of its rulers, while its occupation and destruction would contribute mightily to the mobilisation of public opinion in the national cause. His bluff was therefore likely to be called.

Alexander had received Kutuzov's note announcing a victory on 11 September, just as the Russian army was taking up position outside Moscow. In a flood of relief and gratitude he promoted Kutuzov, sending him a marshal's baton, and awarded him a grant of 100,000 roubles. Bells were rung in all St Petersburg's churches, and that evening the whole city was illuminated. Alexander wasted no time, and despatched Colonel Chernyshev to Kutuzov with a plan he had devised for the final destruction of the French.

The next day, the service in the church of St Alexander Nevsky on the Tsar's name day turned into one of thanksgiving, as Kutuzov's despatch announcing the victory was read out. Alexander walked among the cheering crowd. The capital resounded to artillery salvoes, and in the evening it was again illuminated.

The joy and the relief were unbounded. "Russia rejoice! Raise your head above all the powers on earth!" wrote one inhabitant to a friend. "I am shaking all over from joy. I cannot sleep at night or do anything." The next day he was still too excited, and had to pen another letter. "Everyone is congratulating each other on the victory, hugging each other, kissing. It is impossible to describe the joy and exaltation on every face." The only long faces were those of freshly-minted militia officers like Lieutenant Zotov, who had just proudly donned his uniform and feared he had missed the chance of proving his patriotic ardour. There was much speculation as to whether Napoleon would be brought to St Petersburg in chains or in a cage. But by the third day, the mood had inexplicably changed to one of anxiety and doubt. The streets grew silent, and people noticed that the packing up of state archives and art treasures from the Hermitage Palace was continuing.[37]

On 18 September a courier from Yaroslavl galloped into St Petersburg bringing Alexander a short, breathless note from Catherine, dated 15 September. "Moscow has been taken. There are some things that are beyond comprehension," she wrote. "Do not forget your resolution: *no peace*, and you will still conserve a hope of recovering your honour." Alexander wrote to Kutuzov, saying he had heard of the fall of Moscow through others, and expressing indignation at having been kept in the dark. To Arakcheev he voiced his regret at ever having been prevailed upon to appoint Kutuzov. It was not for another two days that he heard from him directly.[38]

On 20 September, Colonel Michaud appeared at Kamenny Island bearing a letter from the commander-in-chief as well as news of the fire, of which the Tsar still knew nothing. "My God, so much misfortune! What sad news you bring me, Colonel," Alexander exclaimed. The letter was a laconic note from Kutuzov announcing that he had abandoned Moscow. "I make bold, in the most humble terms, to assure you, all-merciful sovereign, that the entry of the enemy into Moscow is not the conquest of Russia," Kutuzov wrote, but this would have been of little comfort to his imperial master.[39]

"From all this I see that Providence expects great sacrifices from us, particularly from me, and I am prepared to bow to her will," Alexander said to Michaud. The Colonel explained that after marching through Moscow the Russian army had disengaged from the enemy and made a flanking march around the south of the city. This had brought it to a point astride the road to Kaluga, where it could rest and repair the damage inflicted by Borodino. He assured the Tsar that morale was good and that the whole army had only one fear – that he might open negotiations with Napoleon.

"Go back to the army and tell our brave warriors, tell my faithful subjects everywhere you go that even when I do not have a single soldier left, I shall put myself at the head of my beloved nobility and my good peasants, I will command them myself and will use all the means of my whole empire!" Alexander replied, going on to say that he would never sign a peace with Napoleon, and would rather end his days as a beggar in Siberia than come to terms with him. He worked himself up into a state of high excitement, finally declaring: "Napoleon or me, him or me – but we cannot reign together."[40]

St Petersburg was by now buzzing with rumour and speculation. Some maintained that Napoleon had been killed in the great battle, others that he had seized Moscow. Those spreading alarmist talk were arrested by the police and forced to sweep the streets in an attempt to stop the rumours getting out of hand, but this did nothing to calm nerves. In the absence of specific news, people began to assume the worst.

The murmuring about treason started up again, and now the finger was being pointed at Alexander himself. His sister Catherine wrote reproaching him with not having stayed in Moscow to defend it. She told him that he was being accused of forfeiting his country's honour, and that feeling was running high against him. "It is not just one class that is blaming you, but all of them in unison," she wrote.[41]

Alexander was particularly stung by the thought that people might think he lacked courage, as he would gladly have faced Napoleon at the head of his army, and had never wavered in his resolve not to negotiate with the enemy. "I would prefer to stop being what I am than to treat with the monster who is destroying the world," he wrote to Catherine in response to her note. Yet there was a whispering campaign, of which he was aware, to dethrone him in favour of his sister. And the small group of those clamouring for coming to terms with France before the whole state disintegrated was gaining ground. Many of the foreign diplomats in St Petersburg thought that a negotiated settlement could not be deferred much longer. John Quincy Adams noted that English residents were making preparations to leave.[42]

"Violent discontent swirled round the capital," in the words of Countess Edling. "The anxious and exasperated populace might rise at any moment. The nobility were loudly accusing the Emperor of all the misfortunes that had befallen the state, and one hardly dared to take his defence in public." It was through a hostile city that Alexander drove to church on 27 September for the customary celebration of the anniversary of his coronation, normally an occasion of joyfulness as well as pomp. He usually rode to church, but his entourage insisted he go in a closed carriage this time. "We drove slowly in our glazed carriages through an immense crowd, whose mournful silence and angry faces were in stark contrast to the holiday we were celebrating," recalled Countess Edling, who was sitting beside the Empress Elizabeth. "I shall never forget the moment when we ascended the steps of the church, between two walls made up by the people who did not utter one cheer. All one could hear at that moment were our steps, and I have never for a moment doubted that it would have taken no more than a spark at that moment to produce a general explosion."[43]

On 29 September there was an official announcement which depicted the fall of Moscow as a minor tactical setback. It was followed up by an imperial proclamation written for Alexander by Shishkov. In tone both angry and proud, it declared that the fall of the ancient capital was the rallying call for all Russians and the turning point in the country's fortunes. Napoleon had climbed into a grave from which he would never emerge, and the Russian nation would triumph.

To Bernadotte, Alexander wrote two days later that although Kutuzov had retreated, Borodino really had been a victory. "I repeat to Your Royal Highness the solemn assurance that more than ever I and the nation at the head of which I have the honour to stand, are decided to persevere and to bury ourselves under the ruins of the empire rather than to come to terms with the modern Attila."[44]

15

Stalemate

Alexander's determination was born of fatalism and inner conviction rather than any kind of calculation. For one thing, he did not really know whether he had an army left, whatever Michaud might say. One of his aides-de-camp, Prince Sergei Grigorievich Volkonsky, who had been at Wintzingerode's headquarters, assured him that "from the commander-in-chief to the last soldier, all are ready to lay down their lives in the defence of the fatherland and your Imperial Majesty," which was encouraging, but it did not accord with what he was hearing from other quarters.[1]

"The soldiers are no longer an army, but a horde of bandits, looting under the very eyes of their commanders," Rostopchin wrote to him from Kutuzov's camp. "One cannot shoot them: how can one punish several thousand people a day?" Alexander might have been inclined to take anything the Governor General of Moscow wrote with a pinch of salt, but he would almost certainly have had similar reports, either directly or from people who received letters from the army. "My heart aches at the disorders and anarchy I see in almost every unit of the army, which is heading for catastrophe," General Dokhturov wrote to his wife; Prince Dmitri Mikhailovich Volkonsky lamented that "our own marauders and cossacks are robbing and killing people"; and there were plenty of other officers who did not hide the truth or their fears. Many of them were in despair at their army's failure, and openly proclaimed that they were ashamed to wear the uniform.[2]

The senior generals were justifying their own conduct by accusing each other of everything from incompetence to treachery. Barclay was the victim of a stream of aspersions, of which he apprised Alexander in hurt tones. Bennigsen informed anyone who would listen that Kutuzov was an imbecile and a coward who had lost the respect of the army. "The soldiers hate and despise him," echoed Rostopchin in a letter to Alexander. Bennigsen wrote to the Tsar complaining that Toll, whom he held responsible for the disaster at Borodino, had insulted him. Rostopchin warned Alexander that General Pahlen hated him and denounced Platov as a traitor who had made arrangements with the French for his future. Like a gaggle of petulant squabbling schoolchildren, they sneaked on each other to the Tsar in letters that must have made baffling as well as painful reading. Egging them on and criticising them in his regular letters was General Wilson, who mistrusted all of them. Particularly distasteful to the somewhat prudish Alexander was the stream of lewd tittle-tattle about Kutuzov's private life. Bennigsen and Rostopchin both gleefully informed him that the old commander-in-chief had smuggled a couple of girls disguised as cossacks into his quarters, where he spent whole days attending to them while his demoralised army seethed with indignation.

Clausewitz considered it fortunate that Alexander did not join the army, as the sight of what it had come to might well have weakened his resolve. He also thought that if the Tsar had been able to contemplate at close quarters the devastation being visited on his land and the effect it was having on society, he might have agreed to negotiate with Napoleon.[3]

The country was in a volatile mood, whatever the later legend of the patriotic war might suggest. Leaving aside questions of patriotism and loyalty, Napoleon's invasion inevitably raised a number of others as to the viability of the nature and constitution of the Russian state. It had never been put to such a test, and Alexander could not be sure that the rapidly expanded structure of the empire could take the strain.

"Over the past twenty years I have been present at the funeral of several monarchies," the old royalist Joseph de Maistre wrote from St Petersburg at the beginning of October, "but none of them struck me as much as what I see now, for I have never seen anything so mighty totter . . . Everywhere I see loaded boats and carriages; I hear the language of fear, of resentment and even of ill will; I can see more than one terrible symptom."[4]

On the face of it, the invasion had provoked a surge of patriotism and devotion to the Tsar in all classes. The nobility were, as Alexander had witnessed in Smolensk and Moscow, apparently eager to sacrifice their lives and their wealth to the cause. When Kutuzov began to organise the St Petersburg militia, he received the following letter, which gives an idea of this:

> *Having had the pleasure of serving under the command of your excellency in the previous Turkish war, in the Bug rifle corps, I took part in many battles including the storming of Izmail, in three victorious encounters beyond the Danube and at Machin at the defeat of the Vizir, where we were always victorious with you. After that I took part in all the encounters against the French in Italy and was wounded very gravely in the leg, when my thighbone was shattered by a bullet, which remained lodged in it, and as I could not walk I was retired from service with the rank of Major General with the right to wear uniform but no pension. For ten years and six months I suffered from this bullet, seeking relief everywhere, but nobody could remove it; at last here, in St Petersburg, Jacob Vasilievich Wille decided to deliver me, and after an operation counselled by him removed the bullet, the wound healed and the bone grew back, and now I have the full use of my leg and complete freedom, as proof of which I have a certificate from him, which I enclose. Having the most passionate desire to serve my fatherland*

> *under the command of your excellency in the militia, I beg most humbly to be admitted into its ranks.*[5]

Young boys ran away from home to join the army, twenty-two pupils at the Kaluga school for nobles volunteered, and from the fringes of the empire Bashkirs, Kalmuiks, Crimean Tatars and Georgian princes declared their willingness to fight. Groups of fashionable young gentlemen clubbed together to form units of volunteers at their own cost, taking the opportunity to design flashy uniforms with death's-head symbols and to call themselves "the immortals" or some such dramatic name. Some showed their patriotic ardour in drastic ways: Sergei Nikolaevich Glinka burnt his entire collection of richly bound French books.[6]

But not all were prepared to make such sacrifices. While some picked their best serfs for the militia and personally led them into the ranks, others refused to serve themselves or would only do so on local order-keeping forces. Most did everything they could to hold on to their workforce. Many small landowners sent tearful letters to the authorities in an attempt to evade the obligation. Others dragged their feet, hoping the war would be over before they were forced to part with their serfs. Others still selected the old, the crippled, the shirkers, drunks, miscreants and the village idiots. As a result, the province of Kaluga, which should have yielded 20,843 men, furnished no more than 15,370, and hardly more than a third of the men raised as a whole by the levy were suitable for active service. With patriotic proclamations appealing for defenders of the fatherland to come forward, some serfs, assuming that if they fought they might be rewarded with personal freedom, actually volunteered, but they were pursued and arrested as fugitives and dealt with harshly by their masters. According to Rostopchin, two aristocrats who had loudly pledged to raise and equip a regiment each during Alexander's visit to Moscow, never contributed a man or a penny between them.[7]

Notwithstanding numerous proclamations issued by the authorities enjoining them to destroy anything that could be of use to the invader and to abandon occupied areas, many landowners stayed put. There were plenty of instances of them providing forage and victuals to the French, taking payment in cash or notes. A foraging party led by Captain Abraham Rosselet of the 1st Swiss Regiment was not only plentifully supplied by the Russian landowner they visited, but put up for the night and in the morning assisted in evading a unit of cossacks which was preparing to ambush them.[8]

When Alexander asked Sergei Volkonsky about the attitude of the nobility in the country at large, he replied: "Sire! I am ashamed to belong to it." But the nobility were not the only ones lacking in patriotic spirit. Grand Duke Constantine himself forced the army to buy remounts from him at inflated prices, and of the 126 horses he sold, only twenty-six were fit for service while the rest had to be destroyed. Arakcheev was taking a cut from suppliers. The civil servants responsible for equipping and supplying the army stole and sold on goods bought for it, inflated prices and took bribes to issue receipts for deliveries that never took place, with the result that the troops never received much of what had been procured for them. Those responsible for caring for wounded officers evacuated out of the war zone deflected to their own pockets the sums destined for the feeding and care of their charges. And according to some sources, members of the clergy showed little courage and abandoned their posts as the French approached.[9]

The merchant class appears to have been more generous, although much of this could be put down to the fact that the war against France was also a war against the Continental System, which was so ruinous to them. And there were many instances of profiteering among them too. They joined with the commissary officials in fixing prices, and some certainly profited from supplying the army. Following Alexander's appeal for volunteers and offerings in Moscow, the city's armourers raised the prices of sabres from six to between thirty and forty roubles; of a pair of pistols from seven or eight roubles to thirty-five or fifty; and of a musket from eleven or fifteen to eighty.[10]

When asked about the attitude of the common people, Volkonsky answered: "Sire! You should be proud of it: every peasant is a hero, devoted to the fatherland and your person." But this is hardly borne out by the evidence. The peasants had no interest in the war, but were understandably keen to preserve themselves and as much of their livestock as possible, usually by taking it off to the woods. The retreating Russian army encouraged this trend, telling the peasants of the horrors that awaited them if they stayed behind. "These rumours are producing a sensation among the peasants, who with the greatest *sang-froid* in the world set their huts on fire so as not to abandon them to the enemy," recorded one Russian officer. But many were not happy to see the retreating Russian army torching their villages, and some put out the fires as soon as the soldiers had moved off. The peasants had also been told, from the pulpit, that this invader was an infidel, and many referred to the French as "*Bisurman*," a traditional term for a Muslim.[11] Hence the fearful and hostile attitude encountered by the French.

Once the fear was dispelled, relations could be perfectly amicable. Michał Jackowski, an officer in Poniatowski's horse artillery, rode into a village accompanied only by one trooper, and was promptly surrounded by some fifty armed peasants. But when he gave them the traditional Christian greeting habitual in Poland and Russia, they lowered their weapons and said that if he was a Christian they had nothing against him. He found that this never failed, and that he always obtained supplies by prefacing his request with the statement that he would only buy what they could spare.

A similar attitude is recorded by other Poles, who were better placed than the other nationalities of the Grande Armée to communicate with the locals. Every French division had a Polish officer seconded to it for this purpose, and there are accounts of peaceful and fruitful foraging expeditions. General Berthézène denied that there was any widespread animosity at this stage. "On the contrary," he wrote, "I saw our servants go off singly and without escorts, foraging around Moscow; I saw peasants warning them of the approach of cossacks or of ambushes; I saw others show us where their masters had hidden their supplies and share them with our soldiers." A number of French and allied officers corroborate this with accounts of amicable foraging expeditions.[12]

One Westphalian soldier recorded that when his unit came to evacuate Mozhaisk after a five-week stay, the man they had pressed into service to work for them bade them farewell with tears in his eyes, making the sign of the cross over them. Lieutenant Peppler, whose Hessians were cantoned outside Mozhaisk, found that by treating the locals politely they had nothing to fear. "We had won the trust and even the friendship of those good people to such a degree that we felt as safe among them as though we had been in a friendly country," he wrote. And when Bartolomeo Bertolini escaped from captivity, he found friendly peasants giving him food as he made his way across country to Moscow.[13]

Even allowing for some exaggeration, such accounts are revealing, and they are corroborated by evidence from the Russian side, where the attitude of the lower orders aroused the deepest fears. "We still do not know which way the Russian people will turn," Rostopchin warned Sergei Glinka.[14]

Soon after the invasion began, there were instances of serfs refusing to carry out their duties and even staging minor revolts, and there was much ransacking of manors abandoned by fleeing nobles. In a letter to a friend, one landowner described how, after a French foraging party had come and taken what they needed from his estate, the serfs rushed in and looted all that remained. Once the local authority had evaporated the peasants began to behave like "bandits," even assaulting priests and torturing them in order to extort supposed Church riches. Peasants also helped French marauders to attack and loot manor houses. Some complained of their condition to the French, and seemed to expect Napoleon to do something about it. Many of those landowners who stayed put, often wives of officers who were away with the army, surrounded themselves with armed servants and asked the French for protection. There were cases of landowners being roughly handled and even killed, but most of the disorders were opportunistic rather than politically inspired.[15]

Pavel Ivanovich Engelhardt, a landowner on the fringes of the area occupied by the French in the province of Smolensk, led his peasants in an attack on some French marauders. Emboldened by the action, they began to question his rights over them and refused to work. He called on a detachment of cossacks hovering in the area to come and restore discipline. The serfs then denounced him to the French authorities in Smolensk, and he was imprisoned. But as the French could find nothing specific to charge him with, they released him. He once more called in the cossacks, and his serfs were whipped into submission. But the moment the cossacks had gone they buried in his park the bodies of a couple of French soldiers they had killed and then denounced him again. This time he was shot by the French.[16]

There were also cases of peasants showing extreme devotion to their masters. Aleksandr Benckendorff recounted how his detachment fell upon a party of French marauders looting the estate of one of the Galitzine family and chased them off. The peasants, who had assembled, asked the Russian officer in charge for permission to drown one of their number, a woman. When asked why they wanted to do this, they replied that she had revealed to the French the place where the Princess's jewels had been hidden. The officer suggested that she might only have done this under duress, and they answered that she had assuredly been flogged to within an inch of her life, but that nevertheless she must be punished.[17]

The Russian army's failures, followed by the loss of Smolensk and then Moscow, inevitably lowered respect for the authorities and for the Tsar, so that it was not uncommon to hear peasants making ribald jokes about the incompetence not only of Barclay and the "Germans," but of Alexander himself. In the general mood of mistrust and paranoia even Russian officers in uniform found themselves arrested by the populace and in at least one case nearly lynched as "spies."[18]

The nationalist Filip Vigel commented, approvingly, that the lower orders had shed their deference and become much more outspoken, while others noted, with alarm, the frequency with which the name of the rebel Pugachov was uttered by them. "The influence of the local authorities, particularly of the police, grew weak, and the common people grew restive," according to the merchant M.I. Marakuev. "It was necessary to treat them with skill and flattery. The decisive tone of authority and mastership was out of place and could be dangerous." Even the authorities recognised this, and proclamations were couched in populist terms and a cajoling rather than commanding style.[19] "The ideas of freedom that have spread through the land, the widespread devastation, the total destitution of some and the selfishness of others, the disgraceful attitude of landowners, the abject example they have set to their peasants – will this not lead to great upheaval and disorder?" noted Lieutenant Aleksandr Chicherin in his diary as he observed the situation around the retreating army.[20]

In a letter to a friend, Maria Antonovna Volkova expressed the conviction that Rostopchin had saved Moscow from social upheaval, even though she had lost her house in the fire. "Only a man like Rostopchin knew how to deal with minds in such a state of ferment and prevent terrible and irreversible things happening," she wrote. "Moscow has always had an influence on the whole country, and you can be sure that if there had been the slightest disorder between groups of her inhabitants, the upsurge would have been universal. We all know with what perfidious intentions Napoleon invaded. It was necessary to counteract them, to turn minds against the scoundrel and thereby contain the common people, who are always thoughtless." She was talking about revolution.[21]

A great deal of effort had been put into influencing the attitudes of the people. Alexander's proclamations and religious sermons were accompanied by a steady trickle of propaganda and rumour. News of the burning of Moscow, universally attributed to the French, of the profanation of churches and of alleged atrocities committed on the population was circulated widely. "It is impossible to imagine the horrors the French are said to be committing," noted Lieutenant Uxküll. "One hears that they're burning and desecrating churches, that the weaker sex – or rather any individuals who fall into their frantic hands – are sacrificed to their brutality and the satisfaction of their infernal lusts. Children, greybeards – it's all the same to them – all perish beneath their blows."[22] Rumours were disseminated among the peasants to the effect that Napoleon would convert them all to Catholicism by force and brand them on the heart. Much was also made of the fact that Napoleon was in league with Russia's historic enemy, the Poles, who were supposedly intent on recapturing parts of Holy Russia.

But according to Yermolov, there would have been no truly national dimension to the war and no way of harnessing the peasants to the Tsar's cause had it not been for the clumsy and increasingly undisciplined behaviour of the French.[23] They did bed down and stable their horses in churches – mainly because these were the only suitable buildings in small towns and villages. They were also undoubtedly rough with the natives. Peasants who brought their produce to sell in Moscow were beaten up and robbed. The more and more widespread depredations of the Grande Armée's foraging parties, often conducted without any regard for the livelihood, let alone the feelings, of the Russians, forced them to take up arms in order to survive. It was a question of self-preservation.

The peasants began to lay ambushes for foraging parties or lull them into a false sense of security and then overpower them. They acquired arms and were able to take on small units. They vented their rage on their captives in acts of barely believable savagery, mutilating them, burying them alive or roasting them over fires. "Approaching a village in order to get some supplies," Lieutenant Uxküll noted in his diary, "I saw a French prisoner sold to the peasants for twenty roubles; they baptised him with boiling tar and impaled him alive on a piece of pointed iron." The French and their allies responded in kind, encouraging a degenerating spiral of horror. "People became worse than wild animals and killed each other with incredible cruelty," noted A.N. Muraviov.[24]

There were exceptions, at every level of society. Édouard Déchy, the thirteen-year-old son of a doctor in Davout's corps, had been brought along by his father since, his mother being dead, there was no one to care for him at home. His father was put in charge of one of the hospitals in Smolensk after the action at Valutina Gora. A local Russian landowner, a countess, seeing the child all alone, begged his father to let her take him off to the country, and the boy spent an idyllic few months being pampered and playing with the Countess's children. In the city of Orel, a Russian woman took pity on some French prisoners who had been brought there and, taking them into her house, ruined herself clothing, feeding and tending to their wounds; when her means ran out she wandered the city begging for money to feed them.[25]

Such acts of charity were by no means restricted to the educated classes. A peasant recalled how his entire village had taken refuge in the forest, along with their cattle and all the food they could carry. One day they were discovered by a couple of Frenchmen, so they tried to kill them, but one got away. Expecting a punitive expedition, they sent one of their number to alert some Russian troops stationed nearby, who duly set an ambush. A couple of hundred French troops came and demanded food, threatening to take it if they were not obeyed, whereupon the hidden Russians attacked and the French were quickly disarmed. "But as they were all begging for bread," explained the peasant, "we felt sorry for them, we cooked some potatoes and brought them bread, and even some beef, and we could see how hungry they were, how they all threw themselves on the food we had given them, and with what eagerness they started to eat it. Some of the Frenchmen were trying to say something with tears in their eyes, evidently thanking us in their language, and we said to them: go on, eat, cheers, we have plenty of bread."[26]

The unaccountability of the peasants was a source of debate and anxiety, as to involve them in the war was, to some extent, to empower them: it was the first time in Russian history that a Tsar had been obliged to appeal to the serfs to defend him and his state. And nobody could be sure how they might use this implicit power.

First attempts to engage the peasants in the war had mixed results. A detachment of Prince Eugène's Army of Italy came across a band armed with pikes, scythes and axes, under the command of their squire. He led them bravely towards the Italians, only to find that all his serfs deserted him and fled. When the selection for the militia began, many wounded themselves in order to avoid being drafted. And not all of those who did end up in the militia displayed quite the right spirit. Nikolai Andreev, a lieutenant in the 50th Jaeger Regiment of Neverovsky's division, noticed that at Borodino the militiamen responsible for carrying the wounded back to the dressing stations relieved the officers of all their valuables as they did so.[27]

Yet the example of Spain, where the *guerrilla* was causing such damage to the French, was an alluring one. A Spanish "national catechism" was translated and published, and many believed that the military value of the peasantry, as distinct from those drafted into the militia, should be harnessed to the national cause. "But a national war is too much of a novelty for us," lamented the populist Fyodor Glinka. "It seems that they are still afraid of unbinding hands."[28] Ultimately, it was left to circumstance.

Soon after the beginning of the war, Denis Davidov, an officer of Hussars in the Second Army, wrote to Bagration suggesting that if he were given a small independent command he could carry out effective partisan operations against the French. It was not until the beginning of September, just before Borodino, that he was given his command, of fifty Hussars and eighty cossacks. He began to operate in the French rear, but was put out to find himself being shot at by Russian peasants, who regarded all soldiers with equal hatred. After wasting time and energy trying to convince them that he was on their side, he exchanged his regimentals for a peasant smock, let his beard grow, and replaced the cross of the order of St Anne on his breast with a small icon of St Nicholas. This permitted him to approach villages without being shot at, and he could then begin to convince the inhabitants to make common cause with him. He scored a few successes against French foraging parties and isolated units, and began to involve the local villagers in his actions. By 24 September his detachment had swelled to some three thousand horsemen, as peasants and Russian stragglers or escaped prisoners joined him, and by the end of October he could muster large numbers of local peasants, many armed with muskets taken from the French, for specific operations.

After the fall of Moscow, Kutuzov sanctioned the formation of more such "flying detachments" whose object was to prey on the French lines of communication and supply. He detached General Wintzingerode with 3,200 men to operate along the Tver road, and General Dorokhov with two thousand to harry French traffic along the Smolensk road from bases around Vereia. He had also formed smaller detachments under Seslavin, Figner, Lanskoy and others. Some of these units did gather peasant recruits to their side, but mostly they only made use of the intelligence and help provided by the population, occasionally arming them and using them for prisoner or supply escort duty.

Davidov was held up as a hero by Pushkin and admired by Walter Scott, who had a portrait of him in his study, before being immortalised by Tolstoy in the guise of Denisov in *War and Peace*. His exploits and those of other "partisan" units have been greatly exaggerated by legend, and the claims made on their behalf are often absurd. One Soviet historian assures us that a detachment of a hundred Russians attacked a village defended by two cavalry squadrons and two companies of infantry. They allegedly killed 124 Frenchmen and took a futher 101 prisoner, at a cost of two wounded men and six horses wounded or killed. A child can work out that in the space of time it would take to kill 124 men a substantial number of Russians must have been killed and wounded – unless, that is, the French surrendered after a couple of shots and were then butchered. Sergei Volkonsky, who commanded one of the partisan units, admitted that most of the heroic stories were nonsense. The whole point of this kind of warfare was that it had to be cautious and low-risk. The trick was to avoid a fight and capture the French detachment while it slept. The reality was not as glorious as the legend either. Figner was a cold-blooded murderer. Sergei Lanskoy was, according to General Langeron, a rapist and a brigand.

There were also a number of what one might term guerrilla bands operating along the fringes of the territory held by the French. But it is difficult to disentangle truth from fiction, as the idea of patriotic peasants was alluring to slavophiles and communists alike. It cannot be supported or challenged by documentary evidence, precious little of which ever existed. All one can do is repeat what has been written by Russian historians, and take it with a pinch of salt.

Yermolai Chetvertakov, a Russian dragoon who had been taken prisoner near Gzhatsk but managed to escape, teamed up with a local peasant and together they began ambushing and killing individual French soldiers. They were joined by others, and the band snowballed. Numbers were never constant, as they might be joined by up to several thousand volunteers from the locality if there was a tempting convoy to attack and plunder, but most of these would then go home. Fyodor Potapov, alias "Samus," a Hussar who had been wounded in a skirmish with the French, had taken refuge in the woods, where some peasants gave him shelter. He enlisted their support and similarly built up a band of partisans. Stepan Eremenko, an infantry private who was wounded and left behind outside Smolensk, followed a similar course.

There were also a number of peasants who formed guerrilla bands, and a couple of women have gone down in history as defenders of Holy Russia. In the village of Sokolovo in the province of Smolensk a peasant woman by the name of Praskovia defended her virtue, or possibly her livestock, so effectively that she allegedly sent six Frenchmen to their deaths with her pitchfork. She was outdone by Vasilisa Kozhina, who supposedly despatched dozens with her scythe.

There must have been countless instances of peasants confronting and killing enemy soldiers or civilian camp-followers, particularly in the latter stages of the campaign. But modern Russian historians are generally agreed that there was no *guerrilla* that could bear any comparison with the Spanish model, and that the contribution of the peasants was largely confined to opportunistic pillage and murder.[29]

Opportunistic or not, the instincts animating the peasants, and Russian society as a whole, were such as to rally them to the cause of the empire, and this would become ever more apparent as the fortunes of war deserted the French. Despite the foot-dragging, a total of 420,000 men were drafted into the militia, of whom 280,000 managed to take part in the fighting. A total of a hundred million roubles was donated to the cause by all classes – a sum equivalent to the entire military budget for that year.[30] And at no stage did the machinery of administration cease to function outside the area controlled by the enemy. Towns such as Kaluga, which were too close to the war zone for comfort, were efficiently evacuated of their institutions, schools, hospitals, archives and so on, which were soon functioning again at a safer distance.

Alexander was not to know any of this. He was unwell, suffering from painful rashes on his leg, and he was sick of the wavering sycophancy of his court, so he had withdrawn into the isolation of Kamenny Island. His only source of comfort was his inner conviction that he was but an instrument of God's will, acting out a higher purpose. This had outgrown the mere defence and liberation of Russia, and among those whom he now saw with the greatest pleasure were the little group of people working under his wing on the liberation of Germany and indeed the whole Continent. He had formed them into a German Committee under the presidency of George of Oldenburg. This was organising a German Legion under Colonel Arentshild and conducting a propaganda campaign throughout Germany, orchestrated by Stein and his secretary the poet Ernst Moritz Arndt. Alexander's manifestoes were translated and circulated throughout Germany, enhancing his position as the champion of all those opposed to Napoleonic rule. The idea was to prepare a fifth column which would rise at the appropriate moment. When this moment would come depended on the performance of the Russian troops.

"The fate of the armies will decide that of Germany," Stein declared before the battle of Borodino.[31]

There is a hackneyed story that, seeing his friend Aleksandr Nikolaevich Galitzine so serene, Alexander had questioned him about the source of his inner peace, to which the Prince is supposed to have answered that it came from reading the Bible, and that a few days later the Prince's wife lent the Tsar her copy, marked at various passages. In fact, Alexander had been reading the Bible for some time, and using it to find scriptural reinforcement of a sense of his own destiny which had already matured. He had also immersed himself in mysticism, and a detailed memorandum on the origins of mystic literature which he sent to his sister Catherine at this time reveals surprising familiarity with it.[32]

It was fortunate for Russia that Alexander found the inner strength to resist the temptation to act at this critical moment. "If fate has condemned your empire to fall, you must perish with it and fight among your faithful subjects, who have decided to die under your eyes on the field of honour, on which you yourself should win or perish with them," Rostopchin urged.[33] The idea of assuming command of his army once more and sallying forth to a fight to the death with Napoleon held strong appeal for Alexander. Had he done so, he would have been beaten and Napoleon would have regained the initiative. As it was, the somnolent Kutuzov was the perfect man for the moment, as time was working for the Russians and against the French.

After leaving Moscow to the French, Kutuzov had marched out in a south-easterly direction. He then veered right and began a flanking march around the city which eventually brought him to a point due south of it at Krasnaia Pakhra. It was a risky move, particularly in view of the state of the army.

Morale had reached a nadir. "Every part of the army is in a state of terrible disorder," noted Dmitry Volkonsky, "and not only has there been a general weakening of obedience, but even the sense of courage has weakened since the loss of Moscow." Tens of thousands had fallen behind or deserted in the retreat from Borodino and the march through Moscow. Some formed bands of marauders. "The saddest thing of all is that our soldiers spare nothing," wrote Lieutenant Uxküll of the Imperial Chevaliergardes. "They burn, pillage, loot, and devastate everything that comes to hand." There were even instances of them looting churches.[34]

Those that were left were hardly a force to be reckoned with. "The soldiers seemed to have taken fright," according to the female cavalry officer N.A. Durova. "From time to time they would come out with a few words, to say that it would have been better to be dead than to have given up Moscow." As they marched round the south of Moscow, the troops could see the city in flames. "Mother Moscow is burning," they murmured incredulously to each other. "The superstitious ones, unable to comprehend what was happening before their eyes, already decided, with the fall of Moscow, that they had witnessed the fall of Russia, the triumph of the Antichrist, soon to be followed by the Final Judgement and the end of the world," in the words of Lieutenant Radozhitsky.[35]

Bennigsen and others were expecting Kutuzov to attack the French advance guard under Murat, which had ventured out on its own, but Kutuzov once more ordered a retreat. This provoked a confrontation with Bennigsen that went beyond their previous disagreements. Bennigsen was convinced that he had saved the day at Borodino. He was horrified by the abandonment of Moscow, and had come to the conclusion that the Field Marshal was an incompetent old fool. He was supported in this view by Wilson and a few others, and accusations of "cowardice" began to be made against Kutuzov.

The Russian army withdrew in a south-south-westerly direction to Tarutino, where Kutuzov set up a fortified camp. Bennigsen began to argue that the position was no good and his tactics were inappropriate, but Kutuzov put him in his place. "Your position at Friedland was good enough for you," he snapped. "Well, I'm quite happy with this one, and this is where we will stay, because it is I who am in command here and I who am responsible for everything."[36]

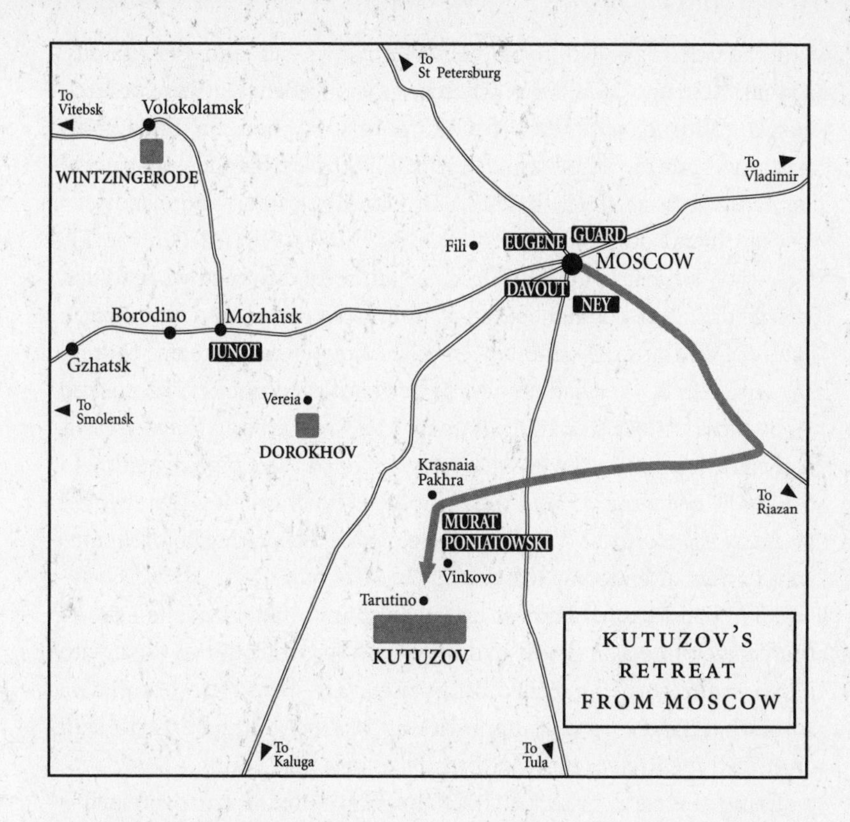

It was a good position. It was far enough from Moscow not to be vulnerable to an attack by Napoleon, it was a good jumping-off point for operations against his lines of communication, and it commanded the approaches to Kaluga and Tula. These were the centres of Russian military production, and they were also the gateways to the fertile south. Once the benefits of this move had been recognised, several of the other commanders ascribed to themselves the merit of having chosen it. In fact, as Clausewitz pointed out, it had been dictated by a logical imperative rather than by any flash of genius.[37]

What Kutuzov needed was time, and he later described every day spent at Tarutino as "golden," since it helped to restore the strength of the army. Supplies of food and equipment began to flow in from Kaluga and Tula. Local peasants brought eggs, milk, bread and pies, while merchants rolled up in their wagons with all manner of goods, so that the troops could buy whatever they needed. Kutuzov ordered winter uniforms, with thick trousers, sheepskin coats, fur-lined boots and gloves for the whole army. The soldiers dug pits and constructed "*banyas*," Russian steam-baths, so they could clean up and relax.

"We spend our time very pleasantly," noted Nikolai Dmitrievich Durnovo, an officer on Bennigsen's staff. "All day long we feed, eat and drink." "We cooked beef stew and often sour soup with cabbage, beetroot and other vegetables," recalled Lieutenant Nikolai Mitarevsky with relish. "We had fry-ups of beef and even poultry; we cooked buckwheat with butter and potatoes." They sat around playing cards and chatting, and in the evenings they smoked their pipes listening to the soldiers singing around the campfires. Every evening there would be prayers before the Virgin of Smolensk accompanied by religious songs, often attended by Kutuzov.[38]

The Field Marshal had set up quarters in a cottage at the edge of the village of Letashevka. It had one room, in which he worked, with a bed in the corner screened off by a curtain. Bennigsen occupied a somewhat larger cottage opposite, and other officers of the staff crammed themselves into nearby huts as best they could.

Using the field press provided by Alexander, Kutuzov issued a stream of propaganda in regular bulletins, *Izvestia iz Armii*, which reported every skirmish, magnifying its significance and inflating figures of captured French soldiers and guns. More importantly, the bulletins represented the Russian soldiers as happy, brave and keen to fight, with well-fed horses. The wounded were apparently being lovingly cared for by wives and mothers, and every peasant was a true son of the fatherland ready to support the army in its struggle. The French were represented as hungry, sad and isolated. It was clever psychology, as it gave comfort and emotional support to soldiers who had just suffered not only defeat but also the shock of seeing their revered capital invaded and burnt.

The units were reinforced and the new levies given elementary training. But there was none of the parade-ground discipline that made the Russian army such hell. Nobody bothered to pipeclay their crossbelts. Men wore those elements of their uniform that suited them, and supplemented them with overcoats or cloaks that kept them warm and comfortable. Shakos were jettisoned in favour of soft forage caps. The junior officers developed a swashbuckling swagger. "There was no sparkle, no gold or silver; epaulettes and sashes were rare; the only things that gleamed were muskets, bayonets and artillery pieces," recalled Mitarevsky. "There were no rich or fashionable uniforms, only felt cloaks, thick capes, dirty, torn greatcoats, crumpled forage caps . . ."[39]

Many of these young officers had known nothing of soldiering, nothing of the common soldier and nothing of the peasants. Prince Piotr Andreevich Viazemsky, a Moscow aristocrat, had volunteered after Alexander's visit. "I was a middling rider, and had never taken a gun in my hand," he wrote. "At school I had learned to fence, but my acquaintance with the rapier had grown distant. In a word, there was nothing warlike about me." At a dinner he met General Miloradovich, who took him on as an aide-de-camp. He felt confused and out of place as he followed his General about the battlefield of Borodino. But that suddenly changed. "When my horse was wounded under me, an inexplicable feeling, of joy, of pride, welled up inside and enveloped me."[40]

A large number of young men such as him found themselves for the first time connecting, through the solidarity of war, with each other and with the mass of the Russian people as represented by the common soldier. In the heat of battle and the rigours of the bivouac they were able to see their serf-soldiers as human beings. Their shared experience over the next two years was to give rise to a kinship and a new vision of Russia, one that would perish on the gallows and in the exile that followed the failed Decembrist rising of 1825, but would live on through its enormous influence on Russia's cultural life.

The shock of Borodino and the destruction of Moscow, followed by the boyish idyll of Tarutino, had produced an extraordinary effect. "We were in a state of bliss!" recalled Dushenkievich, a fifteen-year-old

Lieutenant in the Simbirsk Infantry regiment. "What had happened to the sorrow, from where did we get the sense of security and self-assurance which now flooded over us, while we grieved over Moscow and the Fatherland?"[41]

16

The Distractions of Moscow

"I spent the evening with the Emperor yesterday," Prince Eugène wrote to his wife on 21 September. "We played *vingt-et-un* to pass the time; I foresee that we will find the evenings very long, as there is not the slightest distraction, not even a billiard table." The prospect of staying in Moscow did not fill Napoleon's entourage with enthusiasm. "Napoleon was never more than a man of genius, and it was not in his nature to know how to amuse himself," remarked *Commissaire* Henri Beyle, alias the novelist Stendhal, adding that his court was a dreary zone.[1]

The Emperor had once again taken up residence in the Kremlin, where he occupied the same apartment overlooking the river Moskva and part of the city as Alexander had a few weeks before. It consisted of one vast hall with great chandeliers, three spacious salons and a large bedroom, which doubled as his study. It was here that he hung Gérard's portrait of the King of Rome. He slept on the iron camp bed he always used on campaign. His campaign desk had been set up in one corner and his small travelling library laid out on shelves – but his copy of Voltaire's history of Charles XII was always within reach, on either his desk or his bedside table. He instructed his valet to place two burning candles at his window every night, so that passing soldiers would see that he was watching and working on their behalf.

Napoleon had hoped to set up a Russian civil administration, but there was a dearth of Russian citizens of any calibre, and most of those available did everything to wriggle out of collaborating with the French. He therefore fell back on the expedient of appointing Jean-Baptiste de Lesseps, a former French Consul in St Petersburg, who gathered together all those Russian inhabitants prepared to serve in a provisional administration. Aside from restoring order in most parts of the city, this body made housing available to those Muscovites who had lost their homes in the fire, and tried to encourage peasants from the surrounding countryside to come and sell their produce in the city. But those who did come forward were mostly beaten up and robbed by the soldiery.

A semblance of normality was established in other respects. People travelled "as easily between Paris and Moscow as between Paris and Marseille," according to Caulaincourt, although it took a little longer. The post, carrying thousands of letters from the men to their families and sweethearts, took up to forty days. But the Emperor did not have to wait that long. Every day an *estafette* would arrive from Paris, having covered the distance in only fourteen days. This was the high point of Napoleon's day, and he would grow restless if, as happened on one or two occasions, it arrived a couple of days late.[2]

News from Paris was always welcome, particularly if it caressed Napoleon's vanity. He read with pleasure that his birthday, which he had spent before Smolensk, had been celebrated in his capital by the laying of foundation stones for the Palais de l'Université, a new Palais des Beaux-Arts and a monumental building to house the national archives. He was informed that "the enthusiasm of the Parisians, on hearing of the Emperor's entry into Moscow is tempered only by their fear of seeing him march out of it in triumph on a conquest of India." News that Wellington had taken Madrid was less welcome.

If he felt any anxiety about his position he kept it well hidden, and attended to affairs of state as well as those of his army with a punctiliousness that probably helped him avoid facing up to the realities of his situation. He badgered Maret, pressing him to put pressure on the

American Minister, the poet Joel Barlow, who had just arrived in Vilna, to forge a closer alliance with the United States against Britain. He gave instructions for 14,000 horses to be sent from France and Germany. He ordered the purchase of large quantities of rice in Trieste which was to be shipped across Europe to Moscow. He also held frequent parades on the great Krasnaia Square before the Kremlin, at which he awarded crosses of the Légion d'Honneur and promotions earned at Borodino.[3]

But he was not looking forward to a winter away from home. "If I cannot return to Paris this winter," he wrote to Marie-Louise, "I will have you come and see me in Poland. As you know, I am no less eager than you to see you again and to tell you of all the feelings which you arouse in me."[4]

His soldiers felt much the same. "Another winter will go by without the happiness of being able to press you in my arms, for it is said that we are going to take winter quarters, though where exactly has not yet been decided," Captain Frédéric Charles List wrote to his wife on 22 September. "I am very tired of this campaign and I do not know when God will give us peace," the simple ranker Marchal wrote to the *curé* of his village. General Junot was no less depressed. "Enough said about the war, I now want to tell you, my darling L—e, that I love you more every day, that I am bored to death, that I desire nothing in the whole world as much as to see you again, that I am stuck in the most unworthy country in the world, and that I will die of sorrow if I do not see you soon and die of hunger if I remain here much longer," he wrote to his mistress from Mozhaisk. A *commissaire* who had come out on campaign at the age of fifty because he found his desk job dull and thought he might make his fortune, poured out his regret and disgust to his wife, adding, somewhat insensitively, that there were not even any pretty girls in Russia. Marie-François Schaken, a nineteen-year-old surgeon in Davout's corps, complained to his sister that he was eating poorly, while his horses were gnawing at their manger, but affirmed his unbounded faith in Napoleon, who would undoubtedly lead them home safely. "Find me a pretty little mistress for my return, for there are none here," he begged her. "Tell her I will love her very much."[5]

Although he may have turned his nose up at them, there were in fact plenty of women to choose from in Moscow. For one thing, most of the whores seem to have stayed. "This class of person was the only one which drew some profit from the sack of Moscow, as everyone, in their eagerness to have a woman, welcomed these creatures with pleasure, and once they had been introduced into our dwellings, they straightaway became the mistresses of the house, and squandered everything the flames had spared," according to Jean-Pierre Barrau, quartermaster of Prince Eugène's corps. "There were others who really deserved consideration on account of their birth, their upbringing, and above all their misfortune; hunger and poverty forced their mothers to bring them to us."[6]

Louis Joseph Vionnet de Maringoné, a senior officer in the Grenadiers of the Guard, was shocked to see young women reduced to the extremity of selling their sexual favours to French officers in order to be able to feed themselves, and indeed to protect themselves from the attentions of unruly soldiers. "I often found during my walks through the city old men weeping to see this awful immorality," he wrote. "I did not know their language well enough to be able to console them, but I would point to the heavens and then they would come and kiss my hands and conduct me to where their families were huddled in the ruins, moaning from hunger and misery."[7]

The worst disorders had largely died down with the fire. Looting became a clandestine activity, carried on at night or in out-of-the-way burnt-out quarters of the city. The frenzied need to save things from the flames had given way to more methodical rummaging. The French soldiers carried it on jointly with abject locals who found a role for themselves as guides and procurers. Violence against citizens and rape also declined, and in several Russian accounts there are instances of young girls pushing away would-be molesters with impunity.[8] Moscow was a huge and sprawling city, and parts of it remained dangerous, particularly at night. Yet a somewhat bizarre *modus vivendi* had evolved between the various groups living side by side in the ruined city.

The best guarantee of safety for the inhabitants was to have a high-ranking officer in residence. One servant girl recalled that there was no trouble of any sort while a French officer took up quarters in the house, but the moment he left the place was looted thoroughly by Russians. Another Muscovite would send a servant to alert the aides of a French marshal who had quarters nearby whenever a gang tried to loot his house, and an armed patrol would immediately be despatched to arrest the miscreants.[9]

G.A. Kozlovsky, the son of a landowner from Kaluga who was stranded in Moscow when the French arrived, made friends with some French officers, ate and played chess with them. The only risks he ran were at the hands of the city's inhabitants who had stayed behind. "In those days, one feared the Russian peasants more than the French," he recalled. "In almost all the houses we went into there were still women, children and old people, mostly servants it is true, as the masters had left," remembered Jean Michel Chevalier. "Not only were they respected and protected by us, but even fed, for we shared with them anything we could get." The painter Albrecht Adam moved in with a Russian whom he treated politely, and they made common cause of finding food and the other necessities of life. A group of Italian soldiers became so fond of their "hosts" that when the time came for leavetaking, there were tears on both sides. One French soldier who found a poor Russian woman squatting in some ruins about to give birth, brought her to his lodgings and fed her. And the hardly belligerent Stendhal actually drew his sword against a drunken French soldier who was mistreating a Russian civilian.[10]

A mounted grenadier of the Guard named Braux came across a great visitors' book of the city council, and wrote in it, in such bad French that it would be impossible to reproduce its tone: "There is not one Frenchman who is not desperately saddened by the misfortune which has befallen your lovely Moscow. I can assure you that as far as

I am concerned, I weep for it and regret it, for it was worthy of being preserved. If you had stayed at home it would have been preserved. Weep, weep, Russians, over the misfortune of your country. You alone are the author of all the ills that it endures."[11]

The French were very impressed by the city and its many fine buildings. Dr Larrey thought the hospitals "worthy of the most civilised nation on earth," and was of the opinion that the foundling hospital was "without argument the grandest and the finest establishment of its kind anywhere in Europe." And they all wrote admiringly of the fine palaces, many of which succumbed to the fire. "Even the French, so proud of their Paris, are surprised at the size of Moscow, of its magnificence, of the elegance of life here, of the wealth we have found here, even though the city was almost entirely evacuated," a Polish officer wrote to his wife. Louis Gardier, Adjutant Major in the 111th of the Line, also thought it very fine, but was shocked by Muscovite morals. "As an eyewitness, I can say that I have never seen so many indecent pictures and furnishings," he wrote, "and lewdness was on display in particularly disgusting ways in the houses of the great."[12]

Although a large part of the city had been destroyed, those troops stationed in Moscow itself managed to make themselves quite comfortable. "I found quarters in the palace of Prince Lobanov," recalled Dezydery Chłapowski of the Chevau-Légers of the Guard. "General Krasiński took up his quarters opposite, in the house of the merchant Barishnikov. Both of these houses were very well appointed, everything was in order, both upstairs and down there were very comfortable wide beds with morocco-covered mattresses. Behind the palace were outbuildings, haylofts, a garden with an orangery and, beyond, a field and a kitchen garden. The front of the palace was in town, the back seemed to be in the country. There were about a hundred Muscovites in the two ranges of outbuildings, including servants, craftsmen and peasants, whom we found very helpful in everything. Our soldiers gave them work, which they needed. The behaviour of these people towards us was very calm and civil."[13]

"In spite of the disasters, the fire of Moscow and the flight of the inhabitants, the army is quite comfortable here and has found immense supplies of victuals and even wine," General Morand, who was recovering from the wound received as he stormed the Raevsky redoubt, wrote to his "*Émilie adorée.*" "My division is quartered in a very large building, and I have a very fine and very comfortable house nearby on a large square . . . I await with impatience news of your confinement, may the good Lord protect you as he has protected me in battle . . ."[14]

Baron Paul de Bourgoing found billets in Rostopchin's palace, and spent happy hours browsing in the Count's magnificent library. One day he came across an edition of a book written by his father. "It is with real pleasure that the son of the author has found one of his father's books so far from his fatherland," he wrote in the flyleaf. "He only regrets that it should be war that brought him here."[15]

B.T. Duverger, paymaster of the Compans division, installed himself in the house of some German inhabitants of Moscow, and lived quite happily, with the Italian Guard parading outside his windows to good regimental music in the mornings. "I was rich in furs and paintings; I was rich in cases of figs, in coffee, in liqueurs, in macaroons, in smoked fish and meats," he noted, "but of white bread, fresh meat and ordinary wine, I had none." There were twelve of them in the house altogether, and as they sat down to dinner they would drink a toast to next year's campaign and their entry into St Petersburg.[16]

"The grenadiers went out and found us some table linen and household items; others furnished us with provisions of every kind; the flocks of cattle which have rejoined the army are providing us with meat; our bakers are making bread with flour found under the ashes; in a word, the army has everything it needs in spite of Rostopchin," wrote Captain Fantin des Odoards. In order to provide themselves with vitamins through the winter, the more provident set about making sauerkraut out of the cabbages in which the city's numerous kitchen gardens abounded.[17]

The soldiers employed the various cobblers and tailors left in the city to repair their uniforms or make new boots. They also stocked up on essentials at the markets that had sprung up, where they could buy things salvaged or looted by others. The Grenadiers of the Guard, who entered into the city early on and had had ample opportunity to lay hands on every manner of goods when they were detailed to extinguish the fire of the principal trading bazaar, had set up a market outside the Kremlin where an astonishing array of victuals and goods could be had. But although they had managed to corner the market in some types of commodity, stalls sprang up all over the city. "The streets which had been spared by the fire resembled real markets, with the peculiarity that all those taking part, merchants and customers, were all soldiers," noted Lubin Griois. Another peculiarity was that the troops found it more convenient to barter than to use money, so everyone involved was wandering about with an extraordinary array of objects and delicacies. Frenchmen could sample finer French wines and cognacs than they would ever be able to afford at home, and one had his first taste of a pineapple in Moscow.[18]

Much of the mercantile activity was driven on the one hand by the need of soldiers to make some money or to provide themselves with objects that would be saleable back home, and on the other by the desire to find presents for wives, mistresses and sisters. What they all wanted was fine furs, for which Russia was famous, and the woven cashmere shawls imported from Persia and India that were a fashionable and indeed essential accessory to the high-waisted but low-cut empire-style dresses. The Continental System had sent the price of both rocketing in Paris. But furs and shawls were not normally stored in cellars, so a large part of the city's stock had gone up in flames.

General Compans, who was recovering from the wound he received while leading his division's attack on Bagration's *flèches* at Borodino, was newly married and eager to shower his young wife with presents. But, as he wrote to her, he was finding it very difficult, even though he had several people on to the job. On 14 October he was at last able to write:

> *Here,* ma bonne amie, *is what I have been able to procure in the way of furs:*
>
> *One large fur of black and red foxes, in alternating bands;*
>
> *One large fur of blue and red foxes, in alternating bands;*
>
> *That is how they assemble fox furs in this country, when they are not using them merely as trimming. These two furs are new and are considered to be very fine.*
>
> *One large collar of silver-grey fox;*
>
> *One collar of black fox;*
>
> *Both of them are very beautiful, but too small for you to make much use of for yourself, but I could not find anything else in that line;*
>
> *Enough sable for two or three trimmings for furs as large as the one in chinchilla which you bought in Hamburg;*
>
> *A large muff in grey-black fox made up of choice pieces sewn together in little bands of an inch and a half in width. This muff is highly regarded here; it must have taken quite a few fox furs, much silk and a great deal of work to make up such a muff. I think you could probably use it either as a trimming or as a cape. All of this, my dearest Louise, will be packed in a trunk, and I will seize the first possible opportunity to have it delivered to you.*[19]

Whether the furs ever reached her is doubtful – the letter did not, as it was picked up by marauding cossacks after they attacked a courier.

Fur fever gripped men of every station. "I have made the acquisition of an extremely fine pelisse in fox fur backed with a very beautiful violet satin," Lieutenant Paradis of the 25th of the Line wrote to his mistress. "I would very much like to send it to you, but I do not know how to go about it. As you can imagine, the object is rather voluminous." Colonel Parguez, chief of staff to the 1st Division in Davout's corps, suggested his wife send one of her maids over to collect the "six dozen fine sables, all ready and perfect to trim at least six pelisses." The girl could be back in Paris with them by 1 January, in time for her to wear them in the New Year.[20]

Guillaume Peyrusse, paymaster to Napoleon's household, encountered terrible difficulties in getting hold of any of the things his wife longed for. "Try as I might, I have been able to find neither piqué, nor muslin, nor cashmere shawl . . . Nothing delicate in the way of lady's furs . . . Not a print, not a view of Moscow, not a medal, not the slightest curio of any sort." This was particularly galling, as he had been given a whole list of items by not only his wife, but his sister-in-law and various other members of his family. Many others, including Marshal Davout himself, complained of the difficulty of getting hold of good stuff for their womenfolk. "In Moscow, even at court, the conversation turned on nothing except foxes, rabbits and sables," as Eustachy Sanguszko put it.[21]

The more culturally curious explored what was left of the city, visiting the Kremlin and the tombs of the Tsars, which had been ripped apart by looters. Vionnet de Maringoné found a functioning "*banya*" which he frequented with much pleasure. Colonel Louis Lejeune met his sister, who had been living in Russia for twenty years. Others struck up acquaintance with the French residents of Moscow, though some old revolutionary soldiers sneered at them as "*émigrés*," and with various other foreign residents, including Germans, Italians, and even some English, who did their best to entertain the invaders.[22]

A troupe of French actors resident in Moscow had stayed behind, and they gave a series of performances of light comedies by Marivaux and others. They played not in any of the public theatres, which had been reduced to ashes, but in the private court theatre of an aristocratic palace. "You cannot imagine through what magnificent salons we passed in order to get to the theatre," Major Strzyżewski wrote to his wife in Warsaw. "I was entranced by everything I saw. In one of the drawing rooms, I thought particularly of you, since it was filled with the most beautiful flowers." He judged the actors "passable." In the interval the spectators were served refreshments by grenadiers of the Old Guard.[23]

Some arranged their own entertainments. Napoleon did not go to the theatre, but did attend a recital given for him at the Kremlin by the singer Signor Tarquinio. The twenty-seven-year-old Sergeant Adrien Bourgogne and his messmates had, in the process of providing themselves with the necessities of life, amassed quite a wardrobe of rich court clothing, some of it dating back to the previous century, which they had found in abandoned palaces. One evening they and the Russian trollops who had moved in with them prinked themselves up in this finery, with the regimental barber dressing their hair and making them up. They then held a ball and danced to the sound of fife and drum, the whores dolled up as eighteenth-century marquises high-kicking and causing a great deal of mirth.[24]

Although Moscow boasted a French Catholic church, St Louis des Français, whose parish priest, the Abbé Surrugues, had remained at his post, churchgoing did not figure among the activities of the soldiers. A handful of officers, mostly from aristocratic backgrounds, came to mass or confession, and the Abbé was only asked to give Christian burial on two occasions. He went around the hospitals to talk to the wounded, but found them interested only in their physical wants, not their spiritual needs. "They do not seem to believe in an afterlife," Father Surrugues wrote. "I baptised several infants born to soldiers, which is the only thing they still care about, and I was treated with respect."[25]

While Napoleon held frequent reviews, at which his troops looked their best, he had not once since reaching Moscow inspected their bivouacs or quarters, with the result that he had no idea of their real state of mind and body. At Petrovskoie, where a large part of Prince Eugène's 4th Corps was stationed, the generals had installed themselves in the summer residences of wealthy Muscovites, the officers in various pavilions, follies and summer houses scattered around their parks, and the soldiers in the surrounding fields. They sat around their campfires on fine furniture rescued from some gutted palace, eating their gruel off silver plate and drinking the finest wines from precious goblets. "Our actual poverty was masked by an apparent abundance," observed an officer on Prince Eugène's staff. "We had neither bread nor meat, and our tables were covered with preserves and sweets; tea, liqueurs and wines of every kind, served in fine porcelain or in crystal vessels, showed how close luxury was to poverty in our case."[26]

Junot's Westphalian 8th Corps, stuck out at Mozhaisk, also suffered from a lack of decent housing and continual shortages of food. The men would come into Moscow whenever they could in order to buy necessary supplies, but as they were obliged to purchase them from the looters, they had to pay high prices.

Undoubtedly the worst off in every way were Murat's cavalry and Poniatowski's 5th Corps, stationed to the south of Moscow, around Voronovo and Vinkovo, in close proximity to Kutuzov's camp at Tarutino. It was an unusual situation. An unspoken armistice had come into existence, with both sides merely keeping an eye on each other. On one occasion, some French foragers came across a herd of cattle in the no-man's-land between them, and divided the booty up amicably with the Russians. On another Murat himself rode over to some Russian pickets and told them it would be more convenient if they moved a few hundred yards further back, which they obligingly did. One day he had a chat with Miloradovich, who was inspecting his outposts. Whenever Murat rode out in his operatic costume, the cossacks would greet him with shouts of "The King, the King!" As a sign of respect for his reckless courage, they never fired on him, and in his naivety Murat seems to have fancied that he could subvert these wild children of the steppe. Officers on outpost duty would pass the time of day talking to their opposite numbers, exchanging prognoses about the war and debating whether they would soon all be off to India together. The French believed that it was only a matter of time before peace was signed, and the passage of Lauriston through their camp on his way to see Kutuzov only confirmed them in their conviction.[27]

But the conditions in which the French stationed here waited for the hoped-for peace were terrible. They were mostly camped out in the open fields, with no shelter to protect them from the rain and the cold. They slept on improvised beds of straw or branches under the stars, or under a *caisson* or carriage. The autumn days were cold, even if it was sunny, and at night there was always a frost. There was a severe shortage of food, and unlike their comrades stationed in or around the city, the men here could not go to Moscow, some eighty kilometres distant, to stock up.

The bivouac of the Polish Chevau-Légers at Voronovo was better than most. They had taken over the ruins of Rostopchin's magnificent country house, which he had left with a large notice stating that although he had spent years building and planting his estate, he had personally burnt it down lest it provide shelter or comfort for the French invaders. Some of the officers had made makeshift tents in the ruins or squeezed into peasant huts in the village, while the men sheltered wherever they could find even a wall to shield them from the wind. The regimental *cantinières* created a café with surviving pieces of furniture, including one fine sofa from the palace, and the men sat around drinking coffee from a bizarre array of gold, silver and china vessels, discussing the campaign and listening to General Colbert, who commanded the division, and his two aides-de-camp sing airs from Paris vaudeville in the evenings.

Men and horses wasted away at an alarming rate in these conditions, and the words "corps," "division" and "regiment" are highly misleading when considering the state of the French cavalry by the middle of October. The 3rd Cavalry Corps, consisting of eleven regiments, could only muster seven hundred horsemen. The 1st Regiment of Chasseurs could only field fifty-eight, and that only thanks to some reinforcements which had reached it from France. Squadrons in the 2nd Cuirassiers, usually 130 strong, were down to eighteen or twenty-four men. General Thielmann's Saxon brigade was down to fifty horses.[28]

The condition of the horses was dreadful, and by mid-October many of them were "entirely spoiled," in the words of Lieutenant Henryk Dembiński of the 5th Polish Mounted Rifles. "It was so bad that, even though we had folded blankets to the thickness of sixteen, their backs had rotted through completely, so much so that the rot had eaten through the saddlecloth, with the result that when a trooper dismounted, you could see the horse's entrails."[29]

What is truly extraordinary is the degree to which even in these conditions Napoleon inspired absolute confidence in his men. As they sat around in camp with nothing to do, they endlessly discussed the situation. "We could see that we were slowly perishing, but our faith in the genius of Napoleon, in his many years of triumph, was so unbounded that these conversations always ended with the conclusion that he must know what he is doing better than us," recalls Lieutenant Dembiński.[30]

Most of them were anxious – at being so far away from home, at the state of the army, at the lack of food, at the general turn events had taken. "But all our reflections did not give us the slightest fear: Napoleon is there," as Captain Fantin des Odoards put it. Among the letters found strewn in the road after a courier had been ambushed by cossacks was one from the Comte de Ségur, dated 16 October, telling his wife in the tenderest tones how much he loved and missed her, and discussing the progress of the tree-planting programme he had initiated in the park of his château.[31]

Many of them were convinced that Napoleon was bent on a march to India. "We are expecting to leave soon," noted Boniface de Castellane on 5 October. "There is talk of going to India. We have such confidence that we do not reason as to the possibility of success of such an enterprise, but only on the number of months of marching necessary, on the time letters would take to come from France. We are accustomed to the infallibility of the Emperor and the success of his projects." Others fantasised about liberating girls from the Sultan's seraglio, one dreaming of a Circassian girl, another of a Greek, another of a Georgian. "After a good treaty of alliance with Alexander, who willy-nilly will be dragged along with us like the others, we will go to Constantinople next year and from there to India," one officer wrote home. "It is only loaded down with the diamonds of Golconda and the cloths of Kashmir that the Grande Armée will return to France!"[32]

At the beginning of October, Murat sent his aide-de-camp General Rossetti to Moscow to inform Napoleon personally of the critical condition of the cavalry and of his exposed position. But Napoleon dismissed his report, saying that the Russians were too weak to attack. "My army is finer than ever," he told Rossetti. "A few days of rest have done it the greatest good." That was perhaps true of the troops which paraded before him in Moscow, but certainly not of the cavalry. On 10 October Murat wrote to General Bélliard on Berthier's staff, urging him to get the truth through to the Emperor. "My Dear Bélliard," he wrote, "my position is atrocious. I have the whole enemy army in front of me. Our advance guard is reduced to nothing; it is starving, and it is no longer possible to go foraging without the virtual certainty of capture. Not a day passes without me losing two hundred men in this way."[33]

Napoleon was far too astute not to realise that his strategy had gone badly wrong, and that Caulaincourt had been right all along. But he did not like to admit it. And he recoiled from the only logical next step, which was to withdraw. He liked neither the idea of retreat, which went against his instincts, nor the implications of such a withdrawal on the political climate in Europe. He also had an extraordinary capacity for making himself believe something just by decreeing it to be true. "In many a circumstance, to wish something and believe it were for him one and the same thing," in the words of General Bourienne.[34] So he hung on, believing that Alexander's nerve would break or that his own proverbial luck would come up with something.

He had studied weather charts, which told him that it did not get really cold until the beginning of December, so he did not feel any sense of urgency. What he did not realise, in common with many who do not know those climates, was just how sudden and savage changes of temperature can be, and how temperature is only one factor, which along with wind, water and terrain can turn nature into a viciously powerful opponent.

The unusually fine weather at the beginning of October contributed to his complacency. He teased Caulaincourt, accusing him of peddling stories about the Russian winter invented to "frighten children." "Caulaincourt thinks he's frozen already," he quipped. He kept saying that it was warmer than Fontainebleau at that time of year, and dismissed suggestions that the army provide itself with gloves and items of warm clothing. He was not alone in his delusions. "We have been having the most wonderful weather for the past few days, which could not have been finer in France at this time of year," Davout wrote to his wife. "In general, people exaggerate the harshness of the climate here."[35]

With every day Napoleon spent in Moscow, the harder it was to leave without loss of face, and the usually decisive Emperor became immobilised by the need to choose between an unappealing range of options on the one hand, and stubborn belief in his lucky star on the other. He fell into the trap of thinking that by delaying a decision he was leaving his options open. In fact, he only really had one option, and he was reducing the chances of its success with every day he delayed.

On 12 October the daily *estafette* from Moscow to Paris was attacked and captured between Moscow and Mozhaisk, and on the following day the one coming from Paris was intercepted. General Ferrières, who had travelled all the way from Cadiz, was captured almost at the gates of Moscow. These events shook Napoleon, and the gravity of his position was underlined by the first light shower of snow, on 13 October, which covered the ruins of Moscow and the surrounding countryside in a blanket of brilliant white.

"Let us make haste," he said on seeing the snow. "We must be in winter quarters in twenty days' time."[36] It was a bit late, but by no means too late. Smolensk, where he had some supplies, was only ten to twelve days' march from Moscow, and his well-stocked bases at Minsk and Vilna were only another ten and fifteen respectively from there. Having reached these, his army would be well fed and supplied, safe in friendly country and able to draw on reinforcements from the depots he had built up in Poland and Prussia. In the spring he would be able to march on St Petersburg or any other point he chose.

A withdrawal is always a risky enterprise, as it can easily turn into a flight, but there are ways of limiting the damage, and in this case the first imperative was to ensure mobility by travelling as light as possible. Only this could have given Napoleon the initiative, even as he withdrew. And the need to jettison things along the way tended to lower the morale of the retreating force while raising that of the pursuers. Expediency therefore demanded that he send on ahead or leave behind as much as possible, in terms of people and equipment.

But as with every other aspect of this campaign, political imperatives prevented him from taking the course dictated by military considerations, not to say common sense. His original assumption that his occupation of Moscow would produce peace meant that instead of regarding the city as a forward position he treated it as a base. The transportable wounded of Borodino were not sent back to Smolensk and Vilna to recuperate, but brought forward to Moscow. It was only on 5 October that he gave orders that the movable wounded still at Mozhaisk, Kolotskoie and Gzhatsk be gradually taken back to Smolensk, and not until 10 October that a first convoy of wounded left Moscow. Had he begun this process a week earlier, thousands of men of all ranks would have survived. Those who did get sent back in the first week of October travelled unmolested in perfectly good conditions all the way to Paris. Stendhal, who left Moscow with a convoy of wounded as late as 16 October, got through to Smolensk without problem. They were harassed by cossacks, but not enough to disturb him in his reading of Madame du Deffand's *Lettres*. Even the trophies – banners, regalia and treasures from the Kremlin, the great silver-gilt cross Napoleon ordered to be wrenched from the dome of the tower of Ivan the Great, which he intended to erect in Paris – were not sent on ahead.

He kept ordering all available reinforcements forward, rather than building up reserves along his line of retreat. It was only on 14 October, the day after the first snowfall, that he gave orders that no more troops were to be sent forward to Moscow, but ordered back to Smolensk, and that the remaining wounded in Moscow be evacuated immediately, those from Mozhaisk and Kolotskoie by 20 October, those from Gzhatsk two days later.

The more seriously wounded, of whom there may have been as many as 12,000, should have been left where they were, which was what Dr Larrey had intended – he even left medical teams, supplemented by French inhabitants of Moscow. Dr La Flise was horrified when the order came to transport them, realising that they would mostly perish from the buffetings on the road even if they escaped the pikes of marauding cossacks.[37]

Napoleon fixed on 19 October as the date he would leave Moscow, later rescheduling it for 20 October. But even then, various political considerations vitiated sensible preparations. He could retreat straight back down the road along which he had come, which had the advantages of being familiar, guarded by French units and punctuated with supply depots, as well as being the most direct. The only disadvantage of this road was that the country alongside it had been ravaged by the advance and would not provide much in the way of sustenance. Napoleon therefore asked General Baraguay d'Hilliers, stationed at Smolensk, to identify two side roads running parallel to the main road, so that some elements of his army could march through virgin country.

But going back the way he had come would be tantamount to an admission that he was retreating. He considered marching northwestward, through Volokolamsk, where he could crush Wintzingerode's detachment, to join up with Victor and St Cyr at Vitebsk, where he could attack Wittgenstein and whence, if necessary, he could withdraw to Vilna. This option had the merit that it would threaten St Petersburg, which might just cause Alexander's nerve to snap. Or he could march southwards, strike a blow at Kutuzov, and then march back to Minsk via Kaluga or Medyn. This option had the disadvantage that if he fell back on Smolensk even after defeating Kutuzov, it would look like flight. So he entertained the possibility of returning to Moscow after defeating him. Instead of evacuating the city, he therefore gave orders for Davout's, Mortier's and Ney's corps to gather up and stockpile three months' worth of rations and six months' worth of stewed cabbage, to improve the defences of the Kremlin and turn all the monasteries into strongpoints, which were to be held by horseless cavalrymen armed with muskets "during the absence of the army." A lasrge part of his household was also to remain in Moscow when he moved out.[38]

"It is possible that I may return to Moscow," Napoleon wrote, as late as 18 October, to General Lariboisière, inspector-general of the artillery, who was worried at the vast quantities of equipment stockpiled there. "So nothing that could be of use must be destroyed." When, in the end, Moscow did have to be evacuated, Lariboisière would have to burn five hundred *caissons*, 60,000 muskets and several hundred thousand measures of powder before leaving. In the absence of a sufficient number of horses to pull the guns, he also wanted the useless four-pounders destroyed, but Napoleon felt this smelt of defeat.[39]

Napoleon should have sent back all his dismounted cavalrymen. There were several thousand of these by the time he reached Moscow, and the number grew daily. Instead, he ordered them to be formed up into dismounted units and issued with carbines, which was pointless. The men did not know how to and did not wish to fight as infantry, they did not know the drill, and they could feel no *esprit de corps* in these units. "The worst infantry regiment is much more effective than four regiments of dismounted cavalry," wrote Boniface de Castellane. "They bleat like donkeys that they were not made for this work."[40]

Colonel Antoine Marbot, commanding the 23rd Chasseurs à Cheval under St Cyr, ignored orders to keep all dismounted men near the front, and sent his all the way back to Warsaw, where he knew they would find horses; in this way he had 250 well-mounted men ready for action at the end of the campaign, while all the dismounted cavalrymen who remained with St Cyr were taken prisoner. "It would have been so easy throughout the summer and autumn to send men to Warsaw, whose remount depot had plenty of horses but no riders," he wrote.[41] Had Napoleon evacuated his horseless cavalrymen just one week ahead of the army, he would have had the cavalry whose lack was to rob him of victory in 1813 and 1814.

Colonel Marbot also made his men acquire rough sheepskin coats from the local peasants at the beginning of September, thereby saving the lives of a great many of them. The Colonel of the Polish Chevau-Légers of the Guard did the same, and as Master of the Horse, Caulaincourt made all the riders, grooms and drivers under his command provide themselves with not only sheepskin coats, but also gloves and fur hats.[42]

Other officers showed similar prescience, but usually only with respect to themselves, investing in good fur coats (as opposed to fancy items for the ladies back home), fur-lined overboots, gloves and fur caps. Lieutenant Henckens of the 6th Chasseurs bought some small pieces of fur which he got one of his men, a tailor by trade, to make up into a vest, to be worn under his shirt. Colonel Parguez proudly informed his wife in a letter that he had had a pair of bearskin boots made up with the fur on the inside.[43]

Captain Louis Bro of the Chasseurs à Cheval of the Guard was taking no chances. "I bought two little cossack horses used to surviving off straw and the branches of pine trees. They carried my personal effects and a hundred kilograms of reserve victuals, principally chocolate and *eau-de-vie*; I foresaw that my exhausted French horse would not go far. The two horses, shod with steel, would take me all the way to the Niemen. I also furnished myself with a fur-lined cloak, a fox fur, a fur-lined cap, felt boots, and resin bricks which would allow me to light a fire at any moment."[44]

Louis Lagneau, a surgeon with the Young Guard, had taken the precaution of having a small tent made in Moscow, in which he and three colleagues would be able to sleep in relative warmth and shelter even in the coldest conditions – it got perfectly warm inside with the four of them. And Antoine Augustin Pion des Loches, recently promoted to the rank of colonel in the foot artillery of the Guard, prepared himself against all eventualities. In his small wagon he packed a hundred large dry biscuits, a sack of flour, three hundred bottles of wine, twenty to thirty bottles of rum and other spirits, ten pounds of tea, ten pounds of coffee, a large quantity of candles, and "in the event of winter quarters east of the Niemen, which I felt to be inevitable, a case containing quite a fine edition of the works of Voltaire and Rousseau; a *History of Russia* by Le Cler; and that by Levesque, the plays of Molière, the works of Piron, *de l'Ésprit des Lois* and a few other works, such as Raynal's *Histoire philosophique*, all bound in white calf and gilt-edged."[45]

Yet while people such as these were clever enough to equip themselves with the means of survival, there was not a single order given from the top, not even at corps command or divisional level, to take appropriate measures to protect the troops during the forthcoming operations. Good commanders such as Davout made sure the soldiers' uniforms and boots were repaired, but that was as far as it went. And whatever other measures they might have taken would in any case have been largely nullified by one great omission, which was to cost tens of thousands of lives and turn a potentially orderly retreat into a tragic rout.

The moment they had come to rest at Moscow, all Polish units set up forges and began making horseshoes with sharp crampons in preparation for winter. They told their French comrades to do likewise, but their advice bounced off a wall of Gallic unconcern. "The stubbornness and arrogance of the French, who felt that having been through so many wars they knew better than everyone else and did not need their advice, did not allow them to sharp-shoe their horses," wrote Józef Grabowski, a Polish officer attached to imperial headquarters. Luckily for Napoleon, Caulaincourt, who had seen several Russian winters, took it upon himself to have all the horses of his household properly shod. But when it was suggested the same measures be ordered throughout the army, the Emperor dismissed it – with fatal consequences for him and his whole army.[46]

17

The March to Nowhere

Napoleon's military success in the past had rested on his capacity to make a quick appraisal of any situation and to act intelligently and decisively on its basis. Yet from the moment he set out on his "Second Polish War" he displayed a marked inability either to make the correct appraisal or to act decisively. There were probably many reasons for this, and without doubt one of them was a difficulty in comprehending what his opponents were trying to achieve.

The Russians had spent a year and a half deploying for an offensive, only to retreat the moment operations began. This at first led Napoleon to expect a trap, and then to assume that they were avoiding battle out of fear of losing. He was not to know that most of it was the result of chaos and intrigue at Russian headquarters. When they did stand and fight, at Borodino, he defeated them, and since they then gave up their capital, he had assumed that they were beaten. Kutuzov's passivity over the next few weeks appeared to confirm this, and by giving him a false sense of security ultimately contributed to his defeat.

The more romantically minded historians have tried to make out that the Russian Field Marshal's inaction was a clever ploy to lull Napoleon into staying in Moscow as long as possible so as to ensure that he would be caught by the dread Russian winter. This may indeed have been the case. Or it may have been that Kutuzov simply did not know what to do, and was afraid of doing the wrong thing. That is certainly what many at his headquarters thought.

Soon after his arrival at Tarutino he began to receive reinforcements. Over the summer months 174,800 regulars, 31,500 irregulars, mostly cossacks, and 62,300 militia were fed into the armies operating against Napoleon, and Kutuzov received the lion's share. His army was growing stronger with every day that passed, and in the four weeks it spent at Tarutino the force of no more than about 40,000 tired and dispirited troops had grown to 88,386 regulars, with 622 guns, supplemented by 13,000 Don cossacks and 15,000 irregular cossacks and Bashkir cavalry.[1] Kutuzov knew that as his forces grew, those of the French dwindled. Every day his patrols brought in scores of French foragers, marauders and deserters, from whom he knew that the Grande Armée was suffering from an acute shortage of fodder and fresh food, and that morale was sinking. He was contributing to this process through the "flying detachments" he had organised. These hovered on the fringes of the area occupied by the French, catching any who ventured out and occasionally swooping in to snatch a small detachment or supply train. In all, the French lost some 15,500 men in this manner during their five-week stay in Moscow.[2]

But Kutuzov was in no hurry to engage in regular warfare. Alexander's masterplan, brought from St Petersburg by Colonel Chernyshev, which envisaged a great encirclement of Napoleon as he fell back from Moscow by the combined armies of Kutuzov, Tormasov, Chichagov, Wittgenstein and Steinheil, was politely discussed and shelved on the grounds that the Russian army was not strong enough to face the French yet.[3]

This was true enough, as, apart from giving the troops time to rest and drilling new recruits, Kutuzov had done little to prepare for offensive operations. He amalgamated the First and Second Armies, leaving Barclay with the title of commander but no actual role. He continued to foster chaos by his idiosyncratic style of command, in which he was ably seconded by his new chief of staff, General

Konovnitsin, who spent his days smoking his pipe and chatting with brother-officers, but refused to sign a single order, according to Captain Maievsky, one of Kutuzov's duty officers. "Fearful for his reputation, Konovnitsin wanted, it seems, to appear highly active; but not being up to taking in the whole, he correctly assumed that everything that was good would be ascribed to him, and everything that was bad would be attributed to the Field Marshal," according to Maievsky. "But it seems that Kutuzov employed the same tactic, with the opposite aim." Konovnitsin was heartily loathed by Yermolov, who was still chief of staff to Barclay, who commanded the combined armies, and who found the atmosphere at headquarters so poisonous that he moved to a village some way from the camp and only came in when summoned.[4]

At the end of September, unable to stand the situation any longer, Barclay wrote a long and tearful letter to the Tsar in which he warned him that the army was being commanded by a nonentity and a "bandit," and resigned his command. "My health is ruined and my moral and physical strength have run out," he explained. He felt hard done by: his strategy had been fully vindicated, yet he had been denied credit for it. He had saved the day at Borodino, yet this had been acknowledged by nobody. To add injury to insult, his carriage was pelted with stones by the mob as he drove through Kaluga after leaving the army.[5]

Barclay's departure did nothing to ease tensions at Russian headquarters. Bennigsen, hoping to be appointed to command in Kutuzov's place, was ringleading a cabal which accused him of idleness, dissipation and cowardice. And although the old Field Marshal still enjoyed huge respect and love from the rank and file, even junior officers were beginning to ask themselves whether Bennigsen and his supporters might not have a point. After a couple of weeks' rest they were bored and ready for action, while the gossip swirling around the camp further unsettled them. "I have listened for so long to so many opinions and disturbing rumours that I no longer know whom or what to believe," Lieutenant Aleksandr Chicherin noted in his diary.[6]

News of the fall of Madrid to the British forces reached Tarutino at the beginning of October, but the wild rejoicings were followed by questions about why they were standing by while others were fighting the French. The old spectre of treason was ever-present. And it began to assume a degree of substance when a French staff officer presented himself at the Russian pickets with a letter from Berthier to Kutuzov. This informed the Field Marshal that General Lauriston had reached Murat's camp and requested an interview. Kutuzov agreed to meet Lauriston in the no-man's-land between the two armies at midnight, no doubt hoping that this would ensure secrecy.

But rumours were soon flying around the Russian camp, and Bennigsen, who had convinced himself that his superior was about to do some kind of deal with the French, alerted Wilson to what was going on. The English busybody needed little incitement and went straight to Kutuzov's quarters, where he subjected the Field Marshal to a harangue about not betraying the Tsar's wishes and the Russian cause. Whatever he may have thought of the importunate Briton, the Russian commander could not now proceed with his plan, so he sent in his place the Tsar's aide-de-camp Prince Piotr Volkonsky, who had recently arrived from St Petersburg. Lauriston was disappointed not to find Kutuzov at the appointed meeting, and protested that he was the bearer of a letter from Napoleon to the Russian commander which he could only hand him in person. This caused further delay, and in the end Lauriston was brought into the Russian camp and into Kutuzov's log cabin.

The purpose of Lauriston's mission was to ask for a safe conduct so that he might go to see the Tsar in St Petersburg. Kutuzov answered that he had no authority to grant one, and could do no more than write to his sovereign passing on the French request. Lauriston and Napoleon would have to wait patiently for an answer. Lauriston asked whether a ceasefire could not be arranged while they waited. He also suggested an exchange of prisoners, and that measures be taken to curb the violence developing on both sides. Kutuzov was polite but negative, pointing out that it was not the invader's business to complain about the behaviour of the ravaged population. The Frenchman's mission came to nothing. But it had set tongues wagging against the Field Marshal – so much so that he actually had an officer arrested for spreading the rumour that he was in talks with the French – and it was to earn him a stern rebuke from Alexander, who had given strict instructions not to enter into any kind of communication with the enemy.[7]

Alexander had been urging Kutuzov to attack exposed units of the French army, and could not understand why the Field Marshal had not moved against Murat at Voronovo and Vinkovo.[8] Bennigsen, Yermolov, Platov, Baggovut and others had also been begging Kutuzov to seize the opportunity to destroy this force, numbering no more than about 25,000, camped right under their noses, and which had been lulled into dispensing with most of the precautions observed when in the face of the enemy. Considering that the Russian army now numbered more than 100,000, this should have been about as risky as a field day. Kutuzov had resisted them all, but he could no longer oppose what had become the general wish of the whole army, so he finally agreed.

He fixed the date of the attack for 17 October, but he did so at short notice, with his usual disregard for correct channels. Yermolov could not be traced in time to pass on orders to some of the units (possibly because it was Konovnitsin who had summoned him), with the result that when Kutuzov rode out to take command of the operation he found that half of the relevant units were going about their daily business of foraging and cooking rather than standing to for battle. He flew into a violent rage and threatened terrible retribution on Yermolov and various others, before rescheduling the attack for the next day.

But on 18 October it was Kutuzov who failed to turn up on time, and Bennigsen therefore took charge of the assault. Baggovut moved against Murat's left flank, while Orlov-Denisov appeared on his right. The surprise was complete, and Orlov-Denisov's cossacks swept into General Sébastiani's camp while its occupants were mostly still asleep, taking hundreds prisoner while the rest took to their horses and fled. Instead of driving home their advantage, the cossacks then fell to looting the camp, which allowed the French to rally and counter-attack. They were nevertheless surrounded and outnumbered. But the Russians failed to take advantage of the situation. Miloradovich's corps stood idly by while Bennigsen's pleas to Kutuzov for reinforcements went unanswered, and the various Russian units manoeuvred without purpose while the French, having rallied under cover of a strong stand by Poniatowski's corps, made an orderly retreat.

What should have been a walkover turned into a fiasco, and while the French lost some 2500 men, thirty-six guns and one regimental colour, the Russians lost over a thousand men, including General Baggovut. They boasted that the King of Naples had fled so fast that he had been obliged to leave behind even his wardrobe and his plate.* Yet they had failed to destroy a single French unit. And the episode ratcheted up the conflicts within the Russian command. Kutuzov called Bennigsen an "imbecile" and a "red-headed coward" (the word for "red-head," *ryzhy*, also means "clown"), and the other responded in kind.[9]

The disappointing outcome of the engagement did not prevent Kutuzov, who had never actually shown up on the battlefield, from trumpeting to all and sundry that he had taken on Murat's army of 50,000 men and routed it. The French had "run like hares," he wrote in one letter. The account he sent to St Petersburg was so boastful that the victory was honoured with two days of gun salutes and illuminations, and earned him a ceremonial sword with laurel leaves.[11]

Yet there was no follow-up. Officers cursed and soldiers wondered why they were still not being allowed to get at the French, but as it was getting colder they began to build dug-outs for themselves. Many began to think that Kutuzov really was frightened of confronting Napoleon. Nikolai Durnovo, an officer on Bennigsen's staff,

---

* The French had also left behind a large number of Russian women who had come out from Moscow to comfort them.[10]

complained that Kutuzov "is afraid of making the slightest move . . . sitting in his den like a bear who does not want to come out" while the French were vulnerable to attack. "It is driving us all mad with rage."[12]

As the Russian attack on Murat was unfolding at Vinkovo, Napoleon was at the Kremlin, reviewing Ney's 3rd Corps. "The parade was as fine as the circumstances permitted," wrote Colonel Raymond de Fezensac, whose 4th of the Line was taking part. "The colonels surpassed themselves in order to present their regiments in the best light, and nobody seeing them could have imagined how much the soldiers had endured, and what they were still enduring." Shortly after midday, Murat's aide-de-camp General Bérenger galloped onto the parade ground and informed Napoleon of what had happened at Vinkovo. "Without being frightened, Napoleon was nonetheless very agitated," according to Bausset. He hurried through the review, giving out crosses of the Légion d'Honneur and promotions, and then returned to his rooms, where he began issuing orders to begin the evacuation planned for the morrow. "He kept opening the door to the room where those on duty waited, calling now for one person, now for another, spoke rapidly and could not remain still for a single instant," wrote Bausset. "He had hardly sat down to lunch when he was up again, and he put so much urgency into all his ideas and plans that I believe the fatal consequences of his long sojourn in Moscow were suddenly revealed to him on that day." But when he actually left the Kremlin that evening, Napoleon had regained his usual calm, and seemed more awake than he had for some time.[13]

Fezensac's regiment marched out that very night. On the way he noted with anger that another unit which was also moving off had set fire to a store of flour and fodder which it had no means of transporting. "We had space in some of our *fourgons*," wrote Fezensac, "but we had to watch as flames consumed victuals that might have saved our lives." Such organisation boded ill for what lay ahead.[14]

But many of the troops were too happy to be leaving or too preoccupied with the preparations to think of anything else. "We hasten back to our quarters, we fold up our dress uniforms and with pleasure put on our marching uniforms," noted Cesare de Laugier in his diary. "Everything is in a state of commotion; one can read joy on every face at the prospect of leaving. The only thing that bothers us is to have to leave behind comrades who are incapable of walking. Some of them are making superhuman efforts to follow us. At five o'clock, with drums beating and loud music, we march through the streets of Moscow . . . Moscow! which we had been so desirous of reaching, but which we leave without regret. We are thinking only of our native land, of Italy, of our families, which we will soon be seeing at the end of this glorious expedition."[15]

The commotion in the city as news of the impending departure spread was indescribable. Large numbers of civilians who feared being lynched by the mob once the French army was gone decided to follow it. While they were mostly from the French colony they included other foreigners, among them at least one English woman, the wife of a Polish merchant. They also included a considerable number of Russians, particularly women and petty criminals, who had thrown their lot in with the French or otherwise become attached to them. Their last-minute preparations only added to the confusion as the troops packed and loaded up. The city quickly filled with small traders anxious to buy anything the French could not carry off with them, and with peasants from the surrounding countryside eager to rifle through whatever was left behind. But the soldiers of the Grande Armée were not going to leave behind anything they did not have to.

One of the most damaging consequences of the fire was the large number of soldiers who had accumulated a stock of booty which they hoped would, on their return to western Europe, make their fortune. Probably as many as eight thousand had left the ranks to attend to the vehicles or pack animals on which they loaded their treasures. But they could only make the journey home under cover of the army, whose movements they would impede at every step.

Even disciplined soldiers were determined to hold on to their hard-gotten gains, as the case of Sergeant Adrien Bourgogne shows. His knapsack contained: several pounds of sugar, some rice, dry biscuit, half a bottle of liqueur, a Chinese silk dress embroidered with silver and gold thread, several small silver and gold items, his dress uniform, a lady's riding coat, two *repoussé* silver plaques, one representing the Judgement of Paris, the other Neptune in his chariot, several medallions and a diamond-studded Russian decoration. In a large bag slung over his shoulder he had a silver medallion representing Christ, a Chinese porcelain vase, and a number of other objects. This was all in addition to his full equipment, a spare pair of shoes and sixty cartridges. "Add to that health, gaiety, good will and the hope of paying my respects to the ladies of Mongolia, China or India, and you will have some idea of a sergeant of Vélites in the Imperial Guard," he concluded. After a few miles, he stopped by the roadside and sifted through his pack to see what he could jettison in order to lighten it. He decided to throw out the breeches of his dress uniform.[16] Others would jettison cartridges and musket-cleaning equipment. Artillerymen would throw out shells to make room in their gun carriages, and company wagons would be emptied of anvils, horseshoes and nails in favour of the farrier's booty.

"Anyone who did not see the French army leave Moscow can only have a very weak impression of what the armies of Greece and Rome must have looked like when they marched back from Troy and Carthage," wrote Pierre-Armand Barrau. Ségur thought it looked more like a Tatar horde returning from a successful foray. Every soldier groaned under a bulging knapsack and clanked with gold and silver items suspended from his crossbelts. The regimental baggage wagons were laden with non-regulation bundles. And the intervals between the marching units were crammed with wagons and carts of every kind, including handcarts and barrows piled with loot being pushed or pulled by Russian peasants who had themselves become part of the booty, and the most elegant carriages and landaus, often drawn by scraggy little *cognats*. These vehicles rolled along, three or four abreast, amid a crowd of soldiers without arms or uniforms, servants and camp followers. "It was no longer the army of Napoleon but that of Darius returning from a far-flung expedition, more lucrative than glorious," in the words of Adrien de Mailly.[17]

The mood was effervescent, and the scene reminded many of a masquerade or a carnival. Mailly watched in amazement as a family of French merchants from Moscow drove out. "These ladies were dressed just like Parisian *bourgeoises* off for a picnic in the Bois de Vincennes or Romainville," he wrote, adding that one was wearing a pink hat trimmed with silk, a pink satin doublet and white dress trimmed with lace, and white satin slippers.[18]

Colonel Griois could hardly believe his eyes as he watched. "Dense columns, made up of soldiers of various arms, advanced without order; the weak, skinny horses dragged the artillery along with difficulty; the soldiers, on the other hand, were full of health and strength, since they had enjoyed six weeks of abundant victuals. Generals, officers, soldiers, commissioners, all had employed every means for taking with them everything they had amassed. Carriages of the greatest elegance, peasant carts, *fourgons*, drawn by the little local horses and overloaded with luggage, trundled in the middle of the columns, higgledy-piggledy with saddlehorses and draught animals. The soldiers were bent under the weight of their packs. To abandon any part of their booty would have been too cruel. But they had to face up to it, and from that first day baggage was abandoned by the roadside, along with carriages whose horses could no longer draw them. This mass of men, of horses and of vehicles resembled rather the migration of a people on the move than an organised army."[19]

Estimates of the number of non-military vehicles vary from 15,000 to 40,000, and the number of civilians, including servants, may have been as high as 50,000.[20] This was an extraordinary amount of impedimenta for an army to drag along in its wake. Quite apart from slowing down its movements and cluttering the roads, it had a profoundly demoralising effect. The men's thoughts were on their booty rather than on the fighting ahead, and this introduced an element of anxiety and distraction which undermined the cohesion of even the best units.

The actual armed forces at Napoleon's disposal as he left Moscow numbered no more than 95,000, and probably less. According to Lieutenant-Colonel de Baudus, aide-de-camp to Bessières, most of the young soldiers had fallen by the wayside or died by the time the Grande Armée had reached Moscow, with the result that those who were left were a fine lot. This was certainly true of the infantry. The men had not only had a rest, they had restored their strength with regular food, and had repaired their uniforms and boots. Dr Bourgeois thought the troops were in fine fettle. "They marched along gaily, singing at the tops of their voices," noted Mailly.[21]

Captain François Dumonceau of the Lancers of Berg was impressed by the aspect of Prince Eugène's Italians as they marched out of Moscow, laughing and singing. "These troops had a fine martial look," he wrote. "It was evident that they had recovered well from their previous exertions; the soldiers appeared to be gay and ready for anything." His impression was confirmed by Prince Eugène himself, who thought they were in good shape, even though, with 20,000 infantry and two thousand cavalry left out of an original total of nearly 45,000 men, their numbers had been drastically reduced.[22]

Those troops that had been stationed furthest from Moscow were in the worst shape. After Vinkovo, Poniatowski's 5th Corps was down to no more than four thousand worn-out men. Junot's Westphalians, who had been quartered at Mozhaisk, were no longer a fighting force. And while the cavalry of the Guard was in reasonable order, even if its horses were not in prime condition, and some of the light cavalry attached to the various corps was still serviceable, Murat's once fine corps was by now a phantom force.

A few miles outside Moscow, General Rapp saw Napoleon standing by the roadside, and rode up to him. "Well, Rapp, we are going to withdraw to the frontiers of Poland by the Kaluga road," the Emperor said. "I shall take good winter quarters; I hope that Alexander will make peace." Rapp said that the winter would be cold and conditions harsh, but Napoleon dismissed this. "Today is 19 October, and look how fine the weather is," he said. "Do you not recognise my star?" But Rapp felt this was no more than bravado, and noted that "his face bore the mark of anxiety."[23]

The high spirits of the troops as they left Moscow soon flagged as they struggled along the crowded road, trying to avoid being forced into the ditch or crushed by the thousands of vehicles. Every time they came to a bridge or a defile there would be fierce struggles for precedence, with fists and even swords and bayonets coming into play as well as curses in a dozen languages. Most of the bridges in this part of the world were made of long pine logs laid across the ravine or stream, possibly supported by an upright or two, with shorter round logs laid across them to form the causeway, which was then covered with straw and compacted earth to make a smooth surface. If one of the transversal logs rotted or snapped and fell through, the others would roll back and forth, catching and breaking the legs of horses and men, particularly if there was a crush of people struggling to get across. On 22 October the heavens opened and the road became a sea of mud. More and more vehicles had to be abandoned, more and more cumbersome objects jettisoned from knapsacks, and the line of march lengthened as stragglers failed to keep up.

Napoleon moved southwards, down the old Kaluga road, making straight for Kutuzov's camp at Tarutino, while Prince Eugène and his 4th Corps marched down the new Kaluga road, a little to the west. One can only speculate as to Napoleon's intentions. He may have wanted to avenge the defeat of Vinkovo by attacking the Russian army full on while Prince Eugène outflanked it, but if he did he changed his mind, for two days out of Moscow, on 21 October, he marched his main forces across country in a westerly direction onto the new Kaluga road and joined Prince Eugène at Fominskoie. He ordered him to move fast and seize the little town of Maloyaroslavets, commanding the crossing of that road over the river Luzha.

He also sent orders to Mortier in Moscow to abandon the city and fall back on Mozhaisk. He instructed him to evacuate all the wounded left in the city, including those who had previously been deemed too sick to move. "I cannot recommend strongly enough that all the men still remaining in the hospitals should be loaded onto the wagons of the Young Guard, of the dismounted cavalry and on all those that come to hand," he wrote. "The Romans gave civic crowns to those who saved their fellow citizens; the Marshal Duke of Treviso [Mortier] will merit as many of those as he saves soldiers." He told him not to be afraid to overload the wagons, as he would find fresh horses and empty supply wagons at Mozhaisk. He was to give preference to officers and NCOs, and to Frenchmen. As a final exhortation, Napoleon reminded Mortier of his own feat of bringing home his wounded from Acre in 1799.[24]

He had previously instructed him to blow up the Kremlin before leaving, and to torch the town houses of Rostopchin and Count Razumovsky, a diplomat for whom he had a personal dislike. At 1:30 on the morning of 23 October the units that were not too far from Moscow could hear the dull thuds as the charges went off. "This ancient citadel, which dates from the foundation of the monarchy, this first palace of the Tsars, no longer exists!" proclaimed the twenty-sixth Bulletin.[25] Fortunately, many of the fuses failed, and although the damage was extensive, the Kremlin was not actually destroyed.

This did nothing to soften the feelings of the rabble who began to rampage around the capital the moment the last French troops had marched out. They murdered any sick Frenchmen they found and attacked the Foundling Hospital, where Captain Thomas Aubry, a Chasseur wounded at Borodino, helped organise a defence along with three wounded Russian generals. In spite of being in the throes of a raging fever, Aubry took up his post in a ward full of soldiers of both armies and, sword in hand, gave the firing orders to keep the attackers at bay until regular Russian troops arrived.[26]

News of the French evacuation of Moscow was brought to Tarutino at night by Lieutenant Bolgovsky. He informed Konovnitsin and Toll, who went to wake the Field Marshal. After a while the Lieutenant was called into Kutuzov's room, where he found the old man sitting on his bed in his frock-coat and his decorations. "Tell me, my friend," Kutuzov said to him, according to the Lieutenant's perhaps fanciful account, "what is this news that you have brought me? Can it really be true that Napoleon has left Moscow and is retreating? Speak, quickly, do not torture my heart, it is trembling." When the young officer had told him everything he knew, Kutuzov began to sob and, turning to the icon of the Saviour in the corner of the room said: "God, my creator, at last you have heard our prayers, and from this moment Russia is saved."[27]

Meanwhile General Dokhturov, who had been sent out by Kutuzov to reconnoitre the area in front of Tarutino, located Prince Eugène's 4th Corps near Fominskoie and decided to attack it. Luckily for him, his scouts picked up a couple of prisoners from whom he learned that he was about to take on not just Eugène but the whole Grande Armée. He also learned that it was headed for Maloyaroslavets. Realising that if it were allowed to occupy the town it would outflank the Russian army and threaten its supply lines, he sent word to Kutuzov and himself made a forced march in the hope of getting there before Prince Eugène. But when he reached the little town at dawn on 24 October, Dokhturov found that the French had got there first.

Maloyaroslavets is perched atop a curving ridge running along the southern bank of the Luzha in a wide semicircle. It had been occupied by two battalions of Delzons' division, the rest of which was camped on the low ground on the north bank of the river. Dokhturov stormed the town and managed to throw out the two battalions, but was then forced to give it up by a counterattack mounted by Delzons, in which the Croats of his 1st Illyrian Regiment distinguished themselves.

Kutuzov, who was far closer to Maloyaroslavets than Dokhturov when he received news of the French movements, was so slow to react that it was not until mid-morning that the first of his units, Raevsky's corps, turned up on the scene. In a dashing attack, Raevsky

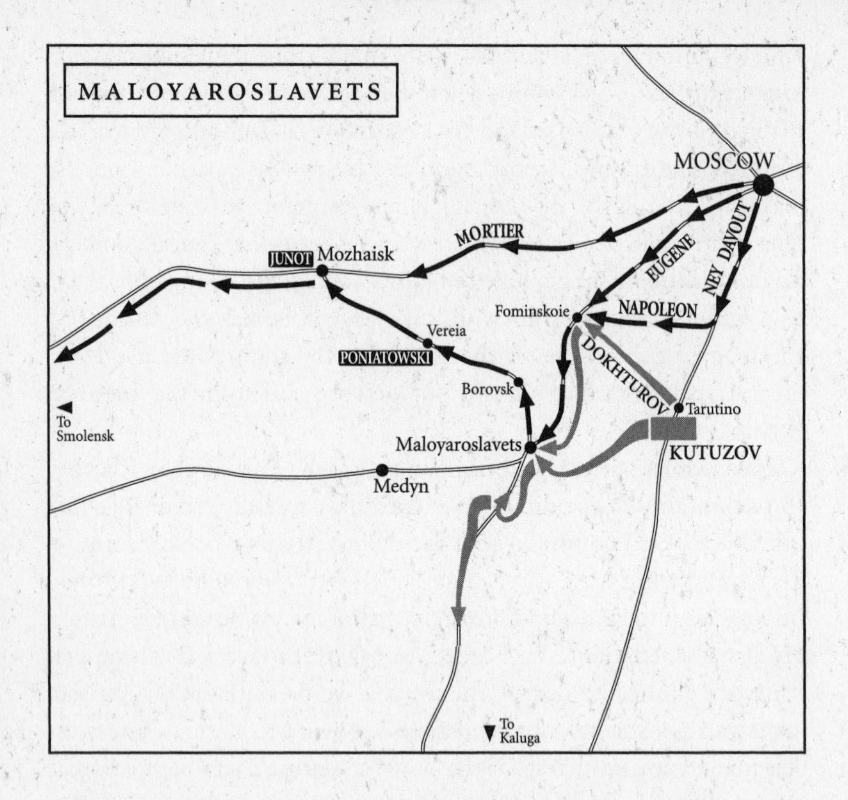

recaptured the town from the Italians, who had been reinforced by Broussier's division, but was himself thrown out by a spirited attack by Pino's division. Maloyaroslavets changed hands no fewer than eight times as more and more reinforcements came up on both sides. Most of the fighting took place at very close quarters, much of it in the streets of the by now flaming town, with burning houses crashing down on the living, the dead and the wounded. Many of those who managed to crawl free were crushed under the hooves and wheels of the artillery, which was constantly changing position.

The fighting continued long after nightfall, when the spectacle became "indescribably magnificent and interesting" in the words of General Wilson, who was watching from the Russian lines. "The crackling flames – the dark shadows of the combatants flitting amongst them – the hissing ring of the grape as it flew from the licornes – the rattling of the musketry – the ignited shells traversing and crossing in the atmosphere – the wild shouts of the combatants, and all the accompaniments of the sanguinary struggle formed an ensemble seldom witnessed."[28] The Italians fought like lions, and at nightfall they remained masters of the town.

"This battle was one of the finest feats of arms of the whole campaign," in the opinion of General Berthézène. Napoleon was full of praise for Prince Eugène and his troops. So was Wilson. "The Italian army had displayed qualities which entitled it evermore to take rank amongst the bravest troops in Europe," he wrote. The Delzons, Broussier and Pino divisions, as well as the Italian Royal Guard, had in the latter part of the action been supported by the Compans and Gérard divisions of Davout's 1st Corps, bringing the total number involved on the French side that day to about 27,000, with seventy-two cannon. They had triumphed over some 32,000 Russians with 354 cannon. "Yes, this was indeed one of the most glorious victories," wrote Captain Bartolomeo Bertolini, "which, both on account of the disasters by which it was followed, and on account of the ingratitude and malice of the French historians who principally write about Napoleon's battles, is hardly mentioned."[29]

But the cost had been heavy. French losses came to about six thousand dead and wounded, and included General Delzons, killed by a bullet in the head while leading the attack. The Russians had lost more, but they could afford to. "Another victory like this one, the soldiers are saying, and Napoleon won't have an army left," recorded the chief of staff of Prince Eugène's corps.[30] And tactically, the French victory was meaningless.

The entire Russian army, numbering at least 90,000 men, was by now in position behind the town, commanding the area before it with over five hundred guns. Napoleon could not muster 65,000 (Mortier was between Moscow and Mozhaisk, Junot was at Mozhaisk,

Poniatowski at Vereia). If he wanted to carry through his plan of taking Kaluga or even of just getting back to Smolensk through Medyn and Yelnia, he would have to defeat the Russians in a pitched battle. His instinct told him to do just that, and he was supported by Murat, but most of his entourage were against such a course.

That night at his headquarters, a squalid cabin in the village of Gorodnia whose single room was divided in two by a dirty canvas sheet, Napoleon asked his marshals for their opinion. He listened in silence, staring at the maps spread before him. While some were for going ahead, or at least crossing the Luzha elsewhere and making for Smolensk through Medyn, the majority favoured what they felt was the more prudent course of rejoining the main Smolensk–Moscow road and retreating along that. "Thank, you, gentlemen, I will make my own decision," he said as he broke up the meeting and retired for the night.[31]

Before dawn on the following morning, Napoleon mounted up and set off to reconnoitre the situation. He had only ridden a short way when a swarm of cossacks assailed his little party. The Chasseurs and Chevau-Légers of his escort, along with some of his staff officers and entourage, chased them off while he rode to a place of safety. But it had been a narrow escape. He then rode across the battlefield and through Maloyaroslavets to assess the Russian positions. The ruins of the little town presented a gruesome sight. "The streets were strewn with corpses, many of them hideously mutilated by the wheels of guns and *caissons*," recalled Colonel Lubin Griois. "What added to the horror of the scene was the large number of those who had fallen victim to the fire, and who were more or less blackened and foreshortened, depending on whether they had just been slightly burned or completely calcinated." According to Baron Fain, Napoleon was very much affected by the sight.[32]

What he saw beyond the battlefield left him undecided. In the early hours Kutuzov had fallen back a couple of kilometres and deployed in a strong defensive position. This suggested to Napoleon that he had a chance of defeating Kutuzov decisively, which would not only avenge Vinkovo but transform the march back to Smolensk into a victorious progress. But the strength of the Russian positions meant that a victory would be costly, and he would have to leave behind the heavily wounded.

In falling back, Kutuzov had actually opened the road to Medyn, leaving Napoleon free to pursue his retreat to Smolensk that way, but the retreating French would then have had the entire Russian army on their heels. Napoleon discussed the situation further with various members of his entourage without reaching a decision.

While Kutuzov lost no time in proclaiming Maloyaroslavets a Russian victory, he was far from confident. His forces were so overwhelmingly made up of raw recruits and inexperienced officers that although they were not lacking in spirit, they had made a poor showing in the fighting.[33] How they would behave in the context of a pitched battle was anybody's guess. Conversely, the fight put up by the French had demonstrated that their stay in Moscow had not affected their morale enough to diminish their fighting fitness. So while Kutuzov's forces comfortably outnumbered and decidedly outgunned those facing him, Napoleon would have had an enormous advantage in terms of quality.

"Very strange things were taking place at Russian headquarters that night," according to Colonel von Toll. Bennigsen, Konovnitsin, Wilson and Toll himself tried to goad Kutuzov into action, asserting that victory was well within their grasp. But Kutuzov was not about to stake everything even on such odds. If he were defeated now, the army would not be capable of reconstituting itself as it did after Borodino, and his whole argument, that he had given up Moscow in order to preserve the army which would chase Napoleon from Russia, would be made to look very hollow. As the French were not moving away but evidently reconnoitring the terrain, Kutuzov can only have assumed that they were preparing to give battle on the following day. And when he received intelligence that French units had been spotted making for crossings over the Luzha further west and for Medyn, he became alarmed that he might be cut off from Kaluga. He therefore decided to avoid battle, and ordered a retreat for that night. "The extent of his poltroonery exceeds the measure permitted even to a coward," Bennigsen fumed in a letter to his wife.[34]

18

Retreat

Had Napoleon known what was going on in Kutuzov's mind, he could have advanced boldly and made for Medyn, where he would have found victuals and forage, and from there to Yelnia, where he had a division under General Baraguay d'Hilliers waiting to meet him, and on to Smolensk, which he would have reached in fairly good shape on 3 or 4 November. But taking into account the strong positions the Russians had taken up, he decided otherwise. That evening, he ordered the retreat, through Borovsk and Vereia to Mozhaisk, and from there along the main road to Smolensk. In one of the more bizarre episodes in military history, the two armies were now moving away from each other.

Two days later, as he reached Mozhaisk, Napoleon met Mortier coming from Moscow with the Young Guard. He also had with him two prisoners, General Ferdinand von Wintzingerode and his aide-de-camp Prince Lev Naryshkin, who had unwisely ridden into Moscow to verify reports that the French had left, only to be captured by a patrol. On seeing Wintzingerode, a native of Württemberg in Russian service who seemed at that moment to epitomise the *internationale* that was forming against him, Napoleon erupted into a violent rage, the like of which none of his entourage had ever witnessed. "It is you and a few dozen rogues who have sold themselves to England who are whipping up Europe against me," he ranted. "I don't know why I don't have you shot; you were captured as a spy." He took out all his frustration and mortification on the unfortunate General, accusing him of being a renegade. "You are my personal enemy: you have borne arms against me everywhere – in Austria, in Prussia, in Russia. I shall have you court martialled."[1]

Even this tirade did not succeed in venting all his pent-up anger, and on seeing a pretty country house that had somehow escaped destruction, Napoleon ordered it to be torched, along with every village they passed through. "Since *Messieurs les Barbares* are so keen on burning their own towns, we must help them," he raged. He soon countermanded the order, but that hardly made much difference.[2] As they stopped for the night the troops would dismantle houses to feed their campfires or crowd into them for warmth. They would light fires inside or overheat the mud stoves, which often led to them catching fire, and in villages or small towns in which every building was of wood, this usually led to a general conflagration.

The order to retreat had a depressing effect on the army, which instinctively felt that something had gone wrong with the infallible Emperor's calculations. But, ironically, it was when they began to feel threatened that the troops rallied to him and took comfort in his perceived greatness. On the day the retreat began, General Dedem de Gelder reported to the Emperor for orders. "Napoleon was warming his hands behind his back at a small bivouac fire which had been laid for him on the edge of a village one league beyond Borovsk on the road to Vereia," he recalled. The General disliked Napoleon, partly because of the way he had treated his native Holland, but he could not help being impressed by him now. "I have to do justice to this man hitherto so spoilt by fortune, who had never yet known serious setbacks; he was calm, without anger, but without resignation; I believed he would be great in adversity, and that idea reconciled me to him . . . I saw then the man who contemplates disaster and recognises all the difficulties of his position, but whose soul is in no way crushed and who says to himself: 'This is a failure, I have to quit, but I shall be back.' "[3]

The spirits of the army were further lowered when, shortly after rejoining the Moscow–Smolensk road at Mozhaisk on 28 October, they found themselves marching across the battlefield of Borodino. It had never been cleared, and the dead had been left where they lay, to be pecked at and chewed by carrion crows, wolves, feral dogs and other creatures. The corpses were nevertheless surprisingly well preserved, presumably by the nightly frosts. "Many of them had kept what one might call a *physiognomy*," recorded Adrien de Mailly. "Almost all of them had large open staring eyes, their beards seemed to have grown, and the brick-red and Prussian blue which marbled their cheeks made them look as though they had been horribly sullied or luridly daubed, which made one wonder if this were not some grotesque travesty making fun of misery and death – it was odious!"[4] The stink was indescribable, and the sight cast a pall over the passing troops.*

At Mozhaisk and at Kolotskoie they saw thousands of emaciated wounded, barely surviving in dreadful conditions. Colonel de Fezensac went into the Kolotsky monastery to see if there were any men from his regiment. "They had left the men there without medicines, without rations, without any form of succour," he wrote. "I was barely able to get in, so encumbered were the stairs, corridors and the middle of the rooms with ordure of every kind."[6]

Napoleon was annoyed to find so many wounded still there, and grandly determined that they should all be taken along. Against the advice of Larrey and other doctors, who had left medical teams to care for them, he gave instructions for them to be placed in carriages,

---

* Eugène Labaume, an officer on the staff of Prince Eugène, was the first to print the story that as they were passing the battlefield they heard a man call out and discovered a soldier who had lost both legs during the battle and been left for dead but managed to survive by finding shelter in the belly of a dead horse and feeding off scraps found in the pockets of dead men. This story, incredible and almost certainly untrue, was subsequently repeated by countless other chroniclers, some of whom claimed to have seen or spoken to the man, and provides a good example of how strong the power of suggestion can be when old men try to remember.[5]

on *fourgons*, on the wagons of the *cantinières*, gun carriages and every other possible conveyance. The result was predictable. "The healthiest people would not have stood up to such a method of transport, or been able to remain for long on the vehicles, given the way they had been loaded on," wrote Caulaincourt. "One can therefore judge the state these unfortunates were in after a few leagues. The jolting, the exertion and the cold all assailed them at the same time. I have never seen a more heart-rending sight." The owners of the carriages in question were far from happy to have extra weight laid on vehicles which their horses could barely draw as it was, and faced with trepidation the prospect of having to feed their new charges. Realising that they were unlikely to survive anyway, they mostly decided to precipitate the inevitable. "I still shudder as I relate that I saw drivers purposely drive their horses across the roughest ground in order to rid themselves of the unfortunates with whom they had been saddled and smile, as one would at a piece of luck, when a jolt would rid them of one of these unfortunates, whom they knew would be crushed under a wheel if a horse did not step on him first."[7]

After giving the orders for the evacuation of the wounded on the afternoon of 28 October, Napoleon rode on to Uspenskoie, where he stopped for the night in a devastated country house. But he could not sleep. At two o'clock in the morning he called Caulaincourt to his bedside and asked him what he thought of the situation. Caulaincourt replied that it was much graver than Napoleon thought, and that it was unlikely that he would be able to take up winter quarters at Smolensk, Vitebsk or Orsha, as he still hoped. Napoleon then said that it might be necessary for him to leave the army and go to Paris, and asked him what he thought of such a plan and what he thought the army would make of it. Caulaincourt replied that going back to Paris was the best course of action, though he would have to choose his moment well, and that what the army thought was of no consequence.[8]

Napoleon's position was indeed very bad. Ten days after leaving Moscow, he was only three days' march down the Smolensk road.

This not only represented a dangerous delay, it also meant that his army had used up ten days' rations. At the rate it was now moving, Smolensk was still over ten days' march away, and the only sustenance to be found before that was a small magazine at Viazma. And, with no intelligence and not enough cavalry to send out scouting parties, Napoleon had no idea of what the Russians were up to.

When Volkonsky had reached St Petersburg and handed Alexander the letter Napoleon had sent through Lauriston, the Tsar had hardly bothered to read it. "Peace?" he said. "But as yet we have not made war. My campaign is only just beginning."[9] In fact it would be some time before it began.

It was only after a couple of days' hurried retreat that Kutuzov turned about and began gingerly to follow the retiring French. He sent Miloradovich on ahead, and himself followed at a more leisurely pace. Having marched north to Mozhaisk, the French were now marching west along the Moscow road in a wide arc that curved southwards. Kutuzov was therefore excellently placed to cut across their line of retreat. But while he could not resist writing to his wife that he was the first general who had ever made Napoleon run, he made no attempt to intercept him.

The only enemy the French saw were cossacks, who followed them at a respectful distance, like hyenas stalking a wounded animal. The regular cossack regiments they had met hitherto were now outnumbered by irregulars from the Don and the Kuban. "Dressed and hatted in a variety of styles, without any appearance of uniformity, dirty-looking and scruffy, mounted on mean raw-boned little horses with unkempt manes which kept their necks stuck out and hung their heads, harnessed with no more than a simple snaffle, armed with a crude long pole with a sort of nail at its point, milling around in apparent disorder, these cossacks made me think of teeming vermin," remembered François Dumonceau.[10] The cossacks were supplemented by bands of Bashkir horsemen armed with bows, who amazed the French by firing arrows at them.

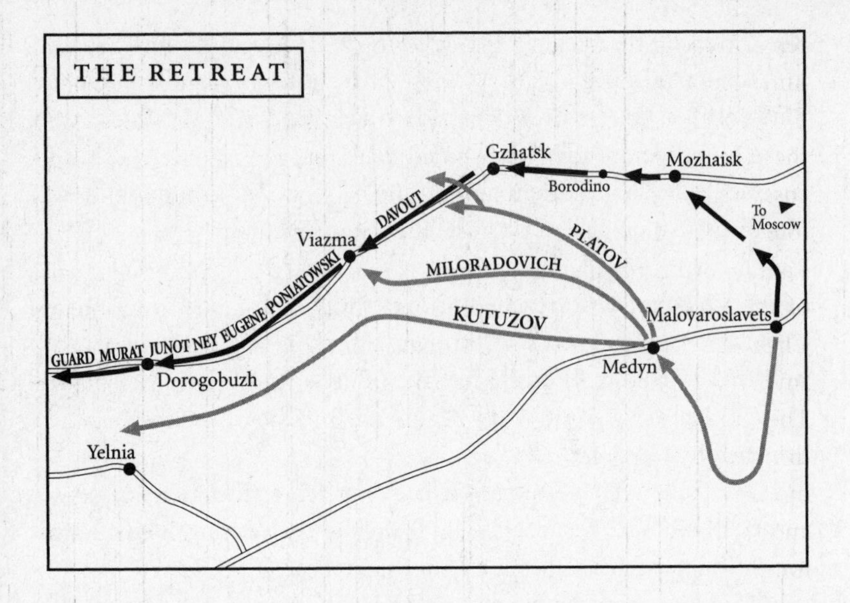

These wild horsemen were of no military value in themselves. Their principal tactic was to rush forward yelling "Hurrah!," hoping to terrify the enemy into flight, at which point they would catch a few fugitives and pick through whatever booty the others had left behind. If a soldier stood his ground and levelled a musket at them they would invariably run, but he was wise not to fire it, as they would return and get him while he was reloading. The cossack pike had a thin round point which could only prick and not sever tendons or muscles, so unless it found a vital organ its wound was not serious.

On the advance, the French had ignored the cossacks, making fun of their shameless unwillingness to expose themselves to the slightest danger. "If one were to raise a regiment of French girls they would, I believe, show more courage than these famous cossacks with their long pikes and their long beards," commented one soldier. But in the conditions of a retreat, and in the absence of adequate numbers of cavalry on the French side, they were to exert an influence quite beyond their potential. "The French soldier is easily demoralised," remarked Lieutenant Blaze de Bury. "Four Hussars on his flank terrify him more than a thousand in front."[11]

On 2 November Marshal Lefèbvre harangued the Old Guard on the subject with his usual directness. "Grenadiers and Chasseurs, the cossacks are there, there, there and there," he said, gesturing to the four points of the compass. "If you do not follow me, you are f—d. I am no ordinary general, and it is with good reason that in the army of the Moselle I was known as the Eternal Father. Grenadiers and Chasseurs, I say to you again: if you do not stay with me you are f—d. And anyway, I don't care a f—k. You can all go and f—k yourselves."[12] The Guard did not disappoint, and the ranks remained steady throughout; but the same could not be said of other troops. Once morale began to crack on the retreat, an irrational fear took over, and the mere shout of "Cossacks!" would send old soldiers scurrying for cover.

The French were retreating in echelons, with Napoleon leading the way accompanied by the Old Guard, the Young Guard, the remains of Murat's cavalry and Junot's corps, and reaching Viazma on 31 October. Next came Ney, followed by Prince Eugène's Italians and what was left of Poniatowski's Poles. Bringing up the rear was Davout with his 1st Corps.

Progress was slow, mainly due to lack of horsepower. Shortage of fodder had debilitated the horses, which were growing too weak to pull the guns and *caissons*. Guns normally drawn by three pairs were now having teams of twelve or fifteen horses hitched to them, and even these could not manage to pull the heavy pieces over the muddy rivulets and up the many inclines in the road. Passing infantry would be enlisted to help push the guns, but the exhausted footsloggers did not relish this task and did everything to avoid it. Powder wagons were blown up and surplus shells jettisoned to lighten the load. The private carriages and booty-laden wagons of individuals were seized and burnt by the artillery, who commandeered the horses. At Gzhatsk on 30 October Henri-Joseph Paixhans, an aide-de-camp to General

Lariboisière, passed a column of wagons laden with wounded men whose horses had been taken. "These poor unfortunates implored our pity with their hands joined in prayer," he recalled. "They called out to us in heart-rending tones that they too were Frenchmen, that they had been wounded fighting at our side, and they begged us tearfully not to abandon them."[13]

Part of the problem was that Napoleon saw himself as carrying out a tactical withdrawal rather than a retreat. Several of the corps commanders wanted to abandon a proportion of their guns, which were of no use to them. This would have liberated horses with which to draw the rest and saved much time, but Napoleon would not hear of it, maintaining that the Russians would claim the abandoned guns as trophies. This determination not to lose face would cost him dear.[14]

Along with other unnecessary impedimenta, the French had with them some three thousand Russian prisoners. Even though their presence cost nothing in terms of supplies – the unfortunate wretches were given no food at all, so they fed off the dead horses they found by the wayside and ended up, by some accounts, eating their own dead – they were an aggravating encumbrance to the Portuguese infantry detailed to escort them, and took up valuable space on the road.[15] And space was at a premium.

A major drawback of retreating in echelons down the same road, as Napoleon had elected to do, was that only the leading unit had a clear field of march, while all the others had to move through the mess left behind by the preceding ones. Their path was laboured by tens of thousands of feet, hooves and wheels – into a stormy sea of mud if it was wet, and into a skating rink of compacted snow and ice when it began snowing. Such supplies as there might have been along the way were devoured, and even the available shelter was dismantled for firewood by those who had gone before. The road was littered with abandoned carriages and wagons, dead horses and jettisoned baggage; and, worst of all, the following columns kept coming up against a slow-moving mass of traffic.

Apart from the tens of thousands of civilians following the army there were *commissaires* and other functionaries attached to it, and officers' servants. They were mixed up in a throng of booty-laden deserters, some on foot, some in wagons; *cantinières* with their laden vehicles; and wounded officers travelling in carriages, tended by their servants. There were also some lightly wounded from the transports which had left Moscow in the days before the evacuation who were caught up and eventually overtaken by the retreating army. Their numbers were swelled daily by those wounded in the fighting along the way.

There were a large number of soldiers who had fallen behind and become separated from their units, which they sought, and occasionally managed, to rejoin. But it was difficult for them to catch up, as they had to push their way through a compact mass of people, horses and vehicles. There were others who, having fallen behind, threw away their weapons and were absorbed into the mass of stragglers, demoralised and guided more and more by herd instinct.

This swelling throng of people moved along the same road as the army, using up whatever resources were left and cluttering its path. It encumbered the approaches to every bridge and defile, as the absence of discipline coupled with a desperation verging on panic invariably produced chaos at such places. "Men, horses and vehicles would press forward pell-mell, pushing and shoving without any mutual consideration," wrote Dumonceau. "Woe betide those who allowed themselves to be knocked over! They could not get up, were trodden underfoot and caused others to trip and fall on top of them. In this manner mounds of men and horses, dead and dying, gradually piled up, blocking the way. But the crowd kept coming, banking up and cluttering the approaches to the obstacle. Impatience and anger would come into play. People quarrelled, pushed each other away, knocked each other over, and then one could hear the cries of the unfortunates who, knocked over, trampled, were caught and crushed beneath the wheels of carriages or other vehicles."[16] And if a cry of "Cossacks!" went up, the ensuing panic would multiply the number of those crushed to death.

As well as slowing their progress, all this had a demoralising effect on the following troops, who marched down a devastated road and saw only abandoned equipment, human and equine corpses, and men who had thrown away their weapons. The situation was worst for the rearguard, which not only had to march over a veritable obstacle course, but also to roll before it a snowballing mass of stragglers who impeded its movements and even impaired its ability to fight. Colonel Raymond de Fezensac, who found himself in the rearguard with his 4th of the Line between Viazma and Smolensk, would have his bivouacs crowded by cadging or thieving stragglers, who refused to make use of the night to move ahead but would try to march with his force when it set off in the morning. He would chase them off with rifle butts and warn them that he would not let them take refuge inside his squares if he was attacked. But still they hung about his regiment, getting in his way and making it easier for his men to desert.

The constant sight of disbanded men thinking only of themselves weakened the resolve of those who were still trying to do their duty. "The soldier who remained with the colours found himself in the role of a booby," explained Stendhal. "And as that is something the Frenchman abhors above all, there were soon only soldiers of heroic character and simpletons left under arms."[17]

On the evening of 2 November Miloradovich, who was full of fight and keen to get at the French, tried to cut the road in front of Davout's corps, which was bringing up the rear, at a defile near Gzhatsk. His guns wreaked havoc in the pile-up of field guns, *caissons*, private carriages and stragglers. A convoy of civilians and wounded was caught up in this chaos, and many of those who were not able to abandon their vehicles and make a dash for it perished. But Miloradovich did not have enough infantry to attack the French, and had to back off when Davout deployed troops against him.

Two days later, with a full complement of some 25,000 men, he made a second attempt to cut off Davout, just east of Viazma. This time he came between Davout's 14,000 or so exhausted men and the preceding echelons, while Platov attacked Davout from the back and Figner's and Seslavin's irregulars harried his flanks. The French rearguard was thus caught between two fires and found itself in a perilous situation.

Prince Eugène and Poniatowski heard the guns and promptly turned about. Mustering about 13,000 and 3500 men respectively, they mounted a determined attack which repulsed Miloradovich and opened the road, while Ney, who had also turned about, covered the approaches to Viazma. The Russians were reinforced by the arrival of Uvarov's cavalry, but Davout was nevertheless able to beat an orderly retreat and, when the Russians tried to harry him too closely, he even sallied out and captured three guns. In the late afternoon two fresh Russian divisions, Paskievich's and Choglokov's, attacked the outskirts of Viazma, and Ney withdrew across the river, burning the bridges.

Losses on the French side were about six thousand dead and wounded, and two thousand taken prisoner, while Russian casualties were no higher than 1,845, and possibly less. Poniatowski's horse fell while he was jumping a ditch, crushing his knee and shoulder and causing severe internal injuries, which put him out of action. But the most depressing aspect of the battle for the French was that two standards had been lost, and that at one point towards the end of the day some of Davout's men had broken into a panicked flight.[18]

The Russians, however, had nothing to rejoice over. If Miloradovich and Platov had squandered an opportunity to destroy Davout's corps, Kutuzov had missed an even greater one. The Field Marshal with his 65,000-odd men had spent the day a couple of miles to the south of Viazma in a position from which he could, without any trouble at all, have taken Ney's corps in the rear, thus nullifying Prince Eugène's and Poniatowski's efforts, and wiping all four of the enemy corps off the chessboard, leaving Napoleon with little more than his Guard. Although he did despatch some reinforcements to Miloradovich, the old man had resolutely opposed any and every suggestion to make an offensive move. He was not even on speaking terms with Bennigsen by now, having suspended him from his duties and told a staff officer he had sent to him: "Tell your General that I do not know him and do not wish to know him, and that if he sends me one more report I shall hang his messenger." Bennigsen, Toll, Konovnitsin, Wilson and others were beside themselves. That night Wilson wrote to Lord Cathcart, British Ambassador in St Petersburg, asking him to use all his influence to get Kutuzov sacked. On 6 November he wrote to the Tsar himself, saying that Kutuzov was a tired old man who should be replaced with Bennigsen.[19] In the event it hardly mattered who was in command of the Russian army, as on that very day a new element had come into play.

Accounts of the retreat vary a great deal, depending on who the memoirist was, which part of the army he was with, and what befell him. The distance between the head of the column and the rearguard was rarely less than thirty kilometres, and at times stretched to a hundred, which meant that various units often marched through different weather on the same day. By the same token, the one who writes that the retreat was an orderly one up to Smolensk and the one who paints a picture of chaos on the first day can both be right.

Captain Hubert Biot, a Chasseur incapacitated by a shell at Borodino, left Moscow on 18 October in a carriage with two other wounded officers, and the three of them rolled without mishap all the way to Paris, because they were always ahead of the army. Madame Fusil, one of the French actresses in Moscow who decided to return to Paris with the Grande Armée, was perfectly comfortable in an officer's carriage until 7 November, when his horses died. She then had a very difficult time, but eventually managed to find a place in a marshal's carriage and rolled on comfortably enough in the first echelon. The aristocratic young Adrien de Mailly and his friend Charles de Beauveau, both of them wounded, shared a comfortable carriage and sang songs or read to each other as they drove. "To support the misfortunes of war with courage and gaiety, there is nothing like being French, being young, and also, perhaps, being a nobleman," he wrote.[20] Those who trudged further back had a rather different picture of events.

But most were still in relatively good spirits in the last days of October, glad to be heading for home. "It was the 29th or the 30th, the weather was magnificent, and in the course of the morning a regiment which was marching by me was singing joyfully and continually," recalled Lubin Griois, Colonel of artillery in Grouchy's corps. "I was struck by this: there had been no singing at our bivouacs for a long time, and this was the last I was going to hear." Colonel Jean Baptiste Materre who was on Ney's staff, marching in the middle echelon, noted signs of a general flagging of spirits on 31 October. The process gained over the next couple of days. "The position of the army is beginning to look rather unfortunate," noted Cesare de Laugier on 2 November.[21]

The weather had something to do with it. On 31 October, at Viazma, Napoleon again compared it favourably with that at Fontainebleau at the same season, and derided those who had been attempting to scare him with stories of the Russian winter. The nightly frosts did not bother anyone unduly. Colonel Boulart, writing on 1 November to his wife from a camp outside Viazma, summed up the mood. "I am writing to you, my darling, on the most beautiful day with the most beautiful frost, sitting on the most beautiful hummock, feeling cold all over, which also means at the tips of my fingers, in order to tell you that you should not be anxious on my account."[22] The army's postal service still functioned, erratically it is true, and even when the probability of their letters getting through had diminished, the men still wrote, clinging to that tenuous link with home.

The weather remained fine during the first days of November. "The days are as warm as in summer, and the nights are cold," noted Boniface de Castellane on 3 November. "I can remember seeing fields carpeted with pansies of every hue, which I amused myself by making into bouquets," wrote the bluff Colonel Pelet of the 48th of the Line. But 3 November was to be the last warm day. The new moon on the night of 4–5 November brought with it a sharp drop in temperature, and on 6 November the retreat entered a new phase. "That day has remained deeply engraved on my memory," continued Pelet. "After we had passed through Dorogobuzh it started raining quite hard, and it began to get cold; the rain turned to snow, and in a very short time it lay two feet thick on the ground."[23]

Sergeant Bourgogne, two days' march further west, would not forget that day either. It had already started getting cold on the eve, which unfortunately for her was the moment their *cantinière* Madame Dubois went into labour. The grenadiers built her a shelter out of branches and the Colonel himself lent his cloak to lay over it, but the poor woman nevertheless had to give birth in sub-zero temperatures.[24]

Another who could not forget that first cold night of 6 November was François Dumonceau. "Our campfires, which we could only keep going with difficulty, did not succeed in warming us," he wrote. "The biting north wind came and found me even under the bearskin rug I was covered with. Frozen on one side, scorched on the other, suffocated by the smoke, alarmed by the roar of the wind as it tore at the trees of the dense wood, I could not bear it and, like the others, ran this way and that in order to warm myself, spending a night without rest and experiencing suffering the like of which we had never known."[25]

While he stamped his feet in the wood, further east, at Dorogobuzh, a group of Italian officers huddled together in the ruins of a roofless hovel watching their comrade, Lieutenant Bendai, die from his wounds, malnutrition and cold. "I only regret two things," the Lieutenant murmured before breathing his last. "Not to be dying for the freedom and independence of our Italy . . . and not to be able to see my family again before I go."[26]

The following night Colonel Pelet gallantly invited the actress Madame Fleury, who was sitting in her carriage while her coachman had gone in search of fodder for the horses, to share his dinner and his fireside. But when morning came it turned out that her horses had died of cold in the shafts.[27] The day after that he saw for the first time a man who had frozen to death.

It was not just the wounded, lying still on the top of some wagon and unable to seek the warmth of a fireside, who died. On the morning of 7 November, Faber du Faur, a Württemberger serving in Ney's 3rd Corps, caught up with some fellow countrymen who had been a day's march ahead of him. He approached the camp of makeshift huts constructed out of pine branches in which, to his surprise, they seemed to be still fast asleep. In actual fact, they were frozen stiff. Colonel von Kerner, chief of staff to the Württemberg division, came out of the barn in which he and his companions had spent the night in order to muster the troops, but came running back after a while. "I have just seen the most appalling sight of my life," he said. "Our men are there, sitting around their campfires just as we left them last night, but they are all dead and frozen." This sight became commonplace until the men learned to keep their fires going and to sleep only for short periods. "When we got up in order to move out," recalled Marie Henry de Lignières, "many would remain seated; we would shake them to wake them up, thinking they were asleep; they were dead."[28]

The drop in temperature had not been that great – certainly not more than −10°C (14°F). But the French army was not dressed for cold weather. There was no such thing as a winter uniform, since in those days armies did not fight in winter. Most of the uniforms were cut away and did not even cover the stomach, which was protected only by a waistcoat, and while the infantry had proper greatcoats, the officers had only tailored overcoats which ended well above the knee. The cavalry had cloaks, but these were not lined, so they provided little shelter from the weather. While the bearskin bonnets of the grenadiers and the fur *kolpaks* of Chasseurs afforded some protection, most of the headgear, and the helmets of cuirassiers and dragoons in particular, had quite the opposite effect. To this has to be added the fact, easily verifiable by a visit to the Musée de l'Armée or any other repository of surviving uniforms, that the quality of the cloth and other materials used was generally poor, and the uniforms were flimsy. "The greatcoats of our infantry are probably the worst in Europe," observed Henri-Joseph Paixhans.[29]

As it grew colder, the men began to supplement their kit in various ways. Scarves were wrapped around the head under the regulation shako, knitted shawls worn around midriffs, and muffs and mittens brought into play to protect the hands. Those who had furnished themselves with sheepskins or fur coats donned these over their regimentals. Those who had not thought of acquiring such winter clothing were forced to put to use the furs (usually lined with feminine shades of silk or satin), shawls, bonnets and other items of clothing they had laid hands on meaning to sell them in Paris or give them to sweethearts.

As falling temperatures vanquished self-consciousness, even ladies' dresses and richly embroidered liturgical vestments came into use. Women's voluminous garments had the advantage of creating a kind of tent, thereby insulating the wearer from the cold. Cavalrymen whose horses died turned their *shabracque*, the sheepskin or embroidered cloth that went under the saddle, into a poncho by cutting a hole in the middle. All sorts of ingenious stratagems were thought up to cover every extremity – some even passed their legs through the sleeves of sheepskin coats and strapped them round their waist in order to keep their legs warm.

"One could see all the provisions and loot of Moscow being gradually brought out; the most elegant dresses and the crudest clothes, headgear of every kind, round caps, some trimmed with silver or gold, the jerkins and fur-lined blouses of the peasant women, the silk pelisses of ladies, dressing gowns; in a word, everyone appeared in whatever he had brought," recalled Colonel Pelet. "It was a hilarious sight to see these tanned faces, these moustaches, these fearsome miens enveloped in the most tender colours, these huge bodies barely covered by the most frivolous raiments. It was a continuous masquerade, which I found highly entertaining, and ribbed them about it as they passed."[30] Amusing as it may have been, this kind of get-up did not facilitate movement, and often impeded the men in wielding their weapons.

Many agree that the first serious snowfall on 6 November, accompanied as it was by a sharp drop in temperature, had a profound effect on the cohesiveness of the army. "It is from that point that our misery began," wrote Colonel Boulart, "and that misery was to grow and to last for another six weeks! Luckily, we could not see into the future; the present sufferings absorbed all our faculties, we thought only of ways of attenuating them and we thought little of the sufferings of the morrow; each day held quite enough affliction."[31]

The snow was soon compacted by the tramp of tens of thousands of feet into a rock-hard and slippery surface. As the horses were having difficulty in pulling wheeled vehicles, many coachmen and wagoners took the wheels off their vehicles and improvised runners. On 8 November there was a thaw and the roadway turned boggy. Those who had thrown away their wheels could do nothing but abandon their vehicles. But on the following day there was a hard frost, and the roadway now turned into a sheet of ice.

It was difficult to stay upright even while walking along the level, and Lieutenant Marie Henry de Lignières counted that he made over twenty falls a day. "Whenever we came across steep slopes we had to descend, which happened frequently, we would sit down and allow ourselves to slide down to the bottom; which meant that those behind would fall on top of one with their arms and luggage," he wrote.[32] Carts and gun teams had to be secured by men pulling on ropes from behind to stop them sliding down inclines, but if those holding the ropes slipped, the whole lot, gun, limber, horses, men and all, would go crashing down, taking with it anyone and anything in its path. As it became more difficult to walk, many more soldiers fell behind.

The cold made it painful to touch the muskets, and below a certain temperature the skin would stick to the frozen steel and come away from the hand. Those who had no gloves and could not improvise some form of protection for their hands were obliged to throw away their arms, and many more took the cold as an excuse to do so.

The cold also proved the last straw as far as many of the horses were concerned. Tens of thousands of the undernourished and exhausted beasts died in the space of three days, partly from the cold and partly because they were improperly shod. The ordinary shoes with which they were for the most part shod gave no purchase on compacted snow or ice, and acted rather like skates. Some French units did have studded shoes, and the artillery began fitting these after the first snows fell, but they wore down quickly to a smooth surface. What was required was shoes with sharp crampons, and these had only been fitted by the Poles, by Caulaincourt to the horses of the imperial household, and by a few sensible officers. The rest of the Grande Armée's bloodstock did not stand a chance once the snow fell and the cold set in. They would slip and fall, often breaking a leg in the process, and the terrible thrashing efforts involved in getting them to their feet again exhausted and distressed them further.

Some people tried wrapping rags round the horses' hooves, others realised that it was better for the horses to have no shoes at all than the standard ones, and prised theirs off. The little local *cognats* with their broad hooves and their low centre of gravity were at a premium, as they could still trot without shoes. Jacob Walter acquired one that even knew how to sit down on its hindquarters at the top of an icy incline and slide down without his having to dismount.[33]

But there was no substitute for proper sharp-shoeing. "As we Poles mounted on sharp-shod horses passed French generals at a gallop, they looked at us with surprise and envy, while at every hillock their guns got into difficulty and could only be pulled to the top thanks to the shoulders of infantrymen," wrote Józef Załuski of the Polish Chevau-Légers.[34]

Prince Eugène's 4th Corps lost 1,200 horses in two days. Albert de Muralt, a Swiss lieutenant serving in the Bavarian Chevau-Légers, recorded that his brigade, numbering two hundred horsemen when it reached Viazma, was down to about thirty to fifty the following day, and ceased to be a fighting unit on the next. The same story was repeated throughout the army.[35] The loss of the cavalry and a large part of the artillery radically reduced the fighting potential of the Grande Armée and made it vulnerable to the ubiquitous cossacks, who circled the retreating columns like blowflies.

But it was the loss of thousands of draught animals that had the greatest impact on the army and its chances of survival. Hundreds of vehicles had to be abandoned, some with much-needed supplies and equipment, as well as the personal effects and booty of the troops. Many threw away their arms in order to carry their belongings. "The road was strewn with precious objects, such as paintings, candlesticks and many books," recalled Sergeant Bourgogne, "and for the best part of an hour I would pick up books which I would look through, and which I threw away in turn, to be picked up by others who, in their turn, threw them away." Prince Józef Poniatowski, who trundled past, his mangled body laid out in his carriage, asked a passing soldier to hand him something to read from the selection at the roadside, and this book, which absorbed him, was to be his only booty of the campaign.[36]

On 8 November Commandant Vionnet de Maringoné of the Grenadiers of the Guard realised that his horses would be unable to draw his carriage much further. He therefore transferred all the most essential items, beginning with vital rations of food and some spare clothes, into one portmanteau and left all unnecessary luxuries in the carriage, which he abandoned. By not overloading it, he was able to keep one horse alive to carry this portmanteau. Major Claude-François Le Roy of the 85th of the Line in Davout's corps sat down with needle and thread at Viazma and sewed two huge pockets onto the inside of his coat, into which he stuffed all his most vital possessions, thereby ridding himself of the need to carry a bag.[37] Others were not as prescient and, when forced to make the choice, ditched the sack of grain or the bag of rice and kept the gold and silver vessels. It is easy to condemn them, but it must have been difficult to part with a lifetime's chance – of being able to afford to marry, to buy a house, to start a business. And they did not realise what lay ahead. They struggled on as best they could, hoping that things would improve.

As often as not, they had to bed down for the night in open countryside, with no cover of any kind, too exhausted to think even of building shelters out of branches. They would use their sabres to cut down saplings to burn. But the green, resinous wood produced clouds of acrid smoke before giving off any warmth, and quickly burnt out, so the fire had to be fed regularly all night. Even when they did get a good fire going it could only provide warmth for the hands and face, while their backs remained exposed to the temperature of the night. They would lay down branches around the fire to sit or lie down on, and huddle round the flames in tight groups of eight or ten men, hoping to create a small circle of warmth. But the fire would melt the snow, and they would soon find themselves sitting or lying in wet mud.

If they were lucky, they might find a half-ruined deserted village, but it usually contained unwelcome relicts. "Under its still-warm cinders, which the wind drove into our faces, would be the bodies of several soldiers or peasants," wrote Eugène Labaume, "and sometimes one could also see murdered children and young girls who had been slaughtered in the very place they had been raped." The generals and senior officers would usually take possession of the best of the remaining huts, but disagreements about precedence sometimes led to duels. The men would crowd into whatever huts, barns, sheds, styes or other shelter they could find. If there were large numbers of them they would trample each other in the process, on some occasions even suffocating those who had got in first and were pressed harder and harder by an incessant flow of new arrivals desperate to get out of the cold.[38]

Once a group occupied a hut, it would defend the entrance by force. But the thatch would soon be stripped off the roof by others eager to feed their horses. Those who could find no shelter would settle down in the lee of the hut and start tearing slats off the roof, shutters and any other accessible elements in order to build fires, with the result that those inside would find their shelter gradually dismantled around them. All too often those outside would build their campfires too close to the walls, and the huts would catch fire. If they were very crowded or those inside were asleep, they might be burnt alive.

Even without outside intervention, soldiers who found a hut for the night ran the risk of finding their death in it. The Russian huts were heated by stoves, about two metres square, made of wood rendered with clay, which had to be heated up gradually, but the frozen soldiers would stoke them up with every piece of wood they could lay their hands on, and as often as not the stove would catch fire and the hut would go up in flames as they slept.

The misery of having no shelter for the night was compounded by lack of food. Most of the rations brought from Moscow had been consumed by the time the army reached Mozhaisk, and they were now condemned to retreat along a road which had been devastated by the retreating Russians and then bled dry by themselves on the way to Moscow. It was not possible to send foraging parties out on either side of the road, for these would naturally fall behind the main body of the army, and become easy prey for the pursuing enemy.

As supplies ran out and materiel was left behind in abandoned wagons, organised feeding of the troops became impossible. There might be a sack of corn, but there would be no way of grinding it (a large supply of small grinders had been distributed at Dorogobuzh, but most had been left by the roadside as the horses died). There might be some groats or buckwheat, cabbage or scraps of meat, but no pot in which to make a stew.

Conscientious and resourceful officers who managed to keep their companies together ensured that essentials were not discarded, and organised the fairly shared consumption of whatever was available, so soldiers belonging to a disciplined unit had a better chance of survival. When they stopped for the night, one detail went in search of firewood, another built shelters, another prepared food, and so on; others were detailed to feed the pack animals; others still kept the fires burning and stood watch while their comrades slept.

Some units took care of themselves remarkably well. Dr La Flise, who had become separated from his regiment, fell in with a squadron of Polish lancers who would leave the road in the evening, find an inhabited village, surround it, and then strike a bargain with the peasants, promising not to harm them if they would just give them a little food and shelter for the night. They and their horses were thus able to keep in good shape, and as they had a couple of women with them, the officers would even spend the evenings entertainingly.[39]

Another who made sure he lacked for nothing was Colonel Chopin, commander of the 1st Cavalry Corps artillery. "A happy-go-lucky man who believed that the important thing in life was to think of oneself first, Colonel Chopin had, as soon as the retreat began, gathered about him a dozen of his most alert and resourceful gunners," recalled one of his comrades. "A *fourgon* with a good team of horses followed him and every evening this was the rallying point for all the gunners, each of whom brought in what he had managed to procure, either in the villages along the way or from isolated stragglers, from whom they took by guile or by force whatever they might have. In this way the Colonel's band (and one cannot call it otherwise) lacked for nothing, the *fourgon* was amply stocked, and watching and listening to his purveyors, one sensed it would never be empty."[40]

"It was rare for those who had stayed with their unit not to be able to share some kind of stew," wrote Colonel Boulart. "But woe betide those who had become separated, for they found no help anywhere." The only exception was if they had something to offer. Colonel Pelet watched a singular trade taking place around a large fire. "Who's got some coffee? I've got sugar. Who'll exchange some salt for flour? Who's got a pot? We could cook up a *popote* between us. Who's got a coffee pot?" and so on. "The man who had a small bag of salt could count on several days' food, as he could trade it everywhere," he wrote. Albert de Muralt owed his life to the possession of a small iron cooking pot, which he would lend to people who had food to prepare in return for being allowed to share their meal.[41] The only hope for those who had nothing was to team up with others in the same situation, and as a result corporations of eight or a dozen men sprang up, usually owning a horse or a wagon, which operated in much the same way as Colonel Chopin's gunners.

A particularly vulnerable group were the servants of officers. As they were not soldiers, they could not claim rations, and if their master were killed or wounded, or found them surplus to requirements, they were left without resource. By the same token, a good master was, for many, the only hope of salvation. In Moscow, General Dedem de Gelder had been prevailed upon to take on an extra servant, a bright young boy who drove his carriage and cared for his horses, and it was only much later, in the chaos of the retreat, that he realised the boy was a fifteen-year-old French girl who had fallen in love and run away from home to follow an artillery officer, only to see him killed at Borodino.[42]

Reading the accounts of survivors, one is struck by how little food was required to stay alive. But it was essential, for psychological as well as physical reasons, to have a regular supply. Lieutenant Combe had received a packet from home just before the army left Moscow. "What joy! News from Paris, from my father, my beloved mother, my whole family, my friends!" he wrote. "Nothing in the world could compare with what I felt then." It was only later that he would come to realise that this packet saved his life, for it contained little tablets for making hot chocolate and stock cubes to make bouillon. This meant that he could brew up a cup of something nourishing whenever all else failed. Others had the intelligence to load their pockets with tea and sugar, and quite a few claim to have survived for up to two weeks on nothing but tea.[43]

On the retreat as on the advance, thoughts of food never left the soldiers' minds. They would try to distract themselves by imagining that they were sitting down to dinner in one of the best restaurants in Paris. "Each of us would order his favourite dish, we would discuss their relative merits against other dishes, and in this way would distract ourselves for a while from the hunger which devoured us," recalled Victor Dupuy of the 7th Hussars, "but all too soon the horrible reality would assail us in all its power."[44]

The reality was indeed repellent, the principal source of meat being dead horses, but even that was not easy to come by. When a horse fell and could not find the strength to get up, soldiers would rush up and start cutting it up. The most experienced would slit open its stomach in order to get the heart and liver. They would not bother to kill the horse first, and would swear at it for making their job more difficult as it struggled and kicked. Captain von Kurz noted that after the men had finished, the carcase looked as though veterinary surgeons had been carrying out an anatomical investigation.[45]

Many were disgusted by the idea, as well as the taste, of horsemeat, but the taste could be smothered by tearing open a cartridge and sprinkling a liberal dose of gunpowder on it, and most of them soon got used to it. Jacques Laurencin, a geographer attached to Napoleon's headquarters, wrote to his mother explaining that horsemeat was really quite pleasant if sliced thinly and fried. General Roguet of the Young Guard thought it worth recording that the meat of the local *cognats* had a more delicate taste than that of French or German horses.[46]

Horses were not the only source of meat. "At Viazma we treated ourselves to a very good fricassée of cats," Laurencin assured his mother in a letter which would never reach her. "Five of us devoured three fine cats which were excellent." On the evening of 30 October, at Gzhatsk, Christian Septimus von Martens and his comrades cooked their first cat. "In order to allay the disgust which was welling up in us," he wrote, "I assured them that the gondoliers of Venice, who were by no means as miserable as we were at that moment, regarded a *ragoût* of cat as a treat."[47] The marching column was accompanied by dogs from the villages they had burnt, howling and disputing the carcases of horses with the famished men, and these too found their way into the pot if they were not careful. The pet hunting dogs or poodles various officers had brought along with them also began to disappear into cooking pots or onto the straight swords of cuirassiers and dragoons, which made good spits.

Bread was almost impossible to get hold of, but flour and groats of one sort or another could be obtained here and there, so the men would make a paste of these using water and chopped-up straw for binding, and bake it into flat biscuits in a peasant stove or in the ashes of a campfire. But usually they would throw anything they could find into a pot and boil up a pottage, often adding the stump of a tallow candle to provide nourishing fat. Jakob Walter from Stuttgart, who had found it so difficult to adapt to campaign conditions at first, had grown quite resourceful, learning to pick hemp seeds and dig up cabbage stalks, which could be turned into nourishment if boiled for long enough.

"We made our gruel with all kinds of flour mixed with melted snow," explained Captain François. "We would then throw in the powder from a cartridge, as the powder had the virtue of salting or at least of enhancing the bland taste of food prepared in this way." Duverger, the paymaster of the Compans division, wrote down the recipe for what he called "The Spartans' Gruel": "First melt some snow, of which you need a large quantity in order to produce a little water; then mix in the flour; then, in the absence of fat, put in some axle grease, and, in the absence of salt, some powder. Serve hot and eat when you are very hungry."[48]

The conditions under which they had to take their meals did not help. The men were often so hungry that they scoffed the food raw, and even if they did cook it they would swallow it hurriedly, in fear of the enemy. Among the consequences were vomiting, indigestion, colic and diarrhoea. Another reason for wolfing down any food they might come across was that it might otherwise be stolen. "Thieving and bad faith spread through the army, reaching such a degree of brazenness, that one was no more secure in the midst of one's own than one would have been surrounded by the enemy," noted Eugène Labaume. "All day long one heard only: " 'Oh God! somebody's stolen my portmanteau; or knapsack, or bread, or horse,' " recalled Louise Fusil.[49]

For many, particularly those who were on their own, stealing had become the only possible means of survival other than pilfering abandoned wagons, trunks and the pockets of those who had died along the way. Everyone despised these disbanded men, referring to them as *fricoteurs*, from the word *fricoter*, to cook something up, as they were often to be seen pathetically trying to concoct something to eat by the roadside. If they came up to a campfire looking for a little warmth they would be brutally pushed away. Sometimes they would stand just behind those sitting round the fire, hoping to glean at least some warmth from it.

Many of these unattached men walked over to the Russian bivouac fires to give themselves up, in their thousands on particularly cold nights. But their hopes that this would put a term to their sufferings were soon dashed, and their fate was not to be envied. Although they officially subscribed to the code accepted throughout Europe, the Russian attitude to prisoners was generally one of contempt.

There were some shining examples of consideration. When the much-loved young Colonel Casabianca, commander of the part-Corsican, part-Valaisain 11th of the Line, was captured outside Polotsk, his captors spared no effort to keep him alive. When he died of his wounds a few days later they returned his body, escorted by a guard of honour whose officer handed over a note from General Wittgenstein. "I am returning the body of the valorous Colonel of the 11th regiment, whom we mourn as much as you, for a brave man must always be honoured," it ran.[50]

Some officers treated their captured counterparts with courtesy. The partisan leader Denis Davidov went to great pains to trace and return the lover's ring, locket and love letters taken from a young Westphalian Hussar lieutenant who had been stripped of them on being taken by the cossacks. But his colleague Alexandr Samoilovich Figner took sadistic pleasure in slaughtering his prisoners, often when they least expected it. General Yermolov also ill-treated prisoners, particularly Poles, whom he despised as traitors to the Slav cause. After Vinkovo he spat in Count Plater's face and instructed the cossack escorting him to feed him only with lashes of his whip. Yermolov's attitude was not unusual. "Our soldiers took some prisoners among the French," noted a young Russian officer after the fighting at

Smolensk, "but all the Poles fell victim to vengefulness and contempt." When one officer reported in after a patrol in the course of which he had taken some French soldiers who were looting a church, he was told by his senior officer that he should not have bothered to bring them back. So he went out and told his men to bayonet them to death.[51]

The Tsar himself wrote to Kutuzov complaining of reports of ill-treatment of prisoners and insisting that all captured men must be treated humanely, fed and clothed. But the example set by his own brother undermined any chance of his complaints being heeded. General Wilson was riding along with other senior officers behind Grand Duke Constantine when they passed a column of prisoners. Their attention was attracted by one of them, a distinguished-looking young officer, and Constantine asked him if he would not rather be dead. "I would, if I cannot be rescued, for I know I must in a few hours perish by inanition, or by the cossack lance, as I have seen so many hundred comrades do, on being unable from cold, hunger, and fatigue to keep up," he answered. "There are those in France who will lament my fate – for their sake I should wish to return; but if that be impossible, the sooner the ignominy and suffering are over the better." To Wilson's horror, the Grand Duke drew his sabre and killed the man.[52]

There was a set of regulations in existence which laid down not only where prisoners should be held, but how much they were to receive for their sustenance. But it was a dead letter in the reality of this campaign. Sergeant Bartolomeo Bertolini, who had been taken while foraging on the eve of Borodino, could hardly believe the treatment he and his companions were subjected to. They were forcibly relieved of everything, even their uniforms and their boots. "Our misery was so great that I could never adequately convey it in words," he wrote. "They gave us no pay, as happens normally with prisoners among civilised nations, nor did they give us any rations to keep us alive." They were marched quickly, beaten, and killed if they strayed off the path to pick up a rotten potato or scrap of food.[53]

Dr Raymond Faure was taken at Vinkovo. He and other captured officers were brought before Kutuzov, who treated them with chivalry, giving them clothes and some money. The same treatment was not accorded to rank-and-file prisoners, who were robbed, stripped and beaten. And as soon as the convoy of prisoners left the Tarutino camp, under the escort of militia levies, the officers began to suffer the same fate, being robbed by the militia officers of everything Kutuzov had given them.[54]

By the time the retreat started the war had grown more vicious, and captives had become an unwelcome encumbrance: with food and clothing scarce on both sides, there was none to spare for them. As the Russian prisoners being goaded down the road by the French weakened and fell behind, they would be despatched with a bullet to the head. The Russians were no less brutal. Most of the prisoners were taken by cossacks, whose first action was invariably to strip them and take not only all valuables but also all serviceable items of clothing. They would then hand them over, or preferably sell them, to local peasants, who would massacre them with varying degrees of sadism.

Some would be buried alive, others would be tied to trees and used for target practice, others would have their ears, noses, tongues and genitalia cut off, and so on. General Wilson saw "sixty dying naked men, whose necks were laid upon a felled tree, while Russian men and women with large faggot-sticks, singing in chorus and hopping around, with repeated blows struck out their brains in succession." In one village the priest told his flock to be humane and drown the thirty prisoners under the ice of a lake rather than torture them. At Dorogobuzh, Woldemar von Löwenstern was horrified to see Russian troops stand by while the locals massacred unarmed camp followers with axes, pitchforks and clubs. "It was a ghastly spectacle," he wrote, "they looked like cannibals and a fierce joy lit up their faces."[55]

Common humanity did occasionally triumph in the midst of all this savagery, as in the case of Lieutenant Wachsmuth, a Westphalian wounded in the hip at Borodino. He was in the process of relieving himself by the roadside when some cossacks overran the group he was travelling in. Seeing him squatting helplessly with his trousers around his ankles, they burst out laughing and subsequently treated him well. Julien Combe had strayed off the main road with five other officers in search of fodder for their starving mounts, and got lost. After spending a cheerless night during which they were nearly buried under the snow, they found a hamlet where the peasants gave them shelter and food. "The snow was falling in thick flakes, and the aspect of this miserable countryside, seen through the small panes of dull yellow glass, the danger of our position, the uncertainty of our future, all seemed to conspire to plunge us into the most sombre reflections," he wrote. "But I was suddenly awakened from my musings by an exclamation of *Mama! Mama!* distinctly uttered by a child, whose cradle, suspended like a hammock by four ropes from the roof beams and hanging in a dark corner, had escaped our notice.

"Nothing could convey the impression that this word, almost a French one, made on us," he continued. "It brought everything back to us; it seemed to contain in itself all our memories of family, of happiness and of home." He took the child in his arms and wept. The mother was so touched that she looked after them and alerted them when cossacks were signalled in the area, giving them directions on how to escape and food for the journey.[56]

At Viazma, Lieutenant Radozhitsky, who was following the retreating French, came across a Russian woman who had been hired by a French colonel and his wife as a wetnurse for their baby. They had been killed in the fighting, but she had saved herself and the child. "He's only a little Frenchman, why bother with him?" the Lieutenant asked. "Oh, if you only knew how good and kind these masters were," she replied. "I lived with them as with my own family. How can I not love their poor orphan? I will not abandon him, and only death can separate us!"[57]

19

The Mirage of Smolensk

On 18 October, as Napoleon was setting out from Moscow, Marshal Gouvion St Cyr, who had taken over command of the 2nd Corps from the wounded Oudinot, was attacked outside Polotsk by overwhelming Russian forces under General Peter von Wittgenstein. In a fierce battle lasting two days his emaciated force of 27,000 French, Bavarian, Swiss, Italians, Poles and Croats held off Wittgenstein's 50,000 Russians, inflicting heavy losses. But when the city was set alight by the Russian artillery bombardment, it became indefensible. "No battle has ever appeared more awful," wrote Captain Drujon de Beaulieu of the 8th Lancers. "It made me think of the fall of Troy, as it is recounted in the *Aeneid*." Fearing encirclement, St Cyr abandoned Polotsk and fell back to the river Ula, along which he took up defensive positions.[1]

Napoleon did not hear of this until he reached Viazma on 2 November, but he was confident that Victor, who was marching to St Cyr's support, would assist him in retaking the city. He was more preoccupied with the slowness of Davout's retreat, complaining that he was deploying for battle every time a few cossacks appeared on the horizon, and himself marched briskly on towards Smolensk. But when he heard of the fighting outside Viazma and realised that Kutuzov was hovering a couple of miles to the south, he decided to give battle himself.

As he began to muster his forces on 4 November, he became aware of just how disorganised they were. "You want to fight, yet you have no army!" protested Ney, who had replaced Davout in the rearguard. Since Davout had now got free of Miloradovich and joined up with the preceding echelons, Napoleon decided to make for Smolensk and take winter quarters. He ordered Junot and Poniatowski to head for Smolensk itself, Davout to take up positions outside the city in the Yelnia area – "They say the country is rich and abundant in victuals," he assured him – and Prince Eugène to march to Vitebsk and take winter quarters there. He dictated these orders at Dorogobuzh on 5 and early on 6 November, before setting off towards Smolensk.[2]

He soon found himself driving through a blizzard, and as the temperature dropped he was forced to accept that he had got his timing dangerously wrong. That was not the only disagreeable reality he had to face that day. When he reached Mikhailovka that afternoon, he found an *estafette* from Paris waiting for him with the astonishing news that a couple of obscure officers, headed by General Malet, had attempted to seize power in a *coup d'état*. Napoleon could hardly believe it. The plot had been far-fetched in the extreme, but the very fact that it had got off the ground at all raised alarming questions about the solidity of Napoleonic rule in France. "With the French," he quipped to Caulaincourt, "as with women, one should never stay away too long." But he was shaken by this revelation of the fragility of his authority.[3]

The following morning he wrote to Victor, instructing him to join up with St Cyr and retake Polotsk. A note of real alarm is detectable in the letter. "Take the offensive, the salvation of the army depends on it," he wrote. "Every day of delay is a calamity. The army's cavalry is on foot, the cold has killed all the horses. Advance, it is the order of the Emperor and of necessity."[4] He himself made for Smolensk with all possible speed.

The cold had become so intense that Napoleon abandoned the traditional grey overcoat and small tricorn which made him instantly recognisable to all at a distance, and from now on wore a Polish-style fur-lined green velvet frock-coat and cap. He had also taken to warming himself by getting out of his carriage at intervals and tramping alongside his grenadiers, with Berthier and Caulaincourt at his elbow. It was while he was walking unsteadily on the slippery ice at noon on 9 November, with a temperature of −15°C (5°F) accentuated by a bitter north wind, that he caught sight of Smolensk. The thick blanket of snow that lay across the city, concealing the charred ruins, allowed him to forget what it had looked like when he had left it, and to entertain for a while the feeling that he had reached a safe haven.

As soon as he had set up quarters in the city he began dictating orders detailing the reorganisation of the cavalry into two divisions, one of light cavalry and one of cuirassiers and dragoons, each of which was to be divided up into picket regiments which were to cover the Grande Armée's winter quarters. He then ordered every unit to concentrate at specified assembly points in order to allow stragglers and detached elements to rejoin. But within a few hours grim reality had begun to bring home to him the futility of his plans, with a succession of painful blows.

Napoleon had given orders for large stores of food and equipment to be built up at Smolensk. But those who tried to implement them found that obtaining food and forage from the surrounding countryside was an unrewarding struggle, while supplies coming up the road from Vilna had to be sent on to Mozhaisk and Moscow. There were some 15,000 sick and wounded soldiers left over from the storming of the city and from Valutina Gora who had to be fed, while a constant stream of reinforcement echelons moving through on their way to Moscow, as well as Marshal Victor's 9th Corps which had been operating in the area, had been drawing on the stores as well.[5]

At the beginning of October Napoleon had issued urgent orders for the magazines to be restocked. One of those entrusted with carrying out these orders was Stendhal. "They expect miracles," he complained to a colleague as he set about the business, adding that he wished he could be sent to Italy.[6] Substantial stores were in fact built up, and there was certainly enough there to feed the Grande Armée for some time. But not enough to last through the winter for more than a division, and the idea of even a single corps taking winter quarters in the city was out of the question.

A more serious blow to Napoleon's plans was the news brought by Amédée de Pastoret, whom he had named intendant of White Russia, based at Vitebsk. Pastoret had built up a magazine there which could have fed one corps through the winter, and Napoleon had already assigned it to Prince Eugène's 4th Corps. But following the fall of Polotsk the Russians had moved down the Dvina and thrown Pastoret and his insignificant garrison out.

Another unwelcome piece of news waiting for him at Smolensk was that General Baraguay d'Hilliers, who had been sent out with his division to meet Napoleon's intended retreat along the Medyn road at Yelnia, had met not Napoleon but Kutuzov's main forces, and one of his brigades, General Augerau's, 1,650 strong, had been surrounded and forced to surrender.

As his own columns trudged into Smolensk from Viazma Napoleon could see how depleted they were. Estimates of the forces at his disposal in Smolensk vary wildly, but most sources agree that he had lost at least 60,000 men since leaving Moscow three weeks earlier, and that there were no more than about 40,000 left with their colours.[7] And this included several thousand cavalrymen who were of no use without their mounts. "Horses, horses and more horses, whether for cuirassiers, dragoons, or light cavalry, or artillery or military *caissons*, that is the greatest of our present needs," Napoleon wrote to Maret in Vilna on 11 November.[8] On the same day he heard of the disaster that had befallen his stepson.

He had ordered Prince Eugène to leave the main road at Dorogobuzh and make in a more or less straight line for Vitebsk. After a day's march he came to the Vop, an insignificant river no more than fifteen or twenty metres wide at this point, and his sappers set about building a bridge across it. The best they could do with the materials to hand was not good enough, and the bridge collapsed. The entire 4th Corps had by now come up, and a two-mile-long tailback formed as the troops waited for it to be rebuilt. As they stood patiently in driving snow and freezing temperatures, Platov's cossacks had time to come up and unlimber their guns, and began shelling the queuing Italians. With no possibility of rebuilding the bridge Prince Eugène decided to ford the river, which was nowhere deeper than a metre and a half. The Royal Guard led the way, and although the water came up to the chins of the shorter men, they got across without much difficulty.

Prince Eugène himself followed, and ordered the artillery to be brought over so it might deploy on the western bank and cover the crossing with its fire. But although it is not deep, the Vop flows between steep banks some three metres high, made slippery by the snow. After only two guns had been dragged across and up the opposite bank, a *caisson* got stuck and then overturned. The vehicle behind it also became wedged in the river's bed, and the one behind that slammed into the back of it. Other guns and *caissons* trying to bypass the jam also got stuck in the softened ooze, and soon the riverbed was a mass of vehicles whose wheels had sunk into the mud, and of horses desperately thrashing about trying to free themselves from the freezing water. "I can still see those brave soldiers of the train, obliged to spend whole hours with their teams in the water and, after having managed to drag one cannon or *caisson* out, go back in and double the team on another vehicle and start the struggle all over again," wrote Colonel Griois, who spent the whole day trying to get the guns across.[9]

He succeeded in dragging a dozen to the other side, but as night began to fall and the cossacks crept nearer, he realised that he would have to spike the rest. As soon as it became clear that the carriages and wagons would have to be abandoned, pandemonium broke out. Trunks were hauled down and broken open as men hurriedly transferred their most precious possessions and as much food as they could carry onto the backs of the unharnessed horses or their own before plunging into the river. Others seized the opportunity to rifle through the abandoned luggage of others before following. As they struggled to get across, many of the men and horses, gripped by the shock of the icy water, went under and drowned. Many more died of hypothermia as they huddled round bivouac fires in their wet clothes that night. "It is impossible to describe the situation of the men after this crossing, or the physical torments endured and the pain resulting from this icy bath," wrote one of them, and the Italians dubbed it "*la notte d'orrore.*"[10]

Prince Eugène lost around 2,500 men at the crossing, about a quarter of his force, as well as a large number of civilians and stragglers who had balked at the cold water. He also left behind fifty-eight spiked guns and his baggage train, which meant virtually all his rations and ammunition. He was now in no position to march as far as Vitebsk, and had to make a dash for Smolensk. This was just as well, since Vitebsk had fallen to the Russians. But the experience of the Vop crossing had demoralised many of his men and, notwithstanding his fine leadership qualities, there was not much he could do about it. "I ought not to hide from Your Highness," he reported to Berthier, "these three days of suffering have so crushed the spirit of the soldier that I believe him at this moment to be hardly capable of making an effort. Many men have died of hunger or of cold, and others, out of despair, have gone off to get themselves taken by the enemy."[11]

In Smolensk, Napoleon vented his frustration at the course events had taken by blaming all his marshals and accusing them of not carrying out his orders. "There's not one of them to whom one can entrust anything; one always has to do everything oneself," he complained to Pastoret in a long diatribe which covered many subjects. Everything was somebody else's fault, even his presence in Russia. "And they accuse me of ambition, as though it was my ambition that brought me here! This war is only a matter of politics. What have I got to gain from a climate like this, from coming to a wretched country like this one? The whole of it is not worth the meanest little piece of France. They, on the other hand, have a very real interest in conquest: Poland, Germany, anything goes for them. Just seeing the sun six months of the year is a new pleasure for them. It is they that should be stopped, not me. These Germans with all their philosophy don't understand a thing."[12]

Rant as he might, the retreat would have to go on. And it would have to be rapid, for St Cyr and Victor would not be able to hold back Wittgenstein for much longer, while Kutuzov was already overtaking Napoleon on his other flank. And an altogether new threat was developing in the south, where Schwarzenberg and Reynier had been obliged to give ground before the combined forces of Tormasov and Chichagov: instead of falling back towards Minsk, where they would have joined forces with Napoleon, they had gone off westwards, back into Poland, leaving Napoleon's line of retreat through Minsk dangerously exposed.

Napoleon's disappointment on reaching Smolensk was as nothing compared to that of his troops. The last stages of the march had sapped not only the physical strength but also the spirits of the bravest soldiers. "Morale nevertheless held," according to Dedem de Gelder, "the majority of the army believing that Smolensk would be the term of their misfortunes." On 7 November the front echelons passed a substantial convoy of food moving the other way destined for Ney's rearguard and this lifted their spirits, as it seemed to endorse the image of plenty at Smolensk. Soldiers hurriedly rejoined their units in the expectation of regular distributions of food. They somehow managed to forget that the last time they had seen the city it had been a smouldering ruin, and as they approached they had a picture of warmth and abundance in their minds. "The idea that the end of our travails was nigh lent us a kind of gaiety," wrote one, "and it was with many a joke about our prolonged slitherings and frequent falls that my comrades and I came down the hill and up to the city walls."[13]

But while the Guard, which entered the city with Napoleon, received a distribution of food and spirits, and settled into the ruins for a welcome rest, the units marching behind it were less fortunate. The Guard had been preceded by a rabble of fleeing deserters who

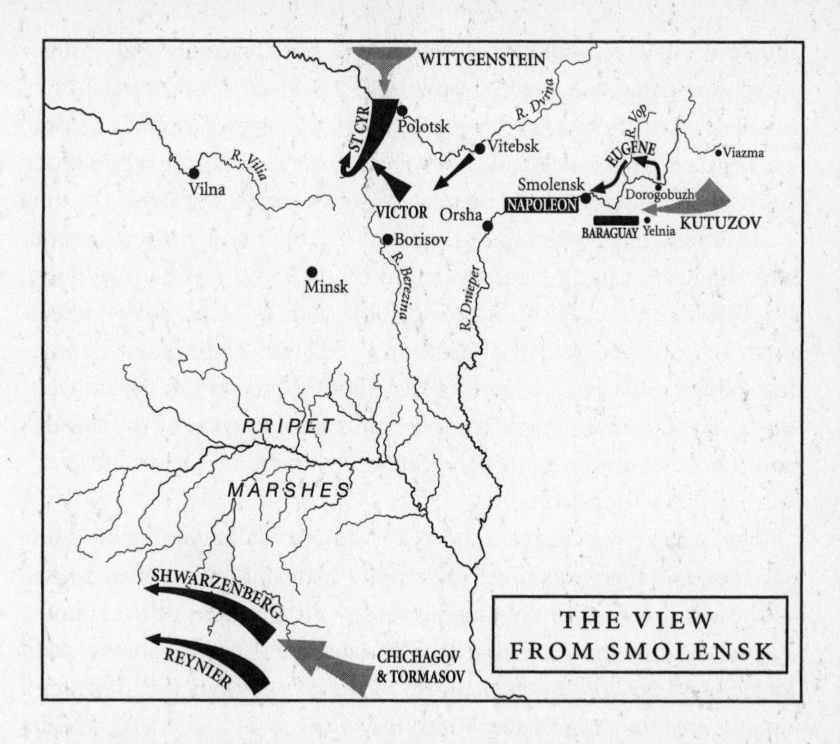

had tried to storm the stores, with the result that those in charge of distributing them became even more fussy than normal in following procedure. After the Guard had entered, the gates of the city were shut, and the gendarmes manning them admitted only armed units marching under the command of an officer. But as well as excluding the stragglers, this measure punished those who had fallen behind through no fault of their own, the wounded, and the cavalrymen whose units had quite simply dissolved through the death of their mounts.

Even those who managed to regroup outside the town and present an organised appearance received a less than satisfactory distribution. As Napoleon did not want news of his setbacks to spread, he had not warned the authorities in places such as Smolensk of his impending arrival, let alone of the real state of affairs. With prior warning, the local administration could have baked bread and divided stores up into rations which could have been distributed quickly and easily. As it was, companies were simply issued with sacks of flour which, as they lacked the means of baking bread, they boiled up into a thin gruel, an ox which they had to set about slaughtering, and a barrel of spirits, half of which would be wasted as it was decanted.[14]

All attempts at maintaining order were nullified by the deserters and stragglers who managed to infiltrate the city and set up dens of brigands in the cellars of burnt-out houses, from which they sallied forth to steal and raid the magazines. Fights kept breaking out in the stores, officials in charge of distributing them were beaten up, those carrying away rations for their units were waylaid by those who could not obtain food through regular channels, and a vast amount was wasted in the process.

The Guard was accused by other troops of having stolen the supplies, and there was much grumbling against it, but in effect most of those still with their colours did receive distributions of rice, flour, spirits and in some cases beef.[15] The Guard also aroused envy and anger as it appeared to take control of the great bazaar which sprang up at one of the main crossroads in the town.

The conditions of the retreat had turned out to be very different from those envisaged as they left Moscow, and as a result everyone was trying to adapt their arrangements by trading one kind of booty for a more manageable or transportable variety. "Here a suttler-woman would be offering watches, rings, necklaces, silver vases and precious stones," recalled Amédée de Pastoret. "There a grenadier was selling brandy or furs. A little further on a soldier of the train was hawking the complete works of Voltaire or the letters to Émilie by Desmoustiers. A *voltigeur* had horses and carriages on offer, while a cuirassier had set up a stall with footwear and clothing."[16] Those who had failed to get a regular distribution of food sold whatever they had in order to buy some.

The civilians, who did not qualify for military distributions, had no other way of obtaining it, and when they ran out of money or things to sell they were reduced to begging. In this, the women had an unenviable advantage, as Labaume records. "Mostly on foot, shod with cloth *bottines* and dressed in thin dresses of silk or percale, they wrapped themselves in pelisses or soldiers' greatcoats taken from corpses along the way. Their predicament would have wrenched tears from the hardest of hearts, if the rigours of our position had not been such as to strangle every feeling of humanity. Among these victims of the horrors of war, there were some who were young, pretty, charming, witty, and who possessed all the qualities capable of seducing the most insensitive man, but most of them were reduced to begging for the slightest favour, and the piece of bread they were given often required the most abject form of gratitude. While they implored our help, they were cruelly abused, and every night belonged to those who had fed them that day."[17]

The misery was compounded by the fact that on 12 November the temperature fell sharply, with readings as low as −23.75°C (−10.75°F). On the night of 14 November it was so cold that the men on picket duty around Ney's bivouac had to be threatened with the direst consequences to keep them from coming in to find shelter. Marshal Mortier took a more relaxed view. Seeing a sentry standing outside his lodgings, he asked him what he was doing and received the reply that he was on guard. "Against whom and against what?" Mortier asked. "You won't prevent the cold from coming in or hardship from attacking us! So you may as well come in and find a place by the fireside."[18]

A large proportion of the army was camped out in the open outside the city, and they struggled desperately to escape the cold. "Around our bivouac there were some huts in which officers and men had sought shelter from the cold and in which they had lit fires," recalled Sergeant Bertrand of the 7th Light Infantry in Davout's corps. "One of my good friends had gone inside as well. Foreseeing what was bound to happen, I begged him to come out. At my insistence, the officers and several soldiers, who were already numbed by the warmth and incapable of making a decision, did come out, but he would not hear of it and found his death there. As I had foreseen, crowds of other men soon began to assail these huts while those inside tried to defend their haven, a terrible struggle began and the weaker men were crushed mercilessly. I ran to the bivouac to get help, but I had barely reached it when flames engulfed the huts with all those inside them. In the morning there were only ruins and corpses." Sergeant Bourgogne, who had himself tried to get into one of the buildings, stood by helplessly as he watched screaming comrades being devoured by the flames.[19]

What made the conditions so hard to bear was the blow morale had suffered from the disappointed hopes. "A bivouac set up in deep snow in the ruins and the courtyard of a burnt-out house, a few meagre victuals, for the possession of which we had to come to blows at the entrance to the stores with thousands of ghosts enraged by hunger, and one single day of rest, with a temperature of [−22.5°C (−8.5°F)]: that was all we found in Smolensk, in those much-vaunted winter quarters," recalled an artillery officer of Ney's 25th Württemberg Division.[20]

"In an attempt to prevent the men from losing heart, the Emperor affected impassivity in the face of all this bad news, in order to make himself seem above all the adversity and ready to face any eventuality," noted Louis Lejeune. "But this was wrongly interpreted as indifference." The fatherly concern the troops used to sense in Napoleon was not in evidence. Auguste Bonet, a simple soldier, wrote to his mother from Smolensk on 10 November. "*Ma chère maman*, write to me often and at length, it is the only pleasure, the only consolation that remains to me in this wild country that the war has turned into a wilderness."[21]

Perhaps the most unfortunate were Prince Eugène's Italians who, having lost all their possessions and supplies at the crossing of the Vop, survived their icy bath and finally struggled into Smolensk, only to find the gates closed. After three hours of pushing, shoving and arguing they were at last admitted, only to discover that the supplies had been thoroughly pillaged. They camped in the streets, and the few wounded they had managed to bring along on their remaining wagons died in the night without shelter. "Many of us lost what was left of our spirit, of that spirit that kept hope alive," wrote Cesare de Laugier, while Bartolomeo Bertolini felt that "every soldier had lost the hope of ever seeing his motherland again."[22]

The Italian Guardia d'Onore, a kind of cadet force made up of the scions of the nobility of northern Italy who held officer's rank but served as simple soldiers, elicited general pity, for they lacked all the skills of the regular soldier. They had lost their mounts and tramped awkwardly in their ungainly top-boots instead of cutting them down, they had been too pampered to know how to fix their footwear or sew up a tear in their uniforms, let alone how to cook up a stew from whatever might be on offer; and they had been too well brought up to stoop to pillaging or even pilfering from dead men. Only eight of them survived out of a total of 350, which was low even by the standards of this campaign.[23]

Cavalry were particularly vulnerable, as every time a horse died another man was left behind. They were gradually dispersed, and therefore denied any mutual support system. So even while they had plenty of able-bodied men, cavalry units tended to disintegrate. On 9 November General Thielmann wrote to the King of Saxony that he must regard the two cavalry regiments which had been under his command as completely lost. But there were exceptions, and the lancers Dr La Flise had teamed up with rode into Smolensk with unfurled colours and music, and managed to get food for themselves and fodder for their horses.[24]

It required a strong hand to keep any regiment together, as the kindly but gruff Colonel Pelet of the 48th of the Line in Davout's corps attested. Not without effort, he had managed to obtain a quantity of flour, a barrel of vodka and four live oxen from the stores, but before he could set about feeding his men he was ordered to turn them out on parade before Davout. He was determined not to let his precious victuals out of his sight, so he took them along to the parade. Luckily, Davout was late. "I kept an eye as constantly as I could on the regiment and the barrel," Pelet wrote, "and suddenly

I noticed that it had been broken open. I ran over to it, but it was too late; nearly all the spirits had been pillaged, or at least distributed without measure or order. I hastened to overturn the barrel, but my men were already tipsy, and a number of them dead drunk. In order to hide this accident from the severe eye of Davout I tried to make the regiment manoeuvre, but this proved beyond them." He managed to lead them out of sight of the dreaded Davout's quarters and then came back to clear up. "More than eighty knapsacks, muskets and shakos were strewn about as after a battle," he added.[25]

Despite the general demoralisation, there was still a nucleus of disciplined men in most units, and many regiments found reinforcements in Smolensk, in the shape of echelons sent from depots in France, Germany or Italy. Pelet's regiment, for instance, had shrunk to six hundred men but found a couple of hundred uniformed and armed men waiting for them. Raymond de Fezensac's 4th of the Line was down to three hundred, but was joined by two hundred fresh men. The only problem with these men was that they had not been through the same tempering process as their comrades, and they were not up to dealing with the conditions. The 6th Chasseurs à Cheval received 250 recruits from their depot in northern Italy, but the shock to their system was such that not one of them was alive a week later.[26]

The loss of up to 60,000 men and possibly as many as 20,000 camp followers since leaving Moscow could, theoretically, have been to Napoleon's advantage. Caulaincourt was one of those who believed that if a couple of hundred cannon had been thrown into the Dnieper, along with the wagons carrying the trophies from Moscow, and all the wounded left in Smolensk with medical attendants and supplies, liberating thousands of horses, the slimmed-down but more mobile force of 40,000 or so men could have operated in a more aggressive manner and fed itself more easily. He blamed Napoleon for failing to take stock of the situation. "Never has a retreat been less well ordered," he complained.

It is certainly true that Napoleon's unwillingness to lose face prevented him from taking drastic measures and making a dash for Minsk and Vilna. He put off every decision to fall back further until the very last moment. "In that long retreat from Russia he was as uncertain and as undecided on the last day as he was on the first," wrote Caulaincourt. As a result, even the march could not be organised properly by the staff.[27]

But the real problem vitiating any attempt to reorganise the Grande Armée was that at every stop along the line of retreat it picked up fresh troops, who were often more of a liability than an asset, as well as *commissaires*, local collaborators, wounded and sick who had been left behind on the advance, and all the riff-raff who had been infesting the area under French occupation. As the Grande Armée retreated, it pushed all this dead weight before it, and had to march through it, losing resources and gaining chaos in the process.

Napoleon still entertained hopes of halting the retreat at Orsha or, failing that, along the line of the river Berezina. After four days in Smolensk, he sent the remnants of Junot's and Poniatowski's corps ahead, and left the city himself on the following day, 14 November, preceded by Mortier with the Young Guard and followed by the Old Guard. Prince Eugène, Davout and Ney were to follow at one-day intervals.

The going was hard, through deep snow which became slippery when compacted by the tramp of feet and hooves. There were many slopes in the road testing men and horses, and a number of bridges over small ravines causing bottlenecks. On the evening of the first day out of Smolensk, Colonel Boulart with part of the artillery of the Guard got stuck at a bridge which was followed by a steep rise. There was the usual jam of people, horses and vehicles, all vying for precedence, and every so often cossacks would ride up and cause panic. The Russians had now placed light guns on sleighs, which meant they could be brought up, fired and pulled away before the French had time to unlimber their cannon and fire back. Boulart realised that if he did not take decisive action, his battery would disintegrate in the midst of the jam. He therefore forced a passage for himself, by overturning civilian vehicles or pushing them off the road. He got his men to dig under the snow on either side of the road until they found earth, and to sprinkle this on the icy surface of the road leading up the slope, which he also broke up with picks. It took him all night to get his cannon across the bridge and up the slope. "I fell heavily at least twenty times as I went up and down that slope, but, sustained as I was by the determination to succeed, I did not let this hinder me," he wrote.[28]

While Boulart struggled with his guns, Napoleon, who had stopped at Korytnia for the night, called Caulaincourt to his bedside and again talked of the necessity of his going back to Paris as soon as possible. He had just heard that Miloradovich had cut the road ahead of him near Krasny. He could not rule out the possibility of being taken, and his close encounter with the cossacks outside Maloyaroslavets had unnerved him. In order to arm himself against capture he bade Dr Yvan prepare him a dose of poison, which he henceforth wore in a small black silk sachet around his neck.[29]

The following morning, 15 November, Napoleon fought his way through to Krasny, where he paused to allow those behind him to catch up. But Miloradovich had closed the road once more behind him, and when Prince Eugène's Italians, now not much more than four thousand strong, came marching down it the following afternoon they in turn found themselves cut off. Massed ranks of Russian infantry supported by guns barred the road in front of them, while cavalry and cossacks hovered on their flanks. Miloradovich sent an officer under a white flag to inform Prince Eugène that he had 20,000 men and that Kutuzov was nearby with the rest of the Russian army. "Go back quickly whence you came and tell him who sent you that if he has 20,000 men, we are 80,000!" came the reply. Prince Eugène unlimbered his remaining ten guns, formed up his corps into a dense column and forged ahead.

The Russians, who could see how few of them there were, once again summoned them to surrender. When this was rejected, they

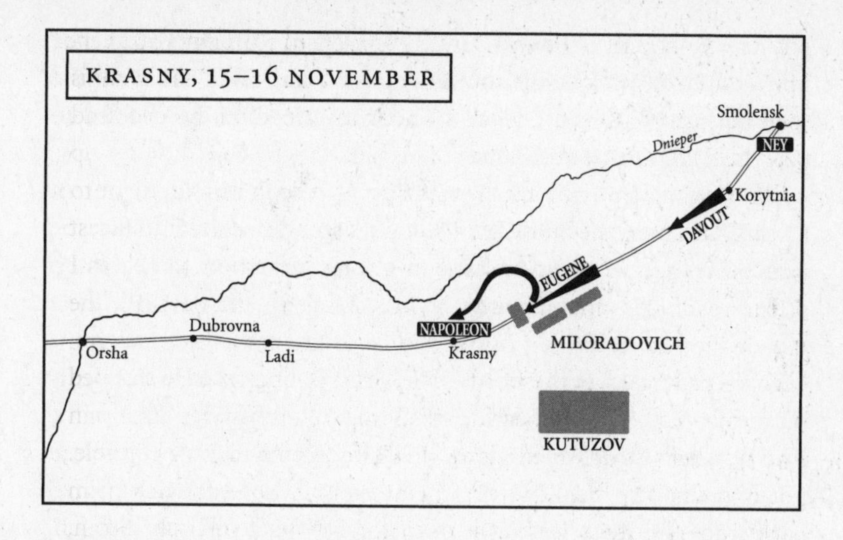

opened fire, and a fierce and bloody fight ensued. "We fought until nightfall without giving ground," recalled one French officer, "but it fell just in time; one more hour of daylight and we would probably have been overpowered." The Russians were nevertheless still between them and Krasny, and would easily crush them on the following day. In the circumstances, Prince Eugène could see no way out other than to fall in with the plan of a Polish colonel attached to his staff. When darkness fell, he formed up his remaining men in a compact file and, leaving behind all unnecessary impedimenta, marched off the road, into the woods, and across country round the side of the Russian army. When challenged by Russian sentries, the Polish colonel marching at the head of the column brazenly replied in Russian that they were on a special secret mission by order of His Serene Highness Field Marshal Prince Kutuzov. Unbelievably, the ploy worked, and in the early hours, just as Miloradovich was preparing to finish it off, the 4th Corps marched into Krasny behind his back.[30]

Napoleon was relieved to see his stepson, but he was now in something of a quandary. He ought to wait for Davout and Ney, in case they too had difficulty in breaking through Miloradovich's roadblock, but he was in peril of being stranded himself, as Kutuzov had turned up a couple of miles to the south of Krasny, and could easily cut the road between him and Orsha. In order to gain time, he decided to take the field himself at the head of his Guard.

Walking in front of his grenadiers, Napoleon led them out of Krasny back onto the Smolensk road and then turned them to face the Russian troops who had massed in a long formation to the south of the road. "Advancing with a firm step, as on the day of a great parade, he placed himself in the middle of the battlefield, facing the enemy's batteries," in the words of Sergeant Bourgogne. He was vastly outnumbered, but his bearing, standing calmly under fire as the Russian shells struck men all around him, seems to have impressed not only his own men but the enemy as well. Miloradovich moved back from the road, leaving it open for Davout to march through. And Kutuzov resisted the entreaties of Toll, Konovnitsin, Bennigsen and Wilson, who could all see that the Russians were in a position to encircle Napoleon and overwhelm him by sheer weight of numbers, ending the war there and then.[31]

Napoleon was alarmed to discover that Davout had hurried on westwards without waiting for Ney, who was still some way behind. But he could not afford to wait any longer himself, as Kutuzov had by now turned his wing and threatened his line of retreat to Orsha. He left Mortier and the Young Guard to hold Krasny and cover Davout's retreat, and himself marched through the town and out onto the Orsha road, at the head of the Old Guard.

It was not long before he came up against a horde of civilians and deserters who had gone on ahead and, finding the road cut by the Russians, come rushing back in a panic. Napoleon steadied them, but not before they had caused chaos in the ranks and among the wagons following the staff, with the result that some careered off the road and sank in the deep snow covering the boggy ground on either side of it.

As the French resumed their march, they were caught in a murderous enfilading fire from the Russian guns. The last of Latour-Maubourg's cavalry struggled to keep cossacks and Russian cavalry at

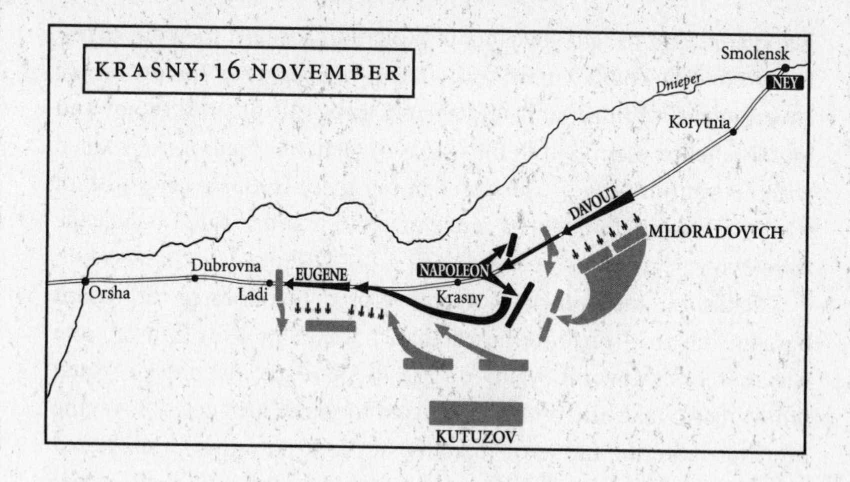

bay, while the dense column of men and vehicles made its way down the cluttered road. Colonel Boulart, who had managed to keep all his guns thus far, had a terrible job getting them through here too. The civilians and men who had left the ranks were getting in the way, and their skidding vehicles obstructed the road. Boulart cleared some ground at the side of the road, and, one by one, led his gun teams round the jam. But the chaos increased as the Russian artillery were now shelling the bottleneck, and when he went back for his last gun he found it impossible to move it among the exploding shells, so he spiked and abandoned it. As he struggled free of the mass of civilians with his last team, he saw a harrowing sight. "A young lady, a fugitive from Moscow, well-dressed and with striking looks, had managed to free herself from the mêlée and was moving ahead with great difficulty on the donkey she was riding, when a cannonball came and shattered the poor animal's jaw," he wrote. "I cannot express the feeling of sorrow I carried away with me as I left that unfortunate woman, who would betimes become the prey and possibly the victim of the cossacks."[32]

In an effort to push back the Russian guns, the infantry made a number of exhausting bayonet attacks through the deep snow, in which hundreds perished. Colonel Tyndal's Dutch Grenadiers, whom

Napoleon used to call "the glory of Holland," lost 464 men out of five hundred. The Young Guard was virtually sacrificed in the process of covering the withdrawal. The Russians kept out of musketshot and merely shelled them, but in the words of General Roguet, "they killed without vanquishing . . . for three hours these troops received death without making the slightest move to avoid it and without being able to return it."[33]

Luckily for the French, Kutuzov refused to reinforce the troops barring the road once he heard that it was Napoleon himself who was marching down it. Many on the Russian side felt a deep-seated reluctance to take him on, and preferred to stand by in awe. "As on the previous days, the Emperor marched at the head of his Grenadiers," recalled one of the few cavalrymen left in his escort. "The shells which flew over were bursting all round him without his seeming to notice." But this heroic day ended on a less solemn note as they reached Ladi late that afternoon. The approach to the town was down a steep icy slope. It was utterly impossible to walk down, so Napoleon, his marshals and his Old Guard had no option but to slide down it on their bottoms.[34]

The Emperor struck a more serious tone the following day at Dubrovna, where he assembled his Guard and addressed the dense ranks of bearskins. "Grenadiers of my Guard," he thundered, "you are witnessing the disintegration of the army; through a deplorable inevitability the majority of the soldiers have cast away their weapons. If you imitate this disastrous example, all hope is lost. The salvation of the army has been entrusted to you, and I know you will justify the good opinion I have of you. Not only must officers maintain strict discipline, but the soldiers too must keep a watchful eye and themselves punish those who would leave the ranks." The grenadiers responded by raising their bearskins on their bayonets and cheering.

Mortier made a similar speech to what was left of the Young Guard, which responded with shouts of "*Vive l'Empereur!*" A little further back in the marching order, General Gérard applied more summary methods when a grenadier of the 12th of the Line dropped out of the ranks announcing that he would not fight any more. He rode up to the man, drew his pistol from the saddle holster and, cocking it, announced that he would blow his brains out if he did not return to his place at once. When the soldier refused to obey, the General shot him. He then made a speech, telling the men that they were not garrison troops but soldiers of the great Napoleon, and that consequently much was expected of them. They responded with shouts of "*Vive l'Empereur! Vive le Général Gérard!*"[35]

Later on that same day, 19 November, Napoleon reached Orsha, where he hoped to be able to rally the remains of his army. The city was reasonably well stocked with provisions and arms. "A few days' rest and good food, and above all some horses and artillery will soon put us right," he had written to Maret from Dubrovna the previous day. He issued a proclamation giving assembly points for each corps, warning that any soldier found in possession of a horse would have it taken away for the use of the artillery, that any excess baggage would be burnt, and that soldiers who had left their units would be punished. He himself took up position at the bridge over the Dnieper leading into the town, ordering excess private vehicles to be burnt and unauthorised soldiers to give up their mounts. He then posted gendarmes there to carry on in his place and to direct incoming men to their respective corps and inform them that they would be fed only if they rejoined the colours.[36]

Watching the men streaming into town can only have heightened Napoleon's anxiety over Ney, who seemed irretrievably lost. That evening he paced the room he had occupied in the former Jesuit convent, cursing Davout for not having waited for Ney and declaring that he would give every one of the three hundred million francs he had in the vaults of the Tuileries to get the Marshal back. His anxiety was shared by the whole army, which held the brave and forthright Ney in high esteem. "His rejoining the army from beyond Krasny seemed impossible, but if there was one man who could achieve the impossible, everyone agreed, it was Ney," recorded Caulaincourt. "Maps were unfolded, everyone pored over them, pointing out the route by which he would have to march if courage alone could not open the road."[37]

Ney had been the last to march out of Smolensk, amid harrowing scenes, on the morning of 17 November. He had been ordered by Napoleon to blow up the city fortifications, and his unfortunate aide-de-camp Auguste Breton was given the job of setting the charges and then visiting the hospitals in order to inform the inmates that the French were leaving. "Already the wards, the corridors and the stairs were full of the dead and dying," he recorded. "It was a spectacle of horror whose very memory makes me shudder." Dr Larrey had put up large notices in three languages begging for the wounded to be treated with compassion, but neither he nor they had any illusions. Many of them crawled out into the road, begging in the name of humanity to be taken along, terrified at the prospect of being left at the mercy of the cossacks.[38]

Ney's corps by now numbered some six thousand men under arms, and was followed by at least twice as many stragglers and civilians. He marched along a road strewn with the usual traces of retreat, but beyond Korytnia the following morning he found himself crossing what was patently the scene of a recent battle. And that afternoon, 18 November, he himself came face to face with Miloradovich, who, having failed to capture Prince Eugène and then Davout, was determined not to miss his third chance.

He sent an officer with a flag of truce calling on Ney to surrender, to which the latter answered that a Marshal of France never surrendered. Ney then drew up his forces, opened up with the six guns he had left, and launched a bold frontal assault on the Russian positions. It was carried out with such *élan* that it nearly succeeded in overrunning the Russian guns barring the way, but the French ranks were raked with canister shot and a countercharge by Russian cavalry and infantry sent them reeling back. Not to be deterred, Ney mounted a second attack, and his columns advanced with remarkable determination under a hail of canister shot. It was "a combat of giants" in the words of General Wilson. "Whole ranks fell, only to be replaced by the next ones coming up to die in the same place," according to one Russian officer. "*Bravo, bravo, Messieurs les Français*," Miloradovich exclaimed to a captured officer. "You have just attacked, with astonishing vigour, an entire corps with a handful of men. It is impossible to show greater bravery."

But before long the French were beaten back once again. Colonel Pelet, who was in the front rank with his 48th of the Line, was wounded three times and saw his regiment decimated. The neighbouring 18th of the Line was reduced from six hundred men to five or six officers and twenty-five or thirty men, and lost its eagle in the attack. Fezensac's 4th lost two-thirds of its effectives. Woldemar von Löwenstern, who had been watching the proceedings from the Russian positions, galloped back to Kutuzov's headquarters and announced that Ney would be their prisoner that night.[39]

But this forty-three-year-old son of a barrel-maker from Lorraine was not so easily accounted for. Touchy and headstrong, Ney was furious when he realised that he had been left to fend for himself by Napoleon. "That b— has abandoned us; he sacrificed us in order to save himself; what can we do? What will become of us? Everything is f—ked!" he ranted. But it would take more than that to shake his loyalty to Napoleon. And if he was not the most intelligent of Napoleon's marshals, he was resourceful and certainly one of the bravest. After some discussion with his generals, he decided to try to give the Russians the slip by crossing the Dnieper, which flowed more or less parallel with the road some distance away, and then making for Orsha along its other bank, thus bypassing Miloradovich and putting the river between himself and the Russians.

While he made a show of settling down for the night, Ney sent a Polish officer to reconnoitre the banks of the Dnieper in search of a place to cross. A place was found, and that night, after having carefully stoked up enough bivouac fires to give the impression that the whole corps was camping there, Ney led the remainder of his force – not much more than a couple of thousand men – off the Smolensk–Orsha road and into the woods to the north of it. It was

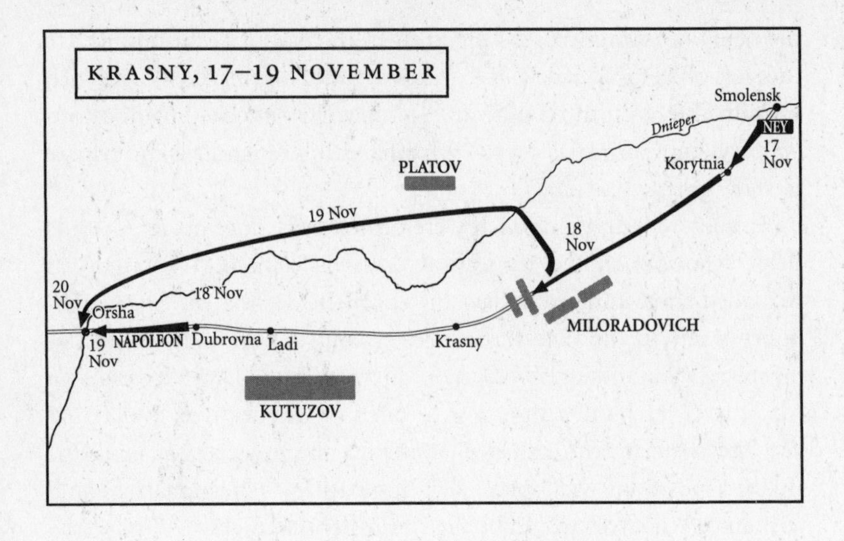

an exhausting and difficult march, particularly as he was still dragging his last few guns and as many supply wagons as he could through the deep snow. “None of us knew what would become of us,” recalled Raymond de Fezensac. “But the presence of Marshal Ney was enough to reassure us. Without knowing what he intended to do or what he was capable of doing, we knew that he would do something. His self-confidence was on a par with his courage. The greater the danger, the stronger his determination, and once he had made his decision he never doubted its successful outcome. Thus it was that at such a moment his face betrayed neither indecision nor anxiety; all eyes were upon him, but nobody dared to question him.”[40]

They soon got lost and disoriented, but Ney spotted a gully which he assumed to be the bed of a stream. Digging through the snow they found ice, and when they broke that they saw from the direction of flow which way they must follow it. They eventually came to the Dnieper, which was covered with a coating of ice thick enough to take the weight of men and horses spaced out, but not to support large groups or cannon drawn by teams of horses.

The men began to cross, leaving spaces between each other,

prodding the ice in front with their musket butts as it groaned ominously. "We slithered carefully one behind the other, fearful of being engulfed by the ice, which made cracking sounds at every step we took; we were moving between life and death," in the words of General Freytag. As they reached the other bank, they came up against a steep and slippery incline. Freytag floundered helplessly until Ney himself saw him and, cutting a sapling with his sabre, stretched out a helping limb and pulled him up.

Some mounted men and then a few light wagons did get across, encouraging others to try but weakening the ice in the process. More wagons ventured onto it, including some carrying wounded men, but these foundered through the ice with sickening cracks. "All around one could see unfortunate men who had fallen through the ice with their horses, and were up to their shoulders in the water, begging their comrades for assistance which these could not lend without exposing themselves to sharing their unhappy fate," recalled Freytag; "their cries and their moans tore at our hearts, which were already strongly affected by our own peril."[41]

All of the guns and some three hundred men were left behind on the south bank, but Ney had got over with the rest and soon found an unravaged village, well stocked with food, in which they settled down to rest. The following day they set off across country in a westerly direction. It was not long before Platov, who had been following the French retreat along the north bank of the river, located them and began to close in. Ney led his men into a wood, where they formed a kind of fortress into which the cossacks dared not venture. Platov could do no more than shell them with his light field-pieces mounted on sleigh runners, but this produced little effect.

At nightfall, Ney moved off again. They trudged through knee-deep snow, stalked by cossacks who sometimes got a clear enough field of fire to shell them. "A sergeant fell beside me, his leg shattered by a carbine shot," wrote Fezensac. "'I'm a lost man, take my knapsack, you might find it useful,' he cried. Someone took his knapsack and we moved off in silence." Even the bravest began to talk of giving up, but

Ney kept them going. "Those who get through this will show they have their b—s hung by steel wire!" he announced at one stage.[42]

Unsure of his bearings, Ney sent a Polish officer ahead. The man eventually stumbled on pickets of Prince Eugène's corps outside Orsha, and as soon as he was informed of Ney's approach, Prince Eugène himself sallied forth to meet him. Eventually, Ney's force, now not much more than a thousand men in the final stages of exhaustion as they stumbled through the night, heard the welcome shout of "*Qui vive?*," to which they roared back: "*France!*" Moments later Ney and Prince Eugène fell into each other's arms, and their men embraced each other with joy and relief.[43]

20

The End of the Army of Moscow

"Yet another victory!" Kutuzov wrote to his wife with touching swank the day after he let Napoleon slip through to Orsha. "On your birthday we fought from morning till evening. Bonaparte himself was there, and yet the enemy was smashed to pieces." His report to Alexander was more measured, but it did state that he had wiped out Davout and cut off Napoleon at Krasny, and it was backed up by his despatch to St Petersburg of the defeated Marshal's baton.[1] Alexander had the splendid velvet-covered and eagle-studded baton paraded in public as a trophy, but neither he nor anyone else was particularly impressed – a real victory would have been followed by the arrival of at least one captive Marshal of France, not just a few of his baubles.*

Vassili Marchenko, a civil servant who arrived in St Petersburg from Siberia in the first week of November, had found the city eerily quiet and gloomy. Many people had fled and the streets were empty. "Whoever was able to, kept a couple of horses at the ready, others had procured themselves covered boats, which waited, cluttering the canals," he wrote. "This sad state of affairs, uncertainty about the

---

* Davout's ceremonial uniform fell to a couple of officers who happened to be brothers, and they sent it home to their mother, who in turn donated it to the local church, where the gold braid was unpicked and used for a new chasuble.[2]

future, and the autumnal weather itself tore at the heart of good Alexander."[3]

This was all the more unwarranted as good news had been pouring in for the past four weeks. On 26 October there had been a solemn ceremony of thanksgiving for the victory at Polotsk. The following day the city resounded to the sound of gun salutes and church bells announcing the victory of Vinkovo and the reoccupation of Moscow. On the twenty-eighth Alexander and the Empress Elizabeth, accompanied by the Dowager Empress and the Grand Dukes Constantine and Nicholas, had driven in state to the cathedral of Our Lady of Kazan for a solemn celebration, and he had been cheered wildly by the crowds. "Courage is returning at the gallop, people have stopped sending away their effects, and I believe that some are even unpacking them," Joseph de Maistre had noted.[4]

But there was still much uncertainty. Kutuzov's rivals and their supporters implied that he had bungled the operations and that in his place they could easily have defeated and captured Napoleon. As the various commanders in the field had their partisans at court, St Petersburg was the scene of endless debates and recrimination. "To the foreign observer," recorded de Maistre, "it appears as either a farcical tragedy or an embarrassing comedy."[5] Alexander himself was by now mistrustful of anything he heard from Kutuzov, and he was receiving contradictory reports.

As he surveyed the map and digested the information coming in from his various armies it seemed clear to him that Vinkovo had not been followed up properly, that Maloyaroslavets had been a missed opportunity, and that a number of chances of cutting off and destroying Davout, Ney, Prince Eugène, Poniatowski and, finally, Napoleon himself had been thrown away. "It is with extreme sadness that I realise that the hope of wiping away the dishonour of the loss of Moscow by cutting the enemy's line of retreat has vanished completely," he wrote to Kutuzov, barely concealing his anger, and complaining of the Field Marshal's "inexplicable inactivity." But he could also see that Napoleon was now stumbling straight into a trap, with

Chichagov and Wittgenstein poised to cut off his retreat and Kutuzov coming up to destroy him from behind.[6]

Kutuzov had sent news of Vinkovo to the Tsar through Colonel Michaud, who also delivered the Field Marshal's invitation to Alexander to come and take command of the army himself. The Tsar had declined, but after four more weeks of apparent failures on Kutuzov's part, he was growing increasingly anxious that Napoleon might actually manage to get away if he did not take a serious hand in the matter. He bestowed the title of Prince of Smolensk on the Field Marshal for the alleged victories of Viazma and Krasny, but also summoned Barclay to St Petersburg.

Alexander briefly thought of going to take command of Wittgenstein's army, bringing about a juncture with Chichagov's and putting himself in a position to deal the final blow. On paper it looked as though he could not fail to destroy the Grande Armée and capture Napoleon, a tempting prospect for the frustrated warrior-Tsar lurking in Alexander.[7] But on balance it is probably a good thing as far as his reputation was concerned that he did not, for the situation as seen on paper and that on the ground were very different.

The long march that had all but destroyed the Grande Armée had also taken its toll of the pursuing Russians. They did enjoy various advantages over the French, as they were better clothed, received fairly regular distributions of food and forage, and had the initiative, so they could stop and rest when they needed to. But although they were better equipped to endure it, they were not immune to the weather. They were deft at building shelters, as Lieutenant Radozhitsky recorded: "It was hard and sad to watch as, having stopped near some small village, each regiment would send out a detail to fetch firewood and straw. Fences would shatter, roofs fall in and whole houses disappear in a flash. Then, like ants, the soldiers would carry their heavy loads to the camp and proceed to build a new village." But when there was no village nearby or they did not have the time, half of the men would spread their greatcoats on the snow and the other half use theirs as overblankets as they lay down, pressed together for warmth.[8]

After the first weeks of the pursuit, distributions of food became more erratic. The men could only expect hard-tack or the occasional thin gruel which they brewed up themselves. They had to rely more and more on sending out foraging parties. "It is wrong to blame only the French for burning and looting everything along the road," wrote Nikolai Mitarevsky, who had trouble feeding his artillery horses. "We did the same . . . When we went out foraging, those soldiers who in time of peace passed for scoundrels and cheats became extremely useful. Nothing escaped their eagle eye." By the time they had crossed the Dnieper they felt little compunction about taking everything from the locals, whom they did not regard as Russians and suspected of having sided with the French.[9]

If the French were moving along a devastated road, the Russian units were mostly marching cross-country, which made progress difficult, particularly for the artillery. Bennigsen suggested to Kutuzov that they leave behind four hundred of their six hundred guns, which would have speeded up the advance, but Kutuzov dismissed this along with all the General's other advice.[10]

"Men and horses are dying of hunger and exhaustion," noted Lieutenant Uxküll on 5 November. "Only the cossacks, always lively and cheerful, manage to keep their spirits up. The rest of us have a very hard time dragging on after the fleeing enemy, and our horses, which have no shoes, slip on the frozen ground and fall down, never to get up . . . My underclothes consist of three shirts and a few pairs of long socks; I'm afraid to change them because of the freezing cold. I'm eaten up with fleas and encased in filth, since my sheepskin never leaves me."[11]

Under such conditions the army melted away quickly. By the end of the fighting around Krasny, Kutuzov had lost 30,000 men, and as many again had fallen behind, leaving him with only 26,500 available for action. Mitarevsky found that by the time they had reached Krasny almost all the reinforcements, men and horses alike, that he had received at Tarutino had died or fallen behind. Their morale was inevitably affected by the horrors they witnessed as they followed in the wake of the Grande Armée. "Despite their being our enemies and destroyers, our desire for revenge could not stifle feelings of humanity to the point where we could not pity their sufferings," wrote Radozhitsky. But he found Kutuzov's bombastic proclamations a source of comfort and strength in his weariness and misery, likening them to "manna from heaven."[12]

Kutuzov certainly had a better understanding of his soldiers' needs than the parade-ground martinet Grand Duke Constantine, who had rejoined the army. Appalled at the sight of the dirty men wrapped in sheepskins, he held a review with the intention of smartening them up and appeared in full dress uniform without an overcoat in order to make his point. "He wanted to set an example, but we felt cold just looking at him," recalled Captain Pavel Pushchin of the Semyonovsky Life Guards.[13]

But even those soldiers who worshipped him "started saying that it would be good if our Field Marshal grew a little younger," according to Mitarevsky. After allowing the French to slip through at Krasny, he held a service of thanksgiving instead of pressing on with the pursuit. The more merciful pointed to his age and his poor health – he was often bent double with lumbago and suffered from acute headaches; others speculated as to his true motives. Several in his entourage, perhaps repeating each other's testimony (or possibly just their assumptions), report him as saying that his intention was not to destroy Napoleon but to provide him with "a golden bridge" out of Russia. Wilson recorded a conversation in which he claimed the Field Marshal told him that by toppling Napoleon Russia would gain little, while Britain would take over as the dominant power in Europe, which would not be in Russia's interests.[14]

Others, like Yermolov and Woldemar von Löwenstern, believed that what Kutuzov feared was a cornered and desperate Napoleon, even if he only had 30 or 40,000 fighting men with him. He knew that the Emperor was a better general than him, that his marshals and generals were superior to his own bickering subordinates, and that the soldiers of the Grande Armée would outfight his own, a huge proportion of whom were peasants who had been conscripted a couple of months before. Napoleon's orders, written out by Berthier with the usual formality, still mentioned corps, divisions and regiments as though they were fully operational fighting forces, and since many of these orders were now falling into the hands of the Russians, Kutuzov could only have formed the impression that they were.[15]

The most probable explanation for Kutuzov's behaviour is a combination of these motives. He hoped to wear Napoleon down further before taking him on. When he gave Yermolov a unit to command, he begged him to be prudent. "Little dove," he said with his usual familiarity, "be careful and avoid any actions in which you might suffer losses in men." He instructed Platov to harry the French and to "create incessant night alarms."[16] He himself would, by marching alongside them, force the French to hurry lest he cut them off, which would prevent them from regrouping.

Kutuzov was not the only one to err on the side of caution. Denis Davidov, a confident commander with a well-tried unit, would not take on anything more organised than a band of stragglers or an isolated platoon. Even when the retreating column looked temptingly disorganised, other Russian commanders took nothing for granted. "Groups of ten or twenty men would form up and refuse to let us scatter them," wrote Woldemar von Löwenstern. "Their bearing was admirable. We would let them continue their march and fall instead on other groups which did not put up any resistance." There was little point in the Russians exposing themselves, as they could take baggage and cannon without a fight, and as thousands of starving soldiers would come to their bivouacs at night to give themselves up anyway.[17]

Kutuzov probably preferred to take no chances and to wait until he could be certain of success. He was counting on the relatively fresh armies of Chichagov and Wittgenstein to cut off the French line of retreat along the Berezina. Napoleon's forces would be weakened by fruitless attempts to break through, which would allow Kutuzov to come up from behind and take the Emperor himself. And if Napoleon

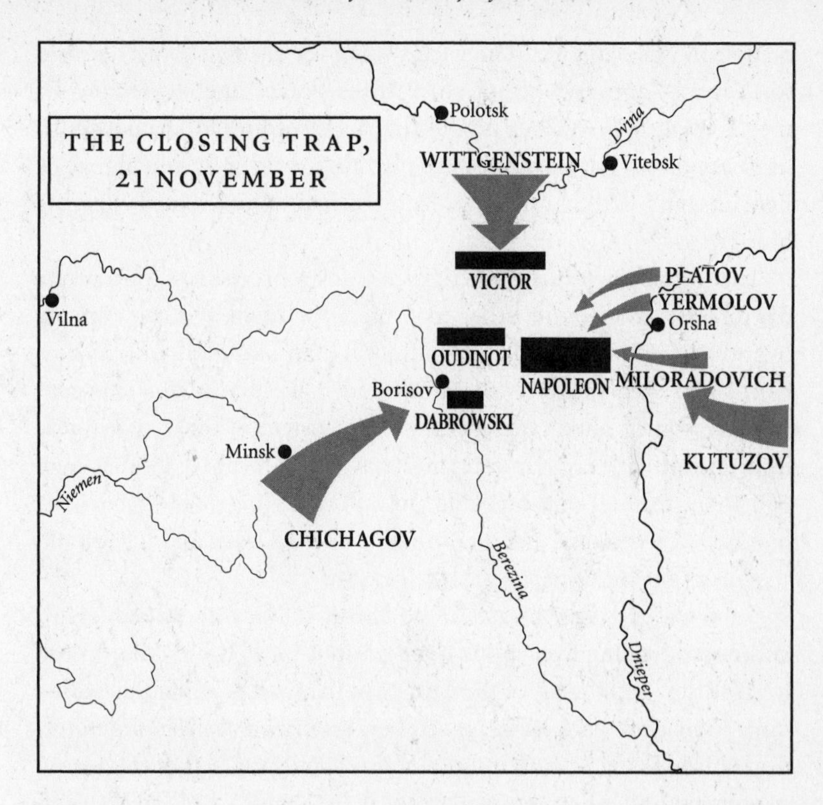

should get away, it could be blamed on one or both of the other generals.

Chichagov, whose army of Moldavia had been swelled by Tormasov's forces to a total strength of some 60,000 men, was moving fast to meet Napoleon head-on. On 16 November, the day Napoleon entered Krasny, the Admiral seized Minsk, Napoleon's best-stocked supply base. He then carried on towards Borisov, where a Polish division under General Dąbrowski was guarding the only bridge over the river Berezina. A couple of days' march to the north, Wittgenstein hovered threateningly with his 50,000 men over Napoleon's line of retreat, about halfway between Orsha and Borisov.

Although he still knew nothing about the fall of Minsk and assumed that Schwarzenberg was at least keeping Chichagov in check,

Napoleon was nervous. "Things are going very badly for me," he said to General Rapp, whom he called to his side at one o'clock on the morning of 18 November at Dubrovna. According to Caulaincourt, he guessed that the Russians were planning to encircle him along the Berezina, and that that was why Kutuzov had so far avoided engaging him.[18]

That same day he sent urgent orders to Dąbrowski to concentrate his forces at Borisov in order to protect the town and the crossing over the Berezina. He ordered Oudinot to join him there with his 2nd Corps and then move on to Minsk and make that safe. Victor was to make some feint attacks against Wittgenstein in order to give the impression that the Grande Armée was about to move against him. Napoleon realised that he could not afford to linger in Orsha as he had hoped to do, and decided to fall back on Minsk and try to hold the line of the Berezina.

"Will we get there in time?" he rhetorically asked Caulaincourt, and began turning over in his mind various plans for making a dash for it with what was left of the cavalry of the Guard. As if anticipating him, Chichagov had, on 19 November, published a physical description of Napoleon, with an injunction to all loyal subjects of the Tsar to apprehend him if he attempted to sneak through.[19]

Eager to cover the hundred kilometres that still separated him from Borisov, Napoleon moved from Orsha to Baran in the afternoon of 20 November, and it was there that he heard of Ney's miraculous escape and appearance at the French outposts that morning. The news electrified the army. "Never has a victory caused such a sensation," recalled Caulaincourt. "The joy was universal; we were intoxicated; everyone was in motion, coming and going in order to announce the news; telling everyone they met . . . Officers, soldiers, everyone felt that neither the elements nor fortune could hurt us any more, and we felt that the French were invincible!"[20] Although Napoleon is unlikely to have been quite so carried away, the news was welcome, and he appreciated its value as a morale-booster.

The distribution of rations at Orsha had brought a number of men back to the colours and the two-day pause had allowed stragglers to catch up. Those who had lost or thrown away their muskets were issued with fresh ones from the stores, which also contained sixty-two cannon. The remains of Ney's corps were thus able to replace the equipment they had left behind on the bank of the Dnieper. As luck would have it, there was also among the supplies waiting at Orsha a convoy of wagons carrying a long pontoon bridge. This was of no apparent further use to Napoleon, but the hundreds of fresh horses were invaluable. Napoleon ordered the pontoons to be burnt and the horses given to the artillery.

The weather was fine, with a light frost and a blue sky, as the Grande Armée marched out of the town on 21 November, and the road was straight and even. While Caulaincourt's assertion that Ney's escape had "restored to the Emperor all the faith in his lucky star" is perhaps a little wide of the mark, the whole army had been cheered, and "we set off once more with more gaiety."[21] It could hardly have been more misplaced.

The first leg of the retreat, between Maloyaroslavets and Smolensk, had been disastrous because it was unprepared in every way, and the arrival of the cold weather on 6 November had taken everyone by surprise. As the retreating columns struggled into Smolensk over the next three or four days, tens of thousands of men and horses died as much from undernourishment and exhaustion, both physical and moral, as from hypothermia: the temperature varied from −5°C (23°F) to not much lower than −12°C (10.4°F), which should have presented no problem to an organised army.

Even though the temperature dropped drastically while they were there, the army's short stay in Smolensk did give the men an opportunity to adapt to the circumstances. They adjusted their clothing as best they could, jettisoned some of the more ambitious booty they had set out with, and in most cases tried to provide themselves with personal reserves of food and drink. The first hardships had killed off the least resistant and prompted the weak-willed to give themselves up to the Russians, leaving the more determined and resilient to continue the march. And these gradually grew more used to the cold and the lack of food, becoming more resistant with every day.[22]

The next stage of the retreat, the five-day march from Smolensk to Orsha, was executed in far more difficult conditions than the first, with the temperature varying from −15°C (5°F) to −25°C (−13°F) and regular Russian forces harrying every step. It was dominated by the fighting around Krasny, with each unit having to run the gauntlet. And although the French were generally victorious, the five days of fighting around Krasny had emasculated the army of Moscow. Possibly as many as 10,000 of the best soldiers had been killed or wounded, over 20,000 (many of them civilians) had been taken prisoner and more than two hundred guns had been lost.[23]

As it marched out of Orsha to continue its retreat, the remnant of the army of Moscow was left alone by regular Russian forces. But stragglers were harried by the ubiquitous cossacks and the detachments of Davidov, Figner and Seslavin, and even by bands of French deserters who had settled in this part of the country and were now withdrawing alongside the army. Conditions remained difficult, with weariness and uncertainty about the future sapping the will. In spite of this, a nucleus kept going, displaying an astonishing degree of resilience.

They would set off at first light, as the short winter days of the north gave them little marching time. "We were always in a hurry to leave the frozen bivouacs where we had spent the night, and the hope of being more comfortable on the following night gave us the strength to bear the fatigues of the day," according to Colonel Griois. "It is in this way that for almost two months the hope of an improvement which never came kept us from succumbing to exhaustion."[24]

"We pursued our road in silence; one could hear only the sound of horses being struck and the sharp but frequent curses of the drivers when they found themselves on an icy incline which they could not climb," wrote Cesare de Laugier as he left Smolensk. "The whole road is covered in abandoned *caissons*, carriages and cannon that nobody has even thought of blowing up, burning or spiking. Here and there dying horses, weapons, effects of every sort; broken-open trunks, disembowelled bags mark the way taken by those who precede us. We also see trees at the feet of which people attempted to build fires and, around these trunks, which have been transformed into funerary monuments, the bodies of those who expired while trying to warm themselves. At every step there are dead bodies. The drivers of wagons use them to plug ditches and ruts, to even out the road. At first, we shuddered at such practices, but we soon became accustomed to them."[25]

"With his head bowed, his hands dug deep into his clothes and his eyes fixed on the ground, each one sullenly and silently followed the unfortunate who walked ahead of him," recalled Adrien de Mailly. "The plaintive screech of the wheels on the hardened snow and the croaking of the swarms of crows, of northern rooks and other birds of prey which always followed our army were the only sounds we heard." B.T. Duverger, paymaster to the Compans division, draws a similar picture. At Krasny he had tried to sell the paintings he had looted in Moscow, all neatly rolled up, but there were no takers, so he dumped them in the snow next to a fine collection of books beautifully bound in red morocco which a friend had also tried to sell. He then followed the flow passively. "I was neither gay nor sad," he wrote. "I had become quite indifferent to the circumstances and had decided to accept whatever destiny held in store."[26]

As they moved slowly, they did not cover much ground. But as they had to prepare shelter and fire for the night, they did not get much time to sleep either, and when they did their rest was interrupted by the need to keep the fire going or to move in order to keep warm. Dr Heinrich Roos noted that the younger soldiers, who needed more sleep, suffered this deprivation keenly, and that they were also prone to fall into such a deep slumber when they did get time to rest that they were more likely to freeze to death where they lay than older men.[27]

Every effort was made to keep the men together and under the colours. As the command of a given regiment stopped for the night, the drummers would start beating its signal. "This beating of the drum, dull in tone but audible a long way off, with its particular pattern of rolls and individual beats, slowing and quickening, made up a cadenced melody which was etched on the memory of the foot-soldier as distinctly as the sound of the village church bell on the ear of the rural inhabitant," explained Lieutenant Paul de Bourgoing of the Young Guard. "In time of war, the soldier has no other parish than his regiment, no church steeple other than his colours; when, lost in the night, exposed at every step to come up against enemy patrols or stumble into the midst of one of his columns, he can hear from far away the sound of the drum he recognises, and it is as though he heard a friendly voice egging him on through the murk and the distance."[28]

Sensible men realised that their best chance of survival lay in staying with the colours, and even when regiments were all but destroyed, a kernel stuck together, sometimes no more than a couple of dozen men clustered around their colonel and their eagle. When the number fell below that, they would generally take measures to safeguard the colours. Dr La Flise watched as, just after Krasny, the handful of officers and men of the 84th of the Line left alive unscrewed the eagle from the top of the staff and, wrapping it carefully, strapped it to the Colonel's back. Then, detaching the flag, they folded it and he buttoned it up under his uniform over his chest. After this they ceremonially embraced and set off, with the Colonel in the middle.[29]

Even cavalrymen who had lost their mounts and were obliged to follow on foot did everything to rejoin their mounted comrades for the nightly stop, although it meant making superhuman efforts, as they knew that they would find sustenance, both physical and emotional, among them. Some cavalry units decided to branch out and march parallel to the main road, as this made it easier for them to stay together. General Hammerstein took his remaining hundred Westphalian troopers off the road, and thanks to that kept them together successfully.[30]

Sergeant Bourgogne, who developed a fever after Krasny and fell behind his unit, provides a good example of what could happen to stragglers. He suddenly found himself walking alone along the road, in a gap between marching echelons, and although he was lucky enough not to encounter any cossacks, he saw many corpses of men who had evidently just been killed and stripped of their possessions. When a blizzard engulfed him he got lost and floundered despairingly through knee-deep snow, stumbling over the corpses of men and horses. He was famished, but could not hack away any part of the horse carcases he came across, since they were frozen rock-hard, and had to content himself with a handful of snow which had some horse's blood in it. He was soon reduced to a whimpering wreck, and would have perished if he had not been rescued by a comrade.[31]

Even small gaps between the marching columns were dangerous, as the hovering cossacks were ready to pounce wherever there was no danger to themselves. A wagon whose harness broke and required a pause for repairs was virtually doomed if these could not be completed before the tail of the column it was marching in passed. A soldier or a small group who stopped to chop up a dead horse or make a fire were similarly liable to be taken.

The fate of prisoners grew more dire as the retreat continued into its second month. On their capture they would be robbed. The large numbers of irregular cossacks had no interest in the war beyond looting, and they took anything and everything that might possibly have some value. Towards the end of the campaign one could see cossacks with a couple of dozen fob-watches strung around their necks, wearing several rich uniforms and coats, bedecked with gold epaulettes, a variety of resplendent plumed hats, with an array of booty of every kind strung from their cushion-like saddles. In the baggage left behind at Krasny one cossack found Ney's dress uniform, which he promptly donned, and thereafter French pickets were occasionally treated to the spectacle of what looked like a hirsute Marshal of France trotting up on a cossack pony and sticking his tongue out at them before galloping off.[32]

Regular cossacks, militia and even troops of the line also saw the war as a unique opportunity to make some money, and this included officers. They could not carry cumbrous booty, and would content themselves with money and valuables. So when the French surrendered to regular troops they were merely robbed of these. But this was of little comfort, as they would be relieved of everything else when they were handed over to the cossacks whose duty it usually was to escort them back into Russia.

Sub-Lieutenant Pierre Auvray of the 22nd Dragoons and four comrades had fallen behind on the march, and were captured by cossacks. "First they took my horse, which I had been fortunate enough to preserve from the rigours of the winter and which served to carry the personal effects of myself and my comrades," he wrote. "They looted our possessions and got hold of my portmanteau, which held some precious underclothes and a small box of jewels I had managed to procure in Moscow. Then they searched us and, finding no money on my person, they presumed that my wound must be concealing things that might satisfy their cupidity. They tore away my dressing with such violence that it caused me horrible pain. But so much suffering did not soften their hearts, and they undressed us and beat us with the wooden staves of their lances; we remained in this dreadful position in the snow, in this icy climate for some time, until the cossacks took fright at the approach of some French troops marching in force."[33]

Major Henri Everts of the 33rd Light Infantry in Davout's corps was taken prisoner at Krasny. He was stripped to the skin on the very battlefield by the Russian infantry to whom he had surrendered, and every single item of value was taken from him – they came to blows over his watch. When he and other officers were brought into the Russian camp they complained to an officer, and General Rosen found him an overcoat and gave him a drink in order to comfort him. The following day he set off in a column of 3,400 prisoners, of whom no more than about four hundred reached the provincial town in which they were to be held. The escorting cossacks did not give them any food, and let the local peasants torment them whenever they stopped for the night.[34]

Colonel Auguste Breton, one of Ney's aides-de-camp, was also taken at Krasny, having been wounded, but he was lucky in that he was taken under Milaradovich's gaze. The Russian General actually bound his wound up himself before sending him off to Kutuzov's headquarters, where the Field Marshal treated him amiably. But the moment he left headquarters he and his comrades were stripped and robbed by the escorting cossacks, who pocketed the money meant for the prisoners, gave them little food and took pleasure in gratuitous cruelty, such as not allowing them to drink when they reached a stream or not permitting them to make fires at night. Prisoners were force-marched, and if a man stopped to tie up his leggings or answer the call of nature he would be beaten and, if he did not rejoin the column fast enough, killed. They were sometimes given food and clothing by sympathetic landowners and even peasants as soon as they moved out of the area affected by the war, but this would often be taken from them by their escorts. Of one convoy of eight hundred, only sixteen were alive in June 1813.

As a rule, the later they were taken the worse the lot of the prisoners. When the cold became more intense, the cossacks found it amusing to strip prisoners and stragglers to the skin and leave them naked in the snowbound wilderness. The survival rate had never been good, but it grew much worse as the men were now weaker when taken and there was less food, clothing and shelter at the disposal of the Russians. And the further west they were taken, the longer the march back to the place of detention in Russia.

For many, the only hope of salvation was in establishing some connection that would take them out of the regular convoy. L.G. de Puybusque was captured by Platov's men, not far from Orsha. Platov was impressed by Ney's daring escape (and secretly delighted that Miloradovich had been made an ass of), and therefore treated him well. He was sent on to Yermolov's headquarters, which was a stroke of luck. "I had met him in the drawing rooms of Paris," recalled Puybusque. "And although, by order of our respective sovereigns, we had become enemies, and I found myself among the vanquished, he was the first to remind me of the circumstances in which we had met, which so many others in his position would have pretended to have forgotten. If he allowed me to see the extent of his authority, it was only by giving many orders to ensure that I should enjoy while in his company all the advantages of the most generous hospitality, and that I should have all my needs and those of my companion in misfortune catered for."

General Pouget, who had been the French military governor of Vitebsk, was fortunate enough to be taken not far from there, so that although he was robbed and beaten up by cossacks in the normal way on capture, he was then sent back into the city, where the inhabitants, to whom he had been fair and kind during his reign, interceded for him and recovered most of his possessions.[35]

For those remaining with the army the greatest affliction was now the cold, which added to their troubles at every level. Those fortunate enough to own a horse or carriage could not ride for long periods but had to dismount and walk in order to keep their circulation going. The roadway, churned up on a warmer day, turned into a terrifying ankle-twisting obstacle course and a lacerater of feet and worn boots when the ruts froze hard into jagged-edged canyons.

An unexpected consequence of the hard frost was that there was no water. Ice had to be melted over a fire, which meant that the labour involved in getting something to drink could be prodigious, requiring as it did both a fire and a vessel of some sort. Men and horses became dehydrated, which weakened them and contributed to their death – all the more effectively as they did not expect it at such a temperature.

This affected every function, as fingers grew clumsy and troops struggled with leather straps, harnesses and other pieces of equipment, which were stiff with cold. Even the soles of their boots had to be gradually softened, or they might snap. Men who walked to the side of the road and unbuttoned their pants in order to answer the call of nature, a frequent one since many of them had diarrhoea, would find to their horror that they were unable to button them up again. "I saw several soldiers and officers who could not button themselves up," wrote Major Claude Le Roy. "I myself helped to dress and button up one of these unfortunates, who was weeping like a child."[36]

Another consequence of the cold was a plague of frostbite. Most of the men who made up the Grande Armée came from climates where this phenomenon is entirely unknown, and they could therefore neither take elementary precautions, recognise the signs nor react in the appropriate way. They would try to get as warm as possible in a cottage, or heat their hands and faces at a fire, before setting off into the cold, little realising that this only made the exposed parts more vulnerable. When they did feel a numbness or someone pointed out that their nose had gone a telltale white, they would naturally seek the warmth of a house or rush up to a fire. This would induce instant gangrene. The affected part of the body would go a livid hue of purple and snap off as the sufferer tried to rub it. The only way of preventing damage of this sort is by rubbing the afflicted part vigorously with snow until circulation is restored, attended by excruciating pain. But few, apart from the Poles, the Swiss and some of the Germans, knew this, with terrible consequences. "To amuse the ladies, you can tell them that very probably half their acquaintances will return without noses or ears," Prince Eugène wrote to his wife.[37] But it was no laughing matter.

Captain François watched in disbelief as one of his friends unwrapped his feet from his improvised footwear as they settled down for the night. "As he took off the cloth and leather in which they were wrapped, three toes came away," he wrote. "Then, removing the rags from the other foot, he took the big toe, twisted it and tore it off without feeling any pain."[38] Once a man had lost all his toes, he could no longer walk without assistance; once he had lost his fingers, he was not only incapable of handling weapons, he could not get at any food, other than by tearing at the carcase of a horse with his teeth and sucking its blood.

The freezing temperatures made even this impossible. The horses, which had managed to keep going by eating tree bark and any bush or weed that pierced through the snow and by munching snow where there was no water, could not tear off the frozen bark or crunch the ice, so they died in their thousands. But a dead horse became rock hard in minutes, and its meat could not be cut up. So it was essential to find one that was still alive in order to be able to cut meat out of it.

It was but a short step from there to slicing steaks off a horse's hindquarters while its owner was not looking. The beasts did not feel the pain on account of the cold, and the blood froze instantly. They could carry on for days with such gashes in their hindquarters, but eventually the wounds would fester and start oozing pus, which itself froze. Another resource was to cut a horse's vein and suck out the blood, or collect it in a vessel and boil it up with melted snow to add nourishment to some thin gruel. Some would cut out and devour the tongue of the still-living horse. But the best nourishment was to be had by ripping open its belly and tearing out its heart and liver while they were still warm, and this was what increasingly befell those which, unable to go any further, were abandoned by their owners.[39]

From the moment the retreating army passed Ladi it was in former Polish lands, which meant the towns and villages were inhabited. Even in these dire times the ubiquitous Jewish shopkeepers could be relied on to lay their hands on the necessities of life – but only at a price. And the inhabitants were fearful of accepting anything which, if found in their possession later, might lay them open to reprisals on the part of the Russians.

Most currencies had lost their value. Dr Heinrich Roos remembered seeing a Württemberg soldier sitting by the roadside outside Orsha with a silver ingot in his lap, begging to exchange it for the slightest scrap of food, but nobody was prepared to part with life-saving rations in order to acquire a heavy piece of currency which would only have worth once he had got home. The only reaction he elicited from the men shuffling past him was a litany of cruel jokes. Even the last resort of the women – prostitution – was proving worthless in the circumstances. "There was no amorous intent in my action," Boniface de Castellane noted after having given a woman some chocolate. "We are all so tired that everyone is saying they would rather have a bottle of bad Bordeaux than the prettiest woman on earth."[40]

Those who did not manage to remain with their colours and had no money were obliged to pilfer. With the loss of so much of the baggage, the scope for this was greatly reduced, so the thieving often turned into violent robbery, and many a man was killed for a horse's liver, a crust of bread or any other kind of food. An isolated man with something to eat would carefully choose his moment to consume it when nobody was watching, otherwise he might get a bayonet in the back. But by the time Colonel Lubin Griois had found a safe place to consume the wonderful little loaf of white bread he had miraculously managed to acquire, it was frozen rock hard, and he wept as his teeth scraped ineffectually on the crust.[41]

Outside the units which had stayed together, all sense of camaraderie and solidarity had vanished. The various nationalities grew more resentful of each other, with the Germans cursing the French for having dragged them into the war, and the French cursing the Poles for supposedly being its cause. In the struggle for survival another's life meant nothing. "Many times, the implacable, frenzied rabble would shoot at each other when looting, when foraging, over a place to sleep, over a bowl of milk, over a shirt, over a pair of worn-out shoes," wrote Lieutenant Józef Krasiński.[42]

Anything left unguarded for a moment would vanish. An officer walking along with his horse's rein passed through his arm would look around to find that the rein had been cut and the horse taken. Men had their hats stolen from under their heads and the cloaks which covered them purloined as they slept. The length of the retreat and the loss of the baggage meant that their clothing, inadequate from the start, had reached a critical condition.

"My socks had given out a long time ago; my boots were worn out and all but sole-less; I had swathed them in straw which, with the aid of pieces of string, held the whole thing just about together," wrote Julien Combe. "My grey trousers and my uniform jacket were holed and worn threadbare, and I had been wearing the same shirt for the past month." To ward off the cold, Jean-Michel Chevalier, an officer in the Chasseurs à Cheval of the Guard, wore a flannel vest under his shirt, four waistcoats, one of them of sheepskin, his uniform, a frock-coat and a large cloak, four pairs of trousers or breeches over his underpants, and a bearskin busby.[43] But so many layers made walking, let alone fighting, difficult.

Colonel Griois was more sensible. Under his shirt he wore a flannel vest. His uniform consisted of a red woollen waistcoat, woollen trousers with no underwear, a tailcoat of light wool, and a light overcoat. On his feet he wore a pair of calf-high boots and cotton socks. He managed to get hold of an additional overcoat, but this was stolen. He tried to wear the bearskin cloak he had acquired in Moscow, but while this was excellent for sleeping in, it was far too heavy to march in. He therefore let his horse carry it during the day. But he did cut a strip off it to fashion into a muff, which he suspended from his neck by a string. He used another strip of the bearskin as a muffler, attached by string at the back of his head. "It is in this singular array, my head barely covered by a hat in shreds, my skin chapped by the cold and blackened by smoke, my hair covered in hoar frost and my moustaches bristling with icicles, that I covered the two or three hundred leagues from Moscow to Königsberg, and, in the crowd which flowed along the road, I stood out as one of those whose costume still conserved something of the uniform; the majority of our unfortunate companions looked more like ghosts dressed up for a masquerade. If they had kept elements of their military dress, one could not see it, covered as they were by the warmest clothing they could find. Some, lucky enough to have preserved their greatcoat, had turned it into a kind of habit with a hood, tied around their body with a piece of string; others had used woollen blankets or women's skirts for the same purpose. Many wore on their shoulders women's pelisses lined with precious furs, relics from Moscow and originally intended for sisters and mistresses. There was nothing at all unusual in seeing a soldier with a blackened and disgusting face dressed in a coat of pink or blue satin, lined with swansdown or Siberian fox, scorched by campfires and covered in grease stains. Most of them had their heads wrapped in filthy kerchiefs under the remains of forage caps, and in place of their worn-out footwear they had strips of cloth, blankets or leather. And it was not only the common soldier whom misery forced to such travesty. Most of the officers, colonels, generals were dressed in ways no less ridiculously beggarly, and one day I saw Colonel Fiéreck wrapped in an old soldier's greatcoat and wearing on his head, on top of his forage cap, a pair of breeches buttoned under his chin."[44]

Much of this carnivalesque frippery was quite useless, and those who survived best were often those who were most sensibly dressed. "I had no fur over my uniform, only a blue wool cloak with a very worn collar," wrote Planat de la Faye. "My boots, which I did not take off after Smolensk, had holes in their soles; and to protect my ears I had tied around my head a cambric kerchief, which had become as black as the shako which I kept over it. It is in this attire that I made the whole retreat, and yet I did not suffer any frostbite."[45]

It goes without saying that these clothes were in most cases in shreds and covered in dirt, as were the men, although some made heroic efforts to shave and keep clean. Their faces were filthy, blackened with smoke and smeared with the blood of the animals they ate, with long beards covered in hoar frost which hid the remains of food and saliva. "The most ragged beggars inspire pity, but we could have inspired only horror," wrote Colonel Boulart.[46]

They were also crawling with lice. "As long as we were out in the cold and walking," wrote Carl von Suckow, "nothing stirred, but in the evening, when we huddled round the campfires life would return to these insects, which would then inflict intolerable tortures on us." Colonel Griois also remembered the unpleasant duty that had to be got through every evening. "While our tasteless pottage was on the fire, we would take advantage of this first moment of rest to hunt down the vermin with which we were covered," he wrote. "This kind of affliction, which one has to have experienced in order to have an idea of it, had become a veritable torture which was made all the more powerful by the disgust which it inspired. In spite of all the precautions of cleanliness available on campaign, it is almost impossible, when one has to remain for several days and often entire weeks on end without leaving one's clothes, to preserve oneself entirely from these inconvenient guests. So from our very entry into Russia few of us had escaped this disagreeable inconvenience. But from the beginning of the retreat it had become a calamity; and how could it have been otherwise as we were obliged, in order to escape the deathly cold of the nights, not only not to take off any of our clothes, but also to cover ourselves with any rag that chance laid within our reach, since we took advantage of any free space by a bivouac which had been vacated by another, or in the miserable hovels in which we were able to find shelter? These vermin had therefore multiplied in the most fearful manner. Shirts, waistcoats, coats, everything was infested with them. Horrible itching would keep us awake half of the night and drive us mad. It had become so intolerable that as a result of scratching myself I had torn the skin of a part of my back, and the burning pain of this horrible and disgusting wound seemed soothing by comparison. All my comrades were in the same condition, and we showed no shame in our dirty searches and could perform them in front of each other without blushing."[47]

Such delicacy marks him out – it should not be forgotten that the overwhelming majority of the men marching down that road had been plucked from the most primitive backgrounds. Whatever feelings of shame they might have entertained vanished as quickly under the strain as their strength of character.

Some became helpless sheep, swept along in the general flow, incapable of helping themselves. In the evenings they would stand behind those who had made themselves campfires and were sitting around them. "Soon they would flag under the weight of fatigue, fall to their knees, then sit and then involuntarily lie down," wrote Louis Lejeune. "This last movement would be for them the precursor of death; their dull eyes would look up to heaven; a happy grin would convulse their lips; and one might have thought that a divine consolation was attenuating their agony, which was betrayed by an epileptic salivating." Hardly had that man died than another would come and sit down on his body, until he too fell into a stupor and died.[48]

Indifference to the suffering of others became general. Jean-Baptiste Ricome, a twenty-three-year-old sergeant, recalled how at the start of the retreat he felt agonies of pity when he heard dying men calling for their mothers, and how the familiarity of such cries gradually bred indifference. The struggle for survival hardened the kindest hearts, and men trudged on as their comrades slipped and fell on the ice. "In the beginning, they would find help," wrote Colonel Boulart, "but as the same fate threatened everyone and the frequently repeated falls suggested the futility of assisting them, one passed by these hapless men, who lay on their bellies on the ice, making vain efforts to get up, or scratching the ground in front of them as they battled with death, and one did not stop!"[49]

"This campaign became all the more frightening as it affected our very nature, giving us vices which had until then been unknown to us," wrote Eugène Labaume. On one occasion several hundred men crammed themselves into a large barn for the night, in the course of which the fires they had made set fire to the thatch of the roof and eventually to the whole structure. The rapidity with which the blaze spread made it impossible to save more than a couple of dozen, and the rest perished, saluted only, as Colonel Lejeune observed, by the discharges of the cartridges going off in their loaded muskets. Comrades who had rushed to their aid could only look on in horror. But within a couple of weeks they would simply come up and warm themselves when this kind of thing happened. Such fires were sometimes started on purpose out of miserable fury by men who could find no shelter, and those who stood around warming themselves would make jokes about the quality of the fire.[50]

One night Davout's headquarters found shelter in a large peasant cottage in a deserted village, only to discover that there were three babies still alive lying in the hay in the stable shed, wailing from hunger. Colonel Lejeune told Davout's butler to give them something to eat, which he did. The babies nevertheless continued to wail, preventing them all from sleeping. Lejeune did fall asleep, and when he awoke it was time to move off. As all was quiet he did not think of the babies, but when he enquired about them of the butler later that day he was told that, being able to stand it no longer, he had taken an axe, broken the ice on the drinking trough and drowned them.[51]

According to Captain François, "Anyone who allowed himself to be affected by the deplorable scenes of which he was a witness condemned himself to death; but the one who closed his heart to every feeling of pity found strength to resist any hardship." It certainly took character, as well as fitness, to survive. "A small number of us, with exceptionally strong characters, supported by youth and a solid constitution, resisted all the elements conspiring to our destruction and came out of it well," wrote Louis Lagneau, a surgeon with the Young Guard. "I was thirty-two years old, my health was perfect, I was very used to walking long distances, and as a result I bore everything without any unfortunate results."[52]

Another who bore it all remarkably well was Napoleon himself. He did have the benefit of a regular supply of food and wine, not to mention other comforts. An officer would ride ahead to select a place for the Emperor to stop for the night, which, be it a devastated country house or a peasant hovel, would be made amenable. The iron camp bed would be set up, a rug spread on the floor and the *nécéssaire* containing razors, brushes and toiletries brought in. A study would also be improvised, in the same room if no other could be found, with a table covered in green cloth, the Emperor's travelling library in its case and the boxes containing maps and writing instruments. A small dinner service would be unpacked so that he could eat off plate.

"He bore the cold with great courage," recalled his valet, Constant, "but one could see that he was physically very affected by it." Even though he did have the luxury of a change of clothes, and despite the resources of the *nécéssaire*, Napoleon also had lice. And despite the comfort of his camp bed, he suffered from insomnia – no doubt caused by uncertainty as to what lay ahead and a feeling of responsibility for his army. "Those poor soldiers make my heart bleed, yet I can do nothing for them," he said to Rapp one evening.[53]

Their faith in him remained unshaken. Many grumbled and cursed him, but while some became insolent and insubordinate with their officers and even with generals, they would fall silent and respectful whenever he appeared. "Soldiers lay dying all along the road, but I never heard one complain," recalled Caulaincourt. "Although this man was, rightly, regarded as the author of all our misfortunes and the unique cause of our disaster," wrote Dr René Bourgeois, who held profoundly anti-Napoleonic political views, "his presence still elicited enthusiasm, and there was nobody who would not, if the need arose, have covered him with their body and sacrificed their lives for him." The degree of their devotion is well illustrated by Sergeant Bourgogne, who watched as an officer accompanied by a couple of grenadiers came up to a bivouac asking for some dry wood for Napoleon. "Everyone eagerly proffered the best pieces he had, and even those men who were dying raised their heads to whisper: 'Take it for the Emperor!,' " he recalled. Such devotion was not universal, it is true. On one occasion when Napoleon wanted to stop and warm himself by a fire surrounded by stragglers, Caulaincourt walked over to them but, after exchanging a few words, came back suggesting that perhaps it might not be a good idea to stop there.[54]

Paymaster Duverger, who not being a combatant felt nothing of the soldier's devotion to his chief, still agreed that "his prestige, that kind of aura that surrounds great men, dazzled us; everyone gathered in confidence and obeyed the slightest indication of his will." It is true that Napoleon represented their best chance of getting out of the mess they were in. "His presence electrified our downcast hearts and gave us a last burst of energy," wrote Captain François. "The sight of our overall chief walking along in our midst, sharing our privations, at some moments even brought out the enthusiasm of more victorious times." Whatever their nationality, and whatever their political attitude to him, men and officers alike realised that only he could keep the remains of the army together, and that only he was capable of snatching some shreds of victory from the jaws of defeat.[55]

But there was more to it than that. A German artillery officer who might have been expected to curse the foreign tyrant who had brought him to this as he tramped past Napoleon, standing at the side of the road, expressed feelings which, surprising as they were, were not uncommon. "He who sees real greatness abandoned by fortune forgets his own suffering and his own cares, and as a result we filed past under his gaze in gloomy silence, partially reconciled to our harsh fate."[56]

Ségur sought a metaphysical explanation of this phenomenon. They should have blamed Napoleon but did not because he belonged to them as much as they to him, he argued. His glory was their common property, and to diminish his reputation by denouncing him and turning away from him would have been to destroy the common fund of glory they had built up over the years and which was their most prized possession. This seems to be borne out by the fact that even when they were taken prisoner, the soldiers of the Grande Armée refused to say a word against Napoleon. According to General Wilson, they "could not be induced by any temptation, by any threats, by any privations, to cast reproach on their Emperor as the cause of their misfortunes and sufferings."[57]

Spirits rose when, as they approached Borisov, the army of Moscow met up with Oudinot's 2nd Corps and other units that had been stationed in the rear and had therefore not suffered all the rigours of the retreat. Lieutenant Józef Krasiński, retreating with the bedraggled remnants of the Polish 5th Corps, burst into tears of joy when he saw Dąbrowski's division near Borisov, properly uniformed and marching behind its band. Reactions on the other side were correspondingly painful.

Grenadier Honoré Beulay, who had only recently marched over from France, was incredulous as he watched the retreating units tramp by. "We stood there with our mouths open, wondering whether we were not mistaken, whether these men who hardly resembled human beings were really Frenchmen, soldiers of the Grande Armée!" he wrote. The appearance of the army of Moscow had an unsettling effect on Oudinot's and Victor's corps. "It had been hoped that our example would exert a salutary influence," noted Oudinot. "Alas! it was quite the opposite that prevailed."[58]

What was far more unsettling, however, were the terrifying rumours that flew through the army to the effect that Borisov had fallen to the Russians and that they were now cut off.

21

The Berezina

On 22 November Napoleon reached Tolochin, where he took up quarters in a disused convent. He had not been there long when he heard, from a rider sent by Dąbrowski, that Minsk had fallen to Chichagov six days before. "The Emperor, who by that one stroke lost his supplies and all the means he had been counting on since Smolensk in order to rally and reorganise his army, was momentarily struck with consternation," according to Caulaincourt.[1]

He had been expecting Chichagov to manoeuvre himself into a position to be able to join up with Kutuzov so they could attack him with overwhelming force, not to move into his rear and attempt to cut him off. As it happens, Chichagov was operating in the dark. He had received only scanty orders from Kutuzov, who had instructed him to move into Napoleon's rear and to prevent the French from linking up with Schwarzenberg. Wittgenstein was supposed to cross the Berezina further north and link up with him, so that between them they covered a long stretch of the western bank of the river.[2]

That night Marshal Duroc and Intendant Daru were on duty at the Emperor's bedside, and the three of them sat up late. They discussed the situation at length, and Napoleon allegedly reproached himself for his own "foolishness." He dozed off for a while, and when he woke he asked them what they had been talking about, to which they answered that they had been wishing they had a balloon. "What on earth for?" he asked. "To whisk Your Majesty away," one of them replied. "The situation is not an easy one, it is true," he admitted, and they discussed the possibility of his falling into Russian hands. General Grouchy was instructed to gather all the cavalry officers who still had good mounts into a "dedicated squadron" whose purpose would be to spirit Napoleon to safety in an emergency. But the Emperor remained sanguine, and if he did order the burning of some state papers before they set off in the morning, that was more to lighten the load than anything else – he also ordered the burning of more non-essential carriages. He appeared confident that he would be able to fight his way through.[3]

What he did not know was that while he was digesting the news of the fall of Minsk, Borisov had also fallen to Chichagov. The Admiral, who had a healthy respect for him, was apparently unaware that he was on a collision course with Napoleon, whose whereabouts he did not know, but whose forces he assumed to be at least 70,000. In the event, his advance guard had moved quickly, surprised and defeated the detachment of Dąbrowski's division holding the bridgehead on the western bank of the Berezina and swept into Borisov itself, which it occupied after a stubborn resistance. The Russians then made themselves at home, and their commander, Count Pahlen, sat down to a copious dinner. He had hardly swallowed a mouthful when the alarm was sounded. An advance unit of Oudinot's corps, consisting of five hundred men of Colonel Marbot's 23rd Chasseurs à Cheval, had burst into the town and fallen upon the unsuspecting Russians. No more than about a thousand of them managed to save themselves by fleeing back across the river, leaving behind up to nine thousand dead, wounded and prisoners, ten guns and all their luggage.[4]

But the fleeing Russians had had the presence of mind to fire the long wooden bridge, the only crossing over the Berezina. Napoleon had reached Bobr when he heard of this, and he must have rued the decision to burn the pontoon bridge at Orsha three days before. The boggy trough of the Berezina ran between him and freedom; cold as it was, a slight thaw had broken up the ice on it, and it represented a considerable obstacle. "Any other man would have been overwhelmed," wrote Caulaincourt. "The Emperor showed himself to be greater than his misfortune. Instead of discouraging him, these adversities brought out all the energy of this great character; he showed what a noble courage and a brave army can achieve against even the greatest adversity."[5]

Napoleon momentarily entertained a plan to gather up all his forces, march northwards, knock out Wittgenstein and then make for Vilna, bypassing the Berezina altogether. But he was advised that the terrain was unfavourable for such operations. Instead, he decided to fight his way across the river at Borisov. This would involve repairing the existing bridge and building new ones under enemy fire. In order to reduce the resistance, he decided to disperse Chichagov's forces by giving him the impression that he was planning to cross elsewhere. He sent a small detachment southwards to make a demonstration of activity at a possible crossing point further downstream, and even managed to misinform some local Jewish traders that he was intending to cross there, expecting them to pass the news on.[6]

Everything depended on speed: Wittgenstein and Kutuzov would be coming up behind him in a couple of days, and what would happen if he were caught in the rear by them while attempting to force a passage across the river did not bear thinking about. Napoleon seemed to be energised by the crisis, and did not appear downcast. "The Emperor seemed to have made his mind up with the calm resolve of a man about to embark on an act of last resort," noted his valet, Constant.[7]

The forward units and large numbers of fugitives poured into Borisov on the night of 23 November. The town was strewn with dead bodies and debris from the previous night's fighting. "This countless mass of wagons, with women, children, unarmed men had packed into Borisov in the conviction that the bridge would be repaired and that the crossing would be made there," wrote Józef Krasiński of Poniatowski's 5th Corps, which had also entered the town. "The streets of Borisov were so jammed with this wagon train that it was impossible to pass through them without pushing and crushing people. As a result the streets were covered in mauled bodies, shattered wagons, smashed baggage, and all one could hear were shouts, calls, wails and lamentation . . . I remember that on one of the streets I pulled from beneath the horses' hooves a baby lying in the middle of the road in its swaddling clothes, and further along I saw, by a small bridge, a *cantinière*'s wagon lying in the water into which it had been pushed by the French troops marching before us, and on that wagon the poor woman with a child in her arms was calling for help which none of us could give her."[8]

When General Eblé and his pontoneers reached Borisov and saw the state of the river they were discouraged. It was wider than they had anticipated, and the recent thaw meant that large blocks of ice were being swept down it by a slow but strong current. General Jomini, who was with Eblé, suggested that they cross further north, at Vesselovo, where there had been a bridge which might still be standing. But Oudinot had already identified a better place. One of his cavalry brigades, General Corbineau's, which had been clearing the western bank of the Berezina of cossacks during the previous week, had just rejoined his corps having found a ford by the village of Studzienka, a dozen kilometres upstream from Borisov.

Oudinot had immediately informed Berthier of the existence of the ford, recommending it as the best place for a crossing. But Napoleon stuck to his intention of forcing a passage at Borisov, meaning to defeat Chichagov and then make a dash for Minsk, from where he hoped to be able to make contact with Schwarzenberg. From Loshnitsa at 1 a.m. on 25 November he repeated his orders to Oudinot, urging him to make haste so they could start crossing that very night. Oudinot, who had already ordered some of his units to Studzienka in anticipation, begged Napoleon to reconsider, and sent Corbineau to see him in person. It was only after he had discussed the matter with Corbineau that Napoleon accepted Oudinot's suggestion, and he set off for Studzienka himself late that night.[9]

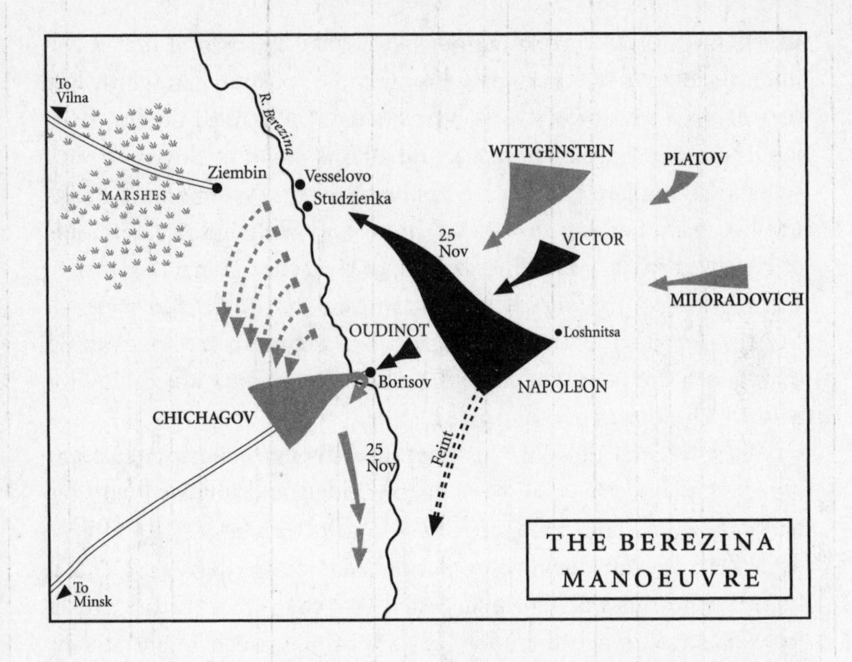

A few hours earlier, Chichagov had moved off with his main forces in the opposite direction along the other bank of the river. He had been anxious about the possibility of Napoleon outflanking him to the south, and the combination of the reports of French activity to the south of Borisov and the information brought to him by three Jews from Borisov convinced him that this was indeed where the French were planning to cross. He left General Langeron with 1,200 infantry and three hundred cossacks at Borisov, and General Czaplic with a few hundred men between there and Vesselovo, while he marched off southwards with the rest of his forces. When the first reports of French activity around Studzienka did reach him on the following day, he assumed this to be a feint meant to deceive him, and continued on his way. The course of the Berezina north of Borisov should in any case have been covered by Wittgenstein, and he had left orders with Czaplic to pull back his outlying units in the area.

But Wittgenstein had no intention of placing himself under

Chichagov's orders, which he would have had to do if he had linked up with him on the western bank. And he too was less than eager to take on Napoleon himself, preferring to spar with Victor, so he ignored Kutuzov's orders to cross the river and cut the French line of retreat. In doing so he not only left the Berezina itself unguarded, he did not, as would have been the case if he had followed his orders, cover the other point at which Napoleon's retreat could have been cut. A few kilometres west of the Berezina, at Ziembin, the road ran through a boggy area along a number of wooden bridges, and could effectively be cut by a platoon of cossacks with a tinderbox.[10]

Oudinot had sent General Aubry with 750 sappers to Studzienka on 24 November to start making struts for a bridge, and followed with his main forces on the evening of the following day. They were joined there by General Eblé with four hundred pontoneers, mostly Dutchmen. Although Napoleon had ordered the pontoon bridge they were accompanying to be burnt at Orsha, Eblé had wisely hung on to six wagons of tools, two field smithies and two wagons of charcoal. The sappers dismantled the wooden houses of Studzienka, sawing the thick logs into appropriate lengths, while the pontoneers forged nails and braces, and turned the logs into trestles.[11]

The riverbed itself, which at this point is less than two metres deep, is no more than about twenty metres across, but its banks are low and boggy, and cut by shallow arms of the main river, so any bridge would need to extend for some distance at either end. A major disadvantage of this as a crossing point was that the western bank, held by the Russians, rose steeply, and any troops occupying it would be in a position to rake the crossing with artillery fire.

Oudinot had placed his men behind a small rise, so they would be out of sight of the cossacks patrolling the western bank, and instructed them to work in silence. But Captain Arnoldi, commanding the Russian field battery of four light guns that had been positioned by General Czaplic to observe the possible crossing points near Studzienka, noticed the French activity on the opposite bank and sent urgent reports to his superior warning that they were preparing to cross the river there. He convinced Czaplic, who came to see for himself and then sent a messenger to Chichagov.[12]

For his part, Oudinot stayed up all night, urging on the sappers and pontoneers, and nervously watching the other bank. "The aspect of the countryside was gripping; the moon lit up the ice floes of the Berezina and, beyond the river, a cossack picket made up of only four men," noted François Pils in his journal. He was a grenadier in Oudinot's corps, but in civilian life he was a painter, which explains his sensitivity to the view. "In the distance beyond, one could see a few red-tinged clouds seemingly drift over the points of the fir trees; they reflected the campfires of the Russian army."[13]

The magnificent sight left Ney, for one, cold. "Our position is impossible," he said to Rapp. "If Napoleon succeeds in getting out of this today he is the very Devil." Murat and others were putting forward various plans to save the Emperor by sending him off with a small detachment of Polish cavalry while the rest of them made a heroic stand. "We shall all have to die," he affirmed. "There can be no question of surrender."[14]

In the early hours of the next morning, 26 November, the troops sitting around the Russian campfires began to withdraw, and Arnoldi's four guns were limbered up and dragged away. Oudinot could hardly believe his eyes. Napoleon, who had reached Studzienka a little earlier, was jubilant: according to Rapp, his eyes sparkled with joy when he saw that his ploy had worked and Chichagov was off on his wild goose chase.

He ordered Colonel Jacqueminot to muster a squadron of Polish lancers and some Chasseurs, each of whom was to take a *voltigeur* riding pillion, and ford the river. Once across, the riders fanned out and, followed by the *voltigeurs*, chased off the few remaining cossacks and took possession of the west bank. Captain Arnoldi, who had clearly seen the French set up a battery of forty guns to cover both banks of the river, had sent a final despairing report to headquarters before withdrawing, expressing his conviction that this was the spot they had chosen for their crossing. But while Czaplic had delayed carrying out the order to withdraw, he did not dare defy it outright. Nor did he have the sense to send a troop of cavalry to hold and, if need be, burn the bridges at Ziembin.[15]

Shortly after the withdrawal of the Russians, at eight o'clock, Captain Benthien and his Dutch pontoneers waded into the icy water and began installing the first trestles. They had stripped down to their pants, and struggled manfully in the strong current, which was carrying with it great blocks of ice up to two metres across. Every so often one of them would lose his foothold on the slimy riverbed and be swept away. They were only allowed to remain in the water for fifteen minutes at a time, but many nevertheless succumbed to hypothermia. They had been offered a bonus of fifty francs per man, but that was surely not the motive that drove them. "They went into the water up to their necks with a courage of which one can find no other example in history," recorded grenadier Pils. "Some fell dead, and disappeared with the current, but the sight of such a terrible end did nothing to weaken the energy of their comrades. The Emperor watched these heroes without leaving the riverbank, where he stood with the Marshal [Oudinot], Prince Murat and other generals, while the Prince de Neuchâtel [Berthier] sat on the snow expediting correspondence and writing out orders for the army."[16]

"At this solemn moment Napoleon himself recovered all the elevation and energy that characterised him," recalled Lieutenant Colonel de Baudus. There are accounts of him looking dejected, and the story of his ordering the eagles of the Guard to be burnt in a fit of despair surfaces here and there. But most witnesses agree that he displayed remarkable self-possession throughout what continued to be a knife-edge situation, and far from ordering the eagles to be burnt, kept enjoining the men to cling to them in order to keep the semblance of a fighting force in existence. Some thought he actually appeared detached as he stood on the riverbank watching the pontoneers at their work.[17]

Major Grünberg, a cavalryman from Württemberg, was struck by this as Napoleon caught sight of him marching past, carrying in the folds of his cloak his beloved greyhound bitch. The Emperor called him over and asked if he would sell the animal to him. Grünberg replied that she was an old companion whom he would never sell, but that if His Majesty so wished, he would give her to him. Napoleon was touched by this and replied that he would not dream of depriving him of such a close companion.[18]

The bridge was completed around midday. It was just over a hundred metres long and about four metres wide, and rested on twenty-three trestles varying in height from one to three metres. There was not enough planking available, so the round logs laid across the top which made up the causeway were covered with flimsy roof slats taken from the houses of Studzienka topped with a dressing of bark, branches and straw. "As a work of craft, this bridge was certainly very deficient," noted Captain Brandt. "But when one considers in what conditions it was established, when one thinks that it salvaged the honour of France from the most terrible shipwreck, that each of the lives sacrificed in the building of it meant life and liberty to thousands, then one has to recognise that the construction of this bridge was the most admirable work of this war, perhaps of any war."[19]

Napoleon, who had hurriedly swallowed a cutlet for breakfast while standing on the bank, walked over to the head of the bridge, where Marshal Oudinot was preparing to march his corps across. "Do not cross yet, Oudinot, you might be taken," the Emperor called out to him, but Oudinot waved at the men drawn up behind him and answered: "I fear nothing in their midst, sire!"[20] He led his corps across, to shouts of "*Vive l'Empereur!*" uttered with a conviction that had not resounded in the imperial presence very often of late. Turning left, he began to deploy his troops in a southerly direction in order to ward off any potential attack by Chichagov. They were quickly lost to sight in the snow that had begun to fall again.

Meanwhile Captain Busch and another team of Dutch pontoneers had been working on a second bridge, fifty metres downstream of the first. This one, built on sturdier trestles and with a causeway of plain round logs, was intended for the artillery and baggage, and it was ready by four o'clock in the afternoon.* While troops continued to trudge across the lighter bridge in an orderly fashion, Oudinot's artillery, followed by the artillery of the Guard and the main artillery park, trundled across the other. At eight o'clock that evening two of the trestles of the heavy bridge subsided into the muddy bed of the river, and the pontoneers had to abandon their firesides, strip off and wade into the water once again. The bridge was reopened at eleven o'clock, but at two in the morning of 27 November three more trestles, this time in the deepest part of the river, collapsed. Once again Benthien's men abandoned whatever shelter they had found for the night and went into the water. After four hours, at six in the morning, the bridge was operational once more.

For the whole of that day the Grande Armée trudged across the Berezina in the lightly falling snow. The Guard began crossing at dawn, then came Napoleon with his staff and household, then Davout with the remainder of his corps, then Ney and Murat with theirs, then, in the evening, Prince Eugène, with the few hundred remaining Italians of the 4th Corps. The bridge was low, barely above the level of the water, and it swayed, so the men crossed on foot, leading their horses. The surface coating of branches and straw had to be firmed up by the sappers from time to time. Even so, the bridge subsided in places, and those crossing it sometimes had water up to their ankles. The sheer weight of numbers and the state of the bridge meant that there was some pushing and shoving, men fell over and horses collapsed, causing obstructions and leading to fights. It was not a pleasant crossing.

Meanwhile a steady flow of guns, *caissons*, supply wagons and carriages of every kind trundled across the other bridge, with a two-hour interruption while the pontoneers repaired two more broken trestles at four o'clock that afternoon. Here too there were jams and

* The original plan had been to construct three bridges, but a shortage of materials prevented the third being built.

outbreaks of violence. The surface of the bridge was scattered with debris and corpses, and a number of horses broke their legs by getting them caught between the round logs making up the causeway. The next vehicles, themselves being pushed on from behind, would try to drive over the struggling and kicking horses rather than stop and wait for them and their vehicles to be heaved over the side. But most of the guns and materiel of the organised units, the treasury, the wagons carrying Napoleon's booty from Moscow, and a surprising number of officers' carriages made the crossing successfully. Madame Fusil, the actress from Moscow, drove across in the relative comfort of Marshal Bessières' carriage.[21]

The approaches to the bridges were guarded by gendarmes who only allowed active units onto them and ordered all stragglers and civilians, and even wounded officers travelling in various conveyances, to wait. A large number of these non-combatants had begun to arrive in the late afternoon of 27 November, cluttering the approaches to the bridge. As they could not cross immediately they settled down, built fires and began to cook whatever they had managed to pick up, scrounge or steal.

Victor's 9th Corps also arrived in the late afternoon and took up defensive positions covering the approaches to the bridges. It had left one division, about four thousand men under General Partouneaux, outside Borisov to mislead the Russians, and this was to follow on under the cover of night.

As most of the army was across by that evening, the gendarmes opened the bridges to the stragglers, *cantinières*, wounded and civilians. But having settled down by their fires, and seeing that their encampment was defended by Victor's men, most did not avail themselves of the opportunity, preferring to spend a peaceful night where they were. Some, like the *cantinière* of the 7th Light Infantry who had gone into labour that evening, had no choice. "The entire regiment was deeply moved and did what it could to assist this unfortunate woman who was without food and without shelter under this sky of ice," wrote Sergeant Bertrand. "Our Colonel [Romme] set the example. Our surgeons, who had none of their ambulance equipment, abandoned in Smolensk for lack of horses, were given shirts, kerchiefs and anything people could come up with. I had noticed not far away an artillery park belonging to the corps of the Marshal Duc de Bellune [Victor]. I ran over to it and, purloining a blanket thrown over the back of one of the horses, I rushed back as fast as I could to bring it to Louise. I had committed a sin, but I knew that God would forgive me on account of my motive. I got there just at the moment when our *cantinière* was bringing into the world, under an old oak tree, a healthy male child, whom I was to encounter in 1818 as a child soldier in the Legion of the Aube."[22]

A remarkable degree of order and even normality reigned over the Grande Armée as it settled down for the night on both sides of the river. A key factor was undoubtedly the presence of the Emperor and the fact that he had visibly taken the initiative, which led everyone to expect great things and kept spirits high. "We are still capable of fun and a good laugh," noted Jean Marc Bussy, a Swiss *voltigeur* sitting around a campfire with his comrades on the western bank of the river. One cannot but admire him. "When night fell, each soldier took his knapsack for a pillow and the snow as a mattress, with his musket in his hand," wrote his comrade Louis Begos of the 2nd Swiss Regiment. "An icy wind was blowing hard, and the men pressed up against each other for warmth."[23]

All that day Napoleon had anxiously listened for the sound of cannon announcing the approach of the Russians. But there was still no sign that Chichagov had realised his mistake. The note he penned to Marie-Louise that evening shows no trace of anxiety.[24]

What he might have heard, had it not been over ten kilometres away, was the end of one of Partouneaux's brigades, which had been holding Borisov. The Partouneaux division, which had only entered Russia recently, had suffered the depressing effects of the conditions in rapid order. The men had been in fine spirits when they had reached Borisov a few days before. At one point they were charged by some Russian cavalry and formed squares. One of the Russian officers, unable to control his wounded mount, had crashed into the middle of the square, where he was pinned to the snow under the thrashing animal. A couple of French soldiers pulled him clear, dusted the snow off his uniform and then went back to their posts in the firing line. The officer bided his time until the French were occupied by another Russian charge and, slipping between them, ran, hopping through the deep snow, to rejoin his own men, at which the entire French square burst into laughter.

But a couple of days later, as they camped out in a windswept spot without fires or food, their mood was very different. "Some wept, crying out plaintively to their parents; some went raving mad; some died under our eyes after a horrible agony," according to one of them. Having held Borisov as long as was necessary, the division had begun to withdraw on the afternoon of 27 November. But one of its brigades lost its way and walked straight into the midst of Wittgenstein's army. After a running battle in which it lost half its number, it was forced to surrender. The men were stripped, beaten and marched off into captivity. One of its regiments, the 29th of the Line, was made up largely of men who had only recently been released from prison hulks in England, having been captured in Saint-Domingue in 1801. "Luck, one has to admit, seems to have abandoned these poor fellows," remarked Boniface de Castellane.[25]

Chichagov had by now realised that he had been duped. Most of his men were still at Borisov and points further south, but he ordered Czaplic to attack the French forces which had already got across the Berezina, promising to send reinforcements. But his men, who had been force-marched some fifty kilometres south and were now ordered to hurry back, made slow progress through the heavy snow. There was much grumbling and even the threat of mutiny. "One of the regiments I had ordered to go and reinforce Czaplic hesitated and then refused outright to move," Chichagov recorded. "My exhortations having produced no effect, I was obliged to have recourse to the threat of firing on it. I had cannon unlimbered and levelled at it from behind." Some of Chichagov's units did however

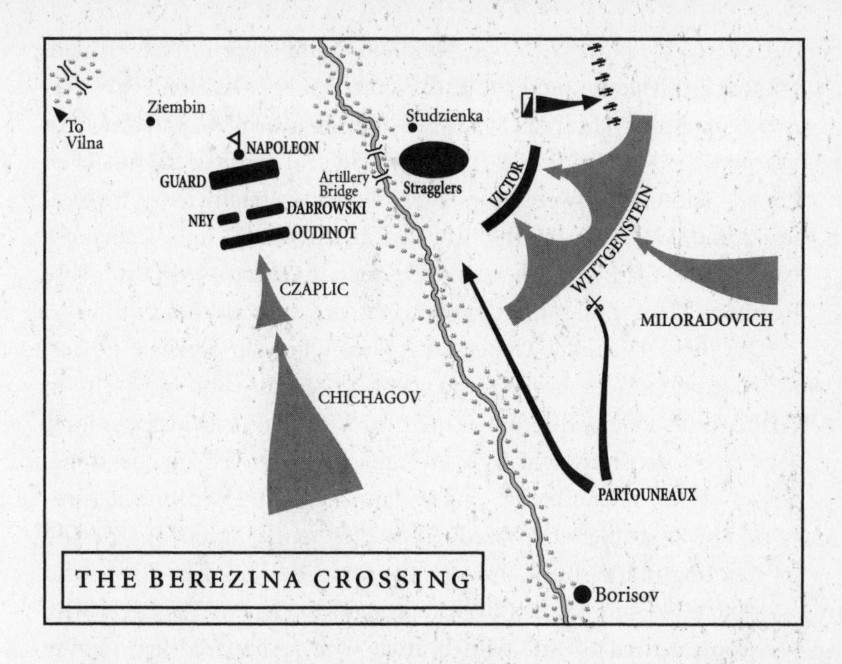

THE BEREZINA CROSSING

come up to reinforce Czaplic that night, and more were on the way.[26]

Before dawn on 28 November, as Oudinot finished gulping down the warming *soupe à l'oignon* his staff had cooked up at their campfire, the first shots resounded on the western bank of the Berezina as a reinforced Czaplic pushed northward under a strong artillery barrage. Oudinot organised a defence, and led his men out under murderous fire from the Russian guns, but he was hit by a shell splinter – his twenty-second wound. Napoleon, who was on the scene, put Ney in command with orders to hold the Russians back at all costs in order to cover the retreat of the remainder of the Grande Armée, the stragglers and, finally, Victor's men.

It was a tall order. Czaplic and Chichagov had over 30,000 fresh troops who had not suffered any serious military losses, and all Ney could put up against them were 12 to 14,000 emaciated and half-frozen remnants: all that was left of Oudinot's 2nd Corps, the Dąbrowski division and a few survivors from Poniatowski's 5th Corps, the Legion of the Vistula, and a handful of other units (his own 3rd Corps had all but ceased to exist, with one regiment numbering forty-two men, another only eleven, and the 25th Württemberg Division's six regiments of infantry, four of cavalry and divisional artillery park down to a grand total of 150 men). Three-quarters of them were not even French. Almost half were Poles, there were four regiments of Swiss, a few hundred Croats of the 3rd Illyrian Infantry, some Italians, a handful of Dutch Grenadiers and Colonel de Castro's 3rd Portuguese Regiment. This motley bunch rose to the occasion magnificently.[27]

The Russians under General Czaplic, a Pole in Russian service, advanced in force through the wooded terrain, but Ney sent in Dąbrowski's Poles, who forced them back to their starting positions. Two more divisions sent by Chichagov then arrived on the scene, Voinov's and Shcherbatov's. They launched a massed attack, supported by an artillery bombardment which sent splinters of pine and fir shooting murderously through the ranks of the Poles. Dąbrowski was wounded and handed over command to General Zajączek, who was soon carried off the field himself with a shattered leg, leaving General Kniaziewicz in command, but he too was put out of action. As the Poles fell back in hand-to-hand fighting among the trees, Ney reinforced them with whatever units came to hand.

Although these were numerically weak, they displayed barely believable spirit. The 123rd Dutch Light Infantry regiment, down to eighty men and five officers, cheered as it went into action. At one point a cannonball shattered the trunk of a huge tree heavy with snow, which came crashing down and buried a dozen men of the French 5th Tirailleurs, but they all clambered out from under the snow laughing like children amidst the bursting shells. When, a few moments later, a shell killed their Colonel's horse, throwing him to the ground, they rushed forward to his aid, but he sprang up and shouted at them: "I am still at my post, so let everyone remain at theirs!"[28]

In order to relieve the pressure on them, Ney sent in General Doumerc with his cuirassiers and three regiments of Polish lancers. They charged the Russians, sowing panic and driving them back.

Czaplic was wounded and General Shcherbatov was captured, along with two thousand others and a couple of standards. A countercharge by Russian hussars and dragoons steadied the situation, but the Swiss regiments, which had now taken over the French front line, supported by the Dutch, the Croats and the Portuguese, held their ground.

The battle raged all day, with the Swiss making no fewer than seven bayonet charges when they ran out of cartridges. "It was worse than a butchery," noted Jean Marc Bussy. "There was blood everywhere on the snow, which had been trampled as hard as a beaten earth floor by all the advancing and retreating . . . One hardly dared to look to right or left, out of fear of seeing that a comrade was no longer there." The fighting was so hot that they forgot about the freezing temperatures, and they kept their spirits up with shouts of "*Vive l'Empereur!*" As death closed in around them in the icy wood, the Swiss broke into the strains of the old mountain *lied* "*Unser Leben Gleicht der Reise.*"[29] The fighting did not stop until eleven o'clock that night, when the Russians, having failed to push the defenders back one step from their morning positions, finally gave up.

It was a magnificent victory for the French, but a bitter one. As they made fires and dragged in their wounded to dress them as best they could, they knew that they would have to leave them behind the following day. The four Swiss regiments had lost a thousand men, and mustered no more than three hundred between them. "We hardly dare speak to each other, for fear of hearing of the death of another of our comrades," recalled Bussy. Of the eighty-seven *voltigeurs* in his company who had laughed around their campfires the previous night, just seven were left alive. The 123rd Dutch Light Infantry had ceased to exist. The Dutch Grenadiers were down to eighteen officers and seven other ranks.[30]

Their heroics were honourably matched by Victor's men defending the crossings on the other bank. They numbered no more than eight thousand men, mostly Badenese, Hessians, Saxons and Poles, and were facing an army over four times that. But here too morale was unaccountably high. They were attacked at nine o'clock in the morning and held their positions until nine that evening against overwhelming odds.

Wittgenstein's first attacks were concentrated on the Baden brigade, commanded by the twenty-year-old Prince Wilhelm of Baden, which was holding the right wing of the French defence perimeter. Prince Wilhelm's men had been greatly cheered when, three days before, they had come across a convoy from Karlsruhe with food and supplies of every sort. The men were able to exchange worn-out uniforms, overcoats and boots for new ones, and to help themselves to food and delicacies. "Every officer had received something from home and everyone jumped on the packages destined for them," wrote Prince Wilhelm. "Thus it was that I saw Colonel Brucker, standing on one of the wagons, open up a large box which I assumed to be full of victuals, and from it he drew a wig and, quick as a flash, he removed the old one he had on his bald head and donned the new one, trying to mould it to his head with his hands." Prince Wilhelm himself was in a good mood that morning as his greyhounds had caught a hare, which he had eaten and washed down with wine that had come in the convoy.[31] Although they were now attacked by overwhelming forces, he and his men stood firm, at the cost of terrible losses.

Hoping no doubt to break the determination of the defenders, Wittgenstein established a strong battery beyond the left wing of Victor's line, and began shelling the area behind it. This was by now occupied by a dense mass of thousands of people, horses and vehicles, up to four hundred metres deep, stretching for over a kilometre along the riverbank. Shortly after midday, Russian shells began to rain down on this vast encampment, bringing hideous destruction as they exploded among the mêlée, killing and maiming people and horses, sending splinters of wood and glass from shattered carriages flying through the air. It was the end of the road for many of the civilians. Captain von Kurz watched in horror as a beautiful young woman with a four-year-old daughter had the horse she was riding killed and her thigh shattered by a Russian shell; realising that she could go no further, she untied the blood-soaked garter from her leg and, after kissing her tenderly, strangled her child and, clutching her in her arms, lay down to await death. Seeing her wagon stuck, Baśka, the *cantinière* of the Polish Chevau-Légers, cut her horse free and, taking her small son in her arms, rode into the Berezina on it. She got more than halfway across before the horse began to drown and sank beneath the surface, throwing her and her son into the water, from which only she was able to wade ashore.[32]

Panic broke out, and a mad rush for the bridges ensued, with people driving their wagons and horses over the corpses of men and beasts, over the wreckage of carriages and abandoned luggage. This merely served to compact the mass pressing around the bridge-ends like a flock of frightened sheep, and now every Russian shell found a target. The massacre continued until Victor managed to mount an attack on the Russian batteries which forced them to pull back out of range.

Although the shelling had stopped, that did not relieve the pressure on the crossings. A mass of people, horses and vehicles converged on the bridges, with those behind pushing forward continuously, so that it was not possible to avoid trampling those who stumbled and fell. "Anyone who weakened and fell would never rise again, as he was walked over and crushed. In this dense mass even the horses were so hard-pressed that they fell over, and, like the men, they too could not get up again," remembered Sergeant Thirion. "By the efforts they made to do so they brought down men who, being pushed from behind, could not avoid the obstacle, and neither men nor horses ever rose again."[33]

Lieutenant Carl von Suckow had become separated from his fellow Württembergers and was caught in the crush. "I found myself being dragged along, jostled and even borne along at some moments – and I do not exaggerate," he wrote. "Several times I felt myself being lifted off the ground by the mass of people around me, which gripped me as though I had been caught in a vice. The ground was covered in animals and men, alive and dead . . . At every moment I could feel myself stumbling on dead bodies; I did not fall, it is true, but that was because I could not. It was only because I was held up on every side by the crowd which pressed in on me. I have never known a more horrible sensation than that I felt as I walked over living beings who tried to hold on to my legs and paralysed my movements as they attempted to raise themselves. I can still remember today what I felt on that day as I placed my foot on a woman who was still alive. I could feel the movements of her body, and I could hear her scream and moan: 'Oh! take pity on me!' She was clinging to my legs when, suddenly, as a result of a strong thrust from behind, I was lifted off the ground and wrenched from her embrace." As he found himself being forced back and forth near the entrance to the bridge, he experienced "the first and only real moment of despair I had felt during the entire campaign." He finally grabbed the collar of a tall cuirassier who was clearing a path for himself with a mighty stick, and was dragged onto the bridge and over the river.[34]

As those caught in the throng could not see in front of them, many found that they came to the river not at the head of one of the bridges, but on the bank. Since they were still being pushed from behind by others they were forced into the water, through which they tried to wade over to the bridges and clamber on from the side. The crush on the bridges themselves was just as great, and those walking in the middle were pressed from both sides as those at the sides moved along facing outwards and pushing inwards with their backs in order not to be thrown into the water.

Those who could not stand up for themselves did not have much of a chance, and many who had somehow managed to make it thus far perished here. A Saxon under-officer named Bankenberg, who had had both legs amputated above the knee after Borodino, had been rescued from Kolotskoie by his comrades. He had been tied onto a horse, and survived all the tribulations of the retreat with courage, but they lost sight of him at the Berezina, and he was never seen again.[35]

In the afternoon Wittgenstein mounted a second assault on Victor's defences, and the Baden brigade was finally forced to give ground. But Victor sent in the Brigade of Berg, made up of Germans and Belgians, and then his remaining cavalry. This, consisting of Hessian chevau-légers and Badenese hussars as well as French chasseurs, no more than 350 men in all, was led into the charge by Colonel von Laroche with such dash that it routed the Russians. A countercharge by Russian cavalry virtually annihilated the Germans, but the French defences had been saved, and as night fell Victor's men were occupying the same positions as they had that morning.

Many of those still hoping to cross found themselves blocked by the barricades of abandoned vehicles, dead horses and human corpses which impeded access to the bridges, and as night began to fall and the fighting died down, they too began to settle down for the night, in the hope that crossing might be easier in the morning.

Victor received the order to withdraw, but seeing the numbers of non-combatants still on the eastern bank, he decided to hold it until daybreak, thus giving them a chance to cross. General Eblé and 150 of his pontoneers cleared the bridges of the corpses, carcases and vehicles that had accumulated on them in the afternoon rush. In order to clear the approaches they dragged many of the abandoned vehicles onto the bridges and then pushed them into the water, and unharnessed and led to the west bank as many of the abandoned horses as they could. They had to drag away or push over carriages and wagons that could not be wheeled away, heaving the carcases of horses and human corpses to the side to create a kind of trench between two banks of dead men and beasts.

At nine o'clock that evening Victor began sending some of his units, his supply wagons and his wounded across, and by one o'clock in the morning of 29 November he had only a screen of pickets and a couple of companies of infantry left with him on the east bank. He and Eblé urged the remaining stragglers to cross, warning them that the bridges would be burnt at first light, but most of them were either too tired or too apathetic. "We no longer knew how to appreciate danger and we did not even have enough energy to fear it," wrote

Colonel Griois, who remained by his fireside along with other comrades from Grouchy's corps. Others were apparently too absorbed by other preoccupations, and the surgeon Raymond Pontier swore that he saw two officers fighting a duel instead of crossing.[36]

At about five o'clock in the morning, Eblé ordered his men to start setting fire to wagons and carriages still littering the eastern bank in order to wake up the non-combatants, and to shout loudly that the bridges would only be open for a couple of hours. A few availed themselves of this, but at six o'clock, when Victor withdrew his pickets and marched across, the remainder began to realise that their last chance had come. A mass of them swarmed onto the bridges, pushing and shoving to get over. Sergeant Bourgogne, who had come back to see if he could pick up any stragglers from his regiment, watched as a *cantinière*, holding onto her husband who had their child on his shoulders, was pushed into the icy water, dragging her family with her, and as a wagon with a wounded officer was tipped over, horse and all, to disappear instantly beneath the ice floes.

Eblé had orders from Napoleon to burn the bridges at seven o'clock, as soon as Victor's last man was across, but he could not bear to leave so many of his countrymen stranded, so he delayed the execution of the order until 8:30. By then Wittgenstein's men could be seen advancing towards the bridges on the opposite bank, and groups of cossacks were already picking over the booty left behind in the wagons and carriages littering the approaches. As Eblé fired the bridges, some of those still on them tried to struggle through the flames, others threw themselves into the water in order to swim the last stretch, while hundreds of others were pushed into it by the pressure of those behind who did not know the bridges now led nowhere.[37]

The morning after the French had marched off, Chichagov rode up to the scene of the crossings. He and his entourage would never forget the grim spectacle. "The first thing we saw was a woman who had collapsed and was gripped by the ice," recalled Captain Martos of the engineers, who was at his side. "One of her arms had been hacked off and hung only by a vein, while the other held a baby which had wrapped its arms around its mother's neck. The woman was still alive and her expressive eyes were fixed on a man who had fallen beside her, and who had already frozen to death. Between them, on the ice, lay their dead child."[38]

Lieutenant Louis de Rochechouart, a French officer on Chichagov's staff, was deeply shaken. "There could be nothing sadder, more distressing! One could see heaps of bodies, of dead men, women and even children, of soldiers of every formation, of every nation, frozen, crushed by the fugitives or struck down by Russian grapeshot; abandoned horses, carriages, cannons, *caissons*, wagons. One would not be able to imagine a more terrifying sight than that of the two broken bridges and the frozen river." Peasants and cossacks were rummaging through the wreckage and stripping the corpses. "I saw an unfortunate woman sitting on the edge of the bridge, with her legs, which dangled over the side, caught in the ice. She held to her breast a child which had been frozen for twenty-four hours. She begged me to save the child, not realising that she was offering me a corpse! She herself seemed unable to die, despite her sufferings. A cossack rendered her the service of firing a pistol at her ear in order to put an end to this heartbreaking agony!" Everywhere there were survivors on their last legs, begging to be taken prisoner. "'Monsieur, please take me on, I can cook, or I am a valet, or a hairdresser; for the love of God give me a piece of bread and a shred of cloth to cover myself with.'"[39]

Estimates of the numbers left behind on the eastern bank of the river vary wildly, from Gourgaud's dismissive assertion that only two thousand stragglers and three guns failed to get across, Chapelle's estimate of four to five thousand along with three to four thousand horses and six to seven hundred vehicles, to Labaume's of 20,000 and two hundred guns, which is certainly too high. Chichagov recorded that nine thousand were killed and seven thousand taken prisoner, which seems closer to the mark. Most are now agreed that during the three days the French lost up to 25,000 (including as many as 10,000 non-combatant stragglers) on both banks, of which between a third and a half were killed in action. Russian losses, all inflicted in the fighting, were around 15,000.[40]

The crossing of the Berezina was, by any standards, a magnificent feat of arms. Napoleon had risen to the occasion and proved himself worthy of his reputation, extricating himself from what Clausewitz called "one of the worst situations in which a general ever found himself." His soldiers had fought like lions. But it was above all a triumph for Napoleonic France, and its ability to create out of the rabble of a score of nations armies which were in every way superior to their opponents, which fought intelligently as well as loyally, and which in this instance did so as though they had been defending their own wives and children. "The strength of his intellect, and the military virtues of his army, which not even its calamities could quite subdue, were destined here to show themselves once more in their full lustre," as Clausewitz put it.[41]

22

Empire of Death

The twenty-two-year-old Captain de la Guerinais was a good swimmer, so when he found himself stranded on the east bank of the Berezina on 28 November he did not bother trying to fight his way onto the bridge but just swam the river. Once across, he found some fellow artillerymen who had got a good fire going. He took off his uniform in order to dry it by the fire, but had the misfortune to fall asleep wrapped in the blanket one of them had lent him. When he awoke, his clothes and boots had gone. He tried to follow the army, wrapped only in the blanket, but it could not protect him from the cold, and he died.[1]

This tale might serve as a parable. The 55,000 or so who had survived the crossing and the fighting of 28 November felt such a rush of relief that they could not help imagining the worst was over. "After the crossing of the Berezina, all faces brightened," in the words of Caulaincourt. Sergeant Bourgogne was cheered by the numbers of men whom he had thought lost who turned up in the course of the following day. "The men embraced, congratulating each other as though we had crossed the Rhine – from which we were still four hundred leagues!" he wrote. "We felt that we had been saved, and, giving vent to less selfish instincts, we pitied and regretted those who had had the misfortune of being left behind."[2] In fact, the worst was still to come.

A vicious wind whipped up a blizzard on the night of 29 November, and even Napoleon found little shelter in the mean hut in which he had taken up his quarters in the village of Kamen. "An icy wind came in from all sides through ill-fitting windows in which almost all of the panes had been smashed," recorded his valet, Constant. "We sealed up the openings through which the wind was blowing with sheaves of hay. A little way off, on a large open space, the unfortunate Russian prisoners which the army was driving along with it were parked in the open like cattle."[3]

The next two days were, according to some, among the worst of the entire retreat. Some could stand it no longer and shot themselves, but most carried on in what had become a mute endurance test. At Pleshchenitse, which Napoleon reached on 30 November, a temperature of −30°C (−22°F) was recorded by Dr Louis Lagneau. Frostbite became even more widespread. Those walking barefoot were so anaesthetised by the cold that they did not notice what was happening to their feet. "The skin and the muscle peeled away like the layers of a waxwork figure, leaving the bones exposed, but the momentary insensitivity allowed them to carry on in the vain hope of reaching their homes," wrote Louis Lejeune. Adjutant Major Louis Gardier of the 111th of the Line noticed a man marching along impassively even though his feet had been lacerated by the jagged surface of the rutted and frozen snow. "The skin had come away from his feet, and trailed like a sole that had become unstitched, so that his every step marked with an imprint of blood the ground he covered," he wrote.[4]

The hundreds of vehicles left behind on the eastern bank contained supplies of every sort and the life-support system of many a soldier. The struggle for survival took on a more vicious character in consequence. As the temperature dropped, people whose clothes or boots had fallen apart or been stolen lost all compunction, and helped themselves to whatever they could. Captain von Kurz remembered seeing a soldier walk up to a colonel who had sat down by the roadside and start pulling off his fur coat. "*Peste*, I'm not dead yet," the Colonel mumbled. "*Eh bien, mon colonel*, I will wait," answered the soldier.' Fezensac saw a man pulling the boots off a general who had collapsed by the roadside. The General protested, begging to be allowed to die in peace, but the soldier carried on. "*Mon général*," he said, "I would be quite happy to, but another will take them, and I prefer it to be me." Von Kurz saw comrades from the same regiment murder each other over a fur coat. "Necessity had turned us into swindlers and thieves, and, without a trace of shame, we stole from each other whatever we required," noted Dr René Bourgeois.[5]

Although they were now moving through inhabited country in which food could be obtained, it was only available to those at the front, and only if they had money. Those further back and the stragglers were left to scavenge. And as thousands of horses had also been left behind at the Berezina, there was, to put it crudely, less meat on the hoof available. "No food was so rotten or disgusting as not to find someone to relish it," wrote Lieutenant Vossler of the Württemberg Chasseurs. "No fallen horse or cattle remained uneaten, no dog, no cat, no carrion, nor, indeed, the corpses of those who died of cold or hunger." There were murderous fights over the carcase of a horse, over the tiniest scrap of food, with men screaming at each other in all the languages of Europe.[6]

Callousness and selfishness reached new heights. "I saw people stubbornly defending access to their fire, not to the half-frozen man who wanted to warm himself for a while . . . that would have been quite natural . . . fire in those moments was life, and nobody shares life – but to him who was begging for a little flame with which he could light his wisp of straw in order to start his own fire," wrote Aleksander Fredro.[7]

It was a bitter moment when officers who considered themselves to be gentlemen were faced with having to admit how low they had fallen, as Carl von Suckow relates. "I had the luck one day, God knows how, to lay my hands on a dozen half-frozen potatoes. Reaching the bivouac, I began cooking them in the ash, and one of my comrades sat down beside me, inviting himself to share my frugal meal. We had come to know each other very well at Stuttgart, where we had been garrisoned together. In spite of this, I had the brutality to refuse his request outright. He got up and walked away, saying in a melancholy voice: 'That is something I shall never forgive you.' It was only then that the ice encasing my heart melted; I called him back and eagerly shared all with him." Colonel Griois, who had procured a small sleigh for himself, encountered a friend who begged to be allowed to share it as he was exhausted, but he brushed him off. "A horrible egoism had taken hold of my heart, and whenever my thoughts go back to that time of my life I shudder at the moral degradation to which misery can make us stoop," he later wrote.[8]

One of the memories that evoked particular revulsion was that of the acts of cannibalism to which some were now driven. There had undoubtedly been instances of it earlier in the retreat, but they had been isolated. Most of the earlier reports are from the Russian side, which is not surprising, since the Russians would, as they followed in the wake of the retreating army, have seen those Frenchmen who had been reduced to the last extremities. They also saw prisoners who, being given no food by their cossack escorts, resorted to eating the flesh of their dead comrades. Nikolai Galitzine's is one of the first accounts to claim to have actually seen French soldiers eating a man at that stage. Wilson relates having seen "a group of wounded men, at the ashes of [a] cottage, sitting and lying over the body of a comrade which they had roasted, and the flesh of which they had begun to eat." In a letter to his wife dated 22 November, General Raevsky reports that one of his colonels saw two Frenchmen roasting pieces of a comrade to eat, and General Konovnitsin wrote to his wife also on the same day affirming that "people have seen them devouring men."[9]

The earliest convincing first-hand account on the French side comes from Lieutenant Roman Sołtyk. Reaching Orsha on his own because he had fallen behind, Sołtyk could not obtain a regular distribution of rations, so he walked up to a group of men standing around a steaming pot and offered them some money in return for being allowed to partake of their stew. "But hardly had I swallowed the first spoonful than I was gripped by irrepressible disgust, and I asked them whether it was horsemeat they had used to make it," he wrote. "They coolly replied that it was human meat and that the liver, which was still in the pot, was the best part to eat."[10]

The practice grew more widespread as psychological barriers broke down under the strain of the conditions on the last leg of the retreat. "I saw – and I do not admit this without a certain sense of shame – I saw some Russian prisoners carried to the very limit by the ravening hunger that possessed them, since there were not enough rations for our soldiers, throw themselves on the body of a Bavarian who had just expired, tear him to pieces with knives and devour the bloody shreds of his flesh," wrote Amédée de Pastoret. "I can still see the forest, the very tree at the foot of which this horrible scene took place, and I wish I could efface the memory as surely as I fled the sight of it."[11]

On 1 December Lieutenant Uxküll noted in his diary that he had seen men "gnawing away at the flesh of their companions" like "savage beasts." Captain Arnoldi of the Russian artillery saw "a small group of [French soldiers] by a fire, carving out the softer parts of a dying comrade of theirs in order to eat them" while he shelled a retreating French column. General Langeron, who was following the retreat between the Berezina and Vilna, did not witness any cannibalism, but did see "dead men who had had strips of meat cut out of their thighs for the purpose."[12]

There are those, such as Daru and Marbot, who deny that any cannibalism took place, and Gourgaud is highly sceptical. But the evidence is against them, as is probability. "One has to have felt the rage of hunger to be able to appreciate our position," wrote Sergeant Bourgogne, who admits that he might well have resorted to the practice. "And if there had been no human flesh, we would have eaten the devil himself, if someone had cooked him for us." Ravening hunger drove people to anything. "It was not unknown even for men to gnaw at their own famished bodies," wrote Vossler, and Raymond Pontier, a surgeon attached to the general staff, also noted this phenomenon.[13]

One of the more interesting things to emerge from the written accounts of the retreat is that there seems to have been a threshold, beneath which the men cheated, killed and even ate each other, and above which they clung to human dignity, a sense of duty and even aspired to happiness. As thousands froze and some were engaged in acts of cannibalism around Pleshchenitse on the night of 30 November, one of Napoleon's orderly officers who happened to have a good singing voice entertained his comrades with a recital of songs as they shivered in the ruins of the manor house. While some died cursing and raging as they gnawed like hungry dogs at some carcase, one young officer was found by his comrades frozen stiff in the act of lovingly contemplating a miniature of his wife.[14] Although circumstances obviously had a major effect, this threshold does not seem to have had anything to do with luck, and everything to do with character.

Sheer determination was a strong driving force. Captain François, who was wounded in the leg at Borodino, walked all the way with the help of a crutch, while Captain Brechtel got home on a wooden leg. Louis Lejeune came across a gunner who had just been wounded in the arm, and, spotting two medical orderlies, asked them to see to the wound. They declared that they would need to amputate, but as they had no table to operate on, asked Lejeune to hold the gunner. "The orderlies opened their bag; the gunner proffered not a word or a sigh; I could only hear the quiet sound of the saw and, a few minutes later, the orderlies telling me: 'It's done! We regret that we don't have any wine to give him to brace him.' I still possessed a half-flask of Malaga, which I was making last by leaving long intervals between taking a drop. I handed it to the amputee, who was pale and silent. His eyes came to life instantly, and, downing it in one, he returned the flask completely empty. 'I've still got a long walk to Carcassonne,' he said, before setting off at a pace that I would have found it hard to keep up with."[15]

Another powerful motive was the shared sense of solidarity within a unit, and men from the same regiment often saved each other from the brink. "In the midst of these horrible calamities, it was the destruction of my regiment which was causing me the keenest pain," wrote Colonel de Fezensac of the 4th Infantry of the Line. "That was my real, or rather my only, suffering, as I do not consider hunger, cold and fatigue to be such. As long as health holds out against physical hardships, courage soon learns to scorn them, particularly when it is supported by the idea of God and the promise of another life; but I admit that courage would leave me when I saw succumbing under my very eyes friends and companions in arms, who are, rightly, termed the colonel's family . . . Nothing binds people together like a community of suffering, and indeed I always found in them the same attachment and the same concern as they inspired in me. Never did an officer or a soldier have a piece of bread without coming to share it with me." According to him, this was so throughout the 3rd Corps, the remains of which were still marching in orderly fashion to the sound of the drum. There were plenty of instances of commanders remaining with their men: both Prince Wilhelm of Baden and Prince Emil of Hesse were exemplary in this respect.[16]

Artillerymen struggled to conserve their guns, which meant making excruciating efforts at every dip or rise in the road; they only spiked them when the last horses gave up the ghost. "It would be difficult to express my heartbreak when I found myself obliged to abandon my last piece," wrote Lieutenant Lyautey.[17]

The anonymous soldiers of the train continued to haul the heavy gold-laden wagons of the *Trésor*, and even those bearing that part of Napoleon's Moscow booty which had made it through Krasny and the Berezina crossing. The man in charge of the convoy, Baron Guillaume Peyrusse, a busybody who saw the whole campaign as an irrelevance next to the punctilious execution of his duty, had a way of seizing the worst moments to lobby various influential people with requests that they put in a word with the Emperor in the matter of his promotion to a higher post. He was certainly the right man for the job, and he managed to get the whole convoy, which included a couple of dozen *fourgons* loaded with gold coins as well as Napoleon's jewels, as far as Vilna without loss.

A more noble object of devotion to duty was Colonel Kobyliński, one of Davout's aides-de-camp, who had his leg shattered by a shell while he was reconnoitring the field on the day following Maloyaroslavets. Fearing that the Colonel would perish in the crowd of wounded trundling along behind the army, Davout entrusted him to a company of grenadiers, with strict orders not to abandon him under any circumstances. The grenadiers took their mission seriously, and carried him all the way. "The Colonel lay on a stretcher constructed like a bier, wrapped in blankets, borne by six soldiers who took it in turns to carry him," wrote another Polish officer. "I often encountered this caravan on the march, and marvelled at their heroic devotion, particularly as its object was not a Frenchman, but one of our countrymen." At one stage, the Colonel begged them to leave him and save themselves, but they would not disobey their orders. The last remaining man of the company dragged the stretcher into Davout's headquarters at Vilna.[18]

Scrupulous observance of discipline, often self-imposed, helped some people through, but few managed to set as high a standard as General Narbonne. "Monsieur de Narbonne was fifty-six years old and had been used to enjoy all the luxuries of life, yet his courage and gaiety in the midst of our disasters were remarkable," wrote Boniface de Castellane. "He wore his hair in the old courtly fashion, and always had it powdered in the mornings at the bivouac, often seated on a log, in the nastiest weather, as though he had been in the most agreeable *boudoir*."[19]

For some, keeping a diary seems to have been a way of reaffirming their humanity as well as performing an act of self-discipline. This is evident from the entry in the journal of Maurice de Tascher, an officer of Chasseurs and a cousin of Empress Josephine, for 4 December, his thirty-sixth birthday, a day on which he might well have fallen below the threshold: "– Bitter cold. silent march. Thoughts to cherish. Anniversary of my birth. – Greetings from my mother . . . tears . . . agony . . . Memories of her. Covered six leagues; stopped in a village, quarter of a league in advance of the general staff. Fever and diarrhoea."[20]

Sergeant Bourgogne noted that women bore the hardships with greater fortitude than men. Dr Larrey made the observation that hot-blooded southern Europeans coped better than the Germans and the Dutch, which was also remarked upon by others. But that did not Prevent General Zajączek's black servant, acquired during the Egyptian campaign, from freezing to death. Albert de Muralt, a Bavarian cavalryman, maintained that officers survived better than soldiers, as they had greater moral resistance and were better educated.[21]

But rank had little to do with perhaps the most vital element in keeping people above the threshold. Devotion to another could be a life-saver. Louis Lejeune encountered a wounded artillery officer waiting by the roadside for his servant to catch up. Two hours later, when he was returning from his errand, he saw the man in the same place and tried to persuade him to go and get some nourishment, which was available nearby, warning him that he was running the risk of freezing, but the man refused, saying: "I agree with you, but my servant, Georges, and I shared the same wetnurse. From the moment I joined the army and particularly since I was wounded he has shown his devotion to me a hundred times. My own mother would not have been more attentive. He was unwell, and I promised I would wait for him, and I prefer to die here than to fail in my promise."[22]

It was not just the officers who gave such proofs of loyalty. One officer of Chasseurs whose feet had been incapacitated by frostbite was dragged all the way to Vilna by a boy bugler of the regiment who had harnessed himself to a little sleigh they had found, and similar examples abound. Corporal Jean Bald of the Bavarian Chevau-Légers gave up his horse to a senior officer who had lost his in battle. "It is far better to save an officer for the King than a simple corporal, who in any case will probably get out on his two strong legs," he said.[23]

Captain Baron von Widemann had managed to get across the Vop with nothing more than what he stood in, but as he huddled by the fire trying to dry his clothes that night, his servant waded back across the river, found his carriage, packed a number of essentials into a portmanteau and brought them back to his master. Paul de Bourgoing's servant, a young Parisian boy, tramped along bravely, carrying on his back as many of his master's possessions as he could, and would be there every evening to attend him as he settled down for the night. One night he failed to show up, and Bourgoing waited on the road for several hours, calling out his name in vain, before lying down to sleep. He woke in the middle of the night to find the boy adjusting the fur rug which had slipped off his feet as he slept. The following evening he did not turn up at all, and was never seen again.[24]

A drummer of the 7th of the Line, married to the company *cantinière*, who had fallen ill, led the horse and cart in which she lay and, when the horse died, dragged the cart himself. When he could go no further, he lay down to die beside her. A *cantinière* in the 33rd of the Line who had given birth to a daughter on the outward march died as she struggled to wade across the Berezina, but with her last strength managed to throw the baby girl onto the bank, where she was picked up and cared for by a stranger, who brought her out of Russia. A fifteen-year-old boy whose parents had died walked along manfully, carrying his three-year-old sister and leading by the hand his eight-year-old brother.[25]

Sergeant Bourgogne met another sergeant of his regiment carrying the regimental dog, Mouton, on his back, since the unfortunate creature had had all four legs frozen and could not walk. Mouton was a poodle they had picked up in Spain in 1808, and had followed the regiment to Germany the following year, been in battle at Essling and Wagram, then accompanied it back to Spain in 1810. It had set off with the regiment for Russia in the spring of 1812, but got lost in Saxony. It had subsequently recognised an echelon of the regiment by the uniform, and followed it all the way to Moscow. Such devotion was not uncommon: General Wilson noted as he followed the last phase of the retreat that "innumerable dogs crouched on the bodies of their former masters, looking in their faces, and howling their hunger and their loss."[26]

Marie-Théodore de Rumigny pampered his favourite horse,

Charles, helping him up if he stumbled and fell, always finding him something to eat and watering him properly – even when this meant stopping, lighting a fire and melting snow in a tin – and as a result he got himself and Charles out of Russia and back to France. The Polish Chevau-Légers made a point of going off in search of fodder for their horses every evening on little local *cognats* they had acquired for the purpose. They even managed to steal a couple of haycarts from some Russian cavalry who were too busy cooking dinner to notice.[27]

Sergeant Bourgogne tells of his friend Melet, a dragoon of the Guard. Melet was devoted to his horse, Cadet, with which he had been through several campaigns, in Spain, Austria and Prussia, and was determined to get it back to France with him. He always went in search of food for Cadet before thinking of himself, and when it became impossible to find any forage at all along the line of retreat of the Grande Armée he went in search of it among the Russians, donning the coat and helmet of a Russian dragoon he had killed in order to get past their pickets. Once inside the enemy's encampment he would help himself to enough hay and oats for a few days and then make his escape. Sometimes he was discovered, but he always got away, and he did return to France with Cadet. A Bavarian Chevau-Léger whose darling mare Lisette fell through the ice of a bog outside Krasny and could not get out simply lay down to die beside her.[28]

Napoleon had originally intended to defeat Chichagov after crossing the Berezina and to make for Minsk, but by the evening of 28 November he realised that his army had given its last in the fighting of the past two days. His only hope now lay in a dash for Vilna. The threat of marauding cossacks had halted regular means of communication, and he had not received an *estafette* for nearly three weeks – a terrible deprivation, as he hated not having news of what was going on in Paris and the outside world. But he was in touch with Maret at Vilna through a number of Polish noblemen who travelled back and forth disguised as peasants, dodging parties of cossacks.

"The army is numerous, but in a state of terrible dissolution," he wrote to Maret on 29 November. "It will take two weeks to bring them back to their colours, but where can we find two weeks? The cold and hunger have dissolved the army. We will soon be in Vilna, but can we make a stand there? Yes, if only we can survive the first eight days, but if we are attacked during that first week, it is doubtful whether we will be able to hold on there. Victuals, victuals, victuals! Without that there is no horror that this undisciplined mob will not visit upon the city. It may be that this army will only be able to rally itself behind the Niemen. In that case, it is possible that I may believe my presence to be necessary in Paris, for the sake of France, the Empire, and the army itself." He instructed Maret to send away all the foreign diplomats so that they should not see the condition of his army, badgered him for news from Paris, demanding to know why no *estafette* had reached him for eighteen days, and begged for news of the Empress. In his next missive to Maret, written the following day, he returned to the subject of food, telling him to bake bread in large quantities and to send convoys of victuals out to meet the army. "If you cannot provide 100,000 rations of bread at Vilna, I pity that city."[29]

But it was the wider political situation that was now uppermost in his mind. He knew that his control over Germany, not to mention other parts of the Continent, would be dangerously impaired if news of the disaster that had befallen him became known. Rumours were already circulating all over Europe, fed by sanguine reports of his defeat coming out of St Petersburg. But if he could rally his forces in Vilna he would still be able to claim some measure of military success, and conceal from the view of Europe the emaciated remnants of his army – which were the most damning evidence of the scale of the catastrophe.

He instructed Maret to trumpet the news of a great victory over the Russians at the Berezina, and ordered Anatole de Montesquiou, one of Berthier's aides-de-camp, to travel to Paris bearing a full report of the six thousand prisoners and twelve guns taken, and carrying with him the eight captured Russian colours. He was to stop in Kovno, Königsberg, Berlin and other cities long enough for the tidings to be disseminated along the way. Ironically, the day after Napoleon dictated these instructions, a service of thanksgiving was held in St Petersburg for the Russian victory at Studzienka.[30]

Napoleon's attempt to manage news could only work if he could hold Vilna and prevent the Russians from moving into Prussia and Poland, and this was looking increasingly doubtful. The relatively fresh corps of Victor and Oudinot on which he had counted quickly became infected by the remnants of the army of Moscow, and within a day or two came to resemble it in terms of condition and discipline.

Yet a skeleton of the army of Moscow, probably fewer than 10,000 men, nevertheless remained operational. And depleted as they were, some units maintained a remarkable degree of spirit. Captain Józef Załuski of the Polish Chevau-Légers recorded an improvement in conditions as they re-entered former Polish lands, and he and his comrades thought nothing of the cold: "We sang our marching songs as usual, particularly during the afternoon march or if it was very cold, when, in order to spare the horses or warm up the riders and prevent them from falling asleep, we would dismount and lead them." He added that many French veterans also bore the conditions remarkably well. "I often marched alongside friends from the Chasseurs who were dressed no more warmly than they would be in France, who were truly astonishing in their endurance and suffered only from the sight of the degradation of the army."[31]

At Molodechno, which they reached on 3 December, they found welcome stores of food. They also encountered a number of *estafettes* and mail from Paris, which meant letters from home, a source of great comfort for these worn-out men, many of whom despaired of ever getting back. The following day, at Markovo, they encountered another convoy of food which included bread, butter, cheese and wine. But none of this could prevent the continuing disintegration of the army, as it could not pause to eat and digest the food properly, let alone tidy up its ranks. The Russians, whom Napoleon thought he had shaken off at the Berezina, kept up the pursuit.

The mess they had made of that operation provoked a surge of self-justificatory recrimination throughout the Russian army. Even Lieutenant Aleksandr Chicherin noted in his diary on 1 December that "the spirit of intrigue has entered everywhere." Kutuzov had been quick to blame everyone concerned for allowing Napoleon to get away. He found it "unbelievable" and "unforgivable" that Chichagov should have allowed himself to be duped. He argued, quite rightly, that even though it had been reasonable for him to send troops south along the Berezina, he should have made Borisov his headquarters and stayed there. Czaplic was "a cow and a fool" who should have fallen back on Ziembin and blocked Napoleon's line of retreat there. Kutuzov had specifically instructed Chichagov to hold the Ziembin causeway, but his order, dated 25 November, only reached the Admiral after Napoleon had slipped through. His worst criticism Kutuzov reserved for Wittgenstein, who had disobeyed specific orders to cross the Berezina and join up with Chichagov on the western bank.[32]

Now that Napoleon was beyond his reach, Kutuzov felt even less inclined to force the pace of the pursuit than before. His army was in terrible condition, most units having lost at least two-thirds of their effectives. His main force, which had marched out of Tarutino 97,112 strong with 622 guns, reached Vilna with no more than 27,464 men and two hundred guns, according to his own figures. The Astrakhan grenadiers were down to 120 men, while the Semyonovsky Life Guards could muster only fifty men per company. "There were no more than twenty or thirty men in a condition to fight in any squadron of our cavalry regiments," noted Woldemar von Löwenstern. "Our horses were in very bad state and almost all suffering from saddle sores, so much so that the stench was appalling and one could smell a cavalry regiment a long way off." The artillery, according to Lieutenant Radozhitsky, was in no condition to go into action. "We are completely disorganised, and we need to be allowed to rest and make up our losses as soon as possible," General Dokhturov wrote to his wife on 4 December. "Our infantry declined into a state of marked disorder," noted Löwenstern. "The cold sapped the soldiers' courage, and once they had managed to find a warm shelter or some heated cottage, it was impossible to get them to leave it. They would snuggle up to the stoves to the point of roasting themselves."[33]

The situation was no better in the other Russian armies. "Our regiments marched pell-mell, our officers were often separated from their troops and could not keep an eye on them," wrote General Langeron, adding that out of a force of 25,000 men who set out under Chichagov after the battle of the Berezina, only about 10,000 reached the Niemen. Wittgenstein's army was hardly in better shape, and could not have put up much of a fight at this stage. And they were still wary of taking on any organised French unit.[34] But the very fact that the Russians were on their tail, constantly threatening, made it difficult for the French to retreat in good order. And it raised severe doubts as to their ability to rally at Vilna.

As usual, Napoleon blamed all his own mistakes and lack of foresight on others. He blamed Victor for "shameful lack of activity," he blamed Schwarzenberg, he blamed the weather, and he blamed the Poles for not having raised large quantities of "Polish cossacks" to replace the cavalry he had so carelessly squandered. But he gave up trying to hide the truth. At Molodechno on 3 December he composed the twenty-ninth Bulletin of the campaign, in which he told the story of the retreat. Although it did not tell the whole truth, it left no doubt as to the magnitude of his defeat. This Bulletin, which ended with the now famous words "His Majesty's health has never been better," was not to be published until 16 December, by which time he hoped to be nearing Paris.[35]

Napoleon had made the only sensible decision in the circumstances. He had resolved to hasten back to Paris, where he would raise a new army in time to sally forth in the spring and not only reassert his control over central Europe but also defeat the Russians. He hesitated as to whom he should leave in command of the remains of the Grande Armée – his preferred choice was Prince Eugène, but he realised that if he placed him above Murat, the King of Naples would probably mutiny, so he chose the latter. Prince Eugène was not happy with the arrangement and asked to be given leave to return to Turin, but Napoleon reminded him of his duty as a soldier. "I have no desire to serve under the King of Naples, who has taken command of the army," the Viceroy wrote to his wife the following day. "But in the present circumstances it would have been wrong to refuse, and we have to remain at our post, be it a good or a bad one." Berthier too begged to be allowed to go back to Paris with the Emperor, but Napoleon would not hear of it. "I know very well that you are of no use to anyone here," he retorted, "but others do not, and your name has some effect on the army."[36]

On the evening of 5 December, at Smorgonie, he called together his marshals and, according to some, apologised for his mistake of having remained in Moscow for too long. He told them of his decision, and after listening to their opinions he climbed into a carriage with Caulaincourt and set off into the night. His carriage, with his Mameluke Roustam and a Polish officer on the box, was followed by a second, with Duroc and General Mouton, and a third, with his secretary Baron Fain and his valet Constant.

Reactions to Napoleon's departure were mixed. There was widespread dismay and a sense of discouragement, but surprisingly little in the way of censure. Officers, and particularly senior officers, generally understood his motives and approved of his decision, and it was only among the lower ranks that imprecations were heard.[37] This was largely because by the time the news broke, on 6 December, they had more vital things to think about.

There had been another sharp drop in temperature. At Miedniki on 6 December Dr Louis Lagneau recorded a temperature of −37.5°C (−35.5°F). "It was really intolerable," he wrote, "one had to stamp one's feet hard while walking along to stop them from freezing." His reading of the thermometer was confirmed, to within a degree or two, by others. François Dumonceau had marched out while it was still dark on that morning. "The air itself seemed to be frozen into light flakes of translucent ice which whirled about," he wrote. "Then we saw the horizon gradually turn an ardent red, the sun rise radiant through a slight aura of vapour enflamed by its rays, and the whole snow-covered plain turn purple and glimmer as though it had been scattered with rubies. It was magnificent to see."[38] But it was hell to walk through.

"The air itself," according to Colonel Griois, "was thick with tiny icicles which sparkled in the sun but cut one's face drawing blood when blown by the wind, which was mercifully quite rare." The phenomenon was recorded by many others. "One could see frozen molecules suspended in the air," noted Ségur. He was astonished by the stillness and silence all around. "We walked through this empire of death like miserable shadows! The dull and monotonous sound of our steps, the crunch of the snow and the feeble groans of the dying were the only sounds that disturbed that vast and mournful taciturnity."[39]

"We were covered in ice; the breath coming out of our mouths was thick as smoke, and it created icicles on our hair, our eyebrows, our moustaches and our beards," recalled Louis Lejeune. "These icicles grew thick enough to obstruct our vision and our breathing." "It frequently happened that the ice would seal my eyelids shut," remembered Planat de la Faye. "I would have to press the lashes between my fingers to make the ice melt in order to open my eyes again." The horses' saliva formed huge icicles at the corners of their mouths, where it dribbled out onto the bit. "I could no longer breathe, as ice had formed in my nose and my lips were stuck together," wrote Sergeant Bourgogne, who felt as though he were walking through "an atmosphere of ice." "Tired and dazzled by the snow, my eyes watered, the tears froze, and I could no longer see."[40]

"There was something sinister, something implacable about the serenity of the sky," records Brandt. "Through a translucent mist of brilliant snowdust, which had the effect of needlepoints on our eyes, the sun looked like a globe of fire, but of a fire that gave no warmth. Houses, trees, fields, all had disappeared under a layer of bright, blinding snow!"[41]

Many succumbed to snow blindness. "The white of the eye would become red and swelled up at the same time as the eyelids, bringing with it a throbbing pain and abundant tears," wrote Dr Geissler. "The sufferer could no longer bear the light and it was not long before he became quite blind." As the retreating column drew closer to Vilna more and more groups could be seen holding each other by the hand as they went.[42]

The men were having such difficulty buttoning their pants in the intense cold that, humiliating and filthy as it seemed to them, they unstitched the back of their breeches or trousers so as to be able to defecate without undressing. But they also had to take care lest their penis froze while urinating, which happened in some cases.

This was also the point at which most of those who kept diaries were forced to give up. Captain Franz Roeder's ink froze, shattering the bottle. Boniface de Castellane's right hand was attacked by frostbite at Miedniki on 7 December, so he had to abandon his journal, to which he could add only a few notes scribbled with his left. As he was leaving the following morning, he found a grenadier on sentry duty standing frozen in death still clutching his musket.[43]

"7 December was the most terrible day of my life," records Prince Wilhelm of Baden. "The cold had reached 30° [−37.5°C (−35.5°F)]. At three o'clock in the morning the Marshal [Victor] gave the order to march out. But when it came for the signal to be given, it turned out that the last drummer boy had frozen to death. I then went among the soldiers, talking to each one, encouraging them, exhorting them to get up, to stand to arms, but all my efforts were vain: I could only assemble fifty men. The others, two or three hundred of them, lay on the ground, dead and stiff from the cold."[44]

"It is during this part of the journey that I saw for the first time numerous examples of men literally struck down by the cold as they walked along," wrote Heinrich Brandt. "They would slow down slightly, totter like drunken men, and then fall, never to rise again." He was not the only one to be struck by this sight of men stumbling around like drunks for a few moments. Either just before or just after they fell, blood would pour from their nose and mouth, and sometimes their eyes and ears.[45]

Weeks of relentless hardship and the succession of blows dealt to their hopes and expectations – at Smolensk, Orsha, Borisov, the Berezina – inevitably also affected the men psychologically. Brandt and others speak of reaching a state of febrile agitation that prevented him and his comrades from sleeping. The fear of dying where they were kept them moving all night. "We were guided by the light of the fires lit in every village, on the edge of every wood, always surrounded by that horrible jumble of the living and the dead," he wrote. "Other corpses marked the road. The dazzling serenity of the sky seemed to be insulting our sufferings; the cold became more and more pervasive, and our little column kept diminishing."[46]

In others, the conditions induced apathy. Dr Larrey observed that "we were all in a state of such despondency and torpor that we had difficulty in recognising each other." Dr Bourgeois noted a similar phenomenon. "A great many were in a state of real dementia, plunged in a kind of stupor, with haggard eyes, a fixed and dazed stare, one could single them out in the crowd, in the midst of which they walked like automata in profound silence. If one hailed them one could get only disjointed and incoherent answers; they had entirely lost the use of their senses and were impervious to everything – the insults and even the blows they received could not rouse them or bring them out of this state of idiocy." Some became so disoriented by the cold that they would walk drunkenly straight into a fire and stand in it with their bare feet, or even lie down in it.[47]

General Langeron, commanding the vanguard of Chichagov's army, was following the French as they made for Vilna. "The Russian army was marching down the middle of the road," he wrote, "and on either side of this road marched, or rather stumbled, two columns of the enemy, without weapons." The Russians ignored them, as they could get nothing out of them. "They knew nothing, remembered nothing, understood nothing," recalled Lieutenant Zotov. The road itself was strewn with frozen corpses, and here and there groups of crazed soldiers gnawed at a carcase, human or animal. "I was born to die in the service of my motherland, and from the beginning I prepared myself to fear neither shells nor other dangers," noted Lieutenant Chicherin in his diary, "but I cannot accustom myself to the horrors and torments that continually present themselves to my eyes along the way."[48]

Another who walked down the same road in the wake of the French retreat was Henri Ducor, who had been taken prisoner on the banks of the Berezina, but, having been stripped and robbed by cossacks and then left to die, had pulled some clothes off a corpse and decided to continue the retreat to Vilna. "Every tree trunk was a support for another victim; sometimes three or four dead bodies were grouped around it in the most bizarre attitudes: some on all fours, others crouching on their heels, others sitting with their arms around their knees and their chins resting on them, others sitting with their elbows on their thighs and their heads bent forward, as though they were sleeping or perhaps eating," he wrote. "But what really aroused my astonishment was a gunner standing erect behind his cannon, his right hand leaning on the breech and facing towards Russia. He still had his uniform. The enemy army had marched past him and left him as he was. In the midst of this ocean of snow he was like a monument commemorating our great disaster."[49]

Those who could simply trudged on, kept alive by the lure of Vilna. "Vilna now became the promised land, the safe haven from every storm and the term of all our misfortunes," according to Caulaincourt.[50] They knew it would not be like Smolensk, that it was a substantial inhabited and friendly city in which they would find shelter and food in abundance. But it would have been better for them if it had been another burnt-out shell like Smolensk.

23

The End of the Road

Vilna was calm. Although Maret had for some time been receiving increasingly desperate letters from Napoleon, demanding more and more horses and men, he had no inkling of the scale of the disaster. He knew the situation was not good, and must have suspected that it was much worse than he was being told, but he was under firm instructions to act as though all was going well. On 2 December he duly celebrated the anniversary of Napoleon's coronation with the usual twenty-one-gun salute, a *Te Deum* in the cathedral and, in the evening, a grand dinner at the former archepiscopal palace for the diplomatic corps and the local grandees, which was surprisingly gay.

The Governor of Vilna, General Dirk van Hogendorp, gave a ball, while the chief *Commissaire* Édouard Bignon gave a humbler reception at his lodgings. It was in the course of this that the owner of these lodgings, who also happened to be one of the Polish noblemen carrying messages between Napoleon and Maret, a Mr Abramowicz, returned from his latest mission. He had left Napoleon at the Berezina and painted a gloomy picture of the situation.

Over the next couple of days strange rumours began to circulate in the city. *Commissaires* and other administrative personnel started leaving, and some of the nobility thought it wise to take themselves off to their country estates. Maret and Hogendorp then received

Napoleon's instructions to bake bread and biscuit, and to send out supplies to meet the army. More alarming were the directives telling them to evacuate all unnecessary personnel and to put the city in a state of military readiness.[1] While Maret asked the ministers of Austria, Prussia, Denmark, the United States and other smaller powers to move to Warsaw, Hogendorp drove out to meet Napoleon.

He found him in a small country house outside Smorgonie on 5 December, and informed him that there were enough rations stockpiled at Vilna to feed 100,000 men for three months, as well as 50,000 muskets, munitions, uniforms, boots, harness and other materiel. There was even, according to him, a small remount depot. He reported that he had ordered the deployment of two fresh divisions that had recently arrived from Germany to form a screen around the city, and three regiments of cavalry along the road from Oshmiana to Vilna. After apparently approving all these measures, Napoleon told him of his intention of leaving for Paris, and asked him to ensure that there would be fresh horses waiting at every posting station on the road to Warsaw.[2]

Hogendorp drove back to Vilna to make the arrangements, and later that evening Napoleon himself set off for France. He did not pass through Vilna, only pausing for an hour on the outskirts in the early hours of 6 December in order to see Maret and give his final instructions. His last orders were that Murat must hold Vilna.

On paper, this was quite feasible. The city was well stocked and there were up to 20,000 fresh troops available to fend off any Russian attack while the 10,000 Bavarians falling back under Wrede and the 30 to 40,000 remnants of the Grande Armée caught their breath. "Ten days of rest and abundant victuals will bring back discipline," Napoleon had assured Maret.[3] The city would not be cut off while Macdonald's 10th Corps held Prussia and Schwarzenberg and Reynier hovered in Poland, where the remains of Poniatowski's 5th Corps would soon be reforming. And the various Russian units approaching Vilna were in no fit state to mount a serious challenge to an organised and determined defence. But no such defence was ever organised, and

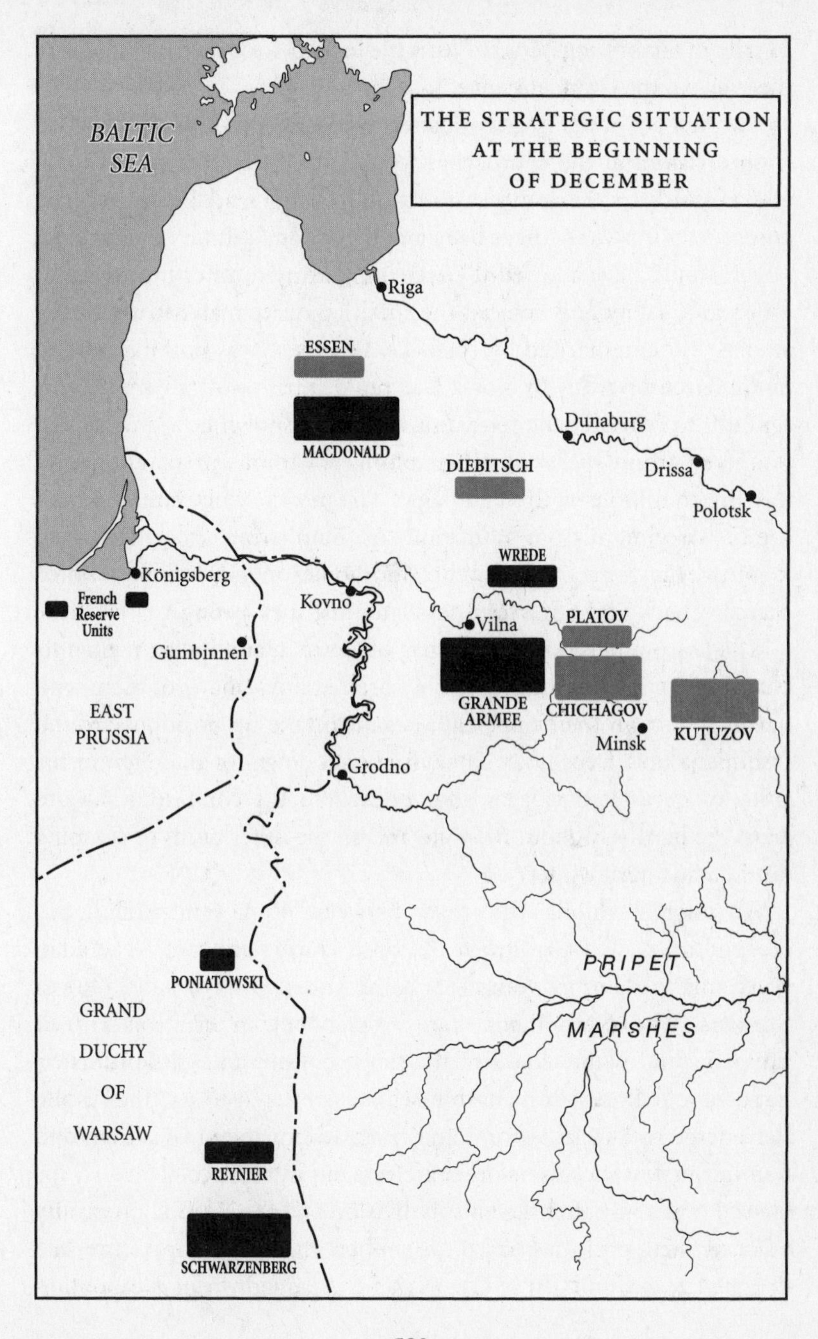

THE STRATEGIC SITUATION AT THE BEGINNING OF DECEMBER
BALTIC SEA
Riga
ESSEN
MACDONALD
Dunaburg
Drissa
Polotsk
DIEBITSCH
WREDE
Königsberg
French Reserve Units
Kovno
Vilna
PLATOV
Gumbinnen
GRANDE ARMEE
CHICHAGOV
KUTUZOV
Minsk
EAST PRUSSIA
Grodno
PRIPET
MARSHES
PONIATOWSKI
GRAND DUCHY OF WARSAW
REYNIER
SCHWARZENBERG

a series of factors conspired to turn the longed-for haven of Vilna into the grave of the Grande Armée.

The first act of the tragedy opened as Hogendorp deployed his two fresh divisions at the approaches to the city, Coutard's to the north and Loison's to the south-east, around Oshmiana.[4] In normal circumstances this would have been an obvious and salutary manoeuvre, which would have allowed the retreating army to filter through into an area of safety and achieve the final leg of its march without the anxiety of being harried by cossacks. But there was nothing normal in the circumstances by now. The rapid attrition of all fresh units sent out to reinforce the retreating army had shown how quickly untempered troops perished when plunged without preparation into the dire conditions of this campaign. The most recent example was of a march regiment from Württemberg, numbering 1,360 men when it joined the retreating army at Smorgonie on 5 December, which marched back into Vilna four days later just sixty strong.[5]

The Loison division, a mixed bag of German and Italian regiments containing large numbers of freshly drafted boys, many of them with hardly a hair on their upper lip, began to take up positions around Oshmiana on 5 December. They hunkered down for the night in the ruins of devastated villages and, unused to the conditions, had to learn the hard way about frostbite and all the other perils of camping out in a northern winter.

At intervals along the road between Vilna and Oshmiana, Hogendorp posted a regiment of Polish lancers and two Neapolitan regiments made up of volunteers commanded by Prince della Rocca Romana. The Neapolitans were resplendent in crimson Hussar uniforms and white cloaks of the finest cotton, and extraordinarily handsome – Rocca Romana himself was referred to as "the Apollo Belvedere" by the ladies of Vilna, with whom they had all enjoyed a short burst of success before marching out into the cold.[6]

And it was just after the Loison division and the Neapolitan cavalry reached their positions, on 6 December, that the temperature had dropped to around −37.5°C (−35.5°F). A squadron of Neapolitans was detailed to complement Napoleon's escort of two squadrons of Polish Chevau-Légers as he left Smorgonie on the first leg of his journey back to Paris. The Imperial Mameluke, Roustam, noted that the wine in Napoleon's carriage froze that night, causing the bottles to shatter. He also noticed that by the time they reached their first halt, there were only Poles surrounding the imperial convoy. The Apollo Belvedere had lost his fingers and all but a handful of his men.[7]

The same icy death met the Loison division. Dr Bourgeois, who passed them on his retreat, could hardly believe the rapidity with which these unprepared men succumbed. "First, one saw them totter and walk for a few moments with an unsteady step like drunken men," he wrote. "It was as though all the blood in their bodies had gone to the head, so red and swollen did their faces become. They were soon gripped entirely and lost their strength, and their limbs were as if paralysed. No longer able to lift their arms, they let them fall under their own weight, their muskets fell out of their hands, their legs gave way beneath them and they fell to the ground after wearing themselves out in vain efforts. As they felt themselves weaken, tears wetted their eyelids, and when they had fallen they would open these several times to stare fixedly at their surroundings; they seemed to have entirely lost all feeling and had a surprised, haggard look, but the whole of their face, the contortions of the muscles in it, were unequivocal evidence of the cruel pain they felt. The eyes became very red, and often blood seeped through the pores, flowing out in drops outside the membrane that covers the inside of the eyelid (the conjunctiva). Thus, one can state without using the language of metaphor, that these unfortunates shed tears of blood."[8]

According to Lejeune, the Loison division lost half of its men in a matter of twenty-four hours, and by the time the retreating army had reached Vilna on 9 December, there was not a single one left. Hogendorp estimates that its 10,000 men dwindled to less than two thousand overnight.[9] To the miseries of the retreating Grande Armée was added the sight of thousands of smartly uniformed hard-frozen soldiers lining the road as they trudged on towards Vilna.

On 7 December, the first individual soldiers and groups began to trickle into town. The shops and cafés were open as usual, and the tattered men could hardly believe their eyes. Every village, town and city they had seen over the past six months had been a ravaged, burnt-out, deserted shell, and they found the aspect of a normal bustling city untouched by war magical. "It was for us the most extraordinary spectacle to see a city where everything was perfectly calm, and one could see ladies at the windows," wrote Colonel Pelet. They revelled in the luxury of being able to walk into a café, sit down and order coffee and cakes. Colonel Griois made for the nearest inn and ordered himself a dinner of bread with butter, meat and potatoes, washed down with a bottle of indifferent Spanish wine. "You will laugh with scorn when I say that this moment, preceded and followed as it was by so much misery and danger, was certainly one of the moments in my life when I felt the sensation of the truest and most complete happiness," he wrote.[10]

Some then went off to find rooms, while others stocked up on provisions. More and more men drifted into the city, and the shops and eating places began to shut as the inhabitants of Vilna realised that the rumours that had been circulating over the past week had been true. "At first they looked at us with surprise, then with horror," wrote Cesare de Laugier, who was one of the early arrivals, with the debris of Prince Eugène's 4th Corps. "They rushed home to their houses, and began barring doors and windows."[11]

Hogendorp had made what he thought would be adequate arrangements to receive the retreating army. After the briefest of consultations with the monks, he designated each of the many monasteries in the city as a barracks for one of the corps, and posted notices at every street corner informing the men that they would find soup and meat there, and giving directions. He detailed an artillery officer to set up a post at the approaches to the city in order to direct the ordnance to places where it could be parked.

On the morning of 8 December he himself went out to greet the incoming army. At around eleven o'clock he saw Murat coming towards him, accompanied by Berthier. "They were on foot, on account of the cold," he wrote. "Murat was enveloped in magnificent large pelisses; a great fur hat, very tall, which crowned his head augmented his height, giving him the air of a giant, beside whom Berthier, whose ample clothing only served to overwhelm his small body, presented a singular contrast." No sooner had Murat taken up quarters than he received a call from Maret, who passed on to him Napoleon's final instructions, which were to hold the city at all costs. "No, I refuse to allow myself to be taken in this pisspot of a place!" Murat is alleged to have answered. And when Berthier asked him for orders, he supposedly told him to write them out himself, since it was obvious what they should be. It is impossible to be sure of this, or of anything else that took place over the next forty-eight hours.[12]

Even if Napoleon's departure had not turned the men against him, it had had an effect, and a profound one, on the course of events. "The presence of the Emperor had helped to maintain the commanders in their duty," noted Eugène Labaume. "As soon as it was known he had left, most of them following his example no longer felt restrained by a sense of shame, and without a thought, abandoned the regiments that had been entrusted to them." This may have been something of an exaggeration, for there are plenty of examples that belie it, but its essence is supported by others.[13]

The effects were felt not only in the army. "The Emperor's passage through the city, which was soon common knowledge in Vilna, was an almost universal signal for people to leave," wrote Hogendorp, who went on to say that military and civil administrators "disappeared in a flash, as if by enchantment," and painted a picture of panic that may have had something to do with the fact that he himself stands accused by some of having bolted on the morning of 9 December.[14]

"In the midst of this extreme disorder, a colossus was needed as a rallying point, and he had just left," wrote Ségur. "In the great void he left behind him, Murat was hardly noticed. It was then that we saw only too well that a great man is irreplaceable." "Murat was not the man who was needed at this moment," concurred General Berthézène.[15]

On 9 December the main mass of the army turned up at the gates of Vilna. The men stationed outside the city by Hogendorp to direct traffic were overwhelmed by the disorderly column of troops. The officer in charge of directing the artillery was confronted by men who refused to listen to orders and just wanted to get into town as quickly as possible.

The road by which they came passed into the city through a medieval gate no more than three or four metres wide and more than twice as long, which created a kind of tunnel. The inevitable jam built up before it, with people pushing from behind. "One could no doubt have found on the left or on the right other roads leading into the city, but we had developed the unfortunate habit of automatically following the path traced by those who went before us," noted Griois, adding that "it was, on a smaller scale, the passage of the Berezina all over again." Some did go off to find other points of entry, but the majority behaved like the sheep they had been reduced to by the experiences of the past weeks.[16]

Pushed and jostled, men and horses went down, those who came after could not hold the pressure from behind and crushed them underfoot. Christian von Martens saw an officer who was pressed so hard against a cannon by the crush that his stomach burst open and his entrails spilled out. "I was swept forward off my feet and finally flung down between two fallen horses, on top of which a rider then stumbled with a third," recalled Captain Roeder. "I gave myself up for lost. Then dozens of people began to pile up on top of us, screaming horribly as their arms and legs were broken or they were crushed. Suddenly the heaving of one of the horses flung me on top, throwing me into an empty space, where I could pick myself up and stagger through the gate."[17]

Once inside, the survivors ignored Hogendorp's notices and made for the nearest eating places, shops and even private houses, knocking on doors and begging to be let in. One cannot but admire the inhabitants who allowed them into their houses. The men were half-crazed, covered in sores and gaping wounds, filthy and crawling with vermin.

"Nothing exhales a more foetid odour than frozen flesh," recalled Sergeant Thirion, and most of the men had at least some affected limbs.[18] The diarrhoea most were suffering from had left traces on their clothes, which cannot have helped, while their breath, after weeks of horseflesh and rotten scraps, was by all accounts particularly foul.

Some of those whose units had disintegrated took the opportunities provided by Vilna to obtain new clothes and stock up with provisions, and promptly moved on along the Kovno road. Many of the officers still with more or less active units made use of their time in the city to prepare themselves for further action. They visited the stores, where they found their trunks with spare uniforms and linen, which had followed the army from Paris via Danzig and then by river and canal to Vilna. Heinrich Brandt was able to have a bath, dress his wounds properly and put on a new uniform; he felt a new man. Major Vionnet de Maringoné also felt transformed by a shave and a change of clothes – which also rid him of the lice infesting his old uniform. Dr Lagneau was delighted to find his trunk, which contained not only fresh clothes, but also some surgical instruments and even books. He took what he needed and gave the rest to the son of the family he was billeted with, who happened to be studying medicine.

But most of the men and officers simply gave in to the luxury of a good meal and a warm, undisturbed night. Colonel Griois took off his boots for the first time in six weeks. A few toenails came away with them, but otherwise his feet seemed in reasonable shape, and he settled down for the night, feeling like a prisoner whose fetters had been removed – but when the time came, he could not get them on again. Marie Henry de Lignières could not resist the temptation of eating a vast amount, after which he climbed into a warm bed for the first time in nearly seven months, but he spent a terrible night and wet himself.

As they relaxed, warm and replete, they felt a sense of security they had not known for six weeks. They were all the more incredulous when they heard the drums beating the stand-to in the morning, and few even considered responding. Even when they heard the sound of cannonfire they assumed that someone else would be dealing with whatever emergency had arisen.[19]

In fact, nobody was by this stage dealing with anything. Murat had tried to call a meeting of all the senior generals, but most were too busy seeing to the needs of their men or themselves, and did not regard a summons from the King of Naples as quite so urgent as one from Napoleon would have been. He spent the rest of that day trying to decide on some plan, but there is no evidence that he actually fixed on one. The only thing he did was to transfer his headquarters to the western end of the city, which gave rise to a rumour that he had left.

As the remains of the Grande Armée had trudged into Vilna on the previous afternoon, the retreating Bavarian division under General Wrede, which was still a fighting force of nearly 10,000 men, had been ordered to take up defensive positions covering the city. Many of the Bavarians could not resist going into town in search of food and a warm night's rest, which disorganised the unit, and when, in the early hours, a few detachments of cossacks appeared to threaten its pickets, these fell back, causing panic among their comrades. Wrede himself seems to have lost his head and was seen running into the city screaming that the cossacks had broken in.

Ney had the call to arms sounded and set off at the head of a detachment of the Old Guard to rally the fleeing Bavarians. He managed to steady the situation and restore order, but returned to headquarters in dispirited mood. "I had the stand-to beaten a while ago and could hardly get five hundred men together," he said to General Rapp. "Everyone is frozen, tired, discouraged, and nobody can be bothered any more."[20]

Murat decided that he would not be able to hold Vilna and resolved to fall back on Kovno. But instead of getting Berthier to issue formal orders to the various corps, instructing them in what order they were to march and at what time they should leave, he simply gave a blanket order for the retreat. He then set off himself without delay. The order to evacuate Vilna flew around the city in somewhat haphazard manner, so that some did not believe it while others never got it. And many of those who did were simply not prepared to carry it out.

Sergeant Bertrand of the 7th Light Infantry in Davout's corps had obediently gone to the designated monastery, where he had found food and shelter. When he heard the bugle sound the alarm in the early hours and began rousing his men, many, even veterans of the Egyptian and Italian campaigns, simply refused to budge. "That night of complete rest around a good fire had been enough to extinguish their courage and their energy," he wrote. "They were overcome by a general drowsiness, a heaviness in the head which seemed to obscure the faculty of thought. Stupefied, and as if drunk, they attempted to get to their feet, only to fall back heavily."[21] The story was the same in other units.

"Instead of staying a whole day in Vilna, it would have been far better to continue the retreat without stopping," wrote Prince Wilhelm of Baden. "Many officers, calling on their last reserves of strength, would have reached the German frontier and would have been saved." When it was time to leave he tried to persuade them to come with him, but after one night of release, these men who had come so far found they had the strength to go no further. "We had for some time been measuring out our forces in order to reach this city in which we believed we would find what we needed to satisfy our most imperious needs; *rest*, *bread*, and *Vilna* had come to form a trinity which was united in our minds as a single hope, and as a result we were clear in our minds that we would go no further than this city," wrote Adrian de Mailly.[22]

In many cases, the psychological as well as physical strain of the past weeks had caused something deeper inside to snap. Planat de la Faye tells of an Italian officer who had inspired all his comrades with his fortitude. "I have never seen a braver man or a gayer one than this Piedmontese," he wrote. "He had lost the toes of both feet to frostbite before the passage of the Berezina. At Smorgonie he developed gangrene and could no longer get his shoes on. Every night as we settled down he would cut away with a knife the gangrenous parts and bind the rest up carefully with rags, and all this with a gaiety which tore at one's heart. The next day he would resume the march, with the help of a stick, only to perform the same operation that evening, so that by the time he reached Vilna, he had not much more than his two heels left." But after a good dinner and a warm undisturbed sleep, he went mad. He was not alone, and one inhabitant of Vilna noted that there were many who had "lapsed into complete idiocy."[23]

Chaos engulfed the city as the evacuation began. Vilna is built on a slope, and the old city is a mass of winding streets. "Naturally, in these narrow streets, covered in ice, the wagons, the sleighs, the carts and the carriages crashed into each other, became entangled, and went over," wrote Adrien de Mailly. "And naturally, the horses knocked each other over, men fell down and got trampled, and the drivers and the crushed screamed as loud as they could, either at their horses or at those who were breaking their limbs."[24] They might have spared themselves the trouble had they known what awaited them a couple of kilometres outside Vilna.

At the small village of Ponary the Kovno road goes up a long incline. Normally, the local authorities would sprinkle sand on such places at regular intervals in winter. But Hogendorp had not thought of doing so. As a result, the compacted frozen snow covering the road became a long sheet of ice, and many of the wheeled vehicles, and even the horses and men on foot, found it difficult to negotiate.

Major Jean Noel, who had come from the opposite direction, bringing two batteries of eight guns each from Germany to supplement the artillery of Loison's division, whose fate he did not yet know, paused to wait for orders when he reached the top of the hill of Ponary on 9 December. He was astonished to see crowds of fugitives coming towards him, and all through that day his men earned good money helping them and their vehicles up the slope. On the following morning, a carriage came up the hill and stopped beside his guns. Murat leant out and asked him who he was and what he was doing there, surprised to see two fine new batteries with well-fed teams.

Having explained himself, Noel asked Murat for orders. "Major, we are f—d," answered the King of Naples. "Get on your horse and run."[25]

Soon large numbers of troops, baggage trains, artillery and carriages with wounded officers were struggling up the increasingly slippery hill. As a wagon stalled or slid backwards, it sent the one behind into a downward slide, and the two would only come to a standstill when a third, fourth or tenth further back overturned. Even those whose horses were properly shod found it increasingly difficult to negotiate what had soon become an obstacle course.

The men on foot either dragged themselves up on all fours, using their bayonets to gain purchase on the icy surface, or floundered through the deep snow on either side of it. Some made a detour along a track around the side of the hill. A number even managed to get their sleighs or carts round this way. But most of the wheeled vehicles and many mounted men persisted in trying to get up the main road. The artillery had no option, as their guns would never have made it through the narrow side track. Some Hessians succeeded in hauling their guns up, but the Bavarian artilleryman Captain von Grawenreuth was less fortunate, and with tears in his eyes abandoned his last and favourite gun, "Mars," an exceptionally accurate piece, at the foot of the slope.[26]

Boulart, who had been unable to get his remaining guns into Vilna when he arrived outside the city on 9 December, had gone out the next day and brought them round by a side road. But by the time he reached the foot of the hill at Ponary the mass of stranded wagons made it out of the question for even the best-harnessed vehicle to get through.

The same went for the convoy carrying the treasury, and all of Baron Peyrusse's superhuman efforts over the past two months were made vain. The gold-laden wagons were too heavy to be hauled up, even if there had been no jam. He began removing sacks of coins from the wagons and transferring them onto the backs of horses. He even managed to get one wagon, which he had emptied, dragged up the slope and refilled, all the way back to Danzig. Marshal Bessières in passing ordered Noel to take some of the gold on his wagons, but the result of this measure was that the wagons disappeared along with the gold. On the other hand, some German officers from Baden and Württemberg allegedly loaded 400,000 francs in gold onto their sleigh and handed it in to the paymaster at Königsberg two weeks later.[27]

It was not long before passing soldiers, seeing the abandoned wagons marked *Trésor Imperial*, began breaking them open and helping themselves, and a free-for-all developed as officers, rankers and civilians fought over sacks of gleaming *Napoléons-d'or*. The ground was littered with silver coins and other booty being cast aside as the men filled their pockets and knapsacks with gold, jewelled icons and other pieces of Napoleon's Moscow booty.

It was a tremendous boon for those who had lost everything at some point along the road. Julien Combe noted that one of his Chasseurs managed to grab a bag containing 20,000 francs, which later permitted him to get married and settle down prosperously in Besançon. But for most, the opportunity to loot proved their undoing. A swarm of cossacks appeared on the scene as soon as the orderly units of the retreating army had passed, and showed no mercy as they too joined in the looting.

The most unfortunate were the wounded whose carriages had been caught in the jam, who were either killed there and then or dragged back to Vilna. As one junior artillery officer pointed out, if only Hogendorp or some other official had sprinkled a little sand on the slope, the French would have saved the entire treasury, several batteries of guns, the papers of the general staff, and hundreds if not thousands of wounded officers and men.[28]

"What took place in Vilna over the few weeks after 10 December is easier to tell than to believe, not that it is easy to speak of either," according to Aleksander Fredro. As soon as the organised units had marched out of the city, swarms of cossacks poured in, hunting down stragglers in the streets and seeking out soldiers and particularly officers who had taken refuge in private houses. They went into the hospitals and the monasteries where the wounded and those who could go no further lay helpless, and began beating and kicking them, tearing off their clothes and their dressings in search of valuables. Those who protested or tried to defend themselves were killed.[29]

The non-Polish population of the city, perhaps out of desire to assert their anti-French credentials and thereby shield themselves from potential reprisals on the part of the Russians, joined in the sport of hunting down French and allied soldiers. Those who had rented rooms to officers or let them take shelter in their houses were quick to kill the inconvenient guests and, after stripping them of their remaining valuables, throw their bodies into the street. There are accounts of them luring starving officers into their homes in order to kill and rob them, of women enthusiastically battering survivors to death, and one of them stuffing ordure into the mouths of prisoners and wounded, saying, "*Le monsieur a du pain maintenant.*" Those who were not killed wandered the streets begging for a crust of bread, and eventually died of cold as they huddled against the wall of a building.

Matters did not improve when regular Russian forces under General Czaplic occupied the city. The soldiers scoured the hospitals in the wake of the cossacks, and the medical staff who eventually took over were little better. Despite the availability of victuals, the wounded were left with no food or water for days, and were ill-treated by the orderlies. Typhus had broken out, and the dead and dying were unceremoniously thrown out of the windows and dumped in the street, where heaps of hard-frozen contorted bodies piled up.[30]

The troubles were far from over for those who had carried on towards Kovno. "The sight which the retreat presented at this stage," wrote Paul de Bourgoing of the Young Guard, "was one of a long stream of men, horses and a few wagons, stretching out of sight like a black ribbon on the uniformly white plain of snow; each man walked on his own, silent and almost crushed by the weight of his thoughts and his fears." The weather continued bitterly cold, with the daytime temperature hovering around the -35°C (-31°F) mark, and frostbite continued to take a toll: "One could see an extraordinary number of soldiers with hands and fingers of bone, as the flesh had fallen away," wrote Vionnet de Maringoné.[31]

A hard core still stuck to their colours, in groups of about fifty. "I remained, with, alas! very few others, with our eagle, whose pole was adorned with no more than a shred of cloth, and which, deprived of one of its wings, carried away by a shell at Eylau, still hovered over the disasters as our sacred rallying sign," wrote Sergeant Bertrand of the 7th Light Infantry. The grim determination of some of the old soldiers is remarkable, and when Marshal Lefèbvre let slip at a moment when they were encircled outside Vilna the despairing comment that they would never see their homes again, one of them turned to him and said: "Shut up, you old fool. If we have to die, we'll die."[32]

Sergeant Bourgogne watched as the pitiful remnant of one regiment turned to face the enemy when its bedraggled colonel called out: "*Allons*, children of France, we must stand again! It must not be said that we hurried our pace at the sound of the cannon! About turn!" Ney, who was commanding the rearguard of some eight hundred men, set a remarkable example of courage and endurance. "He was, at that moment, just as one imagines the heroes of antiquity," noted Bourgogne, who watched him repel a Russian cavalry attack at the head of his troops. "One can fairly say that he was, in the last days of that disastrous retreat, the saviour of the debris of the army."[33]

Bourgogne claims that during this last part of the retreat, as they began to feel that they really were reaching safety, solidarity began to reassert itself and people stopped to help those who had fallen or to assist each other in various ways. Although this may have been so, it seems rather that people revealed themselves in extremes of good and bad.

Captain Drujon de Beaulieu of the 8th Lancers could go no further and sat down by the roadside to wait for death, but a passing trooper from his regiment who still had a horse stopped, gave him a piece of bread and hoisted him onto his mount. Sergeant Irriberrigoyen, a

Provençal cadre serving with the 1st Polish Lancers, was alone with his Lieutenant, having become separated from the rest of their regiment. The Lieutenant turned to him and said that he could go no further. "You can do what you want, my friend, but I'm f—d," he said. "It's hard to have managed to come all this way, from Moscow, to have reached Vilna and to die here . . . But I can't take another step." The Sergeant tried to persuade him to carry on, but the Lieutenant was adamant. At this point they saw a sleigh coming up the road. The Sergeant cried out with joy, but as it drew level with them the Lieutenant recognised the driver: a soldier from his company whom he had disciplined four times for insubordination and looting, having him flogged and even threatening him with the firing squad. The sleigh stopped and the driver got down. He told the Sergeant to climb in, then came up to the Lieutenant. After a while he burst out laughing, took a swinging punch at him, picked him up and, thrusting him into the sleigh, covered him with a fur rug. "You had me punished for a little bit of looting," he said as they drove off, "but you must admit that it has its uses at times, and that at this very moment you are not greatly concerned about the fact that I pinched this well-harnessed sleigh, which will nicely take us out of this damned country."[34]

But Nicolas Planat de la Faye and his superior, General Lariboisière, reached a small hut one evening in which they decided to halt. They found two young Dutch conscripts warming themselves by a fire inside it, and turfed them out, despite the pleadings of one of them, a boy in his mid teens. They could hear him whimpering outside as they fell asleep, and found him frozen to death when they set off in the morning.[35] And in some instances, people found it difficult to tell what was right and what was wrong.

A Belgian soldier of the Guard came across an officer lying in a sleigh, his servant having abandoned him and taken his horse. "Wrapped in a large fur cloak, his hands and feet were frostbitten, and he begged me to kill him, as he was certain that he would not live long in this position," writes the soldier. "I had already primed my musket in order to render him this service which he implored of me, but then

I reflected that he might as well die without me. I left him, but I was some way away before I could no longer hear him begging me to kill him."[36]

Kovno was well stocked with supplies and was certainly defensible, having been fortified with some earthworks. But Murat did not consider stopping there, and sped on towards Königsberg. The organised units received some rations, but the fleeing rabble that streamed into Kovno on 12 December and the following day was in no condition to defend anything. Most of the men made straight for the stores, where they devoured everything they could lay their hands on, without waiting for bread to be baked or distributed in an orderly way. They came across a large supply of spirits, and fights broke out between drunken French and German soldiers. A great many men sat down to drown their sorrows. The alcohol, which warms the spirits but actually reduces body temperature, was to be their undoing, and thousands froze to death as they collapsed, still clutching their bottles, or dozed off, huddled in doorways and porches.

When he reached Kovno with his dwindling rearguard, a ragbag of depleted units, Ney took up defensive positions outside the city in order to allow as many stragglers as possible to pass through it, pick up supplies and get across the Niemen. It was a slow business, as, although the river was now frozen hard and could be crossed anywhere, everyone was crowding onto the bridge, and the ensuing bottleneck caused the usual fights and casualties.

Ney soon found himself threatened with encirclement by cossacks, and was bombarded by artillery brought up by regular Russian cavalry. He had a few guns, including some of those Major Noel had so unnecessarily brought all the way to Ponary, and he kept the Russians at bay for a time. But his troops were melting away. A company of Germans from Anhalt-Lippe gave up when they saw their Captain, who had been wounded, put a pistol to his head and shoot himself. In the end Ney was left with only a handful of French infantry, so he began falling back, carrying out a fighting withdrawal through the town and over the bridge. A soldier's musket in his hand, he remained in the front rank of his diminishing force, commanding them and encouraging them to the last. As he reached the western end of the bridge he discharged one last shot at the Russians and then flung his musket into the frozen bed of the Niemen before turning and trudging off.[37]

The Intendant General Mathieu Dumas had struggled across the river earlier and reached Gumbinnen, where he took shelter in the local doctor's house. The next morning he was just sitting down to a nourishing breakfast and some good coffee when the door opened and in came a man dressed in a brown greatcoat, his bearded face blackened by smoke and his eyes red and sparkling. "Here I am, at last!" the newcomer announced. "What, General Dumas, do you not recognise me?" Dumas shook his head and asked him who he was. "I am the rearguard of the Grande Armée," answered the man. "I am Marshal Ney."[38]

24

His Majesty's Health

The first stages of Napoleon's flight had passed in a gloomy silence as he and Caulaincourt shivered beneath layers of furs in the commodious coach. But that had changed once they crossed the Niemen at Kovno on 7 December. "As soon as we were in the Grand Duchy [of Warsaw] he became very gay and kept talking about the army and of Paris," recorded Caulaincourt. "He would not admit any doubts that the army would manage to hold on at Vilna, and refused to face up to the extent of its losses."[1] They had exchanged their coach for a primitive sleigh made out of an old carriage mounted on runners, and Napoleon chattered on as they sped along, with snow blowing in through the cracks around the ill-fitting doors.

He went over all the events that had led up to the war, which he protested that he had never wanted, and maintained that he had always meant to re-establish the Kingdom of Poland in the interests of world peace. "They do not understand: I am not ambitious," he complained. "The lack of sleep, the effort, war itself, these are not for someone of my age. I love my bed and rest more than anyone, but I have to finish the work I have embarked on. There are only two alternatives in this world: to command or to obey. The policies of every cabinet in Europe towards France have proved that she could only count on her own power, and therefore on force." He kept drifting back to the subject of Britain, which he represented as the one obstacle to the desired peace, and tried to convince Caulaincourt that he was fighting against her on behalf of the whole of Europe, which did not realise that it was being exploited by the fiendish islanders.[2]

He had talked himself into a good mood by the time they reached Warsaw in the early evening of 10 December, and in order to stretch his legs he got out at the city gate and walked to the Hôtel d'Angleterre, where the sleigh had been sent on. He wondered aloud to Caulaincourt whether anyone would recognise him as they walked through the busy streets, but nobody took any notice of the small plump man in his green velvet overcoat and fur-lined bonnet. He seemed almost disappointed.[3]

He continued to talk with animation while dinner was prepared and a servant girl struggled to light a fire in the freezing room they had taken at the hotel. Caulaincourt had been sent to fetch Pradt, who was struck by the jolly mood of the Emperor when he arrived. But that did not make the interview any easier for the Archbishop of Malines. Dismissing his own failure with the phrase "From the sublime to the ridiculous there is but one step," Napoleon laid into Pradt, blaming him for having failed to galvanise Poland, raise money and furnish men. He declared that he had never seen any Polish troops during the whole campaign, and accused the Poles of lacking in courage and determination.

His tone changed with the appearance of the Polish ministers he had summoned. Although he did not attempt to hide the fact that he had been forced into a disastrous retreat which had cost him thousands of men, he told them that he had 120,000 still at Vilna, and would be back in the spring with a new army. They must raise money and a mass levy in order to defend the Grand Duchy. The ministers stood around getting colder and colder as he paced up and down, carried away into a lengthy monologue by his own fantasies.

"I beat the Russians every time," he ranted. "They don't dare to stand up to us. They are no longer the soldiers of Eylau and Friedland. We will hold Wilna, and I shall be back with 300,000 men. Their successes will make the Russians foolhardy; I will fight them two or three times on the Oder, and in six months' time I will be back on the Niemen . . . All that has happened is of no consequence; it was a misfortune, it was the effect of the climate; the enemy had nothing to do with it; I beat them every time . . ." And so it went on, with the occasional self-justificatory "He who hazards nothing gains nothing" and the frequent repetition of the phrase he had just coined, and which he appeared to relish: "From the sublime to the ridiculous there is but one step."[4]

Having had his dinner and impressed upon the Poles that he was not beaten, Napoleon climbed back into his sleigh and sped out of Warsaw at nine o'clock that evening. As they passed through the little town of Łowicz, he realised that he was not far from the country house of his mistress, Maria Walewska. He knew she was at home, and decided to make a small detour and call on her. Caulaincourt was shocked. He told the Emperor that it would be madness to do any such thing. Not only would it delay their arrival in Paris, and increase the danger that some German patriot might hear of their passage and take it into his head to detain or kill them. It would be an insult to Marie-Louise. And public opinion would never forgive him if he were known to have gone off to revel in his *amours* while his army was freezing to death in Lithuania. Napoleon took some time to be convinced.

As the sleigh flew on through the dismal snow-covered landscape, he turned over the whole political situation again and again, as if trying to convince himself that all it had been was a minor setback. "I made a mistake, *Monsieur le Grand Écuyer*, not on the aim or the political opportunity of the war, but in the manner in which I waged it," he said, giving Caulaincourt's ear an affectionate tug. "I should have stopped at Witepsk. Alexander would now be at my knees. The way the Russian army divided after I crossed the Niemen blinded me . . . I stayed two weeks too long in Moscow."[5]

This was perfectly true. Two weeks before Napoleon left Moscow, Kutuzov had no more than about 60,000 men under arms, and the 20,000 cossacks he was counting on were still a long way off.

Napoleon could have stormed his camp at Tarutino, or just withdrawn through Kaluga, Medyn or Smolensk without being molested. He would have been able to evacuate all the wounded and the materiel he needed, and go into winter quarters wherever he wished long before the weather grew cold. And although the cold may not have been the original or even the major cause of the disaster, it was what had ultimately undermined every effort to retrieve anything from it. Most Russians at the time, as well as observers such as Clausewitz and Schwarzenberg, were adamant that the defeat of the French had nothing to do with Kutuzov and everything to do with the weather. "One has to admit," wrote Schwarzenberg, who referred to the Field Marshal as "*l'imbécile Kutuzov*," "that this is the most astonishing kick from a donkey any mortal has ever had the whim to court."[6]

Napoleon's marshals and generals, not to mention his soldiers, were agreed that the Russians could take no credit for what had happened. "In every instance the Russians were beaten, and as soon as the army has had a little rest, they will see their victors again," Davout wrote to his wife from Gumbinnen on 17 December. "The conduct of the troops is splendid; no grumbling. It is as though they all, down to the last soldier, realised that no power and no genius can prevent the damage inflicted by weather." A few days later General Compans wrote to his wife, rebuking her for suggesting that Napoleon had gone mad, albeit admitting that "the calculations of his brain were not as successful in this campaign as in others and that fortune has not been as favourable." The troops were more direct in their appraisal. "We are f—ked, but it doesn't matter . . . we always beat them all the same," muttered a starving Grenadier à Cheval of the Old Guard as he trudged out of Vilna, his uniform in shreds, his bearskin hanging in tatters about his face, with only one boot. "Those little Russkies are no more than schoolboys."[7]

Although Napoleon realised his setback in Russia would give heart to his enemies and threaten his ascendancy, he was confident that he would be able to reassert himself. His aptitude for wishful thinking had not been among the casualties of this war, and he was already preparing the ground for the coming spring campaign as he raced on. He laid great importance on the conflict that had broken out between Britain and the United States of America, assuming that this would distract and weaken his eternal enemy. "He did not doubt that it would turn to [the Americans'] advantage," wrote Caulaincourt. "He saw this moment as that of their total political emancipation and their development into a great power."[8]

He reached Dresden in the early hours of 14 December, and stopped at the French Minister's lodgings. As he dictated letters to his various allies, an officer was sent over to the royal palace, where after much argument he was allowed to wake King Frederick Augustus and inform him that Napoleon was in town. When the bleary-eyed monarch had fully taken in the situation, he dressed hurriedly and had himself carried in a sedan chair to the French Minister's residence. Napoleon, who had managed to snatch an hour's sleep, was sitting up in bed, and it was in this position that he reaffirmed his alliance with Saxony and received assurances that fresh troops would be raised.

Napoleon resumed his journey in a commodious vehicle lent by the Saxon King, pausing only to change horses. At many of the stops he would not even leave the carriage, and he would only go into a coaching inn in order to lunch or dine. At Weimar, where his carriage paused briefly in the middle of the night, he nevertheless found a moment to ask someone to convey his respects to "*Monsieur Gött.*" Four days later he was rolling into Paris.

As the carriage drove up to the Tuileries at a few minutes to midnight on 18 December and the two men wrapped in heavy cloaks with their faces hidden by fur caps got down, the sentries let Napoleon and Caulaincourt pass, taking them to be bearers of urgent news. In the palace itself, the concierge took some persuading that this really was the Emperor and the Master of the Horse. A few minutes later, Marie-Louise heard a commotion and came out to see her ladies-in-waiting trying to forbid entry to two unfamiliar figures. But she could not hide her joy on recognising her beloved husband, and they fell into each other's arms.

Before allowing himself to indulge in marital comforts, however, Napoleon gave a last instruction, one that shows he had lost nothing of his statesmanship. He ordered Caulaincourt to go to the house of the Arch-Chancellor, Cambacérès, to inform him of the Emperor's return and to instruct him that there would be a normal *lever* the next morning.

The twenty-ninth Bulletin had been published three days before, on 16 December. For over a decade every *Bulletin de la Grande Armée* had contained only tidings of victory and glory, and people were stunned to read such an admission of failure. But its closing words about "His Majesty's health" made it clear that the Emperor was not affected by the reverse. And before they could recover from the shock or start drawing conclusions, Napoleon was among them once more, behaving as though nothing had happened. He took the reins of power firmly in his hands and set about raising a new army, to be ready in March. "I am very pleased with the mood of the nation," he wrote to Murat that very day, addressing the letter to Vilna.[9]

By the time Napoleon wrote out that address, Vilna was in Russian hands, and on that very evening Kutuzov was attending a gala at the theatre organised in his honour by the nervous inhabitants. According to Clausewitz, his only contribution to the victory had been the refusal, born of fear, to take on Napoleon, but he was the victor nevertheless. And nobody was more surprised than him. "If two or three years ago anyone had told me that fate would choose me to bring down Napoleon, the giant threatening the whole of Europe, I would have spat in his mug," Kutuzov confided to Yermolov.[10]

Four days later, on 23 December, Alexander himself entered the city. At the gates enthusiastic soldiers unharnessed the horses from his carriage and dragged it up to the archepiscopal palace, where a triumphant Kutuzov was waiting to greet him. The Tsar embraced the Field Marshal graciously, but he was far from satisfied.

At a private meeting on the morning of 26 December Alexander told Wilson that Kutuzov had done "nothing he ought to have done," and that "all his successes have been *forced* upon him." He complained that he could do nothing about it, as the Field Marshal was the darling of the nobility of Moscow. "In half an hour I must therefore (and he paused for a minute) decorate this man with the great Order of St George, and by so doing commit a trespass on its institution; for it is the highest honour, and hitherto the purest, of the empire," Wilson reports the Tsar as saying.[11]

Alexander reproached Kutuzov for losing three days by retreating from Maloyaroslavets, for failing to cut Napoleon off at Krasny and for letting him get across the Berezina. He was annoyed that the French were not being pursued with more vigour. And, like his brother Constantine before him, he was shocked and displeased by the scruffy appearance of the regiments which paraded before him. Kutuzov countered by blaming others for failing to carry out his orders, and stressed his own merit, arrogating all their minor successes to his own credit.[12] He was already dictating legend.

"I mostly humbly beg you, gracious lady, that the fortifications built near the village of Tarutino, fortifications which put fear into the ranks of the enemy and proved a solid bastion at which the rushing torrent of the destroyers which threatened to flood the whole of Russia was stemmed, that these fortifications remain untouched," he wrote to Princess Naryshkin, on whose land the camp of Tarutino was situated. "Let time, not the hand of man, destroy them; let the cultivator, working in his peaceful fields around them, not touch them with his plough; let them in later times become for the Russians sacred monuments of their valour; let our successors, gazing upon them, become enflamed with the fire of emulation and say with rapture: this is the place on which the pride of the predators fell before the fearlessness of the sons of the fatherland."[13]

As far as Kutuzov was concerned, the campaign was over; his army was exhausted and needed a long rest. At the beginning of December he had issued an appeal to the German nation, calling on it to rise.

"The favourable moment has come for you to take revenge for the humiliations you have suffered, and to raise yourself once more to the circle of free nations," it ran. "Your princes are in chains and they expect you all and each of you individually to free them and avenge them." This was followed by a long account of the campaign in German, by a personal address from Kutuzov urging the Prussian forces to leave the Napoleonic camp and join with Russia, and by another appeal to the peoples of Germany.[14] But while he wanted to foment trouble for Napoleon in Germany, he was not about to march out and liberate it. Many Russians felt, like him, that they should take over the whole of the Grand Duchy of Warsaw, thereby settling the Polish problem once and for all, and possibly East Prussia with Danzig as well, leaving the Germans to fight it out with Napoleon further west.

Alexander saw things differently. Although he acceded to Rostopchin's wish for a victory monument to be cast from the bronze of captured French guns, he himself was more interested in building a great church of the Holy Saviour. That his cause had triumphed in spite of all the mistakes he and his commanders had made only confirmed him in the conviction that he was a tool in the hands of the Almighty. "It is God who did everything; it is He who has changed everything so suddenly in our favour, by casting down on the head of Napoleon all the calamities which he had meant for us," he wrote to his sister Catherine on 20 November.[15]

"The fire of Moscow lit up my soul and the judgement of the Lord which manifested itself on our frozen plains filled my heart with an ardent faith I had never felt before," he wrote to another, adding that he realised he must from this moment devote himself to the cause of establishing the kingdom of God on earth.[16] And he was by no means alone in seeing things in such spiritual terms. There was something almost biblical about the scale of the events, and many saw the suffering Russia had endured as punishment for her sins and the fire of Moscow as a purifying expiation. "Her ruins will be a pledge of our redemption, moral and political, and the glow of the embers of

Moscow, Smolensk etc. will, sooner or later, light our way to Paris," A.I. Turgeniev had written to Prince Piotr Viazemsky on 8 October. "The war, which has become a national one, has taken such a turn that it must end with the triumph of the north and a brilliant revenge for the gratuitous wickedness and crimes of the southern nations."[17] He was not the only one who felt the war did not end at Vilna.

"You will place yourself, Sire, at the head of Europe's powers," Stein had written to Alexander on 17 November. "The exalted role of benefactor and restorer is yours to play." He needed little prompting. He was already determined to carry the war into Germany and beyond. On 30 November he had decreed a fresh draft of recruits, eight out of every five hundred souls, who, along with those drafted in July and August, would provide the troops he would lead in the liberation of Europe. This was not to be a mere political liberation from the tyranny of Napoleon – it was to be a crusade. In between the receptions and balls taking place in Vilna that Christmas, Alexander had several intimate conversations with the young Countess Tiesenhausen, whom he had taken such a liking to back in the spring. "The Emperor spoke like a real sage, who only wishes for the happiness of mankind," she noted. "He seemed to be dreaming only of the means by which to bring back the Golden Age."[18]

As Alexander dreamt of bringing heaven to earth and Napoleon laboured at rebuilding his power in Europe, neither spared a thought for the victims of the recent events or those who were still struggling on the fringes of life. While the Tsar and his entourage danced, the final act in the tragedy of Vilna was unfolding beyond the palace windows.

"There was no mound of snow or heap of rubbish that did not have an arm or a leg protruding from it – in uniform, as the stripping of the dead had ceased," wrote Aleksander Fredro. "Throughout the winter in the narrow streets one could see corpses crouching with their backs against the walls, and they had not been spared by mockery. This one had been given a bunch of flowers to hold, that one a pole in the guise of a musket, a third had had a stick resembling a pipe stuck in his mouth."[19]

Cossacks roamed the streets selling their booty, which included children they had taken as they harried the retreating French. "These poor little wretches, who had been torn from their mother's breast by their strange new protectors, could only moan, as they could not even say the name of their parents, who had probably died during the retreat," wrote Countess Tiesenhausen.[20]

The wounded and sick lay in the freezing cold, untended and unfed, in improvised "hospitals," of which there were about forty in Vilna and its environs, situated in monasteries and country houses. They were usually staffed by a couple of medical orderlies assisted by cossacks or militia, who meted out regular doses of brutality but little in the way of food or water, let alone medical attention. Many of the men were afflicted by typhus or other fevers as well as their wounds, but all of them suffered above all from hunger and thirst, with the result that while the military stores left by Napoleon still contained victuals in abundance, the men in the hospitals resorted to cannibalism. Tens of thousands, possibly as many as 30,000, died in the space of a couple of weeks.

In the spring, the corpses would be loaded onto wagons, many of them abandoned French ones, to be taken away and buried or burnt. One survivor never forgot the burlesque sight of twenty or so French *fourgons*, some of them inscribed "*Equipages de S. M. Empereur et Roi*," laden with heaps of stiff bodies piled up like logs.[21]

Conditions were no better elsewhere. Jean Pierre Maillard from Vevey, a sergeant in the 2nd Swiss Regiment and a veteran of the Spanish campaign, was wounded during the battle for Polotsk and found himself lying with hundreds of others in a convent, without food or water. When the Russians arrived, they swarmed into the building and began robbing the wounded men of everything, even unpicking Maillard's sergeant's stripes from his arm. A second wave took his uniform. He lay for days without any assistance, dressing his own wound as best he could, squeezing out the pus, and scraping lice off his skin with a knife. "I thought I was with the damned in hell," recalled Giuseppe Venturini, who had also been dumped on the floor of a "hospital" in Polotsk, where he lay untended and unfed listening to the screams of his comrades. Of the two hundred sick in one hospital at Polotsk when it fell on 20 October, there were twenty-five left alive on 23 November, and two on 12 January 1813.[22]

The fate of the French captives was hardly more to be envied. After being stripped, they were rounded up and either parked out in the open or herded into any available building. At Smorgonie, Henri Ducor was thrown into a grange which was already full of men. The cossacks kept bringing more and more, forcing them to push their way in until they had packed some four hundred men into a space measuring twenty feet by twenty. As men fell over or fainted in the crush they were trampled by others, and Ducor soon found himself standing on a heap of corpses. "We were so tightly pressed together, particularly at the back of the room, that the dead did not even have the space to drop, and, buffeted this way and that, they looked, with their stiffened arms raised, as though they were struggling along with the others," he wrote, "but eventually they would be pushed down, trampled and crushed, as the others climbed onto their corpses in order to relieve their lacerated feet with a trace of human warmth." At Kovno, the cossacks threw forty officers into a cellar after stripping and beating them, and left them there without food or drink for a couple of days, at the end of which only three came out alive.[23]

A physician attached to Berthier's staff, taken at Vilna and sent off in a convoy of three thousand men to Saratov, was subjected to a catalogue of brutality. When he complained to the officer commanding the escort, the latter said he was powerless to do anything, as he feared his own men. Prisoners often had to spend the night out in the open after a long day's march; some would stand in order to avoid sleeping on the snow, but would freeze as they lent against a tree. "Their last sweat would freeze over their emaciated bodies, and they would stand, their eyes open forever, their body fixed in whatever convulsive attitude death had overtaken and congealed them," the doctor wrote. "The bodies stood there until someone removed them in order to burn them, and the ankle would come away from the leg more easily than the sole could be prised from the ground." It was only out of the war zone, where the inhabitants took pity and gave them food and they were escorted by veteran soldiers, many of whom had themselves been prisoners in France, that conditions improved.[24]

Colonel Seruzier of the light artillery of Montbrun's 2nd Cavalry Corps was more fortunate than most. He was taken prisoner by cossacks just as he was about to recross the Niemen. He was wounded, as well as being stripped and robbed, and forced to march through the snow in the intense cold barefoot and naked. He complained of this treatment when he was brought before Platov, but received a supercilious rebuke. Luckily for him, Platov's son took pity on him, and after having his wounds dressed procured him some scraps of clothing to cover himself with. This would have made little difference once he had been sent off under cossack guard, had it not been for the fact that by a stroke of extraordinary good fortune he came across a Russian colonel whom he had taken prisoner at Austerlitz seven years before and treated with chivalry. The Colonel gave him clothes and money, and recommended him personally, with threats, to the commander of the convoy. A further stroke of luck meant that in Vilna Seruzier saw Grand Duke Constantine, whom he had befriended at Tilsit and Erfurt, and after that he was given privileged treatment.[25]

Others who managed to avoid the common fate of prisoners were those who had a useful trade: medical skills were especially valued. As soon as it became clear who he was, Dr La Flise was asked to take over the job of the regimental surgeon of the Russian unit which had captured him. Dr Heinrich Roos was also set to work for the Russian army as soon as he was taken. And tailors, cobblers, blacksmiths and others were in many cases taken along by the advancing Russians. A French bugler somehow contrived to charm the cossacks who captured him by playing for them on his clarinet, and they took him with them as their entertainer. Others with some skill were pulled out of columns of prisoners by the owners of estates they were marched through, who were eager to benefit from free labour.[26]

Many of the soldiers realised that their nationality might help them in some circumstances. Germans often tried to play on theirs, while Dutchmen pretended to be German, as did Poles, in order to avoid being singled out for particularly rough treatment. Spaniards and Portuguese proclaimed their nationality loudly, with some success, as the Spanish *guerrilla* against the French invaders had become a popular legend in Russia. Don Raphael de Llanza, an officer in the Joseph-Napoléon Regiment, called out that he and his men were Spaniards as they surrendered at the Berezina, and they were greeted with enthusiasm by their captors. Their escorting cossacks would shout "Spaniards!" as they passed other Russian units or groups of angry-looking peasants, who in turn hailed them. This did not prevent de Llanza's jaw being so badly frostbitten that he later had to have a silver implant specially made – which he eventually left, along with his sword and his spurs, to his heirs.[27]

Sergeant Bénard was surprised at the reception he and his wounded comrades received as their carts trundled into Moscow. "I must here do justice to the inhabitants of Moscow that they did not let us hear a single shout of hatred, not a single threat; quite the contrary, I received from them unequivocal marks of sympathy, and my little cart was heaped with provisions which anonymous hands had discreetly slipped in," he recorded. But he was saddened to discover when, being well enough to walk, he went to visit the German family with whom he had stayed six months before, that they had been dispossessed and sent into exile for their friendliness to the French.[28]

The prisoners would eventually be handed over to the civil authorities in some provincial town, where they were at last given the subsistence pay due to them. They were able to take jobs, the officers were allowed out on parole, and most of them were treated reasonably. But there were exceptions. "Go to the devil, you French dogs," the Governor of Tambov, a Prince Gagarin, thundered when confronted by some German prisoners who had dared to complain about their conditions. "Heaven is very high and the Tsar is very far away."[29]

Those who had got across the Niemen with the remnants of the retreating army had thought they would be safe once they had crossed the frontier of the Russian empire. It came as a shock to find that bands of cossacks followed them beyond the boundary, and many were killed or taken on Prussian soil. But eventually the survivors made it to Königsberg, where Murat had set up his headquarters and intended to rally the army, and where they felt they could relax at last.

For years afterwards, Bellot de Kergorre could not forget the luxury of his first bath: "I could feel my whole body unwind; it felt as though every part was getting back into its proper place, from which it had been disturbed; all my muscles, all my nerves were shifting and relaxing." The joy of shedding and burning their vermin-infested clothes and putting on clean ones shines through many an account. But the process of readjusting to normal life was not always easy. Marie Henry de Lignières sought out the family he had stayed with in Königsberg on the way out, and they welcomed him with open arms. He spent a delightful evening with them, but reduced them all to tears when he began to play the piano with his frost-damaged fingers. One officer, a German aristocrat, was mortified when he caught himself grabbing at food. "I tended to use my hands instead of a fork," he writes. Some found they could not bear to sit for long in heated rooms, and had to sleep with the window wide open even when it was freezing outside.[30]

A large number of the survivors were in far worse shape than they realised. Just as many had gone mad in Vilna, some now gave way to "nervous fevers" of one sort or another. Others fell victim to the typhus epidemic that had broken out in the last stages of the retreat. Among them were Inspector General of Artillery Lariboisière and General Eblé, whose pontoneers had saved the army at the Berezina.

Murat held reviews and talked of regrouping, but his abandonment of Vilna, which he had been ordered to hold, and then Kovno, which he had himself promised to hold, did not inspire confidence. There was no question of his bravery, but there was also no question of his limited intelligence and strength of character. And his efforts at rallying the army were undermined by the wider political situation.

In Prussia the French came up against the full force of popular resentment. In Königsberg, students would sing belligerent songs and recite poems calling on Germany to rise up against the tyranny of France. As they trudged through the countryside, the soldiers, even if they were Germans, were spat upon as they passed, and the locals often refused to sell them food. Many of the isolated or wounded men were set upon and beaten up.[31]

A more alarming development was taking place to the north, where Macdonald's 10th Corps was falling back to link up with Murat's remaining forces. Macdonald still had nearly 30,000 men in good shape, which was quite enough to hold East Prussia and discourage the Russians from crossing the Niemen. Half of this corps consisted of Prussians under General Hans David Yorck von Wartenburg, a choleric Prussian patriot who bore no love for the French.

At the beginning of November, Yorck had received a letter from General Essen, the Russian commander in Riga, informing him that Napoleon was in full flight and suggesting that Yorck change sides and kidnap Macdonald. Yorck refused to contemplate such a proposal, but after more overtures on the Russian side, he replied that if the Tsar wished to communicate with the King of Prussia, he would be prepared to pass on his letters. Macdonald had noted a "*refroidissement*" taking place, but, confident in the military honour of the Prussian officers under his command, did not fear anything more than a lack of enthusiasm in carrying out his orders.[32]

As Macdonald withdrew at the beginning of December, Yorck and his Prussians fell behind. Macdonald paused in his retreat and sent Yorck orders to catch up before he was cut off, repeating them when he received no response. His entourage began to mutter about treason, but he refused to believe a soldier of Yorck's standing capable of such a thing.

On 20 December the division of General Diebitsch, another Prussian in Russian service, came between Yorck and Macdonald, effectively cutting off the Prussians. There followed a stand-off between the Russians and the Prussians while Major Clausewitz, who was now on Diebitsch's staff, conducted negotiations. Despite a certain amount of suspicion on both sides, they rapidly reached an agreement which, after referral back to the Tsar, was signed as the Convention of Tauroggen on 30 December. By this act, Yorck undertook to cease fighting the Russians, while they granted his corps neutral status.[33]

When news of the Prussian defection reached him, Murat realised that the line of the Niemen could no longer be held and proceeded to evacuate Königsberg. Schwarzenberg also understood that the game was up. He withdrew from his positions in Poland and made for Austria, obliging Reynier and his Saxons to fall back as well so as not to be cut off from Saxony. The whole of East Prussia and the Grand Duchy of Warsaw were now wide open to the cossacks, and thousands more stragglers were taken or killed.

The French held on at Danzig and fortresses such as Küstrin, Toruń and Glogau, and Poniatowski was doing his best to scrape together enough men for a defence of Warsaw. The city presented a curious spectacle, for as emaciated stragglers drifted in from the north and east, a fine new regiment of Italian Vélites marched up from the other direction. "While the heroes were being found places in hospitals, the young Italian gentlemen, beautifully uniformed and groomed, went into society, where they sang with their melodious voices and tasted every kind of pleasure," wrote one Varsavian.[34]

It was only as the French retreat finally came to a stop towards the end of January and the remnants of the Grande Armée reached the towns designated as assembly-points for the various units that a picture of the scale of the disaster began to emerge.

It is impossible to establish with any precision the number of those who perished, since the original figures for the Grande Armée are probably too high, while it was continuously reinforced during the operations in Russia. But one can be fairly confident that the total number of French and allied troops operating beyond the Niemen at some time between June and December 1812 was somewhere between 550,000 and 600,000. Only about 120,000 of these came out in December. Possibly as many as 30,000 had come out earlier, either as lightly wounded, sick, or as cadres sent to form new units in France or Italy. It is also probable that a hefty proportion of the 50,000 or so deserters at the start of the campaign got out while the going was good. The Russians took about 100,000 prisoners, but certainly no more than 20,000 of those survived to be sent home after 1814. Thus it can safely be assumed that as many as 400,000 French and allied troops died, less than a quarter of them in battle. One can only guess at how many more tens of thousands of French and allied civilians, from among those who followed the army into Russia and those who chose to follow it back from Moscow, also perished. And of those, both military and civilians, who did get out, a large number died within a month or two, of typhus, tuberculosis or the nervous after-effects of the campaign.

Losses on the Russian side are no easier to ascertain with any precision. It is now thought that up to 400,000 soldiers and militia perished, around 110,000 of them in battle. The number of civilians who were killed, at the sieges of Polotsk and Smolensk, in the fire of Moscow, in thousands of raids by marauders, or simply of hunger and cold as a result of having to flee their homes, can only be guessed at. But it is safe to say that all in all, between the Grande Armée's crossing of the Niemen at the end of June 1812 and the end of February 1813, about a million people died, fairly equally divided between the two sides.[35]

As far as Napoleon was concerned, the losses were even more serious than they might appear from these figures. Of the 120,000 who had come out of Russia in December, 50,000 were Austrians and Prussians, the former shaky allies, the latter about to become enemies. Of the remaining 70,000, over 20,000 were Poles. If one were to strip away the other allied survivors, one would probably find that no more than about 35,000 Frenchmen had survived, many of them unfit for service. To this must also be added the loss of at least 160,000 horses taken from stock within the Napoleonic empire, and over a thousand cannon.

The extent of the damage varied a great deal from unit to unit. The Austrian contingent had survived largely because, being on its own under Schwarzenberg, it was able to avoid battle on all but one occasion. The Prussians under Macdonald were only involved in light skirmishes, and showed a similar lack of enthusiasm for fighting the Russians – their sector was so peaceful for long stretches of the summer that the officers used to go bathing in the Baltic sea.

Of the 96,000 Poles who took part, as many as 24,000 came out of Russia, a survival rate of 25 per cent. Of the 32,700 Bavarians who crossed the Niemen, the most General Wrede could muster on 1 January 1813 was four thousand, or 12 per cent of the total. The two Illyrian regiments, which had numbered 3,518 men of all ranks, were down to 211, about 6 per cent. Of the 52,000 men of the Army of Italy, only 2637 men and 207 officers reassembled at Marienwerder in January, representing just over 5 per cent. And of the 27,397 Italians who crossed the Alps in the early summer of 1812, only about a thousand returned home, not much more than 3 per cent.[36]

As might be expected, regiments of the Guard came out comparatively well. Lieutenant Marie Henry de Lignières' company of Chasseurs mustered fifty-two men out of 245 at Königsberg. The roll-call in the Lancers of Berg on 3 January 1813 was answered by 370 men out of an original complement of 1,109. The Polish Chevau-Légers, who had crossed the Niemen 915 strong and shed officers and men as cadres for the Lithuanian regiments being raised, still numbered 422 when they recrossed into the Grand Duchy of Warsaw in December.[37]

Line regiments fared less well. According to Sergeant Bertrand, the 7th Light Infantry in Davout's corps opened the campaign with 3,342 men, and numbered only 192 when the roll was called at Toruń on

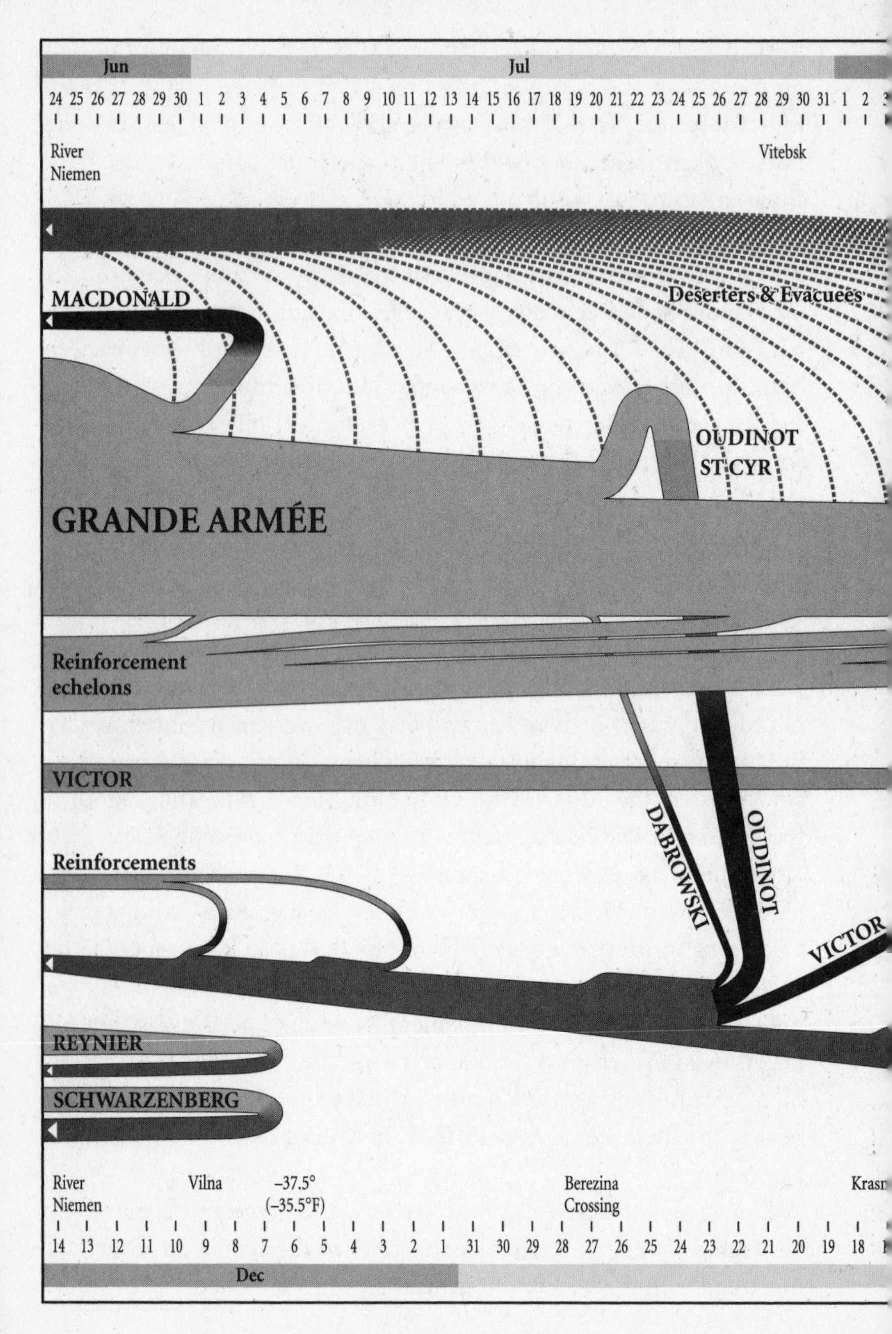
Jun
Jul
24 25 26 27 28 29 30 1 2 3 4 5 6 7 8 9 10 11 12 13 14 15 16 17 18 19 20 21 22 23 24 25 26 27 28 29 30 31 1 2
River
Niemen
Vitebsk
MACDONALD
Deserters & Evacuees
OUDINOT
ST CYR
GRANDE ARMÉE
Reinforcement
echelons
VICTOR
DABROWSKI
OUDINOT
Reinforcements
VICTOR
REYNIER
SCHWARZENBERG
River
Niemen
Vilna
–37.5°
(–35.5°F)
Berezina
Crossing
14 13 12 11 10 9 8 7 6 5 4 3 2 1 31 30 29 28 27 26 25 24 23 22 21 20 19 18
Dec

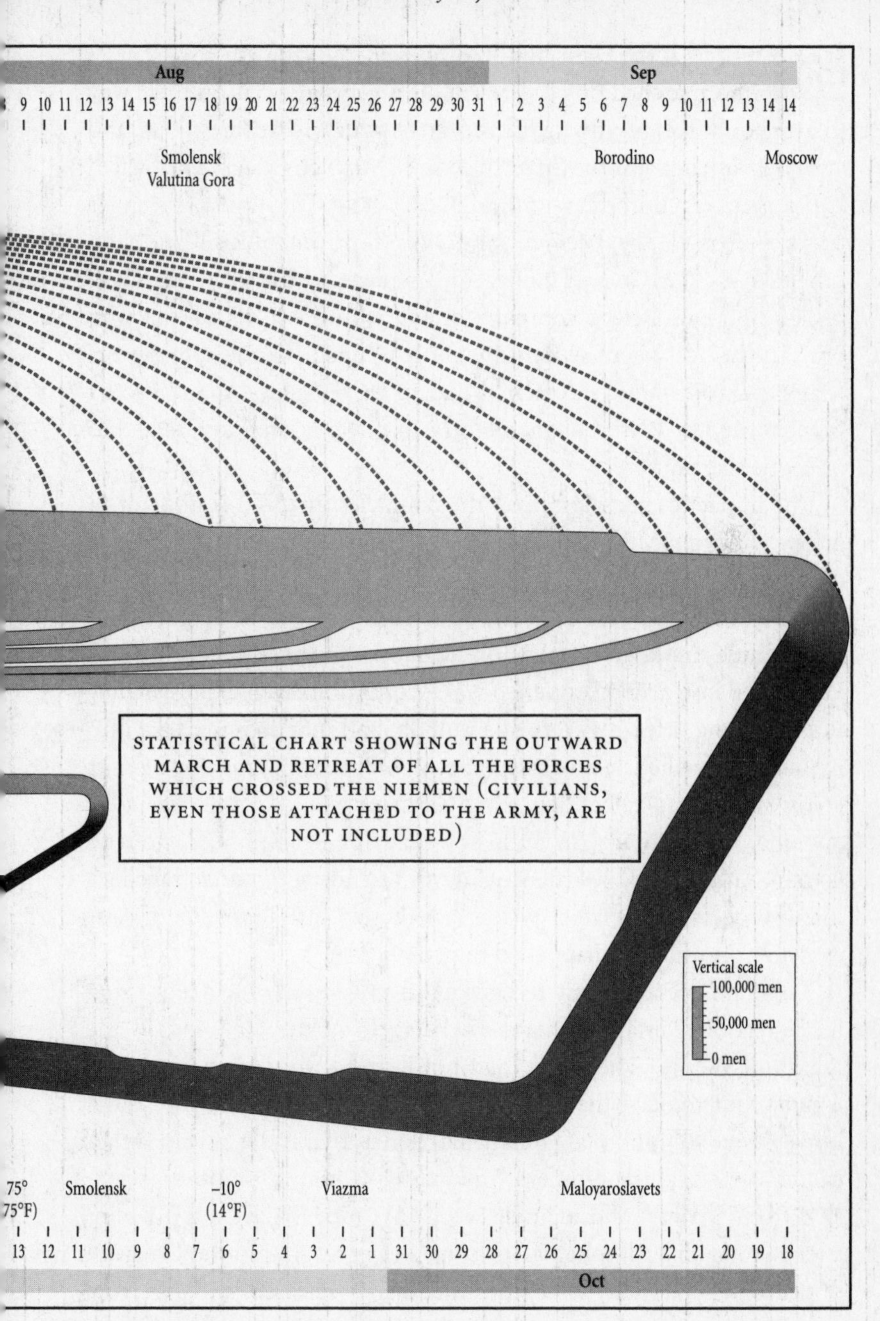
Aug
Sep
9 10 11 12 13 14 15 16 17 18 19 20 21 22 23 24 25 26 27 28 29 30 31 1 2 3 4 5 6 7 8 9 10 11 12 13 14 14
Smolensk
Valutina Gora
Borodino
Moscow
STATISTICAL CHART SHOWING THE OUTWARD MARCH AND RETREAT OF ALL THE FORCES WHICH CROSSED THE NIEMEN (CIVILIANS, EVEN THOSE ATTACHED TO THE ARMY, ARE NOT INCLUDED)
Vertical scale
100,000 men
50,000 men
0 men
75°
75°F)
Smolensk
–10°
(14°F)
Viazma
Maloyaroslavets
13 12 11 10 9 8 7 6 5 4 3 2 1 31 30 29 28 27 26 25 24 23 22 21 20 19 18
Oct

31 December. Of the eight hundred horsemen of the 8th Chasseurs à Cheval who marched out of Brescia on 6 February 1812 on their way to war, only seventy-five gathered at Glogau a year later. The 2nd and 3rd Battalions of the Spanish Joseph-Napoléon Regiment, which had fought at Borodino and Maloyaroslavets in Prince Eugène's corps, recrossed the Niemen with only fourteen officers and fifty other ranks, but Colonel Lopez did manage to carry the standard back with him. Of the company of four hundred pontoneers who built the bridges over the Berezina, only Captain Benthien, Sergeant-Major Schroder and six men returned to Holland. And the only man of the 8th Westphalian Infantry to make it back to Cassel was one lone sergeant.[38]

The impact of such losses on the countries involved is difficult to grasp. For France, the most populous country in Europe, the loss of over 300,000 men out of a population of twenty-seven million would be comparable to a loss of nearly 700,000 today, and that figure does not include the large number of civilians. For the Grand Duchy of Warsaw to lose over 70,000 men was, proportionally, like present-day Poland losing three-quarters of a million, and that does not take into account the civilians who died as a result of the armies moving over its territory. The equivalent figures for Germany would be approximately 400,000, for northern Italy 200,000, and for Belgium and Holland 80,000. And behind these figures lurk thousands of personal tragedies, many of them aggravated by the fact that so little information was available on what had happened to people.

A French peasant wrote to his son on 22 February 1813, addressing his letter to "Captain Flamant of the 129th of the Line, lost in the region of Wilna." The family had not heard a word from him since August, but the fact that he was not a great letter-writer gave them hope. "I like to think that you are a prisoner, that is the only idea that can console us at the moment," he wrote. "Your poor mother is very sick from anxiety, and a single word in your hand would give her back her health." Although all prisoners were set free on the signature of peace in 1814, many were not immediately released, and survivors kept trickling out of Russia for years afterwards. This allowed those who had heard nothing to go on hoping for years, even decades. One peasant woman in Mecklemburg was still making enquiries about her betrothed in 1849.[39]

There were miraculous survivals, and in one case a resurrection. An officer in Dąbrowski's division, Ignacy Dębowski, was wounded and so badly concussed in the fighting outside Borisov that his comrades believed he had been killed. They laid him out on his cloak and buried him under a heap of snow, since the earth was too hard to dig, with full military honours. Dębowski revived after they had marched on, and was taken by the Russians. Like many captured Polish officers, he was pressed into the Russian army and sent to fight as a common soldier in the Caucasus. Years later, having earned his discharge, he reappeared in Warsaw, where he would dine out on stories of his funeral at Borisov.[40]

Some did not return because, having been taken prisoner, they were farmed out to local landowners as cheap labour or took service in order to survive, and never heard of the amnesty or did not have the means to go home. Others did not return because they did not choose to, and made a new life for themselves. They were offered favourable conditions to settle in underpopulated parts of Russia, and even given wives. According to official figures, by 31 December 1814, fifteen senior officers, two medical officers and 1968 other ranks had sworn the oath to become subjects of the Tsar, and another 253 Austrian soldiers were waiting to do so.[41]

In the 1890s a Russian historian discovered Lieutenant Nicolas Savin of the 2nd Hussars, who had been taken at the Berezina, living on the outskirts of Saratov in a small cottage surrounded by flowers which he watered every day. In his study was a bronze statuette of Napoleon and a watercolour portrait of the Emperor done from memory by himself. He lived on, proudly wearing his Légion d'Honneur, until 1894, when he died, apparently aged 127. The reason he had not gone back to France was that he could not bear to see it not ruled by Napoleon.[42]

Guillaume Olive was more pragmatic. He had been born in the United States of America, the son of an émigré, but took service in the French army at an early age. He was taken prisoner during the retreat and decided to remain in Russia. By 1821 he was aide-de-camp to Grand Duke Constantine, and a decade later marshal of the nobility in the province where he had acquired an estate. His son rose to the rank of general of cavalry, married the daughter of a Tolstoy and became a member of the State Council. His grandsons were officers in the Chevaliergardes and gentlemen of the bedchamber, his granddaughters were maids of honour to the Empress. But most did not find it so easy to change their allegiance.

The Russians had hoped to recruit from among the Germans serving with the Grande Armée, many of them unwillingly, and began forming a German Legion to absorb them. In the event, few prisoners volunteered for this, even though it meant immediate release from the dreadful conditions in which they were kept. Astonishingly, most of the German officers and men who had served under Napoleon seem to have remained devoted to him in adversity, and German prisoners throughout Russia solemnly celebrated the Emperor's birthday on 15 August 1813.[43]

The Russians had also been keen to detach the Spaniards serving in the Grande Armée from the Napoleonic cause, and they offered generous terms to any who would surrender. They formed a regiment out of the prisoners, under Don Raphael de Llanza, who had been captured at the Berezina. Although this regiment never fought against Napoleon as had been intended, it did go back to Spain, where, as the Imperial-Alejandro, it took part in the 1820 Riego rising against the very Bourbons Napoleon had dethroned.

Not the least curious outcome of the enmities aroused in 1812 was that the son of Captain Octave de Ségur of the 8th Hussars, who, pierced by two cossack lances and taken prisoner in a skirmish outside Vilna on 28 June, was the first notable French casualty of the campaign, married in 1819 Sophie, the youngest daughter of

Count Fyodor Rostopchin, Governor-incendiary of Moscow. As the Comtesse de Ségur, she went on to write a series of books for children on which generations of French boys and girls were brought up well into the twentieth century.

25

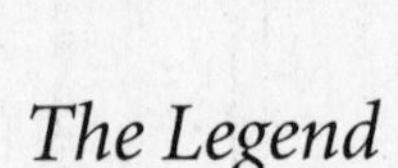

The Legend

The catastrophic outcome of the Russian campaign sealed Napoleon's fate. Not only did it cost him hundreds of thousands of his best soldiers; it punctured the general conviction that he was invincible and tarnished the aura of superiority surrounding his person. "It seems to me that the spell has been broken as far as Napoleon is concerned, and that he is no longer redoubtable as he was in the past," the Dowager Empress Maria Fyodorovna wrote with satisfaction to a friend in the first days of 1813. "He is no longer an *idol*, but has descended to the rank of *men*, and as such he can be fought by men."[1]

She was right. As the master of Europe was seen to stumble and fall, every person who held a grudge against him, every nation which resented his dominion, every group with a dream of change took heart. As the extent of the disaster became known in the first months of 1813, it became apparent that the future of Europe was open to an extent it had not been since the 1790s.

Few saw this as clearly as the German patriots who had been smarting under the humiliation of French dominion. "The first rays of Germany's freedom burst from the east; they were red as blood, but they shed the light of promise," the painter Wilhelm von Kügelgen wrote. General Yorck's defection had been a signal, and young men all over northern Germany prepared to rise up to throw off the French yoke. Stein had set about organising a volunteer militia in Prussia. King Frederick William dithered, unsure of what to do, but was overtaken by events; in the minds of young Germans the war of liberation, the *Freiheitskrieg*, had begun. On 19 February the philosopher Johann Gottlieb Fichte ended a lecture he was giving at the University of Berlin with the words: "This course will be suspended until the close of the campaign, when we will resume it in a free fatherland or reconquer our liberty by death."[2] On 28 February an alliance was concluded between Russia and Prussia, and two weeks later the latter declared war on France.

Napoleon had managed to raise a new army in an astonishingly short time, and took the field at the end of April 1813 with over 200,000 men. He was robbed of the chance to settle scores with Kutuzov, who had fallen ill on the march and died at Bunzlau in Silesia on 25 March. He defeated a combined Russo-Prussian force under Blücher and Yorck at Lützen on 2 May and another under Wittgenstein at Bautzen on 20 May, but the shortage of experienced officers and men in his ranks told, while the lack of cavalry meant that he could not follow up his successes.

Sweden joined the coalition with an army under Bernadotte. Britain contributed money, while her army in Spain under Wellington forced Joseph out of Madrid in May and defeated him at Vittoria on 21 June. And as Napoleon's enemies grew in strength, his allies in Germany began to waver.

Austria had never been an enthusiastic ally of Napoleon, but she did not relish the prospect of a strong Russo-Prussian presence in central Europe, and therefore felt the necessity to save France from annihilation. She offered to mediate and managed to arrange an armistice, which was signed on 4 June. Napoleon was given the opportunity of peace on the basis of a return to France's "natural" frontiers, with no say in the arrangements to be made in the rest of Europe. His situation was desperate. But he felt that if he accepted this he would be placing himself and France at the mercy of the coalition, and he rejected it. Feeling unable to help Napoleon further and fearful for his own future, the Emperor Francis abandoned him and joined the coalition. On 12 August Austria too declared war on France.

Napoleon defeated the Austrians and Russians under Schwarzenberg outside Dresden on 26 August, but Oudinot was defeated by Bernadotte at Blankenfelde and Grossbeeren, and Vandamme by General von Kleist at Külm. On 16 October Napoleon was attacked by the combined allied armies at Leipzig. The "battle of the nations," as it was called on account of the number and size of the armies involved, lasted three days, on the second of which Napoleon's last ally, the King of Saxony, was obliged to change sides. By this stage the French were outnumbered by two to one. Napoleon held his own for as long as he could, giving Barclay de Tolly and Bennigsen, both of whom had been called back to command armies, a very difficult time. But in the end he was forced to withdraw. A nervous engineer blew the bridge across the river Elster prematurely, cutting off Napoleon's rearguard of 20,000 men. Lauriston was taken prisoner with them. Poniatowski perished while trying to swim across. Despite another victory over a force of Austrians and Prussians under Wrede at Hanau, Napoleon's position in Germany was no longer tenable.

He struggled back across the Rhine with no more than about 40,000 men at the beginning of November 1813, and although he fought a brilliant campaign against the invading armies throughout that winter and spring, he could do no more than delay the end. Paris capitulated on 31 March 1814, and Napoleon was forced to abdicate on 6 April. He was exiled to the island of Elba off the Italian coast. A year later, on 1 March 1815, he landed in France and took power once again amid scenes of patriotic jubilation, but on 18 June he was defeated at Waterloo by a combined British and Prussian army under Wellington and Blücher. He was then exiled to the island of St Helena in the Atlantic Ocean, where he would die on 5 May 1821.

As he rode into Paris in triumph on 31 March 1814, Tsar Alexander was cheered by enthusiastic crowds. The French were relieved the war was over. They were also charmed by his person and his manner, which radiated benevolence and spirituality. He seemed uninterested in triumph or retribution, and magnanimously pardoned all the Poles who had fought against him in Napoleon's ranks. He talked of expiation and regeneration, held prayer-meetings and curious semi-religious parades, and represented himself to some as a "Second Abraham" who would bring a new kind of peace to the world.

He was not alone in this exaltation. In Germany, where poets compared Napoleon's Russian débâcle to the fate of the Pharaoh's army in the Red Sea, many believed Alexander to be an instrument of God's will. In London, the Tsar was greeted by Quakers and members of the British Bible Society as though he were some kind of spiritual sage or even avatar. It is not unusual after a long war for people to dream of new beginnings.

None let themselves go to such dreams with greater abandon than the young Russian officers who had taken part in the campaigns of 1812–1814. "Many centuries have passed, and many will follow in the future, but not one of them contained or will contain two such full and wondrous years," wrote Nikolai Borisovich Galitzine, a junior staff-officer. Young men such as he felt they had lived through a national epic that vindicated their country's right to universal respect. "With head lifted proudly, one can at last say 'I am a Russian!,'" wrote the partisan commander Denis Davidov.[3]

It was not just pride in the new power and prestige of their country that animated them. "After the successful conclusion of the patriotic war and our victorious march from the ruins of Moscow to Paris, Russia breathed a sigh of freedom, and came alive with the spirit of renewal and rebirth," reflected Prince Piotr Viazemsky many years later. Another participant believed that the events had "awoken the Russian people to life" and defined them politically. Fyodor Glinka saw the whole episode as a kind of "holy progress" towards a better state of being.[4] It was partly a question of heightened self-awareness: these young officers had experienced a great deal in a very short time, facing death daily, enduring great sufferings, plumbing the depths of despair and riding the crest of triumph. They had discovered a new sense of solidarity with each other and with their men, and dreamt of a fairer and better life for all as they chatted around their campfires.

But as they gave free rein to their better instincts, their more down-to-earth peers back home had busied themselves repairing the damage caused by Napoleon's incursion into the Russian empire. As soon as the French had left Moscow back in October 1812, Rostopchin set up a special commission to investigate all those who had remained in the city under the French occupation. It identified twenty-two, most of them of foreign extraction, who had failed to remain true to their oath of loyalty to the Tsar. They were duly exiled, sent to Siberia or imprisoned. A further thirty-seven were deemed to have been of service to the French but not punished. Others were handed over to civil courts, and some were flogged. Similar investigations were carried out in other provinces. All those who had stayed put under the French occupation were interrogated, along with those who had served the enemy. Very few were penalised, as an amnesty was declared in 1814 before sentences had been passed.[5] Such leniency reflects a degree of relief.

The greatest cause for relief was that the serfs had not taken the opportunity provided by the French invasion to rise up against their masters in any numbers. "I see that this is what most touches the feelings of all the Russians with whom I have conversed on the subject," noted John Quincy Adams in his diary on 1 December 1812. "This was the point on which their fears were the greatest, and that upon which they are most delighted to see the danger past."[6] But many feared the danger was far from over.

While giving fulsome praise to the patriotism of the serfs, people voiced fears of what these peasant heroes might do next, particularly as many had acquired muskets and learnt to use them. The authorities issued proclamations calling on the peasants to hand in weapons, even offering five roubles per musket, but the results were disappointing and the arms had to be confiscated by force. Just how seriously the threat was taken can be gauged by the fact that Aleksandr

Benckendorff, who had commanded a "flying" detachment under Wintzingerode to the north of the Moscow–Smolensk road, received orders to disarm his peasant auxiliaries and to have those who had been most active shot.[7]

"The common people have become accustomed to war and witnessed butchery," Rostopchin wrote to Alexander after a long talk on the subject with Balashov in the autumn of 1812. "Our soldiers pillaged them before the enemy did, and now that Bonaparte has apparently slipped out of our clutches, it would not be a bad idea, as we prepare to do battle with him, to think of the measures to be taken for the struggle within the empire against your enemies and those of the fatherland."[8] These were not groundless fears.

On 9 December, as the debris of the Grande Armée straggled into Vilna, the recently recruited 3rd Militia Regiment of the province of Tula, stationed in the little town of Insar, mutinied. They arrested their officers, beat them up and dragged them about the town before locking them up in the gaol. They began erecting gallows on which to hang them, but were distracted by the lure of booty, and went on the rampage. An officer who managed to escape into the countryside found that the peasants wanted to lynch him and the local landowners. The mutiny spread to other garrisons in Tula before being put down by regular troops supported by artillery. After investigation it emerged that the men had believed they would win their freedom by serving in the militia, and reacted with fury on being told that this was not so. Three hundred were condemned to run the gauntlet, which ended in the death of thirty-four.[9]

The French invasion had indeed altered the attitudes of the peasants affected by it, whether as victims, as partisans, as militiamen or as regular soldiers. They had all suffered or fought alongside their masters in the cause of the common fatherland, and it seemed only right that there should be some acknowledgment of this.

It was the men called up in the course of 1812 who felt most aggrieved. The manifesto under which they had been drafted clearly stated that they were being called upon to defend the fatherland in its moment of need, and that once the enemy had been chased out, they would be free to return home to their families. Instead, they were marched across Europe, fought in two campaigns, and only returned to Russia after three years. On their march to Paris they saw that peasants in every country they crossed enjoyed not only a standard of living but also a degree of freedom beyond their wildest dreams. They felt the least they deserved was some alleviation of their lot. "We shed our blood, and now they want us to go back and sweat for the masters!" they grumbled. "We saved the fatherland from the tyrant, and now the lords want to tyrannise us!"[10]

When the time came for the army to begin its long march back from Paris, many of the soldiers did the sensible thing. "At the first night's stop, twelve of our best soldiers deserted, even more on the second, so that in three days' march the company lost fifty men," wrote Captain A.K. Karpov.[11] His experience was by no means exceptional, and special measures had to be taken to prevent the army from melting away as it marched home.

The young Pushkin, still at school in St Petersburg, wrote an ode to Alexander on his return from Paris, full of joy and yearning for the future. He was expressing the enthusiasm of a generation which hoped that the Tsar would transform their country. For them, the events of the past two years had been a spiritual awakening, and they believed that Russia must realise their promise by breaking down those hierarchies which divided the nation. While they generally rejected foreign values, and French ones in particular, the would-be reformers nevertheless envisaged a process of regeneration that would turn Russia into a progressive liberal state.

The humiliating invasion and the desecration of their country by the French had produced a visceral reaction throughout Russian society, which reached back into its own history and traditions for solace and strength. Ethnic fashions, music and dances invaded the palaces of the aristocracy. Count Ostermann-Tolstoy went as far as tearing out the French décor of the main bedroom in his St Petersburg palace and replacing it with rough-hewn logs in imitation of a Russian peasant's cottage. But such manifestations did not necessarily presage any change in attitude to the peasants themselves.

While the image of the patriotic common soldier was glorified in paintings and prints, and while he was the hero of many a poem and short story, and at least one popular play in which a peasant became an officer, reality remained harshly unaffected. Serfs had to be put in their place and sent back to work. When it was discovered that a lancer who had won the Cross of St George for bravery was a Jew, he was denied the right to wear it.[12]

The majority of Russian society saw the events of 1812–1814 not as a spur to regeneration but as divine vindication of the existing constitution of the Russian state, which alone had been deemed worthy by the Almighty of carrying out His will in the struggle against the evil of revolutionary and Napoleonic France. Abandoning his youthful liberalism, the Tsar himself espoused this view. There would be no more reforms, and the system became in some ways even more stultifying than before. When Denis Davidov attempted to publish a memoir of his heroic deeds, the text was savaged by the censors, and the book was not published for years. It gradually dawned on the heroes of 1812 that, having done their duty, they were now supposed to carry on as before, as though nothing had happened.

In their frustration, they were drawn together. General Orlov, an admirer of the German *Tugendbund*, founded an "Order of Russian Knights," which grew into the "Union of Salvation" and finally the "Union of Welfare." The stated aim of these associations was self-betterment and the regeneration of society, their inspiration a mixture of St Paul and Rousseau, and they did not ostensibly threaten the Russian state. As this was also a period of flowering in Russian literature, a number of purely literary clubs and societies sprang up at about the same time, and there was a certain amount of movement and communication between the two.

In 1822 Tsar Alexander issued a decree banning all such societies, and this was followed by purges in the universities, where subversion was supposedly lurking. The Union of Welfare was driven underground and grew more political, and its members, most of them officers who had fought for their country in 1812, began plotting ways of saving it from autocracy.

They seized their chance in December 1825, when the sudden death of Alexander created a moment of uncertainty, to stage a military coup. But brave as they had been on the battlefield, they had neither the decisiveness nor the ruthlessness necessary to carry through such a venture, and their revolt collapsed under a hail of canister-shot from troops loyal to the new Tsar, Alexander's younger brother Nicholas I.

Among those condemned to a variety of sentences, some of them unwarrantedly harsh, were sixty-five officers who had been at Borodino, and another fifty who had fought in defence of their fatherland in 1812. With them went the hopes of any reform, but they became potent symbols for subsequent generations of Russians, for whom their heroism in 1812 and their self-sacrifice in 1825 stood for all that was best and most worth fighting for.

The effects of Alexander's "liberation" of Europe were no less disappointing to those who had longed for it. While quoting Holy Scripture, he haggled over territory and influence with the best of them, and tried to use the peace settlement as a vehicle to establish an ideological orthodoxy over the European mainland. His Holy Alliance sanctified a system which was designed to invigilate the political life of every state in Europe and intervene militarily if something that displeased him took place anywhere, be it Naples, Belgium or Spain. Contemplating the European scene in 1819, Stendhal came to the conclusion that Russia had achieved such dominion over the continent that only the United States of America could save it.[13]

Such a system could not work for long, and fell apart even before Alexander's death in 1825. But the dominant position Russia had assumed meant that she could prevent any change in the political *status quo* in central Europe which did not meet with her approval. It also strengthened the forces of conservatism in other major states of the area, such as the Austrian Empire and Prussia.

Prussia was perhaps the greatest beneficiary of Napoleon's defeat in Russia. Alexander's use of her as a tool for the liberation of Germany meant that by 1814 her army was one of the best in Europe. He could not reward her by returning all the Polish lands Napoleon had taken from her at Tilsit, since he had claimed most of them for Russia, so he helped her acquire provinces in other parts of Germany and the Rhineland instead. This meant that it was Prussia, not Austria, which would become the dominant German state.

If Russian liberals had been disappointed at the betrayal of their hopes by Alexander and Russian society in general, the German dreamers who had envisaged a great new Germany taking a spiritual lead in Europe were appalled at the way in which their aspirations were quashed. Prussia merely modernised her despotic constitution and, in common with Russia, stamped on even the most innocent indulgences of liberal students and patriotic poets. When a great Germany did arise, in 1870, it was not the chivalric one imagined by Stein and the Romantic poets who had fought in the *Freiheitskrieg*, but the militaristic and autocratic one of Bismarck and the Kaisers.

Far from rolling back those he had called the "northern barbarians," Napoleon had brought them into the heart of Europe. His own defeat and France's resultant eclipse as a Great Power had paved the way for the dominance of both Russia and Prussia. They used that dominance to protect a *status quo* that impeded social, national and religious emancipation, economic enterprise and political development in central Europe, thereby generating militant nationalisms and creating tensions that led to revolution and upheaval in the first two decades of the twentieth century and fed the ideologies which accounted for tens of millions of lives in the third, fourth and fifth decades.

The French reacted to their fall from power in a variety of ways. Some went into denial and dreamt of how things could have been. One, Louis Geoffroy, actually rewrote history in order to comfort himself and his peers, and published his version of events in 1841. According to this, Napoleon did not pause in Moscow in 1812 but marched north. He came up against an army of Russians, Swedes and 25,000 British under the command of Alexander and Bernadotte, and defeated them outside Novgorod on 8 October. A week later he made a triumphal entry into St Petersburg. He forced Alexander to bring the Russian empire into the fold of the Catholic Church, which earned him the forgiveness and wholehearted support of the Pope. He sacked Bernadotte and made Poniatowski King of a restored Poland. He then went to Spain, where, at Segovia on 13 July 1813 he defeated and captured Wellington. The Pope persuaded the Spaniards to accept Joseph Bonaparte as their King, and peace descended on the Iberian peninsula. In April 1814 Napoleon invaded England, landing in East Anglia and defeating the British army under the Duke of York outside Cambridge. He then marched on London, where he stormed into the House of Commons and informed the astonished MPs that Britain was abolished before ordering his troops to clear the chamber. Once this had been done he locked the doors, took the keys and, riding onto Westminster Bridge, cast them into the Thames. While George III was allowed to reign in Glasgow as a vassal King of Scotland and Ireland, England was divided up into twenty-two *départements* and incorporated into the French empire. Napoleon rebuilt Paris as a new Rome to which all flocked. Even Madame de Staël returned, to take up a place in the Académie Française and receive the title of Duchess from Napoleon. He also lavished his attention on Rome, which he embellished, draining the marshes and diverting the Tiber as well as rebuilding St Peter's for his uncle, Cardinal Fesch, who had become Pope. In 1817 Russia, Prussia and Sweden rose up against French rule, but they were crushed; Napoleon wiped Prussia off the map and split Russia into three separate states. As an afterthought, he conquered Constantinople. In 1820 he annexed north Africa and in the following year Egypt. After a great victory outside Jerusalem, he marched east to Baghdad and sent Prince Eugène to Mecca, which he destroyed. Napoleon then abolished Islam. The rest of Asia, China, Japan and the remainder of Africa were easily conquered over the next few years, and, after a great congress at Panama, all the rulers of the Americas, North and South, humbly begged to be allowed to submit to Napoleon's rule. All the world's Jews met in congress in Warsaw and decided to become Catholics. By the time he died, on 23 February 1832, Napoleon ruled the whole world, which spoke French and worshipped Christ.[14]

Preposterous as this fantasy might be, the fact remains that neither Napoleon's final defeat in 1815 nor his death in 1821 succeeded in consigning him to history. Many in France looked back with understandable nostalgia to the days when he made Europe tremble, while old soldiers of other nationalities never forgot what they had experienced when serving under his command. For many he remained a sacred memory, and people would gather on the anniversary of his birth, his coronation or his death. From time to time rumours swept France that he had landed somewhere in Europe and was marching on Paris at the head of a great army. These persisted long after his death, news of which many refused to believe.

The Romantic movement added a further dimension to the fascination Napoleon exerted. Poets such as Alfred de Vigny and Alfred de Musset represented him as a titanic figure, above the common run of man, one who belonged to the heavens rather than to earth, while even writers who hated him, like Châteaubriand and Victor Hugo, saw him as a "poet in action," superhuman and heroic. It was not only the French who were in thrall. To Goethe, Napoleon was the "daimon," representing the darker – but more inspired – element in the dualism of human nature, the ultimate doer of deeds. To Heinrich Heine he was Prometheus, who had stolen the fire of heaven. Byron dwelt at length on his fate, and Walter Savage Landor wove a messianic myth around him. To the Polish Romantic poet Adam Mickiewicz, Napoleon was a kind of demi-god and 1812 a failed incarnation whose memory was to be cherished. Even the Russians were not immune, and it was not uncommon for the heroes of 1812 to have statuettes or prints of Napoleon in their studies. To them he epitomised energy, willpower, action. He was a colossus, an elemental being, exerting the same fascination as an erupting volcano or a raging storm.

In exile on St Helena, Napoleon mused that if only he had been killed in battle at Borodino or on his entry into Moscow, he would have gone down in history as the greatest conqueror and hero of all time.[15] But he was wrong. In classical Greek theatre a hero can only exist in the genre of tragedy, which makes people appear more than they are and lends stature to figures who are not necessarily virtuous or attractive. The more tragic the action, the more terrible the trials to which he is subjected, the greater the hero appears. The same was true for the Romantics, who were fascinated by the concept of man living out a tragic destiny. To be interesting, a man had to be both colossus and victim.

Walter Scott hurried to Waterloo in order to see the field on which the giant had been toppled, and when he first met Stendhal, at the opera in Italy, Byron talked only of the retreat from Moscow, quizzing the French novelist for every detail of the Emperor's comportment. It was the catastrophe Napoleon had met with that turned him into a hero in their eyes. It mattered not that he himself had been the author of this catastrophe, or that he was responsible for the suffering and death of hundreds of thousands of human beings.

Some, like Dostoevsky, were fascinated by the very callousness and monstrosity of the man. "The real *Master*, the man to whom all is permitted, can storm Toulon, stage a massacre in Paris, *forget about* an army in Egypt, *throw away* half a million men in the Moscow expedition and then get away with a witty phrase in Vilna," marvels Raskolnikov in *Crime and Punishment*. "Yet altars are erected to him after his death, for to such a man *all* is permitted. No, such people are clearly not made of flesh, but of bronze!"[16]

One did not have to be a tortured intellectual to sense an element of the numinous in Napoleon. At the first stop after Kovno on his flight from Smorgonie in December 1812, the Emperor took the opportunity to have a wash and a change of linen. When his Mameluke, Roustam, handed the discarded shirt and stockings to the innkeeper to dispose of, they were seized by the locals present, cut up and distributed, to be preserved as holy relics. And relics of the terrible retreat, gathered on the banks of the Berezina and at other scenes of disaster, were surrounded by far greater piety than objects from the fields of Marengo, Austerlitz or Jena.[17]

Sergeant Bourgogne noted that whenever he would get together with old comrades from the Imperial Guard still alive in the 1820s, their talk would invariably turn to the Moscow campaign, which held a fascination for them quite unlike that of any of the other great expeditions in which they had taken part.[18] And when Balzac wanted to engage his readers' nostalgia for the Napoleonic epos, in his novel *Le Médecin de Campagne*, he did not conjure some glorious moment of triumph, but created the character of Gondrin, an old sapper who had helped to build the Berezina bridges and hailed Napoleon from the icy water.

The very name of that muddy river has enormous resonance, even for people who know little history. It is a powerful symbol of the failure and the tragedy that lie at the heart of the Napoleonic myth; the images of heroism and shipwreck it conjures are a fitting nemesis to the hubris not only of Napoleon's crossing of the Niemen five months earlier, but of the whole Napoleonic era.

Notes

Introductory Note

1. Troitskii, *Otechestvennaia Voina*, 3.
2. Orlando Figes, *Natasha's Dance*, London 2002, 81.
3. Pokrovskii, III/181–93; Troitskii, *Otechestvennaia Voina*, 30–1.
4. Troitskii, *Otechestvennaia Voina*, 37–45; Tarle, *Napoleon*, 419–20; Tarle, *Nashestvie*, 3, 4, 292–3.

Chapter 1: Caesar

1. Chevalier, 144.
2. Castellane, I/83.
3. Savary, V/148.
4. Ségur, III/449.
5. Méneval, II/436; Hortense II/127. On the birth of the King of Rome, see also: Las Cases, I, Part 2, 350–2; O'Meara, II/368; Savary, V/146–9; Raza, 202–6; Castellane, I/83; Kemble, 182–4; Rambuteau, 55–6.
6. Dwyer, 129–30.
7. Ségur, III/74.
8. Driault, *Grand Empire*, 126, 127, 231.
9. Herold, 245, 243; Fouché, II/114.
10. Savary, V/149.
11. Comeau de Charry, 440; see also Ségur, IV/78–9.
12. Rambuteau, 70; also Napoleon, *Mémoires*, VIII/150; Las Cases, IV, Part 1, 8–14.
13. Fain, *Mémoires*, 286–7; Kemble, 165, 170. Ségur (III/476) claims that Napoleon was aware of this and was therefore in a hurry to deal with Russia while he could.

Chapter 2: Alexander

1. Deutsch, 17.
2. Madariaga, *Russia in the Age of Catherine*, 231.
3. Hosking, *Russia and the Russians*, 109.
4. Voenskii, *Bonapart i Russkie Plennie*; see also Ragsdale.
5. Kartsov & Voenskii, 7.
6. Alexander I, *Uchebnia Knigi*, 382–3.
7. Hartley, *Alexander*, 1. See also Ratchinski, 55; Deutsch, 43.
8. Ratchinski, 121; Dzhivelegov et al., I/200–1; Fonvizin, 94.
9. Hartley, *Alexander*, 74.
10. Ley, 42; Bazylow, in Senkowska-Gluck; see also Kveta Mejdrička, *Les Paysans Tchèqes et la Révolution Française* in *Annales Historiques de la Révolution Française*, no. 154, Nancy 1958.
11. Hartley, *Alexander*, 76.
12. For Tilsit, see Tatistchev, 379–459, and Vandal, I/224ff.
13. Alexander I, *Corr. avec sa soeur*, 18–19.
14. Many, like Grech, 260, and Davidov, 59, claim that Alexander was not taken in, but Alexander admitted to Mme de Staël that he had fallen under the spell: Staël, 215. See also Palmer, *Napoleon in Russia*, 136–7 & 147–9.
15. Palmer, *Alexander*, 149, 150; Ratchinski, 124; Edling, 60–2; Ley, 32.
16. Vandal, I/217.
17. Caulaincourt, I/242.
18. Vandal, I/195.
19. Roberts, 16; Herold, 48; Pradt, 19.
20. Sokolov, Nov–Dec, 44; Las Cases, III/165; Vandal I/224ff; Talleyrand, 313, 356; Driault, *Tilsit*, 291.
21. Vandal I/242; Tatistchev, 312–13.
22. Vandal I/441, 456–7.
23. Golovin, 391; Shilder, *Nakanunie*, 4–20; Alexander I, *Corr. avec sa soeur*, 20.
24. Caulaincourt, I/273.
25. Ibid., 270; see also Tatistchev, 379–459.
26. Vandal, I/421; Grunwald, *Alexandre Ier*, 176.

Chapter 3: The Soul of Europe

1. Senkowska-Gluck, 274; Ernouf, 105.
2. Ramm, 69.
3. Driault, *Tilsit*, 241; see also Ramm, Servières, Cavaignac, Ernouf, etc.
4. Driault, *Grand Empire*, 188–9.
5. Comeau de Charry, 318–19; Bruun, 174.
6. Bruun, 174; Grunwald, *Baron Stein*, 94.

7. Driault, *Tilsit*, 348.
8. Vandal, II/447.
9. Langsam, 32, 103, 44, 64.
10. Kraehe, I/74; Langsam, 43.
11. Palmer, *Alexander*, 195.
12. Roberts, 83.
13. Bruun, 64; Bismarck, 13.
14. Chernyshev, 205; Vandal, III/201ff; Herold, 182.

Chapter 4: The Drift to War

1. For the planned Russian marriage and the Austrian marriage, see: Vandal; Driault, *Grand Empire*; Caulaincourt, I/293–316.
2. Nicholas Mikhailovich, IV/50–7.
3. Napoleon, *Correspondance*, XX/149–54.
4. Ibid., 157.
5. Ibid., 159.
6. Ratchinski, 21–2, 33.
7. Volkonskii, 4.
8. Ibid., 55; Davidov, 56.
9. Dzhivelegov et al., II/205.
10. Ibid., 194–220; Ratchinski, 284.
11. Scott, 5–6.
12. Kartsov & Voenskii, 50–1.
13. Czartoryski, II/221, 227, 231; Askenazy, 218–20; Dzivelegov et al., III/138.
14. Czartoryski, II/231, 323–4, 225.
15. Fain, *Manuscrit*, I/3.
16. Palmer, *Alexander*, 199.
17. Czartoryski, II/248–53; Josselson, 77; *1812 god v Vospominaniakh sovremennikov*, 78; Fabry, *Campagne de Russie*, I/iff; Czartoryski, II/271–8.
18. Bestuzhev-Riumin, *Zapiski*, 341; Palmer, *Alexander*, 200.
19. Bignon, *Souvenirs*, 46ff; Bonnal, 3, 4–12; Las Cases, II, Part 1, 99; Brandys, II/25.
20. Napoleon, *Correspondance*, XXI/407; XXII/29; Shuvalov, 416; Chernyshev, 21, 72.
21. Chernyshev, 84; Kukiel, *Vues*, 77.
22. Caulaincourt, I/281–96.
23. Ibid., 293.
24. Ibid., 302, 307, 316.
25. Alexander, *Corr. avec sa soeur*, 51.
26. Savary, V/140–1.
27. Vandal, III/209–17; Tatistchev, 572–3.

28. Napoleon, *Correspondance* XXII/266.
29. Ibid., 40–1; Las Cases, II, Part 1, 101.

Chapter 5: La Grande Armée

1. Gary W. Kronk, *Cometography*, vol. 2, 1996, www.Cometography.com
2. Déchy, 364; Butkevicius, 898; Chamski, 61; Butenev, 58; *1812 god v Vospominaniakh sovremennikov*, 172; F. Glinka, *Pisma russkavo ofitsera*, IV/8; S.N. Glinka, *Zapiski*, 261; Palmer, *Alexander*, 206–7; Mikhailovskii-Danilevskii, III/41.
3. Metternich, II/422.
4. Napoleon, *Correspondance*, XXIII/191.
5. Thévenin, 292.
6. Nafziger, 27.
7. Ibid., 85ff; Creveld, 62–3; Dedem, 210; Bonnal, 21; Napoleon, *Correspondance*, XXIII/432.
8. Chernyshev, 116–20; *La Guerre Nationale*, III/97–9.
9. Jerome, V/247; Dedem, 194; Compans, 126; Rapp, 139; Driault, *Grand Empire*, 300–2; Napoleon, *Correspondance*, XXXIII/14–16, 44–6, etc.
10. Chambray, I/162; Funck, 99.
11. Holzhausen, 27.
12. Mikhailovskii-Danilevskii, III/141; Holzhausen, 27.
13. Kukiel, *Wojna*, II/500. Pradt (84) gives a figure of 85,700, but he was probably not counting the Legion of the Vistula.
14. Laugier, *Récits*, 15–16.
15. Combe, 58, etc.
16. Borcke, 161, 171; Jerome, V/188.
17. Comeau de Charry, 428; Sołtyk, 207; see also Sauzey, *Les Allemands*; Wesemann, 25ff.
18. Begos, 175; Boppe *La Légion Portugaise*, 215–27; Castellane, I/178; Boppe *La Croatie Militaire*, 95–120.
19. Calosso, 51; Zelle, 69.
20. Saint-Chamans, 211–12; Fredro, 82.
21. Thirion, 147; Pion des Loches, 27.
22. Lejeune, *Mémoires*, II/170.
23. Funck, 99–100; Berthézène, I/329.
24. Beyle, *Vie de Napoléon*, 227; Fain, *Manuscrit*, I/46.
25. Berthézène, I/328.
26. Funck, 103; see also Baudus, I/336; Ségur, IV/124.
27. Girard, 169–70.
28. Voltaire, 8; Comeau de Charry, 436; Jerome, V/165.
29. Caulaincourt, II/378.

30. Brandys, II/42; Napoleon, *Correspondance*, XXIII/95, 398; Tulard, *Le Dépôt de la Guerre*, 104–9.
31. Blaze de Bury, I/154–5.
32. Brett-James, 57–8.
33. Napoleon, *Correspondance*, XXIII/432.
34. Ibid., 143.
35. Dumonceau, II/5.
36. Blaze de Bury, I/54–5.
37. Elting, 458–73.
38. Bigarré, 297–8.
39. Laugier, *Récits*, 9; Holzhausen, 30–1; Wesemann, 25.
40. Herold, 218.
41. Caulaincourt, II/77–8; Denniée, 15; Castellane, I/102. Some observers thought Napoleon was thinking of founding colonies in distant lands, and there are accounts of bricklayers, carpenters and artisans being drafted into the army or despatched in separate units, but this is probably no more than gossip, based on the fact that prescient commanders such as Davout identified those soldiers who could, for instance, build ovens and bake bread, and made sure that every company had a complement. At the same time, it is certainly true that the army was followed by hordes of people whose purpose was not always clear. See: Begos, 171–2; Bourgeois, 1–2; Chambray, I/164; Puibusque, 9.
42. Saint-Chamans, 213; Paixhans, 33.
43. Compans, 129.
44. Berthézène, 328; Lejeune, *Mémoires*, II/169; according to Bertolini, many of the Italians had their wives with them.
45. Boulart, 239; Walter, 34; Pouget, 184; Fairon & Heuse, 271; Laugier, *Récits*, 10; Pion des Loches, 273; Bourgeois, 2; Denniée, 11; Ganniers, 166; Hausmann, 141.
46. Duverger, 1; Vandal, III/454; Ricome, 47; Meerheimb, 7.
47. Nesselrode, IV/204–5.

Chapter 6: Confrontation

1. Napoleon, *Correspondance*, XXIII/379; Fain, *Manuscrit*, I/43.
2. Villemain, I/155, 162, 163, 167.
3. Ibid., 175.
4. Jerome, V/169.
5. Fouché, II/114; Ségur, IV/74.
6. Kurakin, 360–1.
7. Napoleon, *Correspondance*, XXIII/388.
8. Pasquier, 525.

9. Fain, *Manuscrit*, I/61.
10. Castellane, I/93.
11. Vandal, III/413–16; Baudus, I/338; Garros, 370–1.
12. Beauharnais, VII/340; Méneval, III/25; Savary, V/226; Comeau de Charry, 439; Lejeune, *Mémoires*, II/174.
13. Volkonskii, 147, 148, 149.
14. Voronovskii, 4.
15. Palmer, *Alexander*, 211; Kartsov & Voenskii, 107.
16. Palmer, *Alexander*, 213; Kartsov & Voenskii, 103–4, 107; Dzhivelegov et al., II/221–46; Bulgakov, 2.
17. Dzhivelegov et al., III/64–8; Bogdanov, 78.
18. Volkonskii, 56; *La Guerre Nationale*, VII/333–5.
19. Kharkievich, *Nastavlenie*, 239, 242, 243; Shvedov, *Komplektovanie*, 125.
20. Dzhivelegov et al., III/81; Bogdanov, 61, 65, 72.
21. Shvedov, *Komplektovanie*, 125; Troitskii, *O chislennosti*, 172–3 and *O dislokatsii*. See also: Zhilin, *Otechestvennaia voina*, 95–6; Sokolov, Dec., 36; Bogdanov, 72; Dzhivelegov et al., III/139; also Clausewitz, 12; Shishov, 235; Wolzogen, 87–8.
22. On Barclay, see: Josselson; Wolzogen, 55–6; Toll, I/268.
23. Grunwald, *Baron Stein*, 188.
24. Dubrovin, 9.
25. Ermolov, June/4; Josselson, 77; Muravev, 175.
26. Josselson, 41–2; Dumas, III/416.
27. Caulaincourt, I/291–3; Bignon, *Souvenirs*, 129.
28. Fabry, *Campagne de Russie*, I/iff, x–xxiii, xxviiiff; *La Guerre Nationale*, II/131–44, IV/17–107, V/232–5, VI/264–8, VII/17–27, 37–40; Ermolov, June, 5; Clausewitz, 14; *1812 God v Vospominaniakh sovremennikov*, 79–80; Buturlin, *Byl li u nas plan*, 220; Shvedov, in *Tezisy Nauchnoi Konferentsii*, 32; Marchenko (502, 504–5) was of the opinion that there was a plan to retreat some distance, in order not to fight in Lithuania, where partisan activity was to be expected, but even he is adamant that nobody ever entertained the possibility of giving up Russian territory.
29. On Alexander's various schemes, see: Alexander I, *Corr. avec Bernadotte*, 6–7, 21; Czartoryski, II/281; Askenazy, 231; Volkonskii, 154; Ratchinski, 224; *La Guerre Nationale*, IV/38–55, 413–25, V/359–69.
30. Alexander I, *Corr. avec sa soeur*, 76.
31. *1812 god. Voennie dnievniki*, 77, 81.
32. Benckendorff, 32; Shiskov, 126; Radozhitskii, 21; Simanskii, 1912, No. 2.
33. Bakunina, 396–7.
34. Josselson, 93; *Barclay de Tolly i Otechestvennaia Voina*, August 1912, 197–8.
35. Toll, I/270; see also Vaudoncourt, *Quinze Années*, I/167.

36. Wolzogen, 63.
37. Ley, 45–6; 47, 48.
38. Nesselrode, IV/5–10.
39. Rambuteau, 86; see also Villemain, I/187.
40. Skallon, 450; Nesselrode, IV/35; Voenskii, *Sviashchennoi Pamiati*, 19.

Chapter 7: The Rubicon

1. Denniée, 12–13.
2. Hazlitt, III/398.
3. Fain, *Manuscrit*, I/68.
4. Vandal, III/479; Villemain, I/174.
5. Metternich, I/122.
6. Fain, *Manuscrit*, I/75; Pradt, 56–7; Savary, V/226.
7. Villemain, I/165–6, 163.
8. Fain, *Manuscrit*, I/50.
9. Jomini, *Précis*, I/48; Niemcewicz, I/380; Napoleon, *Correspondance*, XXIII/441.
10. Beauharnais, VII/261, 330, 374; Caulaincourt, I/342; Falkowski, IV/6–7; Lejeune, *Mémoires*, II/180; Potocka, 319–20; Krasiński, 68; Koźmian, II/302; Niemcewicz, 379; Villemain, I/167–70; Las Cases, IV/181–2; O'Meara, I/191; also Kukiel, *Vues*, 77.
11. Chevalier, 175.
12. Brandt, 228; Paixhans, 24–6.
13. Brandt, 228, 230; Dumas, III/417, 419.
14. Dumas, III/418; Castellane, I/101.
15. Everts, 117.
16. Combe, 57.
17. Ogiński, III/114; Falkowski, IV/3; Bignon, *Souvenirs*, 195.
18. Chernyshev, 72, relates that Napoleon himself told him that the Grand Duchy could not support an army of more than 40,000 men; Falkowski, IV/3.
19. Venturini, 218; Dziewanowski, 6.
20. Fredro, 35; see also Blaze de Bury, I/44–5; Brandt, 232.
21. Roy, 130; Hausman, 94.
22. Holzhausen, 31–2; Abbeel, 96.
23. Boulart, 240–1.
24. Walter, 40–1; Le Roy, 130.
25. O'Meara, II/95; Caulaincourt, I/330, 340; Fain, *Manuscrit*, I/87.
26. Fabry, *Campagne de Russie*, IV, Annexe 14–16, 262–348; Vilatte de Prugnes, 249–74, for the official figures from the Dépôt de la Guerre.
27. Beauharnais, VII/276.

28. Berthézène, I/323–6; Dedem, 226–7.
29. Hausman, 94; Fairon & Heuse, 274.
30. Napoleon, *Correspondance*, XXIII/499–50; Garros, 377; Napoleon, *Lettres Inédites* (1935), 41.
31. Napoleon, *Correspondance* (1925), V/435, 396.
32. Garros, 378.
33. Ibid., 378–9; Sołtyk, 9–10; Załuski, 119; Caulaincourt, I/344.
34. Boulart, 241.
35. Napoleon, *Correspondance* (1925), V/435.
36. Dumonceau, II/48; Saint-Chamans, 213; Labaume, 25; Boulart, 242; Ganniers, 181.
37. Lejeune, *Mémoires*, II/175.
38. Planat de la Faye, 71.
39. Lejeune, *Mémoires*, II/175–6; see also Jomini, *Précis*, I/57
40. Fantin des Odoards, 303.

Chapter 8: Vilna

1. Choiseul-Gouffier, 45–8, 58–9; Nesselrode, 45; *1812 god. Voennie Dnevniki*, 76–7.
2. Ley, 49.
3. Shishkov, 127.
4. Buturlin, I/155; *Barclay de Tolly i Otechestvennaia Voina*, Sept. 1912, 327.
5. Altshuller & Tartakovskii, 21; Mitarevskii, 12; Radozhitskii, 35–6.
6. Altshuller & Tartakovskii, 23; Laugier, *Gli Italiani*, 26–7.
7. *Prikaz nashim armiam*, 445.
8. Dubrovin, 13–15.
9. Bloqueville, III/155.
10. Boulart, 243.
11. Bertin, 27; Lecoq, 162.
12. Griois, II/14; Berthézène, I/343; Dumas, III/422.
13. Laugier, *Récits*, 18; Montesquiou-Fezensac, 208; Dumas, III/422; Labaume, 33.
14. Fain, *Manuscrit*, I/200–1; Denniée, 19–21; Napoleon, *Correspondance*, XXIV/33.
15. Castellane, I/110.
16. Caulaincourt, I/354.
17. Dubrovin, 25.
18. Dubrovin, 20–5; Caulaincourt, I/354; Napoleon, *Correspondance*, XXIV/1.
19. Tatistchev, 606; Las Cases, II, Part 1,103.
20. Sołtyk, 36; Dupuy, 166.
21. Brandt, 240; see also Gajewski, 214; Bourgeois, 12; Caulaincourt, I/351.

22. Choiseul-Gouffier, 64; *Dokumenty i Materialy*, CXXVIII, 416; see also Dziewanowski, 6.
23. Falkowski, IV/127, 129; Kukiel, *Wojna*, I/376–88; Załuski, 232.
24. Brandt, 251; Gardier, 32.
25. Chamski, 67; Brandt, 245; Turno, 104; Kołaczkowski, 98; Sołtyk, 59, 65.
26. Choiseul-Gouffier, 65; Płaczkowski, 171–2; see also Faber du Faur, 10.
27. Falkowski, IV/128; Brandt, 241, 243; Fezensac, *Journal*, 13.
28. Dziewanowski, 8; Jackowski, 297.
29. Napoleon, *Correspondance*, XXIV/61; Pradt, 131.
30. *Dokumenty i Materialy*, CXXVIII, 397; Falkowski, IV/135–6; *Dokumenty i Materialy*, CXXVIII, 393.
31. Kukiel, *Wojna*, I/389–390; Falkowski, IV/146, 149–150.
32. Brandys, II/76; Choiseul-Gouffier, 102.
33. Fantin des Odoards, 308.
34. Napoleon, *Lettres Inédites* (1897), II/199; *Correspondance*, XXIV/19.
35. Napoleon, *Lettres Inédites* (1897), II/200; Beauharnais, VII/382; Napoleon, *Correspondance*, XXIV/37.
36. Askenazy, 232; Fain, *Manuscrit*, I/208; Napoleon, *Correspondance*, XXIV/80–1.
37. Josselson, 99; Dubrovin, 36; Voronov.
38. Muravev, 174; Josselson, 93.
39. Grabbe, 827; Ermolov, 11.
40. Shchukin, VIII/165.
41. Chambray, I/210; Kallash, 17–18; Bagration (in Shchukin, VIII/169) gives a figure of 15,000 for all losses; Napoleon's figure of 20,000 deserters, in *Corr. Inédite* (1925), V/492, can be discounted as propaganda.
42. Nesselrode, I/58; Clausewitz, 43; Alexander, *Corr. avec Bernadotte*, 19.
43. Clausewitz, 26–8, 31–4; Grech, 261.
44. Butenev, 69.
45. Shilder, *Zapiska*; Shishkov, 137–8; W.H. Löwenstern, I/209; Palmer, *Alexander*, 232–3.

Chapter 9: Courteous War

1. Espinchal, I/320; Dumonceau, II/123.
2. Bertin, 25–6.
3. Sanguszko, 65; Caulaincourt, I/372.
4. Napoleon, *Correspondance*, XXIV/99.
5. Radozhitskii, 77–8.
6. Bertin, 50–1.
7. Grabbe, 437.
8. Rossetti, 100.

9. Ducor, I/307.
10. Barclay, *Tableau*, 20; Buturlin, *Byl li u nas plan*, 220; Josselson, 109; Kharkievich, *Barclay de Tolly v Otechestvennoi voinie*, 4.
11. Barclay, *Tableau*, 20; Kharkievich, *Barclay de Tolly v Otechestvennoi voine*, 7; Clausewitz (104) thought it would have been madness to fight at Vitebsk.
12. Kharkievich, *Barclay de Tolly v Otechestvennoi Voinie*, 14; Clausewitz, 26–8.
13. Fezensac, *Journal*, 18; Vlijmen, 57; Gruber, 136.
14. Adam, *Aus dem Leben*, 156.
15. Lyautey, 489; Uxküll, 72.
16. Radozhitskii, 106.
17. Mailly, 12.
18. Venturini, 221.
19. Duverger, 4–5.
20. Everts, 123; François, II/765; Brett-James, 53.
21. Brett-James, 54.
22. Walter, 41.
23. Dumonceau, II/96; Brett-James, 67.
24. Laugier, *Récits*, 45; Lagneau, 200; Labaume, 97; Venturini, 220; Brett-James, 52; Radozhitskii, 19; Lejeune, *Mémoires*, II/182–3; Walter, 41; Ducor, I/305–6; Girod de l'Ain, 253.
25. Marbot, 159–60; Suckow, 158; Saint-Cyr, 62; Hausmann (99) says they had 15,000 left; Giesse, 67–8.
26. Dutheillet de la Mothe, 38.
27. Roeder, 95.
28. Holzhausen, 45.
29. Adam, *Aus dem Leben*, 156.
30. Dembiński, I/111; Suckow, 157.
31. Adam, *Aus dem Leben*, 50.
32. Fantin des Odoards, 307; Compans, 160.
33. Walter, 50.
34. Bourgoing, *Souvenirs*, 88–9; Suckow, 156; Ducor, I/310; Everts, 127; C.S. von Martens, 71.
35. Bellot de Kergorre, 57; Brandt, 244; Laugier, *Gli Italiani*, 40; Vaudoncourt, *Quinze Années*, 132.
36. Brett-James, 56.
37. Dedem, 226–7.
38. Ségur, III/205; Villemain, I/198–9; Caulaincourt, I/381.
39. Meneval, III/43; Napoleon, *Correspondance*, XXIV/128, 133; Fain, *Manuscrit*, I/289, 306; Dumas, III/429; Castellane, I/126–7; La Flise, LXXI/465; Bourgoing, *Souvenirs*, 98–100.

40. Dedem, 295.
41. Napoleon, *Correspondance*, XXIV/89.
42. Askenazy, 226; Sanguszko, 75; Jomini, *Précis*, I/85.
43. Sanguszko, 70–1; Caulaincourt, I/407.
44. Fain, *Manuscrit*, I/318–20; Sołtyk, 69; Boulart, 245; Caulaincourt, I/384–5; Bourgoing, *Souvenirs*, 100; Fezensac, *Journal*, 35; Napoleon, *Correspondance*, XXIV/137; Caulaincourt, I/382; Villemain, I/203–4, 208; Fain, *Manuscrit*, I/323.

Chapter 10: The Heart of Russia

1. Dallas, 23.
2. Hartley, *Russia in 1812*, 178, 185–6; Dubrovin, 49; *Akty, Dokumenty i Materialy*, CXXXIX/135ff, 163, 209–68; Simanskii, 1912, No. 4, 176–7; Radozhitskii, 48–9.
3. *Akty, Dokumenty i Materialy*, CXXXIX/17, CXXXIII/173; Hartley, *Russia in 1812*, 196; Butenev, 71.
4. Hartley, *Russia in 1812*; *Akty, Dokumenty i Materialy*, CXXXIX/16–17; Iudin, 27; Shugurov, 253; Benckendorff, 47; Pouget, who was French governor of Vitebsk, was convinced (228) that the Jews spied for the Russians throughout.
5. *Akty, Dokumenty i Materialy*, CXXXIX/269–459; Iudin; Hartley, *Russia in 1812*, 406.
6. Pushkin, 205; Vigel, 43; Golitsuin, 7; Khomutova, 313.
7. Dzhivelegov et al., V/75–81; Troitskii, *1812 Velikii God*, 217; Hosking, *Russia. People and Empire*, 134.
8. Sverbeev, I/62–3; Hartley, *Russia in 1812*, 405; Berthézène, II/38; Dedem, 232.
9. *Prikaz Nashim Armiam*, 446–7; Palitsyn, *Manifesty*.
10. Voenskii, *Russkoe Dukhovenstvo*, 12; Dubrovin, 52.
11. Dolgov, 137.
12. Bakunina, 399, 400, 402; Ogiński, III/179; Tarle, *Nashestvie*, 68.
13. *1812 god v Vospominaniakh, perepiske i raskazakh*, 41–2; Kallash, *Chastnia Pisma*.
14. Voronovskii, 246.
15. Khomutova, 315.
16. *Priezd Imperatora Aleksandra v Moskvu*; Waliszewski, II/72.
17. Kallash, 7–8; Naryshkina, 136; Hartley, *Alexander*, 111; for other accounts of Alexander's visit, see: S. Glinka, *Zapiski*; Grunwald, *Baron Stein*, 195.
18. Viazemskii, 191–2.
19. Edling, 64; Alexander, *Corr. avec sa soeur*, 80–1.
20. Altshuller & Tartakovskii, 32; Ermolov, 29.

21. Mitarevskii, 30; Ermolov, 32.
22. F. Glinka, *Pisma Russkavo Ofitsera*, 18, 22–8.
23. Clausewitz, 111, 113.
24. Shchukin, VIII/167.
25. Simanskii, 1913, No. 1, 155, 156–7.
26. Aglaimov, 41; Altshuller & Tartakovskii, 33; Kharkievich, *Barclay de Tolly v Otechestvennoi voinie*, 14; Josselson 12; Simanskii, 1913 No. 1, 151; Radozhitskii, 98.
27. Fain, *Manuscrit*, I/359.
28. Fantin des Odoards, 317.
29. Askenazy, 235; Jackowski, 297, La Flise, LXXI/461.
30. Voronovskii, 31–3.
31. Clausewitz, 130.
32. Rossetti, 103; Ségur, IV/257.
33. Ermolov, 44; also Raevsky in Fabry, *Campagne de Russie*, IV/69–74.
34. Lyautey, 493.
35. Neverovskii, 79.
36. Thirion, 172–3; see also Faber du Faur 102 & Fanneau de la Horie.
37. Faré, 263; Labaume, 106; Denniée, 49–50.
38. Sukhanin, 277.
39. *1812 god v Vospominaniakh Sovremennikov*, 113; see also F. Glinka, *Pisma Russkavo Ofitsiera*, 35.
40. Uxküll, 74; also Radozhitskii, 111.
41. Bourgoing, *Souvenirs*, 101; Fantin des Odoards, 318; Boulart, 248; Caulaincourt, I/394.
42. Wilson, *Diary*, I/148–9; *Barclay de Tolly i Otechestvennaia Voina*, Oct. 1912, 125; Josselson, 119; Bennigsen, *Zapiski*, July, 102, 114; *Barclay de Tolly i Otechestvennaia Voina*, Oct. 1912, 129, 128.
43. Chevalier, 187–8; Combe, 73–4; Faure, 34.
44. Holzhausen, 62.
45. Ségur, IV/265–6.
46. La Flise, LXXI/472–3.
47. Laugier, *Récits*, 63; Berthézène, II/23; La Flise, LXXI/474; Larrey, IV, 24, 31; Napoleon, *Lettres Inédites* (1935), 62–3.
48. Fantin des Odoards, 321; Dedem, 232.
49. Wilson, *Invasion*, 178–9.
50. Holzhausen, 65.
51. W.H. Löwenstern, I/228–9; Grabbe, 45.
52. W.H. Löwenstern, I/226, 231.
53. Brandt, 258.
54. According to Shvedov, *Komplektovanie*, total Russian losses at Smolensk and

Lubino/Valutina Gora were 20,000; see also Josselson, 127; Troitskii, *1812 Velikii God*, 117.
55. Załuski 241.
56. Brandt, 261; Chevalier, 189; Brandt, 262.
57. Ségur, IV/291; Baudus, II/28.

Chapter 11: Total War

1. Fain, *Manuscrit*, I/394; Chambray, I/332; Caulaincourt, I/393; his intention to halt at Smolensk is confirmed in many other sources.
2. Napoleon, *Correspondance*, XXIV/167, 175, 180–1; Fezensac, *Journal*, 38; Kallash, 32; Caulaincourt, I/406.
3. Berthézène, II/32.
4. Boulart, 250; Fain, *Manuscrit*, I/402. There is some controversy over the stand taken by Davout, as Rambuteau (91) states he was against further advance, while Rossetti (106) and others maintain he was for it. Rapp, 167; Deniée, 62; Lejeune, *Mémoires*, II/199.
5. Fain, *Manuscrit*, I/407–8; Brandt, 252–3; see also Bourgoing, *Souvenirs*, 100.
6. Boulart, 248; see also Labaume, 103; Chevalier, 193.
7. Blaze de Bury, II/324.
8. Clausewitz, 113; Griois, II/9.
9. Abbeel, 110.
10. Combe, 74–5; Brandt, 268; C.S. von Martens, 109.
11. Faré, 261.
12. Pion des Loches, 287.
13. Ségur, IV/320; Bourgeois, 40.
14. Abbeel, 111.
15. Pelleport, II/23; Laugier, *Récits*, 49, 65; Lejeune, *Mémoires*, II/199; Chevalier, 190.
16. Caulaincourt, I/411; Chambray, II/26.
17. Sołtyk, 198–9.
18. Bloqueville, III/167; Roguet, III/474.
19. Muravev, 189; Uxküll, 74.
20. Tarle, *Nashestvie*, 127; Radozhitskii, 129.
21. Koliubakhin, *1812 God. Poslednie dni komandovania . . .*, 470; Radozhitskii, 128; Konshin, 283; Uxküll, 75.
22. Radozhitskii, 130.
23. Ermolov, 48; Simanskii, 159; *1812 god v Vospominaniakh sovremennikov*, 113.
24. Grabbe, 440; Mitarevskii, 41; Dubrovin, 73.
25. Wolzogen, 132–3; Ermolov, 56.
26. Josselson, 115; W.H. Löwenstern, I/240, 244; E. Löwenstern, 113.
27. Clausewitz, 133; Grabbe, 454–5, 455, records this story at Dorogobuzh, but it

seems he is confusing the argument between Bagration and Toll with that between Bagration and Barclay at Dorogobuzh – see Bennigsen, *Zapiski*, July 1909, 115; Koliubakhin, *1812 god. Poslednie dni komandovania* . . ., 468.

28. Gosudarstvenno Istoricheskii Muzei, 176, 188; Dubrovin, 95–6; Josselson, 124.
29. Grabbe, 46; Sukhanin, 279; Mitarevskii, 42; Muravev, 180; Konshin, 283; Grabbe, 47, 455.
30. Grabbe, 47.
31. Koliubakhin, *1812 god. Poslednie dni komandovania* . . ., 470; Clausewitz, 139; Gosudarstvenno Istoricheskii Muzei, 168; Radozhitskii, 128; Koliubakhin, ibid., 471; Barclay, *Tableau*, 38; Ermolov, 61; Beskrovny, *Borodino*, 44.
32. *1812 god v Vospominaniakh, Perepiske i Raskazakh*, 22, 24, 28.
33. Ibid., 77.
34. *1812 god Voennie Dnevniki*, 139; Dubrovin, 64–5.
35. Khomutova, 321; Naryshkina, 141–2.
36. Kologrivova, 340–1; *1812 god v Vospominaniakh sovremennikov*, 96.
37. Naryshkina, 151.
38. *1812 god v Vospominaniakh, perepiske i raskazakh*, 48; Kologrivova, 341; Gosudarstvenno Istoricheskii Muzei, 181; Dubrovin, 102.
39. Khomutova, 319; Beskrovny, *Narodnoe Opolchenie*, 48, 64.
40. Bestuzhev-Riumin, *Zapiski*, 353–7.

Chapter 12: Kutuzov

1. Alexander, *Corr. avec sa soeur*, 82; Palmer, *Alexander*, 237; Bakunina, 409; Wilson, *Diary*, I/155; Butenev, 6.
2. Adams, II/398.
3. Kharkievich, *Barclay de Tolly v Otechestvennoi Voinie*, 49.
4. Bakunina, 403.
5. Palmer, *Alexander*, 239; Chichagov, *Mémoires*, 43; Marchenko, 502–3; Wolzogen, 131–2; Kutuzov, *Dokumenty*, 161–3; Alexander, *Corr. avec sa soeur*, 81, 87; Kharkievich, *Barclay de Tolly v Otechestvennoi Voinie*, 50; Kutuzov, *Dokumenty*, 163–4, 357.
6. Alexander, *Corr. avec Bernadotte*, 55–6; Scott, *Bernadotte*.
7. Wilson, *Invasion*, 111–12.
8. Koliubakhin, *1812 god, Izbranie Kutuzova*, 8–9; Bennigsen, *Zapiski*, Sept. 1909, 492–3; Kharkievich, *Barclay de Tolly v Otechestvennoi Voinie*, 24, 26.
9. Radozhitskii, 120–2; Koliubakhin, *1812 god. Izbranie Kutuzova*, 12; Kutuzov, *Dokumenty*, 360; See also Clausewitz, 136; F. Glinka, *Zapiski Russkavo Ofitsera*, IV/51.
10. Maevskii, 153; Dubrovin, 101; Clausewitz, 138, 139; Toll, II/5–8.
11. Clausewitz, 139; Barclay, *Tableau*, 40.

12. Barclay, *Tableau*, 42.
13. Koliubakhin, *1812 god. Izbranie Kutuzova*, 12; Beskrovny, *Borodino*, 25–6, 45–6.
14. Beskrovny, *Borodino*, 45–6, 55, 59.
15. Toll, II/29, 43.
16. Beskrovny, *Borodino*, 64.
17. Bennigsen, *Zapiski*, Sept. 1909, 495; Beskrovny, *Borodino*, 86.
18. Fezensac, *Journal*, 41.
19. Barclay, *Tableau*, 44–6; Kutuzov, *Dokumenty*, 363, 367–8; Beskrovny, *Kutuzov*, IV/129; Koliubakhin, *1812 god. Izbranie Kutuzova*, 31.
20. Kemble, 188–9; Ségur, V/16–17; Constant, V/61–2; Denniée (74) affirms that Napoleon "*souffrait d'une terrible migraine*" at Borodino; Rossetti, aide-de-camp to Murat, noted (16) that "*il avait l'air souffrant*" on the morning of the battle; Baudus (II/83), who was aide-de-camp to Bessières, records that the latter told him Napoleon was "*très souffrant*" during the battle; Wincenty Płaczkowski, of the Chevau-Légers of the Guard, claims (191) that at one stage Napoleon even "lay down on a coat on the ground and gave his orders from there, and then got up, and, leaning heavily on a cannon, observed the battle from there." Gourgaud (228) is the only one who maintains that Napoleon was in rude health and active throughout the battle.
21. Nafziger, 213; Castellane, I/146; Dumonceau, 129, Dupuy, 176; *Otechestvennaia voina 1812 goda. Istochniki, etc.*, 1998, 75.
22. Troitskii, *Den Borodina*, 195 and Shvedov, *Komplektovanie*, 134; also Beskrovny, *Borodino*, 320; Garting, 76–8; Beskrovny, *Polkovodets*, 204; Tarle, *Nashestvie*, 134; Shishov, 250; the most interesting discussion of the problem can be found in A.A. Abalikhin, *K voprosu chislennosti*, in *Tezisy Nauchnoi Konferentsii* and Troitskii, *1812 Velikii God*, 141.
23. W.H. Löwenstern, I/273.
24. Vionnet de Maringoné, 10.
25. Bausset, II/84.
26. Brandt, 272.
27. Napoleon, *Lettres Inédites* (1935), 69.
28. Combe, 79; Boulart, 252; Holzhausen, 97; Fezensac, *Journal*, 41; see also Labaume, 151.
29. Laugier, *Récits*, 76; Vossler, 60.
30. Mitarevskii, 51, 53–4, 55–6.
31. Sukhanin, 281; see also: Muravev, 193; Golitsuin, 14; F. Glinka, *Pisma Russkavo Ofitsera*, IV/64–5.
32. Rapp, 173.
33. Ibid., 174–5.

Chapter 13: The Battle for Moscow

1. Rapp, 176; Seruzier, 198.
2. Napoleon, *Correspondance*, XXIV, 207; Radozhitskii, 171.
3. Thirion, 180; Vossler, 60–1; Holzhausen, 105.
4. Mitarevskii, 55; *1812 god v Vospominaniakh, perepiskie i raskazakh*, 114.
5. Bourgogne, 7.
6. Laugier, *Récits*, 81, Mitarevskii, 62; Beulay, 56.
7. Kharkievich, *1812 God v dnevnikakh*, 202–3.
8. Rapp, 177.
9. Muravev, 194: Josselson, 141.
10. Kutuzov, *Dokumenty*, 373; Lubenkov, 49–50.
11. Lejeune, *Souvenirs*, II/345.
12. Chambray, II/77, 248; Lejeune, *Mémoires*, II/217; Also: Baudus, II/84; Ségur (IV/382) noted that Napoleon displayed "*un calme lourd, une douceur molle, sans activité*"; Pion des Loches, 290.
13. François, II/791.
14. Beskrovny, *Borodino*, 380; Griois, 40.
15. Toll, II/81–2; Clausewitz, 141; Württemberg, 15–16.
16. Kutuzov, *Dokumenty*, 372–3; Wolzogen, 145; Maevskii, 138.
17. François, II/792, 794.
18. Bennigsen, *Zapiski*, Sept. 1909, 498.
19. Rossetti, Murat's aide-de-camp, relates (119) that when he came up to Napoleon asking for reinforcements, Napoleon ordered General Mouton forward with the Young Guard, but then countermanded the order; Denniée (78–9) agrees that Napoleon had already ordered the Young Guard forward, but then gave way to the counsel of his marshals; Lejeune (*Mémoires*, II/213) records that Napoleon wanted to use the Guard, but "*un conseiller timide*" reminded him that he was a long way from Paris; when Murat sent Bélliard to ask for the Guard, Napoleon answered, "*Je ne vois pas encore assez clair, s'il y a demain une seconde lutte, avec quoi la livrerai-je?*" (Roguet, III/480). Napoleon explained to Dumas (III/440) that he did not use the Guard because he was preserving it for another battle before Moscow. To Rapp (180) Napoleon said: "*Je m'en garderai bien; je ne veux pas la faire démolir. Je suis sûr de gagner la bataille sans qu'elle y prenne part.*"
20. Duffy, 123; Bréaut des Marlots, 17–18.
21. Thirion, 185, 190.
22. Planat de la Faye, 82–3.
23. Meerheimb, 81.
24. Griois, II/38.
25. Holzhausen, 113.
26. Josselson, 139; Grabbe 463.

27. Clausewitz, 166.
28. Dedem, 240; Dumonceau, II/142–3.
29. Brandt, 277; Laugier, *Récits*, 88.
30. Kurz, 90; Faure, 46.
31. Ségur, IV/401; Lejeune, *Mémoires*, II/219; Kołaczkowski, I/126.
32. Bausset, II/99; Bloqueville, III/168; Fain, *Manuscrit*, II/71; Constant, V/83, 64–5.
33. Aubry, 165; Borcke, 187; Vionnet de Maringoné, 10.
34. La Flise, LXXII/45–6; Larrey, IV/49; Roos, 68.
35. Larrey, IV/58; Bourgeois, 51; François, II/793.
36. Larrey, IV/60; Sołtyk, 254.
37. Kallash, 235; Muravev, 199.
38. Muravev, 196.
39. Wolzogen, 145–6.
40. Beskrovny, *Borodino*, 95; Kutuzov, *Dokumenty*, 376; 191; Beskrovny, *Borodino*, 96.
41. Voenskii, *Sviashchennoi Pamiati*, 137.
42. Beskrovny, *Borodino*, 101–7, 111–12. On the Russian side, Kutuzov, Saint-Priest (Kharkievich, *1812 god v Dnevnikakh*, 159) and many others affirmed that the French fell back. Bennigsen (*Zapiski*, Sept. 1909, 500), states that the Russians withdrew, and this is upheld by Shishov (268). On the French side, Berthézène, Labaume and Venturini (Beskrovny, *Polkovodets*, 240) implied that they returned to their morning positions, but Lejeune (*Souvenirs*, II/352) makes it clear that Napoleon's tents were actually pitched at the foot of the battlefield, while Vossen (472), Castellane (I/151), Brandt (279) and many others state that they bivouacked on the battlefield. Either way, Clausewitz (167–8) states quite categorically that the Russians were truly defeated; see also Beskrovny, *Borodino*, 121.
43. Clausewitz, 142. Bennigsen (*Mémoires*, III/87) writes: "That evening, we were still not aware of the huge losses we had suffered during the day; we therefore considered, for a while, retaking our central battery during the night and continuing the battle on the morrow." W.H. Löwenstern (I/278) believed Kutuzov really did want to fight the next day, as did Ermolov (74). Clausewitz (167–8) agreed with Golitsuin that Kutuzov was merely bluffing.
44. Fain, *Manuscrit*, II/47; Denniée, 81.
45. Württemberg, 13.
46. Biot, 34; Holzhausen, 115.
47. Shvedov, *Komplektovanie*, 135. See also: Buturlin, 349; Garting, 78; Josselson, 145; Shishov, 271; Duffy, 139; Troitskii, *1812 Velikii God*, 175–6. Thiry (153) gives the official figures of the Dépôt de la Guerre.

48. Grabbe, 466; Liprandi, 7; Andreev, 192; Shchukin, VIII/110; Simanskii, 1913 No. 2, 165, writes that every company had lost "much more than half" of its men.
49. Voenskii, *Sviashchennoi Pamiati*, 137, 138; Kutuzov, *Dokumenty*, 387; Ermolov, 77.
50. Martos, 489; Kutuzov, *Dokumenty*, 379; Kutuzov, *Pisma*, 339.
51. Beskrovny, *Borodino*, 85; Khomutova, 322.
52. Kallash, 10.
53. Rostopchin, *La Vérité*, 214–16; Naryshkina, 162–4; S. Glinka, *Zapiski o 1812 g*, 54.
54. Ermolov, 78–82; Bartenev, 857; Grabbe, 470; Bennigsen, *Voenny soviet*, 235–8; Barclay, *Tableau*, 56–60; Kharkievich, *1812 god v dnevnikakh*, 173–5 (Saint-Priest); Buturlin, I/357–9; Bennigsen, *Zapiski*, Sept. 1909, 501–3; Bennigsen, *Mémoires*, III/89–93; Beskrovny, *Borodino*, 187–8; Dokhturov, 1098–9; Kharkievich, *1812 god v Dnevnikakh*, 128 (Konovnitsin); Kutuzov, *Dokumenty*, 385 (Raevsky). No official record was made at the time, no doubt because Kutuzov wished to protect his reputation against all eventualities. There are many discrepancies in the various accounts left by the participants, dictated principally by the private agendas of the writers. Ultimately, the details are not that significant, other than for the personal reputations of those involved. On this question, see Josselson, 154.
55. Dokhturov, 1098–9.
56. Kutuzov, *Pisma*, 340; Naryshkina, 164–6.
57. Kharkievich, *1812 god v Dnevnikakh*, 23; Popov, *Dvizhenie*, 518.
58. Buturlin, I/363.
59. Safanovich, 126–8; Kozlovskii, 106; Naryshkina, 146–7; Kutuzov, *Pisma*, 340; Naryshkina, 164–6; Aglaimov, 55; Garin, 18–20; *1812 god. Voennie Dnevniki*, 143, 147; Evreinov, 103; Bennigsen, *Zapiski*, Sept. 1909, 504; *1812 god v Vospominaniakh sovremennikov*, 51; Voenskii, *Sviashchennoi Pamiati*, 139; Shchukin, II/212, V/165.
60. Sukhanin, 482.
61. Ibid.; *1812 god. Voennie Dnevniki*, 144, 147–8.
62. Shchukin, VIII/44–76. Voenskii, *Sviashchennoi Pamiati*, 139; Rostopchin, *La Vérité*, 220.
63. Garin, 24.
64. Sukhanin, 483; also Bennigsen, *Zapiski*, Sept. 1909, 504; Kallash, 92; *1812 god v Vospominaniakh sovremennikov*, 59; Bennigsen, *Mémoires*, III/95; Rossetti, 127–8; Kharkievich, *1812 god v Dnevnikakh*, 205–12.
65. Perovskii, 1033; Roos, 88–90.
66. Uxküll, 87; Chicherin, 14–16; Clausewitz, 192; Aglaimov, 56; Radozhitskii, 165, 172.

Chapter 14: Hollow Triumph

1. Bourgogne, 13.
2. Fantin des Odoards, 331–2; Sołtyk, 261; Shchukin, IV/229–464; Laugier, 94; Combe, 96.
3. A.H. Damas, I/118.
4. Montesquiou-Fezensac, 226–7; Sanguszko, 93; Thirion, 201.
5. The question of how deserted Moscow really was is difficult to answer with any precision. Naryshkina (163) writes that there were still 100,000 inhabitants left on 13 September, and although there was a final exodus on the 14th, this would have left a large number still in the city. Prince Eugène, who was quartered in the city, states (VIII/48) that there were between 80,000 and 100,000. Sołtyk (270) claims that more than half of the inhabitants had remained in the city, but were invisible as they cowered in cellars, at the backs of houses or in out-of-the way areas. Postmaster Karfachevsky (Shchukin, V/165) claimed 20,000 people stayed behind; Ysarn, 41, maintains that many left only on account of the fire, and returned once it had died down. There were also, according to Ségur (V/57), as many as 10,000 Russian soldiers wandering the city. See also Bourgogne, 16.
6. Bourgogne, 16; Fantin des Odoards, 332; Lejeune, *Memoires*, II/222.
7. Sołtyk, 274.
8. Holzhausen, 128.
9. The question of who started the fire has been studied to pieces, mainly because Rostopchin himself (*La Vérité*, 183) decided to deny it at one point, and because a number of Russians wanted to pin responsibility on the French. See also *Pravda o pozhare Moskvy*, in Rostopchin, *Sochinienia*, 201–54. The French contributed to the confusion by overegging the conspiracy: many of them claimed they saw incendiary rockets being fired (Seruzier, 219; Berthézène, II/65), others to have found fuses, explosive devices and other incendiary aids (Bourgoing, *Souvenirs*, 116–17; Berthézène, II/68–9; Caulaincourt, II/16; Laugier, *Récits*, 99; Lejeune, *Souvenirs*, II/365; Castellane, I/162), and others to have seen incendiaries at work (Caulaincourt, II/12–13; Constant, V/93–4; Laugier, *Récits*, 99; Ségur, V/45–6; Rapp, 182), while the commission set up by Napoleon to investigate the fire (Shchukin, I/129–43) conjured a huge conspiracy, involving even Rostopchin's famous balloon. There can be little doubt that Rostopchin did initiate the firing of the city, and in various utterances he revealed how proud he was of it (*La Vérité*, 181). For the facts, see the short version in S.P. Melgunov, *Kto zzheg Moskvu*, in Dzhivelegov et al., IV/162. For a full discussion, see Olivier Daria's exhaustive work. For Rostopchin's order to remove the pumps, see Garin, 21.

10. O'Meara, I/196; Dedem, 255; Larrey, IV/73–4; see also: Lecointe de Laveau, 114; Fantin des Odoards, 335; Boulart, 261.
11. Chambray, II/124.
12. Adam, *Aus dem Leben*, 208; Holzhausen, 129; Boulart, 262.
13. On the looting: Bourgogne, 20ff: Duverger, 10; Brandt 288; Kallash, 37ff, 57, 185ff; Kurz, 95; Surruges, 43; La Flise, LXXII/56; Ysarn; Timofeev; Bozhanov; Pion des Loches, 300–2; *1812 god v Vospominaniakh, perepiske i raskazakh*, 7–10; Barrau, 80, 84–5; Laugier, *Récits*, 103; Garin, 66; Chambray, II/ 122–4; *1812 god v Vospominaniakh sovremennikov*, 51, 61; Shchukin, I/126.
14. Labaume, 226.
15. Duverger, 9; Fantin des Odoards, 337.
16. Dolgova, in *Otechestvennaia Voina 1812. Istochniki etc.* (1999), 30–73.
17. Lecointe de Laveau, 122; Ysarn, 41; Henckens, 134.
18. O'Meara, I/193; Surrugues (39) states that four-fifths of the city was destroyed, but more recent studies have reached a lower estimate. Berthézène (II/74) claims the fire did not deprive the army of anything it needed. Dedem (256) supports this. See also note 29 below.
19. Caulaincourt, II/41.
20. Fain, *Manuscrit*, II/94–7; Rapp, 184.
21. Caulaincourt, II/25–9.
22. S. Glinka, *Podvigi*, 69–71; Fain, *Manuscrit*, II/99–103.
23. Stchoupak, 46; Fain, *Manuscrit*, II/104.
24. Napoleon, *Correspondance*, XXIV/221–2.
25. *K Istorii Otechestvennoi Voiny*, 59–61.
26. Ségur, V/75.
27. Caulaincourt, II/49.
28. Ibid., 22, 24, 57–8, 64–7.
29. According to Dumas (III/446) there were enough stores for a short stay, but not for the whole winter; Davout (Boqueville, III/176–8) claims there were three months' supplies; according to Daru (Segur, V/92) there were enough for the whole winter; Villemain (I/230) reports the same; Bellot de Kergorre (65–6) states that the only area where supplies were short was in fodder for horses; see also Larrey, IV/77; Chambray, II/132ff; Napoleon, *Correspondance*, XXIV/222.
30. Hogendorp, 324.
31. Rapp, 185.
32. Pouget, 204–5.
33. Coignet, 196.
34. Ségur, IV/277, 280; Bellot de Kergorre, 64; Chambray, I/250–1; Hochberg, 69.

35. Roeder, 152; Hochberg, 68, 78; Napoleon, *Dernières Lettres Inédites* (1903), 282; Bignon, *Souvenirs*, 232, 239–40; Beyle, *Corr. Gen.*, II/362ff, 376; Chambray, I/249; Ségur, IV/280.
36. Bellot de Kergorre, 61–4; Napoleon, *Correspondance*, XXIV/225, 226; Napoleon, *Lettres Inédites* (1935), 86–7. Fain, *Manuscrit*, II/134.
37. Dolgov, 171, 290; Zotov, 584.
38. Alexander, *Corr. avec sa soeur*, 83; Adams, II/404–5.
39. Voenskii, *Sviashchennoi Pamiati*, 168,171; Kutuzov, *Dokumenty*, 199–200.
40. Voenskii, *Sviashchennoi Pamiati*, 172.
41. Alexander, *Corr. avec sa soeur*, 83.
42. Ibid., 84, 90; Grech, 279; Adams, II/414, 404–5.
43. Edling, 75, 79–80.
44. Shishkov, 157; Alexander, *Corr. avec Bernadotte*, 37–8.

Chapter 15: Stalemate

1. Kallash, 212.
2. Kutuzov, *Dokumenty*, 204–5, 209; Dokhturov, 1099; *1812 god. Voennie Dnevniki*, 144; Garin, 18; Radozhitskii, 165, 172.
3. A.N. Popov, *Dvizhenie*, Sept. 1897, 623–4; Marchenko, 503; Kharkievich, *Barclay de Tolly v Otechestvennoi voinie*, 34–5; Clausewitz, 195; Simanskii, 1913 No. 2, 168–9.
4. Maistre, I/194–5.
5. Shishov, *Nieizvestny*, 241.
6. See Lazhechnikov, I/181ff; Beskrovny, *Narodnoe Opolchenie*, 459; Hartley, *Russia in 1812*, 401; Butenev, 1883, 6; F. Glinka, *Pisma Russkavo Ofitsera*, IV/74.
7. Dzhivelegov et al., V/43–74, esp. 50, 51, 53; Hartley, *Russia in 1812*, 400; Beskrovny, *Narodnoe Opolchenie*, 132, 45, 60, 62, 65, 132; Sverbeev, 74; *1812 god v Vospominaniakh, perepiskie i raskazakh*, 87.
8. Tarle, *Nashestvie*, 199; Chicherin, 46; Rosselet, 166–7.
9. Kallash, 212; Bakunina, 408–9; Tarle, *Nashestvie*, 71–2; Muravev, 202; Kallash, 212; Kutuzov, *Dokumenty*, 224–5; *1812 god v Vospominaniakh, perepiske i raskazakh*, 104.
10. Bestuzhev-Riumin, *Zapiski*, 349.
11. Uxküll, 75; Compans, 157.
12. Jackowski, 298–9; Gajewski, 239; Chłapowski, 125; Seruzier, 223–4; Roos, 99–100; Berthézène, II/76–7; Vionnet de Maringoné, 21–2; Dedem, 254; Roos, 99–100.
13. Holzhausen, 135; Bertolini, 319.
14. S. Glinka, *Zapiski*, 255.
15. Shchukin, IV/347; Leontiev, 408–9.

16. Voronovskii, 248–9.
17. Benckendorff, 49–51.
18. Dzhivelegov et al., V/81.
19. Vigel, IV/49; *1812 god v Vospominaniakh, perepiske i raskazach*, 36, 102, 104.
20. Chicherin, 47.
21. *1812 god v Vospominaniakh, perepiske i raskazakh*, 62.
22. Uxküll, 75.
23. Dzhivelegov et al., IV/230.
24. Ibid.; Sukhanin, 483; Wilson, *Diary*, I/174, 200, 209; Uxküll, 88; Dolgov, 327; Muravev, 203.
25. Déchy, 369; Dzhivelegov et al., IV/230.
26. *1812 god v Vospominaniakh sovremennikov*, 162.
27. Labaume, 174–5; repeated by Ségur, IV/411; Shchukin, II/202; Andreev, 193.
28. F. Glinka, *Zapiski Russkavo Ofitsera*, IV/32.
29. Volkonskii, 211–12; Beskrovny, *Polkovodets*, 349; Grabbe, 472–3; Langeron, 105; Tarle, *Nashestvie*, 190–1, 247–8, 250; Garin, 94–100, 100–2, 105–6, 109; *Tezisy Nauchnoi Konferentsii*, 66; Tarle, *Napoleon*, 419–20.
30. Bogdanov, 88, 96; Troitskii, *1812, Velikii God*, 223–4.
31. Grunwald, *Baron Stein*, 195–213; Palitsyn, 479.
32. Edling, 75; Grech, 285–6; Ley, 52–4, 55–9.
33. A.N. Popov, *Dvizhenie*, Sept. 1897, 626.
34. *1812 god. Voennie dnevniki*, 144; A.N. Popov, *Dvizhenie*, Sept. 1897, 623, 626; Uxküll, 88; Hartley, *Russia in 1812*; Ermolov, 27.
35. Radozhitskii, 172.
36. Voenskii, *Sviashchennoi Pamiati*, 140–1.
37. Popov, *Dvizhenie*, 519, 525; Bennigsen, *Zapiski*, 507; Clausewitz, 185–6.
38. *1812 god, Voennie Dnevniki*, 95; Mitarevskii, 101.
39. Mitarevskii, 100.
40. Viazemskii, 202, 206.
41. *1812 god v Vospominaniakh sovremennikov*, 117.

Chapter 16: The Distractions of Moscow

1. Beauharnais, VIII/50; Beyle, *Vie de Napoléon*, 219.
2. Bausset, II/113, 183; Saint-Denis, 42; Castellane, I/161; Caulaincourt, II/23.
3. Fain, *Manuscrit*, I/129, 131, 140, 142; Napoleon, *Correspondance*, XXIV/232, 233.
4. *Lettres Interceptées*, 310.
5. Ibid., 25, 34, 84, 106, 59.
6. Barrau, 89.
7. Vionnet de Maringoné, 43.
8. Dolgova in *Otechestvennaia Voina 1812 g. Istochniki, etc.* (1999).

9. Kallash, 189; Shchukin, IX/78–82.
10. Kozlovskii, 113; Chevalier, 208; Adam, *Aus dem Leben*, 213; *1812 god v Vospominaniakh, perepiskie i raskazakh*, 18; Lecointe de Laveau, 125–6; Beyle, *Journal*, IV/209.
11. Shchukin, VII/214.
12. Larrey, IV/65; *Lettres Interceptées*, 80; Gardier, 58.
13. Chłapowski, 127.
14. *Lettres Interceptées*, 67.
15. Bourgoing, *Souvenirs*, 134.
16. Duverger, 11–12.
17. Fantin des Odoards, 339–40.
18. Griois, II/55; La Flise, LXXII/55.
19. Compans, 196–8; *Lettres Interceptées*, 97.
20. *Lettres Interceptées*, 22, 61.
21. Peyrusse, *Lettres Inédites*, 96ff; 103; *Lettres Interceptées*, 80; Bloqueville, III/174; Sanguszko, 107.
22. Płaczkowski, 201; Vionnet de Maringoné, 53; Lecointe de Laveau, 125–7; Sołtyk, 318–19.
23. *Lettres Interceptées*, 80; Combe, 121.
24. Bourgogne, 49–51.
25. Surrugues, 10–11.
26. Labaume, 240; see also Le Roy, 164.
27. Dedem, 250; Dupuy, 185; Ségur, V/79; Simanskii, 1913 No.4, 127; Radozhitskii, 178; Kołaczkowski, 146; Dupuy, 176–7.
28. Castellane, I/168–9; Thirion, 219.
29. Dembiński, I/167.
30. Ibid., 169.
31. Fantin des Odoards, 340; *Lettres Interceptées*, 157–9.
32. Castellane, I/165; Thirion, 215–16; Fantin des Odoards, 321–2.
33. Chambray, II/205; Belliard, I/112.
34. Bourienne, IX/120; see also Caulaincourt, I/315.
35. Caulaincourt, II/26, 42, 56, 65; Bloqueville, III/181.
36. Fain, *Manuscrit*, II/151–2; Napoleon, *Corr. Inédite* (1925), V/595.
37. Beyle, *Corr. Gen.*, II/383; Dumas, III/447, 455, 456; Napoleon, *Correspondance*, XXIV/264; Denniée, 105; Larrey, IV/79; La Flise, LXXII/58.
38. Napoleon, *Correspondance*, XXIV/261, 235–8; Caulaincourt, II/73.
39. Napoleon, *Correspondance*, XXIV/275; Planat de la Faye, 92; Fain, *Manuscrit*, II/162; Ségur, V/90, maintains that Napoleon was afraid the Russians would treat abandoned guns, even if spiked, as trophies.
40. Castellane, I/169.
41. Marbot, III/162–3.

42. Ibid., 161–2; Chłapowski, 128.
43. Henckens, 140; *Lettres Interceptées*, 61.
44. Bro, 119.
45. Lagneau, 219; Pion des Loches, 306.
46. Grabowski, 7; Henckens, 152; Fain, *Manuscrit*, II/157.

Chapter 17: The March to Nowhere

1. Shishov, 288–9; Beskrovny, *Polkovodets*, 274–5; Shvedov, *Komplektovanie*, 127–9, 136.
2. *Otechestvennaia Voina 1812g. Istochniki, etc.* (1998), 20.
3. Toll, II/190–1.
4. Maevskii, 154; Ermolov, 92.
5. Toll, II/204; A.N. Popov, *Dvizhenie*, July 1897, 114–18; Kharkievich, *Barclay de Tolly v Otechestvennoi Voinie*, 34; Dubrovin, 129.
6. Chicherin, 32.
7. Wilson, *Diary*, I/194; Chicherin, 28; Kutuzov, *Pisma*, 359; Wilson, *Invasion*, 182–90; Kutuzov, *Dokumenty*, 231–2.
8. Kutuzov, *Dokumenty*, 226–7.
9. Shishov, 291; Bennigsen, *Zapiski*, 508–22; Maevskii, 156; Tarle, *Nashestvie*, 225–7; A.N. Popov, *Dvizhenie*, August 1897, 366; W.H. Löwenstern, I/303–4; Mitarevskii, 122–3; Kutuzov, *Dokumenty*, 407–10.
10. Radozhitskii, 224.
11. Kutuzov, *Dokumenty*, 228–9, 230; Beskrovny, *Polkovodets*, 295; Altshuller & Tartakovskii, 52–4; Dolgov, 13.
12. Kutuzov, *Dokumenty*, 411; Ermolov, 122; *1812 god. Voiennie Dnevniki*, 98; Württemberg, 21.
13. Fezensac, *Journal*, 64; Bausset, II/126.
14. Fezensac, *Journal*, 68; see also Paixhans, 20.
15. Laugier, *Récits*, 118–19.
16. Bourgogne, 56–7.
17. Barrau, 91; Ségur, V/102–3; Mailly, 72.
18. Pion des Loches, 308; Lecointe de Laveau, 137; Mailly, 71.
19. Griois, II/82.
20. Denniée (107) gives the number of vehicles accompanying the army as 40,000, Bellot de Kergorre (70) as 25,000, Castellane (I/173) as 15,000.
21. Reliable figures are not available. Nafziger (263) says there were 95,000 men; Jomini (239) says there were 80,000 men and 15,000 malingerers; and most are agreed that there were just under 100,000. On the condition of the troops, see: Baudus, II/247; Bourgeois, 85; Mailly, 66.
22. Dumonceau, II/175; Beauharnais, VIII/59; Labaume, 237.
23. Rapp, 192–3.

24. Napoleon, *Correspondance*, XXIV/278, 281.
25. Ibid., 289.
26. Aubry, 167–70.
27. Kutuzov, *Dokumenty*, 415.
28. Wilson, *Invasion*, 229.
29. Berthézène, II/132; Wilson, *Invasion*, 230; Bertolini, I/369.
30. Beauharnais, VIII/22; Labaume, 279.
31. Caulaincourt, II/98–9; Fain, *Manuscrit*, II/248, 251–2, 253, 255; Lejeune, *Mémoires*, II/240; Ségur, V/ 116, 123–8.
32. Griois, II/89; Fain, *Manuscrit*, II/253.
33. Kharkievich, *1812 god v Dnevnikakh*, 45; Mitarevskii, 125.
34. Toll, II/269; Bennigsen, *Zapiski*, 360; Nesselrode, IV/108; Toll, II/270; Dubrovin, 235.

Chapter 18: Retreat

1. Denniée, 118; Volkonskii, 199–203.
2. Denniée, 114–5; Caulaincourt, II/104–5.
3. Dedem, 271–2.
4. Mailly, 78.
5. Labaume, 288; Roos, 115, Ségur, V/152; Dumas, III/127; Bellot de Kergorre, 72; Barrau, 94. François (II/795) claims to have met the man when he was on his way *to* Moscow, three weeks after the battle. Pelet (11) refutes the whole story as nonsense.
6. Fezensac, *Journal*, 75.
7. Caulaincourt, II/109–10, 111–12.
8. Ibid., 112–14.
9. Palmer, *Alexander*, 251.
10. Dumonceau, II/120.
11. *Lettres Interceptées*, 251; Blaze de Bury, I/393.
12. Bellot de Kergorre, 73–4.
13. Paixhans, 39.
14. Chambray, II/367; Nesselrode, IV/116; Pion des Loches, 309.
15. Castellane, I/175; Bourgogne, 63; Kurz, 136; Labaume, 288.
16. Dumonceau, II/190–1.
17. Beyle, *Vie de Napoléon*, 239.
18. Laugier, *Récits*, 133; Kołaczkowski, I/156; Shishov, 298; Prince Eugene of Württemberg, 30, claims his division alone lost 1,000, but Shvedov, *Komplektovanie*, puts Russian losses at no higher than 1,200; Askenazy, 237; Castellane, I/181; Pelet, 21; Fezensac, *Journal*, 79; Voronovskii, 190.
19. Kutuzov, *Dokumenty*, 414; Bennigsen, *Zapiski*, 366; Tarle, *Nashestvie*, 260; Kutuzov, *Dokumenty*, 241, 243.

20. Mailly, 80, 83.
21. Griois, II/96; Bertin, 30; Laugier, *Récits*, 131.
22. Caulaincourt, II/ 117; *Lettres Interceptées*, 184.
23. Castellane, I/180; Pelet, 16; Voronovskii, 185; Pelet, 18.
24. Bourgogne, 66–7.
25. Dumonceau, II/197.
26. Laugier, *Récits*, 137.
27. Pelet, 19–21.
28. Faber du Faur, 244; Holzhausen, 187. Kerner gives a different date to Faber du Faur, but as the latter's account is based on a journal, I give him the benefit of the doubt; Lignières, 118–19.
29. Paixhans, 27.
30. Pelet, 11.
31. Dedem, 279; Labaume, 308.
32. Lignières, 118.
33. Caulaincourt, II/139; Walter, 68; Bellot de Kergorre, 74.
34. Załuski, 251.
35. Laugier, *Recits*, 137; Muralt, 86; Dedem, 276; Griois, II/108; Fredro, 82; Coignet, 213; Vionnet de Maringoné, 64.
36. Bourgogne, 61; Askenazy, 237.
37. Vionnet de Maringoné, 65; Le Roy 205.
38. Labaume, 281; Lignières, 137.
39. La Flise, LXXII/567, 570–1, 574.
40. Griois, II/99.
41. Boulart, 268; Pelet, 76; Muralt, 90.
42. Dedem, 280.
43. Pretet; Płaczkowski, 37.
44. Dupuy, 197; see also Chevalier, 239.
45. Lejeune, *Mémoires*, II/250; Kurz, 144.
46. Roguet, III/508.
47. *Lettres Interceptées*, 227; Holzhausen, 178.
48. Walter, 53; François, II/827; Duverger, 14.
49. Labaume, 294–5; Fusil, *Souvenirs*, 257.
50. Clemenso, 38.
51. Davidov, 119, 134–5; W.H. Löwenstern, I/294–5; Simanskii, 1913, No.3, 142.
52. Dubrovin, 325; Wilson, *Invasion*, 257–8.
53. Bertolini, I/188.
54. Faure, 74.
55. Wilson, *Diary*, I/215 ; Kurz, 145; W.H. Löwenstern, I/228.
56. Wachsmuth, 206; Combe, 145–9.
57. Radozhitskii, 253.

Chapter 19: The Mirage of Smolensk

1. Beaulieu, 33; Shvedov, *Komplektovanie*, 137; see also Zotov, 605.
2. Napoleon, *Correspondance*, XXIV/298–300, 300–2.
3. Rapp, 201; Caulaincourt, II/126.
4. Napoleon, *Correspondance*, XXIV/302.
5. Ibid., 303–6; Jomini, *Précis*, I/173.
6. Beyle, *Corr. Gen.*, II/369.
7. Clausewitz (98) says Napoleon lost 61,000; Berthézène (II/145) maintained that Napoleon had no more than 20,000 fighting men left, but he always gives low figures; Lejeune (*Mémoires*, II/256) states that the Guard consisted of no more than 3–4,000 men under arms; Rossetti (157) that there were only 36,000 under arms in total. See also Nafziger, 305.
8. Alexander, *Corr. avec Bernadotte*, 63.
9. Griois, II/116.
10. Laugier, *Récits*, 141. See also ibid., 138–47; Chambray, II/388; Labaume, 327–31; Zanoli, 202.
11. *Lettres Interceptées*, 318; Shishov, 299, claims that Prince Eugène lost 62 guns; Voronovskii (200) puts the figure at 64.
12. Pastoret, 470–1.
13. Załuski, 252; Dedem, 277; Griois, II/124.
14. Bellot de Kergorre, 76; Caulaincourt, II/131; Griois, II/129.
15. Puibusque, *Lettres sur la guerre*, 109; François, II/815; Lignières, 121; La Flise, LXXII/579; Lecoq, 168; Kurz, 150; Laugier, *Récits*, 150, 153.
16. Pastoret, 472.
17. Labaume, 338.
18. Larrey, IV/91; Voronovskii, 209; see also Angervo; Fezensac, *Journal*, 96; Bourgogne, 81.
19. Bertrand, 147; Bourgogne, 76–7. He may or may not be describing the same incident, but it seems to fit in time and place with Bertrand's account.
20. Faber du Faur, 253–4.
21. Lejeune, *Mémoires*, II/253–4; *Lettres Interceptées*, 251.
22. Laugier, *Récits*, 150; Bertolini, II/10.
23. Labaume, 349–50.
24. Sauzey, III/173; La Flise, LXXII/579.
25. Pelet, 29–30.
26. Ibid., 25; Fezensac, *Journal*, 96; Henckens, 153–4; Dumonceau, II/202–3. Faber du Faur (251) writes that "the last bonds of order and discipline were broken" in Smolensk. Peyrusse (*Mémorial*, 118) claims that discipline was cracking and "our divisions resemble armed mobs."
27. Caulaincourt, II/137, 386–7.

28. Boulart, 270–1.
29. Caulaincourt, II/141; Saint-Denis, 54. Méneval (II/93–4) writes that Napoleon only acquired the poison at Orsha.
30. Griois, II/133; Labaume, 357; Laugier, *Récits*, 155.
31. Bourgogne, 116; Kutuzov, *Dokumenty*, 424–5, 427; Toll, II/321; Wilson, *Invasion*, 272–3.
32. Boulart, 273.
33. Vlijmen, 319; Roguet, III/520.
34. Maevskii, 161; Dumonceau, II/210; Caulaincourt, II/154.
35. Chambray, II/455; see also Fantin des Odoards, 346; Bourgogne, 132; Rumigny, 64.
36. Napoleon, *Lettres Inédites* (1897), II/202; Napoleon, *Corr. Inédite* (1925), V/611.
37. Caulaincourt, II/158, 160, 162; Rapp. 210; Bausset, II/159.
38. Breton, 112–13.
39. Wilson, *Invasion*, 279; Castellane, I/189; Buturlin, II/227–8; Breton, 114; W.H. Löwenstern, I/345–7.
40. Fezensac, *Journal*, 106.
41. Freytag, 169. His description of the events has to be taken with caution, as he mixes up the sequence of events and spreads the action over too many days.
42. Fezensac, *Journal*, 112; Planat de la Faye, 103.
43. Pelleport, II/45, 48; Bonnet, 106; Pelet, 39, 44 (where he claims that he was the one to come up with the idea to cross the Dnieper), 47–52; Podczaski, 110; Chuquet, *Lettres de 1812*, 185ff; Fezensac, *Journal*, 104–18; Chłapowski, 134. There is some doubt about how many made it: Pelet (50) says some 8–10,000 set off, of whom most made it; Materre, who was on Ney's staff, claims (77) that 6,000 crossed the Dnieper; Berthézène (II/157) writes that only 4–500, mostly officers and NCOs, got through; Fezensac (*Journal*, 118) puts the figure at 8–900; Pelleport (II/52), probably the most reliable, says that 1,500 reached Orsha.

Chapter 20: The End of the Army of Moscow

1. Kutuzov, *Dokumenty*, 256, 252.
2. Voronovskii, 228–9.
3. Marchenko, 500.
4. Maistre, I/220.
5. Ibid., 230.
6. Dubrovin, 303; Kutuzov, *Dokumenty*, 249.
7. Marchenko, 503; Palmer, *Alexander*, 254.
8. Radozhitskii, 238; Muravev-Apostol, 36–7.

9. Mitarevskii, 141, 142, 148–9, 153, 154.
10. Bennigsen, *Zapiski*, 369.
11. Uxküll, 100; Radozhitskii, 259.
12. Shvedov, *Komplektovanie*, 136; Mitarevskii, 154; Radozhitskii, 258, 272.
13. Aglaimov, 77.
14. Mitarevskii, 146, 157; Württemberg, 33, 35–6; Kutuzov, *Dokumenty*, 250; Wilson, *Invasion*, 234.
15. Shcherbinin (*Zapiski*), W.H. Löwenstern (I/317), Ermolov (128–9) and Maevskii (161) are among those who believed Kutuzov was afraid of confronting Napoleon. See also Garin (130) and Pokrovskii (III/188).
16. Ermolov, 118; Beskrovny, *Polkovodets*, 311.
17. Davidov, 142–3; W.H. Löwenstern, I/338, 343; Kutuzov, *Dokumenty*, 250.
18. Rapp, 210; Caulaincourt, II/163.
19. Denniée, 141; Napoleon, *Correspondance*, XXIV/310, 311; Caulaincourt, II/166; Voenskii, *Sviashchennoi Pamiati*, 192.
20. Caulaincourt, II/163.
21. Fabry, *Campagne de 1812*, 191; Napoleon, *Correspondance*, XXIV/312; Caulaincourt, II/166; Fain, *Manuscrit*, II/325.
22. Planat de la Faye, 107.
23. Shishov (302) gives the number of prisoners as 422 officers and 21,170 other ranks, along with 213 guns; Bezkrovny (*Polkovodets*, 320) lists 26,000 prisoners and 116 guns, effectively repeating the figures given by Buturlin (II/231); see also Troitskii, *1812 Velikii god*, 279. In a letter to Maret (*Lettres Inédites* [1897], II/202) Napoleon himself writes that he had lost 30,000 men and been obliged to leave 300 guns behind.
24. Griois, II/131–2.
25. Laugier, *Récits*, 154.
26. Mailly, 86–7; Duverger, 15.
27. Roos, 180.
28. Bourgoing, *Souvenirs*, 161.
29. Roguet, III/539; La Flise, LXXIII/55; see also Thirion, 229–31.
30. Holzhausen, 209.
31. Bourgogne, 137–45.
32. Mayer, 342–3; Olenin, 1996.
33. Auvray, 82.
34. Everts, 151, 157–8; Mayer, 347; Wilson, *Invasion*, 256; Płaczkowski, 225.
35. Breton, 114, 126; Chevalier, 249; Holzhausen, 347–8; Rochechouart, 198, 200; Puybusque, 324–5; Pouget, 220; Comeau de Charry, 465.
36. Le Roy, 265.
37. Beauharnais, VIII/112.
38. François, II/813.

39. Thirion, 238–9; Dembiński, I/199–200; Bonneval, 76; Sanguszko, 104; Bourgogne, 68.
40. Roos, 186; Castellane, I/192.
41. Griois, II/129.
42. Krasiński, 98.
43. Combe, 152; Chevalier, 248.
44. Griois, II/174–6.
45. Planat de la Faye, 111.
46. Boulart, 269.
47. Suckow, 206; Griois, II/173.
48. Lejeune, *Mémoires*, II/255.
49. Ricome, 48; Boulart, 267.
50. Labaume, 394; François, II/826; Lejeune, *Mémoires*, II/266–7.
51. Lejeune, *Mémoires*, II/271–2.
52. François, II/826; Lagneau, 237.
53. Constant, V/147–8, 154; Rapp, 210.
54. Caulaincourt, II/189; Bourgeois, 139; Radozhitskii, 263; Bourgogne, 213; Denniée, 143.
55. Caulaincourt, II/189; Roos, 149; Dedem, 275; Labaume, 376; Bourgeois, 139; Duverger, 16; François, II/827; Caulaincourt, II/172; Chambray (II/385) states that there was much grumbling, and Pion des Loches (310) that soldiers openly heckled Napoleon.
56. Faber du Faur, 249.
57. Ségur, V/309; Wilson, *Invasion*, 254; Maistre, I/247.
58. Falkowski, V/85; Hochberg, 106–7; Brandt, 314; Beulay, 67; Oudinot, 214.

Chapter 21: The Berezina

1. Jomini (*Précis*, I/184) writes that Napoleon heard the news on the evening of 19 November, but this is clearly wrong: see Caulaincourt, II/168; Napoleon, *Correspondance*, XXIV/311 ("*rien de nouveau*"), 312, 313.
2. Kutuzov, *Pisma*, 411; Langeron, 55; Czaplic, 515; Chichagov, *Pisma*, 61; Martos, 498; Rochechouart, 182.
3. Chambray, III/15, 25–6; Caulaincourt, II/168–70.
4. Rochechouart, 182; Chichagov, *Mémoires*, 59; Voenskii, *Sviashchennoi Pamiati*, 111ff, 121ff; Chichagov, *Mémoires*, 53; Rochechouart, 188–9; Marbot, III/185–6.
5. Caulaincourt, II/173. Napoleon (*Lettres Inédites* [1935], 102–3) bears this out.
6. Jomini, *Précis*, I/186–8, says that he discussed this plan in Tolochin on 22 November, but at that stage Napoleon still believed he held the crossing at Borisov, so he must have been mistaken; Caulaincourt, II/173, mentions the

plan as having been discussed after he had heard of the fall of Borisov. See also Voenskii, *Sviashchennoi Pamiati*, 191–9.

7. Constant, V/121; Bourgoing (*Souvenirs*, 154) paints a somewhat different picture.
8. Brandys, II/136.
9. Jomini, *Précis*, I/188–90; Fabry, *Campagne de 1812*, 206–7, 208–9, 210, 219, 220, 221, 222, 233–4; Napoleon, *Correspondance*, XXIV/316, 317, 318; see also the (unreliable) account by Korkozevich, 114–17.
10. Voenskii, *Sviashchennoi Pamiati*, 191–9; Langeron, 60; Rochechouart, 192; Volkonskii, 215; Clausewitz, 208–9, 211.
11. Corbineau (48–9) claims that he built the first bridge with the help of some of Oudinot's artillerymen, on 24 November, and Gourgaud (434) also maintains that there was a bridge built by Oudinot's gunners on the 24th, which was swept away by the current. But in their authoritative account, Colonel A. Chapelle, chief of staff to the bridging train, and Chef de Bataillon Chapuis, who commanded one of the battalions of pontoneers, explain that Oudinot's artillerymen and sappers had begun making trestles, but these were inexpertly made, and proved of no use. See Fabry, *Campagne de 1812*, 288.
12. Voenskii, *Sviashchennoi Pamiati*, 103–4; Czaplic, 508–9.
13. Pils, 141.
14. Rapp, 213; Constant, V/127.
15. Voenskii, *Sviashchennoi Pamiati*, 104; Czaplic, 509.
16. Vlijmen, 322; Pils, 143.
17. Baudus, II/274; Boulart, 276; Fusil, 277; Rossetti, 168; Gourgaud, 429; Rosselet, 178; Constant (V/122) maintains there was an eagle-burning; Castellane (I/192) reports that the eagles of most of the cavalry regiments were burned at Bobr.
18. Suckow, 250.
19. Fabry, *Campagne de 1812*, 227–9; Chapelle, 3; Vlijmen, 322; Brandt, 319.
20. Pils, 144.
21. For the bridges and the first day's crossing, Fabry, *Campagne de 1812*, 227–31; Pils, 143–5; Vlijmen, 322–4; Chapelle, 3–6; Fain, *Manuscrit*, II/378–9.
22. Bertand, 152.
23. Bussy, 290; Begos, 191.
24. Napoleon, *Lettres Inédites* (1935), 103.
25. Beulay, 57, 63; Castellane, I/198.
26. Czaplic, 510–12; Rochechouart, 193–4; Chichagov, *Mémoires*, 77.
27. Pils, 146–7. The question of numbers at the Berezina is a vexed one. Most French sources put the total at Napoleon's disposal at 25–27,000, while Russians consistently inflated his strength – according to Clausewitz (208–9),

Wittgenstein thought he had as many as 100,000 men. Russian strengths are not easy to determine either. Chichagov's whole army numbered up to 60,000, but much of this had been left behind to garrison Minsk, to patrol south of Borisov and to keep an eye on potential threats from Schwarzenberg. According to Chichagov (*Pisma*, 54), he only had 18–19,000 infantry at his disposal in the Borisov–Studzienka area; Czaplic (514) affirms that Chichagov had 15,000 infantry and 9,000 cavalry, but my reading of the sources suggests that he had at least 10,000 more than that. See also Tarle, *Nashestvie*, 271; Berthézène, II/160; Faber du Faur, 273.

28. Bourgoing, *Souvenirs*, 160.
29. Bussy, 291; Begos, 192; Chapuisat, 87; Braquehay, 184; Vlijmen, 325–6; Legler, 194.
30. Legler, 198; Bussy, 292; Vlijmen, 325, 326.
31. Hochberg, 113–14, 139.
32. Kurz, 184; Holzhausen, 259; Brandys, II/141.
33. Thirion, 250.
34. Suckow, 256–7.
35. Holzhausen, 180.
36. Griois, II/156; Pontier, 15. On the question of the bridges being free at night: Planat de la Faye, 105 (he was one of those trying to persuade stragglers to cross); Chambray, II/70; Bourgogne, 210, 214; Seruzier, 255; Rossetti, 175 (he did cross with his *fourgon* on the night of the 27th); Turno, 114; Gourgaud, 459; Chevalier, 233; Sołtyk, 452; Marbot, III/199 (he actually went back to look for a lost wagon); Larrey, IV/101 (he went back to pick up some surgical instruments that had been left behind), and many others; only Auvray, 79–80, a less than reliable witness in other respects, claims that there was a terrible jam on the bridges on the night of 27 November.
37. Bourgogne, 215. For the subsequent day's crossings and the burning of the bridges: Fabry, *Campagne de 1812*, 230–2; Chapelle, 7–13; Vlijmen, 326–7; Hochberg, 141–4; *1812 god v Vospominaniakh sovremennikov*, 139–44; Kurz, 177–85; Corbineau, 43–51; Curely, 311–24; Rapp, 213–14; Castellane, I/196–8. Ségur and others who did not witness the worst moments have overpainted the picture of horror, which led others, such as Gourgaud, to belittle it and dismiss much of the writing on the subject as melodramatic.
38. Martos, 502.
39. Rochechouart, 195.
40. Gourgaud, 461; Fabry, *Campagne de 1812*, 234–5; Chapelle, 9; Labaume, 405; Bennigsen, *Mémoires*, III/165; Buturlin, II/386; Langeron, 75; Shishov, 306.
41. Clausewitz, 211.

Chapter 22: Empire of Death

1. Griois, II/164.
2. Caulaincourt, II/192; Bourgogne, 216.
3. Constant, V/133.
4. Lagneau, 234; Vionnet de Maringoné (77) claims that the mercury in his thermometer froze; Lejeune, *Mémoires*, II/286; Gardier, 91.
5. Gardier, 91; Fezensac, *Journal*, 145; Paixhans, 43; Kurz, 194; Bourgeois, 167.
6. Holzhausen, 266; Vossler, 92.
7. Fredro, 44.
8. Suckow, 269; Griois, II/192.
9. For reported cannibalism, see: Maistre, I/246; Olenin, 1986–9; Dubrovin, 301; Nesselrode, IV/120. For evidence of prisoners eating their dead comrades: Roederer, 40; Holzhausen, 271; Cheron, 33; Roguet, III/526. For examples quoted: Golitsuin, 30; Wilson, *Diary*, I/215; Gosudarsvenno-Istoricheskii Muzei, 252; Shchukin, VIII/113.
10. Sołtyk, 415.
11. Pastoret, 497.
12. Uxküll, 105; Voenskii, *Sviashchennoi Pamiati*, 107; Langeron, 93.
13. Ségur, V/448; Marbot, III/215; Gourgaud, 480. For confirmation from the French side, see: Ségur, V/382; Kurz, 199, etc. For quotations: Bourgogne, 78; Vossler, 92; Pontier, 16.
14. Mailly, 101; Fezensac, *Journal*, 139.
15. Planat de la Faye, 108; Lejeune, *Mémoires*, II/272.
16. Fezensac, *Journal*, 146–7; Vionnet de Maringoné, 83.
17. Lyautey, 248.
18. Wybranowski, II/17–21; Lejeune, *Mémoires*, II/293.
19. Planat de la Faye, 107; Castellane, I/206; Ségur, V/348–9.
20. Tascher, 317.
21. Bourgogne, 208; Larrey, IV/125; Planat de la Faye, 108; Muralt, 89, 97.
22. Lejeune, *Mémoires*, II/294; Other examples of officers looking after their servants include: Mailly, 120–1; Chéron, 28.
23. Chevalier, 238; Laugier, *Récits*, 181; Holzhausen, 284.
24. Holzhausen, 201; Bourgoing, *Souvenirs*, 162–3.
25. Fezensac, *Journal*, 146; La Flise, LXXIII/52.
26. Bourgogne, 246; Wilson, *Invasion*, 260.
27. Rumigny, 68; Chłapowski, 135.
28. Bourgogne, 123; Holzhausen, 46.
29. Napoleon, *Correspondance*, XXIV/322, 323.
30. Ibid., 324; Montesquiou-Fezensac, 247 (he mistakenly dates his mission from Smolensk); Maistre, I/266.
31. Załuski, 254, 255.

32. Chicherin, 54–6; Voenskii, *Sviashchennoi Pamiati*, 88ff, 144–6; Kutuzov, *Dokumenty*, 258, 262, 263–4.
33. Chicherin, 63; Kallash, 222; A.H. Damas, I/127; Aglaimov, 78; Tarle, *Nashestvie*, 268; Dziewanowski, 10; W.H. Löwenstern, I/352; Radozhitskii, 284; Dokhturov, 1,107; W.H. Löwenstern, I/356; Chicherin, 63; Aglaimov, 79; Langeron, 91–2, paints a more optimistic picture.
34. Langeron, 104–5; Davidov, 155.
35. Napoleon, *Correspondance*, XXIV/323, 325–7, 331–2.
36. Ibid., 338–9; Beauharnais, VIII/104; Castellane, I/202.
37. Lejeune, *Mémoires*, II/289. Griois (II/177) and Castellane (I/202) are among those who thought it a good thing. Those who thought it discouraged the men, who had seen in him a rallying point (whatever they may have thought of him) and now felt betrayed include: Deniée (168), Labaume (424), Lejeune (*Mémoires*, II/289), Laugier (*Récits*, 181), Bourgeois (171), Mailly (105–6), François (II/835), Dumonceau (II/231), Vionnet de Maringoné (76). Those who felt it made scant impression include: Griois (II/177), Muralt (108), Pelleport (II/58), Bourgoing (*Souvenirs*, 172), Castellane (I/202).
38. Lagneau, 235. Larrey (IV/124) and Sołtyk (454) recorded only minus 28 (Réaumur) at Miedniki. See also Paixhans, 57; Dumonceau, II/231.
39. Griois, II/166; Soltyk, 454; Ségur, V/377.
40. Lejeune, *Mémoires*, II/285; Planat de la Faye, 109–10; Bourgogne, 228.
41. Brandt, 334.
42. Roos, 178; Bourgeois, 190; Holzhausen, 213; Auvray, 80.
43. Henckens, 167; Lagneau, 238; François, II/825; Griois, II/179; Bourgogne, 252–3; Lejeune, *Mémoires*, II/286; Minod, 56–7; Roeder, 173; Castellane, I/203, 205.
44. Hochberg, 181–2.
45. Kurz, 203; Larrey, IV/128; Chevalier, 221.
46. Brandt, 234–5.
47. Larrey, IV/107; Bourgeois, 165; Sołtyk, 440; Fezensac, *Journal*, 145; Vossler, 93.
48. Langeron, 90; Chicherin, 67.
49. Ducor, II/20.
50. Caulaincourt, II/192.

Chapter 23: The End of the Road

1. Jaquemont du Donjon, 106, 107; Hogendorp, 330–1; Bignon, *Souvenirs*, 246–7; Butkevicius, 907.
2. Hogendorp, 332, 338. The military governor of Vilna, Baron Roch Godart (183), writes that there was enough to feed 120,000 for 36 days; see also: Berthézène, II/180; Gourgaud, 484; Fain, *Manuscrit*, II/415; Ségur, V/386;

Dedem, 290. Yermolov, who was put in charge of the French stores when the Russians entered Vilna, confirms (131–5) that there were generous amounts of everything an army could wish for. See also Rochechouart; Czaplic, 521.

3. Napoleon, *Correspondance*, XXIV/330.
4. Hogendorp (336) states that he was acting on Napoleon's orders.
5. Suckow, 286.
6. Hogendorp, 336; von Kurz (202) was struck by the youth of the soldiers; Hogendorp, 327; Choiseul-Gouffier, 129.
7. Raza, 217–19.
8. Bourgeois, 173–4.
9. Lejeune, *Mémoires*, II/291; Hogendorp, 336.
10. Pelet, 70; Chevalier, 242–3; Fezensac, *Journal*, 147; Griois, II/183.
11. Laugier, *Récits*, 182.
12. Hogendorp, 338; Chambray, II/124–5, 126.
13. Labaume, 415; see also Fezensac, *Journal*, 142–3.
14. Hogendorp, 335; Hochberg, 187; Hogendorp himself claims not to have left until instructed to on the following afternoon.
15. Ségur, V/372; Berthézène, II/176.
16. Jacquemont du Donjon, 109; Griois, II/182.
17. Holzhausen, 285; Roeder, 190.
18. Thirion, 267.
19. Brandt, 336; Vionnet de Maringoné, 81; Lagneau, 240; Le Roy, 265; Griois, II/184; Lignières, 130.
20. Chłapowski, 137–8; Berthézène, II/179; Rapp, 218.
21. Bertrand, 165; see also Laugier, *Récits*, 183.
22. Hochberg, 190; Fredro, 45; Lignières, 130; Laugier, *Récits*, 183; Mailly, 123–4.
23. Planat de la Faye, 112; Choiseul-Gouffier, 140; Paixhans, 59.
24. Mailly, 137.
25. Noel, 174–6.
26. Holzhausen, 307.
27. Boulart, 278; Peyrusse, *Lettres Inédites*, 116–18; Peyrusse, *Mémorial*, 118, 136; Duverger, 25; Chuquet, *Lettres de 1812*, 305; Noel, 177; Denniée, 172.
28. Bellot de Kergorre, 103; Jacquemont du Donjon (112) claims to have seen the great cross of St Ivan lying on the ground; Planat de la Faye, 115; Combe, 168–9; Lyautey, 251.
29. Fredro, 45.
30. Bourgeois (180), François (II/837), Vaudoncourt (*Quinze Années*, 155), Kurz (216, 220), Roeder (194), a number of sources in Holzhausen (291–2, 298, 347), Lagneau (240), Chevalier (249), Grabowski (9), Choiseul-Gouffier (138) and many others blame the (admittedly very considerable) Jewish

population of the city. On conditions in the city and the hospitals see: Holzhausen, 300, 302; Rochechouart, 202; Pontier, 17. Bourgeois (190) states that typhus had already broken out in the last stages of the retreat.

31. Bourgoing, *Souvenirs*, 228; Vionnet de Maringoné, 79.
32. Bertrand, 169; Lignières, 131.
33. Bourgogne, 262, 246.
34. Beaulieu, 45; Ginisty, 113–14; Brandt (339–41) tells an almost identical story concerning himself and a soldier he had had flogged for looting in Moscow.
35. Planat de la Faye, 116–17.
36. Bertin, 309.
37. Holzhausen, 319–20; Rumingy, 67; Noel, 180.
38. Dumas, III/485.

Chapter 24: His Majesty's Health

1. Caulaincourt, II/212.
2. Ibid., 230ff.
3. Ibid., 263.
4. Pradt, 207–18; Caulaincourt, II/263–73 (he claims Napoleon said he had 150,000 men in Vilna); Potocka, 331–4; Niemcewicz, 383; Koźmian, II/311.
5. Caulaincourt, II/ 315.
6. Davidov, 172; Marchenko, 503; Alexander, *Corr. avec Bernadotte*, xxxii.
7. Bloqueville, III/193; Compans, 239; Bausset, II/192.
8. Caulaincourt, II/319.
9. Napoleon, *Correspondance*, XXIV/341.
10. Clausewitz, 214; Voenskii, *Sviashchennoi Pamiati*, 146.
11. Wilson, *Invasion*, 356.
12. Voenskii, *Sviashchennoi Pamiati*, 143; Kutuzov, *Dokumenty*, 263–4; Gosudarstvenno-Istoricheskii Muzei, 236.
13. Beskrovny, *Polkovodets*, 272–3.
14. Altshuller & Tartakovskii, 97, 98, 121, 124.
15. Vitberg, 611; Shchukin, I/120; Alexander, *Corr avec sa soeur*, 103.
16. Ley, 54.
17. Garin, 133–4; Kutuzov, *Dokumenty*, 268–70; Shishkov, 168.
18. Kraehe, I/152; W.H. Löwenstern, I/359; Choiseul-Gouffier, 166.
19. Fredro, 46.
20. Choiseul-Gouffier, 147.
21. Ermolov, 134–5; Marchenko, 498; von Kurz, 220–1, 222–3; Minod, 50–2; recent excavations of mass graves in Vilnius have confirmed the cannibalism; Kurz, 223.

22. Maillard, 66–8; Venturini, 227.
23. Ducor, II/57–60; Chéron, 33.
24. Roy, 87; Holzhausen, 83–90.
25. Seruzier, 270ff.
26. La Flise, LXXIII/57; Roos, 197–200; Mitarevskii, 172–3; Holzhausen, 348.
27. Camp, 59.
28. Benard, 145, 146.
29. Holzhausen, 351.
30. Bellot de Kergorre, 112; Lignières, 138; Holzhausen, 323; Bréaut des Marlots, 33–4; Jackowski, I/312; Faré, 269; Noel, 180.
31. Bellot de Kergorre, 110; Bro, 126; Gardier, 96; Holzhausen, 323; Dumonceau, II/248; Bertrand, 179.
32. Seydlitz, 129, 167; *Akty Dokumenty i Materialy*, CXXXIII/329ff; Macdonald, 182; *Akty, Dokumenty i Materialy*, CXXXIII/410.
33. Seydlitz, 197ff; Macdonald, 184–8; *Akty, Dokumenty i Materialy*, CXXXIII/424.
34. Koźmian, II/315.
35. The figures given by early historians of the campaign and even those issued by official military sources on both sides are estimates based on the haziest of data. Meynier, for instance, reveals that most historians have been exaggerating wildly when talking about millions of dead during the Napoleonic wars. Labaume's estimate (437) of 20,000 recrossing the Niemen at Kovno were repeated by Buturlin (II/413) and largely stand, although Buturlin (II/446) maintains that another 60,000, mainly Austrians and Prussians, got out; Gourgaud's (494) fuller estimate errs on the high side, particularly in the case of the 36,000 who allegedly recrossed at Kovno, but his total of 127,000 is not far from present-day estimates. Original Russian figures for the number of prisoners taken varied from Buturlin's 193,000 to Chuikievich's 210,000, even though official Russian figures were only 136,000, and all recent studies have revealed this to be too high (see Sirotkin, in *Otechestvennaia Voina 1812 goda. Istochniki, etc.* [2000], 246ff). The only reliable figures on the deaths during this campaign are those of corpses buried by the Russian authorities on the one hand, and French census figures (Meynier, 21) on the other. Balashov reported that 430,707 human bodies were buried along the road in the spring of 1813. In December 1812 the Russian authorities counted 172,566 corpses and 128,739 animal carcases in the gubernia of Smolensk, 50,185 corpses and 17,050 carcases in the gubernia of Mogilev, 2,230 corpses and 7,355 carcases in the gubernia of Kaluga, a total of 224,981 corpses and 153,144 carcases (Hartley, *Russia in 1812*, 197, 413). But these figures can tell us little, as they do not specify whether the dead were military or civilians, let alone of what nationality. There are also

detailed figures for some units, but to extrapolate from these would be a meaningless exercise, since they vary so wildly. My own estimate is based on the figures given by Vilatte de Prugnes (285–7) and Meynier, those computed by Kukiel, who seems to me the most conscientious of historians of this campaign, on the one hand, and the estimates put forward by more recent Russian historians such as Zhilin, Sirotkin, Shvedov and Sokolov on the other.

36. Kukiel, *Wojna*, II/500; Hausman, 112; Zanoli, 205; Boppe, *Croatie Militaire*, 95, 120, 129.
37. Lignières, 139; Dumonceau, II/257–8; Chłapowski, 137.
38. Bertrand, 4–5; Combe, 178; Boppe, *Les Espagnols*, 157; Vlijmen, 327; Holzhausen, 340; see also: Fezensac, *Journal*, 188; Dupuy, 213; Kołaczkowski, I/167; Pion des Loches, 341; Vossen, 477.
39. *Lettres Interceptées*, 377; Holzhausen, 8.
40. Krasiński, 103.
41. Minod, Roederer; *Otechestvennaia Voina 1812g. Istochniki, etc.* (2001), 20.
42. Voenskii, *Sviashchennoi Pamiati*, 3.
43. Holzhausen, 356–8.

Chapter 25: The Legend

1. Maria Feodorovna, 136.
2. Kügelgen, 136; Bruun, 173.
3. Golitsuin, 5, 22, 23; Davidov, 56.
4. *1812 god v Vospominaniakh sovremennikov*, 3; Kallash, 209; F. Glinka, *Pisma Russkovo Ofitsera*, 154.
5. *1812 god v Vospominaniakh, perepiske i raskazakh*, 84, 86; *Otechestvennaia voina 1812 goda, Istochniki, etc.* (2001), 30ff; Hartley, *Russia in 1812*, 188; Iudin, III.
6. Adams, II/426.
7. Nesselrode, IV/118; Benckendorff, 70–1.
8. *1812 god v Vospominaniakh, perepiske i raskazakh*, 84, 85.
9. Beskrovny, *Narodnoe Opolchenie*, 377–86; Shishkin, 112–51; see also: Dzhivelegov et al., V/98–101; *1812 god v Vospominaniakh, perepiske i raskazakh*, 85; Shchukin, IV/156ff, IX/82–4.
10. Dzhivelegov et al., V/104.
11. Kallash, 224.
12. Davidov, 138.
13. Beyle, *Vie de Napoléon*, 233.
14. Louis Geoffroy, *Napoléon Apocryphe. Histoire de la Conquête du Monde et de la Monarchie Universelle*, Paris 1841.
15. O'Meara, II/107, 156, etc.

16. Dostoevskii, 250.
17. Raza, 220. Such relics are still to be found in collections in Poland.
18. Bourgogne, 281–2.

Sources

Primary sources

1812 god. Voennie Dnevniki, ed. A.G. Tartakovskii, Moscow 1990

1812 god. Vospominania voinov russkoi armii, Moscow 1991

1812 god v Vospominaniakh, Perepiske i Raskazakh Sovremennikov, Voenizdat, Moscow 2001

1812 god v Vospominaniakh sovremennikov, ed. A.G. Tartakovskii, Moscow 1994

Abbeel, Jef, *L'Odyssée d'un Carabinier à Cheval 1806–1815*, ed. René H. Willems, Brussels 1969

Adam, Albrecht, *Voyage Pittoresque et Militaire de Willemberg en Prusse jusqu'à Moscou, fait en 1812*, Munich 1828

Adam, Albrecht, *Aus dem Leben eines Schlachtenmalers*, Stuttgart 1886

Adams, John Quincy, *Memoirs of John Quincy Adams, comprising portions of his diary from 1795 to 1848*, Vol. II, Philadelphia 1874

Aglaimov, S.P., *Otechestvennaia Voina 1812 Goda. Istoricheskie Materialy Leib-Gvardii Semeonovskavo Polka*, Poltava 1912

Akty, Dokumenty i Materialy dlia politicheskoi i bytovoi istorii 1812 goda, ed. K. Voenskii, in *Sbornik Imperatorskavo Russkavo Istoricheskavo Obshchestva*, Vols CXXXIII–IX, St Petersburg 1909–12

Alexander I, *Correspondance Inédite de l'Empereur Alexandre et de Bernadotte pendant l'année 1812*, Paris 1909

Alexander I, *Correspondance de l'Empereur Alexandre Ier avec sa soeur la Grande-Duchesse Catherine*, St Petersburg 1910

Alexander I, *Uchebnia knigi i tetradi velikavo kniazia Aleksandra Pavlovicha*, ed. M.I. Bogdanovich, in *Sbornik Imperatorskavo Russkavo Istoricheskavo Obshchestva*, Vol. I, St Petersburg 1867

Altshuller, R.E., & Bogdanov, G.V. (eds), *Borodino. Dokumenty, Pisma, Vospominania*, Moscow 1962

Altshuller, R.E., & Tartakovskii, A.G. (eds), *Listovki Otechestvennoi Voiny 1812 goda*, Moscow 1962

Andreev, N.I., *Iz Vospominanii Nikolaia Ivanovicha Andreeva*, in *Russkii Arkhiv*, Moscow 1879

Arndt, Ernst Moritz, *Erinnerungen aus dem Ausseren Leben*, Leipzig 1840
Aubry, Thomas Joseph, *Souvenir du 12e Chasseurs*, Paris 1889
Augusta, Duchess of Saxe-Coburg-Saalfeld, *In Napoleonic Days*, London 1941
Bagration, Petr Ivanovich, *General Bagration. Sbornik dokumentov i materialov*, Moscow 1945
Bakunina, Varvara Ivanovna, *Dvienadsaty God*, in *Russkaia Starina*, Vol. XLVII, St Petersburg 1885, pp. 392–410
Bangofsky, Georges, *Les Étapes de Georges Bangofsky, Officier Lorrain*, Paris 1905
Barclay de Tolly, Mikhail Bogdanovich, *Doniesenia Barclaya Aleksandru I i pisma tsaria posle ostavienia tsarem armii v 1812 godu*, in *Voenny Sbornik*, November 1903–September 1904
Barclay de Tolly, Mikhail Bogdanovich, *Tableau des Opérations militaires de la Première Armée*, in *Trudy Imperatorskavo Russkavo Voenno-Istoricheskavo Obshchestva*, Vol. VI, St Petersburg 1912
Barrau, J.P., *Jean-Pierre Armand Barrau, Quartier-Maître au IVe Corps de la Grande Armée, sur la Campagne de Russie*, in *Rivista Italiana di Studi Napoleonici*, No 1, Anno XVI, Pisa 1979
Bartenev, P.I., *Razgovor sc A.P. Ermolovim*, in *Russkii Arkhiv*, Moscow 1863
Baudus, Lieutenant-Colonel K. de, *Études sur Napoléon*, 2 vols, Paris 1841
Bausset, L.F.J. de, *Mémoires Anecdotiques sur l'Intérieur du Palais et sur quelques évènements de l'Empire, depuis 1805 jusqu'au 1er mai 1814*, 2 vols, Brussels 1827
Bawr, Madame de, *Mes Souvenirs*, Paris 1853
Beauharnais, Eugène de, *Mémoires et correspondance politique et militaire du Prince Eugène, annotés et mis en ordre par A. Du Casse*, 10 vols, Paris 1858–60
Beaulieu, Capitaine Drujon de, *Souvenirs d'un militaire pendant quelques années du règne de Napoléon Bonaparte*, Verpillon 1831
Begos, Louis, *Souvenirs des Campagnes du Lieutenant-Colonel Louis Begos*, in *Soldats Suisses au Service Étranger*, Geneva 1909
Belliard, Augustin Daniel, *Mémoires du Comte Belliard, Lieutenant-Général, Pair de France*, 3 vols, Paris 1842
Bellot de Kergorre, A., *Un Commissaire des Guerres sous le Permier Empire. Journal de Bellot de Kergorre*, Paris 1899
Benard, Sergeant, *Souvenirs de 1812. Un Prisonnier Français en Russie*, in *La Giberne*, July 1906
Benkendorf, Aleksandr, *Zapiski Benkendorfa*, Moscow 2001
Bennigsen, General Count Lev, *Mémoires du Général Bennigsen*, Vol. III, Paris 1908
Bennigsen, General Count Lev, *Zapiski grafa L.L. Bennigsena o kampanii 1812 goda*, in *Russkaia Starina*, July–September 1909
Bennigsen, General Count Lev, *Voenny soviet v Filiakh*, in *Voenny Sbornik*, 1903, No. 1

Berthézène, Baron Pierre, *Souvenirs Militaires de la République et de l'Empire*, 2 vols, Paris 1855
Bertin, Georges, *La Campagne de 1812 d'après les témoins oculaires*, Paris 1895
Bertolini, Bartolomeo, *Il Valore Vinto dagli Elementi. Storica narrazione della campagnia di Russia degli anni 1812–1813*, 2 vols, Milan 1869
Bertrand, Vincent, *Mémoires du Capitaine Bertrand*, Angers 1909
Beskrovny, L.G. (ed.), *Borodino. Dokumenty, Pisma, Vospominania*, Moscow 1962
Beskrovny, L.G. (ed.), *Narodnoe Opolchenie v Otechestvennoi Voinie 1812 goda*, Moscow 1962
Bestuzhev-Riumin, Aleksei Dmitrievich, *Zapiski*, in *Russkii Arkhiv*, Moscow 1896
Bestuzhev-Riumin, Aleksei Dmitrievich, *1812 god. Kratkie opisanie proischestviam v stolitse Moskve v 1812 godu*, in *Russkii Arkhiv*, Kniga 2, Moscow 1910
Beulay, Honoré, *Mémoires d'un Grenadier de la Grande Armée*, Paris 1907
Beyle, Henri (Stendhal), *Mémoires sur Napoléon*, Paris 1929
Beyle, Henri (Stendhal), *Vie de Napoléon*, Paris 1929
Beyle, Henri (Stendhal), *Journal*, Vol. IV, Paris 1934
Beyle, Henri (Stendhal), *La Vie de Henri Brulard*, Paris 1982
Beyle, Henri (Stendhal), *Lettres*, in *Campagnes de Russie. Sur les traces de Henri Beyle dit Stendhal*, Paris 1995
Beyle, Henri (Stendhal), *Correspondance Générale*, Vol. II, Paris 1998
Bigarré, Auguste, *Mémoires du Général Bigarré, Aide de Camp du Roi Joseph, 1775–1813*, Paris n.d.
Bignon, Édouard, *Histoire de la France depuis le 18 brumaire jusqu'à la seconde abdication*, Paris 1830
Bignon, Édouard, *Souvenirs d'un Diplomate*, Paris 1864
Biot, Hubert François, *Souvenirs Anecdotiques et Militaires du Colonel Biot*, Paris 1901
Bismarck, F.W., *Mémoires sur la Campagne de Russie*, Paris 1998
Blaze de Bury, E., *La Vie Militaire sous l'Empire*, 2 vols, Paris 1837
Blocqueville, Marie-Adelaide Marquise de, *Le Maréchal Davout, Prince d'Eckmühl, raconté par les siens et par lui-même*, 4 vols, Paris 1880
Bonnet, Guillaume, *Journal du Capitaine Bonnet du 18e de Ligne*, Paris 1997
Bonneval, Armand de, *Mémoires Anecdotiques du Général Marquis de Bonneval*, Paris 1900
Borcke, Johann von, *Kriegerleben des Johann von Borcke, wieland Kgl. preuss. Oberstlieutnants 1806–1815*, Berlin 1888
Boulart, Jean François, *Mémoires Militaires du Général Boulart sur les guerres de la République et de l'Empire*, Paris n.d.

Bourgeois, René, *Tableau de la Campagne de Moscou en 1812*, Paris 1814
Bourgogne, Adrien, *Mémoires du Sergeant Bourgogne (1812–1813), publiées d'après le manuscrit original par Paul Cottin et Maurice Henault*, Paris 1901
Bourgoing, Baron Paul de, *Souvenirs Militaires 1791–1815*, Paris 1897
Bourrienne, Louis Antoine, *Mémoires de M. de Bourienne Ministre d'État sous Napoléon*, 10 vols, Paris 1829
Bozhanov, I.S., *1812 god. Raskaz sviashchennika Uspenskavo Sobora I.S. Bozhanova*, in *Russkii Arkhiv*, Kniga 3, Moscow 1899
Brandt, Heinrich von, *Souvenirs d'un Officier Polonais. Scènes de la vie militaire en Espagne et en Russie (1808–1812)*, Paris 1877
Bréaut des Marlots, Jean, *1812. Lettre d'un Capitaine des Cuirassiers sur la campagne de Russie*, Poitiers 1885
Breton, Auguste, *Lettres de ma cativité en Russie*, in *Mémoires et Lettres de Soldats Français*, Paris 1999
Brett-James, Antony, *1812. Eyewitness Accounts of Napoleon's Defeat in Russia*, New York 1966
Bro, Louis, *Mémoires du Général Bro 1796–1844*, Paris 1914
Broughton, John Cam Hobhouse, Lord, *Recollections of a Long Life*, Vol. II, London 1909
Bulgakov, A. Ya., *Vospominania o 1812 gode i viechiernykh biesiedakh u grafa Fiodora Vasilievicha Rostopchina*, Moscow 1904
Les Bulletins de la Grande Armée, 6 vols, Paris 1841–44
Bussy, Jean Marie, *Notes*, in *Soldats Suisses au Service Étranger*, Geneva 1913
Butenev, A.P., *Vospominania*, in *Russkii Arkhiv*, Kniga 3, 1881; Kniga 1, 1883
Butkevicius, *Napoléon en Lithuanie 1812, d'après des documents inédits*, trs. René Martel, in *Revue de Paris*, August 1932
Buturlin (Boutourlin), Dmitri Petrovich, *Histoire Militaire de la Campagne de Russie en 1812*, 2 vols, Paris 1824
Byl li u nas plan voennykh dieistvii v 1812 godu. Pismo Buturlina k generalu Jomini, in *Voenny Sbornik*, St Petersburg February 1902
Calosso, Colonel, *Mémoires d'un vieux soldat*, Turin 1857
Castellane, Boniface de, *Journal du Maréchal de Castellane 1804–1862*, Vol. I, Paris 1895
Catherine de Wurtemberg, Princesse Jerôme, *Correspondance Inédite*, Paris 1893
Caulaincourt, Armand Augustin Louis, Duc de Vicence, *Mémoires*, 3 vols, Paris 1933
Chambray, Georges de, *Oeuvres du Marquis de Chambray, Maréchal de Camp d'Artillerie. Histoire de l'Expédition de Russie*, 3 vols, Paris 1839
Chamski, Tadeusz Józef, *Opis krótki lat upłynionych*, Warsaw 1989
Chapelle, Colonel A., *Perepravа cherez reku Berezinu v 1812 goda po zapisam*

polkovnikov frantsuzkoi sluzhby Chapelle i Paulin, in *Voenny Sbornik*, Vol. 241, No.5, St Petersburg May 1898
Chernyshev, A.I., *Doniesenia polkovnika A.I. Chernysheva imperatoru Aleksandru Pavlovichu; Doniesenia polkovnika A.I.Chernysheva kantsleru grafu N.P. Rumiantsevu; Pisma A.I. Chernysheva k kantsleru grafu N.P. Rumiantsevu v 1809 godu*, in *Sbornik Imperatorskavo Russkavo Istoricheskavo Obshchestva*, Vol. XXI, St Petersburg 1877
Chéron, Alexandre de, *Mémoires Inédits sur la Campagne de Russie*, Paris 2001
Chevalier, Jean-Michel, *Souvenirs des guerres napoleoniennes*, Paris 1970
Chichagov, P.V., *Mémoires Indédites de l'Amiral Tchitchagoff*, Berlin 1858
Chichagov, P.V., *Pisma Admirala Chichagova k Imperatoru Aleksandru I*, in *Sbornik Imperatorskavo Russkavo Istoricheskavo Obshchestva*, Vol. VI, St Petersburg 1871
Chicherin, Aleksandr V., *Dnevnik*, Moscow 1966
Chłapowski, Dezydery, *Pamiętniki*, Poznań 1899
Choiseul-Gouffier, Comtesse de, *Reminiscences sur l'empereur Alexandre Ier et sur l'empereur Napoléon Ier*, Paris 1862
Chuikievich, Colonel, *Reflections of the War of 1812*, Boston 1813
Chuquet, Arthur (ed.), *Lettres de 1812*, Paris 1911
Clausewitz, Carl von, *The Campaign of 1812*, London 1992
Clemenso, Hyacinthe, *Souvenirs d'un Officier Valaisain au service de France*, Paris 1999
Coignet, Jean-Roch, *Les Cahiers du Capitaine Coignet (1799–1815)*, Paris 1883
Combe, Julien, *Mémoires du Colonel Combe sur les campagnes de Russie 1812, de Saxe 1813 et de France 1814 et 1815*, Paris 1853
Comeau de Charry, Sébastien-Joseph, Baron de, *Souvenirs des Guerres d'Allemagne pendant la Révolution et l'Empire*, Paris 1900
Confalonieri, Federico, *Carteggio del Conte Federico Confalonieri*, ed. Giuseppe Calavresi, Milan 1910
Constant (Louis Constant Wairy), *Mémoires de Constant, Premier Valet de Chambre de l'Empereur, sur la vie privée de Napoléon, sa famille et sa cour*, Vol. V, Paris 1830
Corbineau, Jean-Baptiste, *Passage de la Bérésina*, in *Le Spectateur Militaire*, Vol. III, Paris 1827
Crossard, Jean-Baptiste, *Mémoires militaires et historiques*, 6 vols, Paris 1829
Curely, Jean-Nicolas, *Le Général Curely. Itinéraire d'un cavalier léger de la Grande Armée 1793–1815*, Paris 1887
Czaplic, General Evfemii (Joachim), *Otechestvennaia Voina v Raskazakh Generala Chaplitsa*, in *Russkaia Starina*, Vol. 50, No. 6, 1886
Czartoryski, Adam Jerzy, *Mémoires du Prince Adam Czartoryski et Correspondance avec l'Empereur Alexandre Ier*, Vol. II, Paris 1887

Damas, Ange Hyacinthe de, *Mémoires du Baron de Damas*, 2 vols, Paris 1922
Damas, Roger de, *Mémoires du comte Roger de Damas*, 2 vols, Paris 1914
Davidov, Denis, *In the Service of the Tsar against Napoleon. The Memoirs of Denis Davidov, 1806–1814*, trs. Gregory Troubetskoy, London 1999
Davout, Louis Nicolas, *Correspondance du Maréchal Davout Prince d'Eckmühl*, 3 vols, Paris 1885
Déchy, Édouard, *Souvenirs d'un Ancien Militaire*, Paris 1860
Dedem van der Gelder, Baron Antoine-Baudouin Gisbert van, *Mémoires du Général Baron de Dedem de Gelder, 1774–1825*, Paris 1900
Dembiński, Henryk, *Pamiętnik Henryka Dembińskiego generała wojsk polskich*, 2 vols, Warsaw 1910
Denniée, P.P., *Itinéraire de l'Empereur Napoléon pendant la campagne de 1812*, Paris 1842
Divov, N.A., *Iz Vospominanii N.A. Divova*, in *Russkii Arkhiv*, Kniga 2, No. 7, 1873
Dokhturov, D.S., *Pisma D.S. Dokhturova k evo Supruge*, in *Russkii Arkhiv*, Vol. XII, Kniga 1, 1874
Dokumenty i Materialy, otnosiashchiesia k sobytiam 1812 goda v Litve i Zapadnykh Guberniakh, sobrannie mestnimi gubernatorami, in *Sbornik Imperatorskavo Russkavo Istoricheskavo Obshchestva*, Vol. CXXVIII, St Petersburg 1909
Dolgov, S.O. (ed.), *Sto let nazad. Pisma I. P. Odentalia k A. Ia. Bulgakovu o Peterburgskich novostiach i slukhakh*, in *Russkaia Starina*, July–November 1912
Domergue, A., *La Russie pendant les guerres de l'Empire*, Paris 1835
Dostoevskii, F.M., *Prestuplenie i Nakazanie*, St Petersburg 1884
Dubrovin, N. (ed.), *Otechestvennaia Voina v Pismakh Sovremennikov*, in *Zapiski Imperatorskoi Akademii Nauk*, Vol. 43, St Petersburg 1882
Ducor, Henri, *Aventures d'un marin de la Garde Impériale*, 2 vols, Paris 1833
Dumas, Mathieu, *Souveneirs du Lieutenant-Général Comte Mathieu Dumas de 1770 à 1836*, 3 vols, Paris 1839
Dumonceau, François, *Mémoires du Général Comte François Dumonceau*, Vol. II, Brussels 1960
Dupuy, Victor, *Souvenirs Militaires*, Paris 1892
Durova, Nadezhda, *The Cavalry Maiden. Journal of a Russian Officer in the Napoleonic Wars*, trs. Mary Fleming Zirin, Bloomington 1988
Dutheillet de la Mothe, Aubin, *Mémoires du lieutenant-colonel Aubin Dutheillet de la Mothe, 6 octobre 179–16 juin 1856*, Brussels 1899
Duverger, B.T., *Mes aventures dans la campagne de Russie*, Paris n.d.
Dziewanowski, Dominik, *Przyczyny nięszczęśliwie zakończonej kampanii 1812r*, in Janusz Staszewski (ed.), *Pamiętniki o wojnie 1812 r.*, Poznań 1933
Edling, Comtesse Roxanne, *Mémoires de la Comtesse Edling née Stourdza*, Moscow 1888

Ermolov, Aleksei Petrovich, *Zapiski Generala Ermolova v Otechestvennuiu Voinu 1812g*, in *Russkaia Starina*, June–September 1912
Espinchal, Hippolyte, Marquis d', *Souvenirs Militaires, 1792–1814*, 2 vols, Paris 1902
Everts, Henri Pierre, *Campagne et Captivité de Russie (1812–1813)*, Paris 1997
Evreinov, M.M., *Pamiat' o 1812 gode*, in *Russkii Arkhiv*, Vol. II, Kniga 1, Moscow 1874
Faber du Faur, G. de, *Blaetter aus meinem portefeuille, im Laufe des Feldzugs 1812 in Russland*, Stuttgart 1831–43
Faber du Faur, G. de, & Kausler, F. von, *La Campagne de Russie 1812*, Paris 1895
Fain, Agathon Jean François, *Mémoires du Baron Fain*, Paris 1908
Fain, Agathon Jean François, *Manuscrit de Mil Huit Cent Douze*, 2 vols, Paris 1827
Fairon, Émile, & Heuse, Henri, *Lettres de Grognards*, Liège 1936
Falkowski, Juliusz, *Obrazy z życia kilku ostatnich pokoleń w Polsce*, 5 vols, Poznań 1886
Fanneau de la Horie, Michel, *Notes sur la Campagne de Russie en 1812*, Paris 1906
Fantin des Odoards, General Louis Florimond, *Journal du Général Fantin des Odoards. Étapes d'un Officier de la Grande Armée*, Paris 1895
Faré, Charles A., *Lettres d'un Jeune Officier a sa Mère 1803–1814*, Paris 1889
Faure, Raymond, *Souvenirs du Nord, ou la Guerre, La Russie, et les Russes ou l'Esclavage*, Paris 1821
Fedorov-Davidov, A.A., *Otechestvennaia Voina*, Moscow 1919
Fezensac, R.E.P.J., Duc de, *Journal de la Campagne de Russie en 1812*, Paris 1850
Fezensac, R.E.P.J., Duc de, *Souvenirs Militaires de 1804 a 1814*, Paris 1863
Fonvizin, Mikhail, *Zapiski Fonvizina ochevidtsa smutnykh vremen tsarstvovanii Pavla I, Aleksandra I, Nikolaia I*, Leipzig 1860
Fouché, Joseph, *Mémoires de Joseph Fouché, duc d'Otrante*, 2 vols, Paris 1824
François, Charles, *Journal du Capitaine François (dit le dromadaire d'Égypte) 1792–1830*, Vol. II, Paris 1904
Frantsuzy v Rossii, 1812 god po Vospominaniam sovremennikov-inostrantsev, 3 vols, Moscow 1912
Fredro, Aleksander, *Trzy po Trzy. Pamiętniki*, Kraków 1949
Freytag, J.D., *Mémoires*, Paris 1824
Funck, Karl Wilhelm Ferdinand von, *In the Wake of Napoleon. Memoirs 1807–1809*, trs. Oakley Williams, London 1931
Fusil, Louise, *Souvenirs d'une Actrice*, Brussels 1841
Fusil, Louise, *Souvenirs d'une Femme sur la retraite de Russie*, Paris 1910
Gajewski, Franciszek, *Pamiętniki pułkownika wojsk polskich*, 2 vols, Poznań n.d.

Ganniers, Arthur de, *La Campagne de Russie. De Paris à Vilna en 1812 d'après la correspondance inédite d'un aide-major de la Grande Armée (Socrate Blanc)*, in *Revue des Questions Historiques*, Vol. LXII, Paris 1897
Gardier, Louis, *Journal de la Campagne de Russie*, Paris 1999
Garin, F.A. (ed.), *Izgnanie Napoleona iz Moskvy. Sbornik*, Moscow 1938
Garting, General, *Iz Dnevnika Generala Gartinga*, in *Trudy Imperatorskavo Russkavo Voenno-Istoricheskavo Obshchestva*, Vol. VI, St Petersburg 1912
Geoffroy, Louis, *Napoléon Apocryphe, 1812–1832. Histoire de la Conquête du Monde et de la Monarchie Universelle*, Paris 1841
Gerasimov, P.F., *Raskaz o dvenatsatom godu*, Moscow 1882
Giesse, Friedrich, *Kassel–Moskau–Küstrin. Tagebuch wahrend des russischen Feldzuges*, Leipzig 1912
Ginisty, Paul (ed.), *Mémoires d'Anonymes et d'Inconnus*, Paris 1907
Girard, Just, *L'Enfant de troupe. Souvenirs écrits sous la dictée d'un vieil invalide*, Tours 1858
Girod de l'Ain, General Baron, *Dix ans de mes souvenirs militaires*, Paris 1873
Glinka, Fyodor, *Podwigi grafa Mikhaila Andreevicha Miloradovicha w Otechestvennuiu Voinu 1812 goda*, Moscow 1814
Glinka, Fyodor, *Pisma Russkavo Ofitsera*, 8 vols, Moscow 1815–16
Glinka, Fyodor, *Ocherki Borodinskovo Srazhenia*, Moscow 1839
Glinka, Sergei Nikolaevich, *Zapiski*, St Petersburg 1895
Glinka, Sergei Nikolaevich, *Podvigi Dobrodeteli i Slavy Russkikh v Otechestvenuiu i Zagranichnuiu Voinu*, Moscow 1816
Glinka, Sergei Nikolaevich, *Zapiski o 1812 g*, St Petersburg 1836
Glinka, Vladimir, *Maloyaroslavets v 1812 godu, gdie reshilas' sud'ba Bolshoi Armii Napoleona*, St Petersburg 1842
Godart, Baron Roch, *Mémoires du Général Baron Roch Godart (1792–1815)*, Paris 1895
Goethe, Theodor Daniel, *Ein Verwandter Goethes im Russische Feldzuge 1812*, Berlin 1912
Golitsuin, Nikolai Borisovich, *Ofitserskia Zapiski, ili Vospominania o Pokhodakh 1812, 1813 i 1814 godov*, Moscow 1838
Golovin, Varvara Nikolaevna, *Souvenirs de la comtesse Golovine*, Paris 1910
Golubov, S.N., & Kuznetsov, F.E., *General Bagration. Sbornik Dokumentov i Materialov*, Moscow 1945
Gosudarstvenno Istoricheskii Muzei, *1812–1814*, Moscow 1992
Gourgaud, Gaspard, *Napoléon et la Grande Armée en Russie, ou Examen Critique de l'ouvrage de M. le Comte de Ségur*, Paris 1825
Gouvion Saint-Cyr, Maréchal, *Mémoires pour servir à l'histoire militaire sous le Directoire, le Consulat et l'Empire*, Vol. III, Paris 1831
Grabbe, P.Kh., *Iz Pamiatnikha i Zapisok*, in *Russkii Arkhiv*, No. 5, 1873

Grabowski, Józef, *Pamiętniki Wojskowe Józefa Grabowskiego, oficera sztabu cesarza Napoleona I*, Warsaw 1905
Grech, N.I., *Zapiski moiei zhizni*, St Petersburg 1886
Griois, Lubin, *Mémoires du Général Griois, 1792–1822*, Vol. II, Paris 1909
Grouchy, Émmanuel, Marquis de, *Mémoires du Maréchal de Grouchy*, Vol. III, Paris 1873
Grüber, Carl-Johann, Ritter von, *Souvenirs*, Paris 1909
Guitard, Joseph, *Souvenirs Militaires du Premier Empire*, Paris 1934
Hausmann, Franz Joseph, *A Soldier for Napoleon. The Campaigns of Lieutenant Franz Joseph Hausmann, 7th Bavarian Infantry*, trs. Cynthia Joy Hausmann, London 1998
Hauterive, N., *De l'état de la France à la fin de l'an VIII*, Paris 1801
Henckens, Lieutenant J.L., *Mémoires se rapportant a son service militaire au 6me régiment de Chasseurs à Cheval français de février 1803 à août 1816*, The Hague 1910
Herold, Jean Christopher, *The Mind of Napoleon. A Selection from his Written and Spoken Words*, New York 1961
Hochberg, Count, *La Campagne de 1812. Mémoires du Margrave de Bade*, trs. Arthur Chuquet, Paris 1912
Hogendorp, Dirk van, *Mémoires du Général Dirk van Hogendorp, Comte de l'Empire*, The Hague 1887
Holzhausen, Paul, *Les Allemands en Russie avec la Grande Armée, 1812*, trs. Commandant Minart, Paris 1914
Hortense, Queen of Holland, *Mémoires de la Reine Hortense*, 3 vols, Paris 1927
Jackowski, Michał, *Pamiętniki*, in *Pamiętniki Polskie*, ed. Ksawery Bronikowski, Vol. I, Przemyśl 1883
Jacquemont du Donjon, Porphyre, *Carnet de route d'un Officier d'Artillerie (1812–1813)*, in *Souvenirs et Mémoires*, Paris 1899
Jérôme Bonaparte, King of Westphalia, *Mémoires et Correspondance du Roi Jérôme et de la Reine Catherine*, Vols V & VI, Paris 1864
Jomini, A.H., Baron de, *Vie Politique et Militaire de Napoléon, racontée par lui-même, au tribunal de César, d'Alexandre et de Frédéric*, 2 vols, Brussels 1842
Jomini, A.H., Baron de, *Précis politique et militaire des campagnes de 1812 à 1814*, 2 vols, Lausanne 1886
Jourdain, Armand, *Trente-neuf jours de réclusion dans les prisons de Vilna*, 1858
Kallash, V.V. (ed.), *Dvenadtsaty God v Vospominaniakh i perepiske sovremennikov*, Moscow 1912
Kharkievich, V.I. (ed.), *Barclay de Tolly v Otechestvennoi Voinie: Perepiska Imperatora Aleksandra i Barclaya de Tolly*, St Petersburg 1904
Kharkievich, V.I. (ed.), *Nastavlenie gospodam pekhotnom ofitseram v den srazhenia*, in *Voennii Sbornik*, No. 7, St Petersburg July 1902

Kharkievich, V.I. (ed.), *1812 god v Dnevnikakh, Zapiskakh i Vospominaniakh sovremennikov. Materialy Voenno-Uchennavo Arkhiva Glavnavo Shtaba*, Vilna 1900

Khomutova, A.G., *Vospominania A.G. Khomutovoi o Moskve v 1812 godu*, in *Russkii Arkhiv*, No. 11, 1891

Kicheev, P.G., *Vospominania o prebyvaniu nepriatelia v Moskve v 1812 godu*, Moscow 1868

Kołaczkowski, K., *Wspomnienia Generała Klemensa Kołaczkowskiego*, Vol. I, Kraków 1898

Kologrivova, A.F., *1812 god. Iz semeinykh vospominanii*, in *Russkii Arkhiv*, Kniga 2, 1886

Konshin, M.N., *Iz zapisok N.M. Konshina*, in *Istoricheskii Viestnik*, No. 8, Vol. 17, 1884

Korkozevich, Mikhail, *Vospominania o Nashestvii Frantsuzov na Rossiu v 1812 godu*, in *Viestnik Zapadnoi Rossii*, 1869, Kniga 12

Kozlovsky, G.A., *Zapiski*, in *Russkaia Starina*, Vol. 65, 1890

Koźmian, Kajetan, *Pamiętniki*, Vol. II, Wrocław 1972

Krasiński, Józef, *Pamiętniki od roku 1790–1831*, Poznań 1877

Kügelgen, Wilhelm von, *Jugenderinnerungen eines alten Mannes*, Berlin 1870

Kurakin, A.B., *Doniesenia imperatoru Aleksandru Pavlovichu kniazia A.B. Kurakina; Doniesenia i Pisma kniazia A.B. Kurakina kantsleru N.P. Rumiantsevu*, in *Sbornik Imperatorskavo Russkavo Istoricheskavo Obshchestva*, Vol. XXI, St Petersburg 1877

Kurz, Hauptmann von, *Der Feldzug von 1812. Denkwürdigkeiten eines württembergischen Offiziers*, Leipzig 1912

Kutuzov, M.I., *Pisma, Zapiski*, Moscow 1989

Kutuzov, M.I., *Dokumenty, Dnevniki, Vospominania*, ed. A.M. Valkovich & A.P. Kapitonov, Moscow 1995

Labaume, Eugène, *Relation Complète de la Campagne de Russie en 1812*, Paris 1816

La Flise, N.D. de, *Pokhod Velikoi Armii v Rossiiu v 1812g; Zapiski de la Fliza*, in *Russkaia Starina*, Vols LXXI, LXXII, LXXIII, July 1891–March 1892

Lagneau, L.V., *Journal d'un Chirurgien de la Grande Armée 1803–1815*, Paris 1913

La Guerre Nationale de 1812. Publication du Comité Scientifique de l'état-major russe. Traduction du capitaine de génie breveté E. Cazalas, sous la direction de la section historique de l'état-major de l'armée, 7 vols, Paris 1905–11

Langeron, L.A., Comte de, *Mémoires de Langeron, Général d'Infanterie dans l'Armée Russe*, Paris 1902

Larrey, D.J., Baron, *Mémoires de Chirurgie Militaire et Campagnes du Baron D.J. Larrey*, 4 vols, Paris 1817

Las Cases, Émmanuel Auguste Dieudonné, *Mémorial de Sainte-Hélène. Journal de la vie privée et des conversations de l'Empereur Napoléon à Sainte-Hélène*, 4 vols, London 1823

Lassus-Marcilly, François, *Notes sur ma campagne de Russie*, in *Mémoires et Lettres de Soldats Français*, Paris 1999

Laugier, Cesare de Bellecour de, *Récits de César de Laugier, officier de la garde du Prince Eugène*, trs. Henri Lionnet, Paris 1912

Laugier, Cesare de Bellecour de, *Gli Italiani in Russia*, Milano 1980

Lazhechnikov, I.I., *Novobranets 1812 goda, Polnoe Sobranie Sochinienii*, Vol. I, St Petersburg 1901

Lecointe de Laveau, G., *Moscou, Avant et Après l'Incendie, par un témoin oculaire*, Paris 1814

Lecoq, Adjutant, *Journal d'un Grenadier de la Garde*, in *Revue de Paris*, Tome V, Sept–Oct 1911

Legler, Thomas, *Souvenirs*, Neuchâtel 1942

Lejeune, Louis François, *Souvenirs d'un officier de l'Empire*, 2 vols, Toulouse 1831

Lejeune, Louis François, *Mémoires du Général Lejeune*, Vol. II, *En Prison et en Guerre*, Paris 1895

Leontiev, K.N., *Raskaz Smolenskavo Diakona o Nashestvii 1812 goda*, in *Russkii Arkhiv*, Vol. XIX, Kniga 3, 1881

Le Roy, Claude-François-Madeleine, *Souvenirs*, Dijon 1914

Leslie, A.A., *Raskazy o 1812 gode*, in *Smolenskaia Starina*, Vypusk 2, 1912

Lettres Interceptées par les Russes pendant la campagne de 1812, ed. Leon Hennet & Emmanuel Martin, Paris 1913

Lignières, Marie Henry, Comte de, *Souvenirs de la Grande Armée et de la Vieille Garde Impériale*, Paris 1933

Liprandi, I.P., *Materialy dlia otechestvennoi voiny 1812 goda*, St Petersburg 1867

Lossberg, General Friedrich Wilhelm von, *Briefe in die Heimath geschrieben wahrend des Feldzuges 1812 in Russland*, Cassel 1844

Löwenstern, E. von, *Mit graf Pahlens reiterei gegen Napoleon*, Berlin 1910

Löwenstern, Woldemar Hermann von, *Mémoires du Général-Major Russe Baron de Lowenstern*, Vol. I, Paris 1903

Lubenkov, N., *Raskaz artilerista o dele Borodinskom*, St Petersburg 1837

Lyautey, Hubert, *Lettres d'un Lieutenant de la Grande Armée*, ed. Pierre Lyautey, in *Revue des Deux Mondes*, 15 December 1962, 1 & 15 January 1963

Macdonald, Jacques-Étienne, *Souvenirs du Maréchal Macdonald, Duc de Tarente*, Paris 1892

Maevskii, S., *Moi Viekh, ili istoria generala Maevskavo*, in *Russkaia Starina*, 1873, No. 7–8

Maillard, J.P., *Mémoires*, in *Soldats Suisses au Service Étranger*, Geneva 1913

Mailly, Adrien, Comte de, *Mon journal pendant la campagne de Russie, écrit de mémoire après mon retour à Paris*, Paris 1841

Maistre, Joseph de, *Correspondance Diplomatique*, Vol. I, Paris 1860
Marbot, Antoine-Marcelin, Baron de, *Mémoires du Général Baron de Marbot*, Vol. III, Paris 1891
Marchenko, Vassili Romanovich, *Avtobiograficheskaia Zapiska*, in *Russkaia Starina*, March 1896, pp. 471–505
Maret, Hugues Bernard, duc de Bassano, *Souvenirs intimes de la Révolution et de l'Empire*, 2 vols, Brussels 1843
Maria Feodorovna, Empress of Russia, *Correspondance de sa majesté l'impératrice Marie Feodorovna avec Mademoiselle de Nelidoff, sa demoiselle d'honneur*, Paris 1896
Martens, Carl von, *Denkwurdigkeiten aus dem Leben eines alten Offiziers. Ein Beitrag zur Geschichte des letzten vierzig Jahre*, Dresden–Leipzig 1848
Martens, Christian Septimus von, *Vor funfzig Jahren. Tagebuch meines Feldzugs in Russland, 1812*, Stuttgart 1862
Martens, F.F. (ed.), *Recueil de traités et conventions conclus par la Russie avec les puissances étrangères*, Vols XIII–XV, St Petersburg 1909
Martin, Jean-Baptiste, *Lettres*, in *Mémoires et Lettres de Soldats Français*, Paris 1999
Martos, Inzhinernii Ofitser, *Zapiski*, in *Russkii Arkhiv*, Vol. 8, 1893
Materre, Jean-Baptiste, *Le Général Materre (1772–1843) d'après ses souvenirs inédits*, ed. Georges Bertin, Tulle 1906
Mayer, P.L., *Mémoires d'un Soldat Prisonnier en Russie*, in *Soldats Suisses au Service Étranger*, Geneva 1908
Meerheimb, Franz Ludwig August von, *Erlebnisse eines Veteranen der grossen Armee wahrend des Feldzugs in Russland 1812*, Dresden 1860
Méjan, Étienne, *Lettres du Comte Méjan sur la Campagne de Russie*, in *Miscellanea Napoleonica*, Serie II, Rome 1896
Méneval, Claude-François, *Mémoires pour servir à l'Histoire de Napoléon Ier depuis 1802 jusqu'à 1815*, 3 vols, Paris 1894
Metternich, Klemens Lothar Wenzel, Fürst von, *Mémoires, Documents et Écrits divers laissés par le Prince de Metternich*, 2 vols, Paris 1880
Minod, Charles-François, *Journal des Campagnes et Blessures*, in *Mémoires et Lettres de Soldats Français*, Paris 1999
Mitarevskii, N.E., *Vospominania o Voinie 1812 goda*, Moscow 1871
Montesquiou-Fezensac, Anatole de, *Souvenirs sur la Révolution, l'Empire, la Réstauration et le règne de Louis-Philippe*, Paris 1961
Montigny, L, *Souvenirs anecdotiques d'un Officier de la Grande Armée*, Paris 1833
Müffling, Baron Carl von, *The Memoirs of Baron Carl von Muffling*, London 1997
Muralt, Albert de, *Souvenirs de la Campagne de Russie de 1812*, Neuchâtel 1942
Muravev, A.N., *Avtobiograficheskie Zapiski*, in *Dekabristy, Novie Materiali*, Moscow 1955

Muravev-Apostol, M.I., *Vospominania i Pisma*, Moscow 1922
Napoleon I, *Correspondance de Napoléon Ier*, Vols XXIII & XXIV, Paris 1868
Napoleon I, *Supplément à la Correspondance*, Paris 1887
Napoleon I, *Lettres Inédites de Napoléon Ier*, 2 vols, Paris 1897
Napoleon I, *Lettres Inédites de Napoléon Ier*, Paris 1898
Napoleon I, *Dernières Lettres Inédites de Napoléon Ier*, Paris 1903
Napoleon I, *Supplément à la Correspondance de Napoléon Ier: L'Empereur et la Pologne*, Paris 1908
Napoleon I, *Correspondance Inédite de Napoléon Ier conservée aux Archives de la Guerre, Vol. V*, Paris 1925
Napoleon I, *Lettres Inédites de Napoléon Ier à Marie-Louise, écrites de 1810 à 1814*, Paris 1935
Napoleon I, *Mémoires pour servir à l'Histoire de France sous le règne de Napoléon, écrites à Ste-Hélène sous sa dictée*, Paris 1830
Narichkine, Madame, *neé* Comtesse Rostopchine, *1812. Le Comte Rostopchine et Son Temps*, St Petersburg 1912
Nazarov, Pamfil, *Zapiski soldata Pamfila Nazarova*, in *Russkaia Starina*, 1872, No. 8
Nesselrode, A. de (ed.), *Lettres et Papiers du Chancelier Comte de Nesselrode 1760–1850*, Vol. IV, Paris n.d.
Neverovskii, Dmitri Petrovich, *Zapiski Generala Neverovskavo o sluzhbe svoiei v 1812 godu*, in *Shtenia v Obshchestvie Istorii i Drevnostiei Rossiiskikh*, No. 1, 1859
Nicholas Mikhailovich, Grand Duke, *Les Relations diplomatiques de la Russie et de la France d'après les rapports des ambassadeurs d'Alexandre et de Napoléon 1808–1812*, 6 vols, St Petersburg 1905–8
Niemcewicz, Julian Ursyn, *Pamiętniki Czasów Moich*, Paris 1848
Noel, J.N.A., *Souvenirs Militaires d'un Officier du Premier Empire*, Paris 1895
Nordhof, A.W., *Die Geschichte der Zerstorung Moskaus im Jahre 1812*, Munich 2000
Norov, V.S., *Zapiski o Pokhodakh 1812 i 1813 godov*, St Petersburg 1834
Ogiński, Michał, *Mémoires de Michel Oginski sur la Pologne et les Polonais depuis 1788 jusqu'à la fin de 1815*, 4 vols, Paris 1827
Olenin, A.N., *Sobstvenoruchnaia tetrad'*, in *Russkii Arkhiv*, Vol. VI (1868), 1869
O'Meara, Barry, *Napoleon in Exile, or A Voice from St Helena*, 2 vols, London 1822
Otechestvennaia voina 1812 goda, sbornik dokumentov i materialov, Leningrad–Moscow 1941
Oudinot, Maréchale, *Le Maréchal Oudinot, Duc de Reggio, d'après les souvenirs inédits de la Maréchale*, Paris 1894
Paixhans, Henri-Joseph de, *Retraite de Moscou*, Metz 1868

Partouneaux, Louis, *Adresse du Lieutenant-Général Partouneaux à l'armée française et rapports sur l'affaire du 27 au 28 novembre 1812*, Paris 1815
Pasquier, Étienne Denis, *Mémoires du chancelier Pasquier*, Vol. I, Paris 1893
Pastoret, Amedée de, *De Vitebsk à la Bérézina*, in *La Revue de Paris*, 9e année, Tome 2, 1902
Pelet, Général, *Carnets sur la Campagne de Russie*, Paris 1997
Pelleport, Vicomte de, *Souvenirs Militaires et Intimes du Général Vte de Pelleport*, Vol. II, Paris 1857
Perovskii, V.A., *Iz Zapisok Grafa Vasilia Aleksandrovicha Perovskavo*, in *Russkii Arkhiv*, 1865
Peyrusse, Guillaume, Baron, *Mémorial et Archives du M. le Baron Peyrusse*, Carcassonne 1869
Peyrusse, Guillaume, Baron, *Lettres Inédites du Baron Guillaume Peyrusse écrites à son frère André pendant les campagnes de l'Empire*, Paris 1894
Pfuel, Ernst von, *Retreat of the French Army from Russia*, London 1813
Pils, François, *Journal de Marche du Grenadier Pils (1804–1814)*, Paris 1895
Pion des Loches, Antoine Augustin Flavien, *Mes Campagnes (1792–1815)*, Paris 1889
Płaczkowski, W., *Pamiętniki Wincentego Płaczkowskiego, porucznika dawnej gwardii cesarsko-francuskiej spisane w roku 1845*, Żytomierz 1861
Planat de la Faye, Nicolas Louis, *Vie de Planat de la Faye*, Paris 1895
Pobedonostsev, P.V., *Iz dnevnika 1812 i 1813 godov o Moskovskom razorenii*, in *Russkii Arkhiv*, Kniga 1, 1895
Podczaski, Władysław, *Niektóre szczegóły z życia Władysława Podczaskiego, pułkownika 20 pułku piechoty liniowej polskiej (1812)*, in *Pamiętniki Polskie*, ed. K. Bronikowski, Przemyśl 1883
Pontier, Raymond, *Souvenirs du Chirurgien Pontier, Aide-Major au Quartier Général de la Grande Armée sur la Retraite de Russie*, Paris 1967
Potocka, Anna, *Mémoires de la Comtesse Potocka*, Paris 1897
Pouget, Général Baron, *Souvenirs de Guerre*, Paris 1895
Pradt, Abbé de, *Histoire de l'Ambassade dans le Grand Duché de Varsovie en 1812*, Paris 1815
Pretet, Jean, *Relation de la Campagne de Russie*, in *Revue Bourguignonne*, 1893
Prikaz nashim armiam i dva manifesta Imperatora Aleksandra I v nachale otechestvennoi voiny, in *Russkaia Starina*, June 1912
Puibusque, L.V., *Lettres sur la Guerre de Russie en 1812*, Paris 1816
Puybusque, L.G. de, *Souvenirs d'un invalide*, 2 vols, Paris 1841
Pushkin, A.S., *Roslavlev*, in *Polnoe Sobranie Sochinenii*, Vol. 6, Moscow 1964
Radozhitskii, I.T., *Pokhodnia Zapiski Artillerista s 1812 goda*, Vol. I, Moscow 1835
Rambuteau, Claude Philibert, Comte de, *Mémoires*, Paris 1905
Rapp, Jean, *Mémoires du Général Rapp, Aide de Camp de Napoléon*, London 1823

Raskaz Georgievskavo Kavalera is Divizii Neverovskavo, in *Shtenia v Obshchestvie Istorii i Drevnostiei Rossiiskikh*, No. 1, 1872
Raskazy Krestian-ochevidtsev pro frantsuskoe nashestvie, in *Vestnik Zapadnoi Rossii*, Kniga 12, 1869
Raza, Roustam, *Souvenirs de Roustam, Mamelouk de Napoléon Ier*, Paris 1911
Reguinot, *Le Sergeant Isolé. Histoire d'un soldat pendant la campagne de Russie*, Paris 1831
Réponse de l'auteur de l'Histoire de l'Expédition de Russie, à la Brochure de M. le Comte Rostopchin, intitulée: La Vérité sur l'Incendie de Moskou, Paris 1823
Riazanov, A., *Vospominania ochevidtsa o prebyvanii frantsuzov v Moskve*, Moscow 1862
Ricome, Jean-Baptiste, *Journal d'un Grognard de l'Empire*, Paris 1988
Rigau, Antoine, *Souvenirs des guerres de l'Empire*, Paris 1846
Rochechouart, Louis Victor Leon, Comte de, *Souvenirs sur la Révolution, l'Empire et la Réstauration*, Paris 1889
Roeder, Franz, *The Ordeal of Captain Roeder. From the Diary of an Officer in the First Battalion of Hessian Lifeguards during the Moscow Campaign of 1812*, trs. Helen Roeder, London 1960
Roguet, Christophe Michel, *Mémoires Militaires*, 4 vols, Paris 1862–65
Roos, Heinrich, *Souvenirs d'un Médecin de la Grande Armée*, Paris 1913
Rosselet, Abraham, *Souvenirs*, Neuchâtel 1857
Rossetti, Marie-Joseph, *Journal d'un Compagnon de Murat*, Paris 1998
Rossia i Shvetsia. Dokumenty i Materialy 1809–1818, Ministerstvo Inostrannykh Del SSSR, ed. A.A. Gromyko, Moscow 1985
Rostopchin, Count Feodor Vasilievich, *Sochinienia*, St Petersburg 1853
Rostopchin, Count Feodor Vasilievich, *La Vérité sur l'Incendie de Moscou*, in *Oeuvres inédites du Comte Rostopchine*, Paris 1894
Rostopchin, Count Feodor Vasilievich, *Extrait de Lettres à l'Empereur Alexandre Ier écrites pendant l'année 1812*, ibid.
Rostopchin, Count Feodor Vasilievich, *Fragments des Mémoires sur l'Année 1812*, ibid.
Rostopchin, Count Feodor Vasilievich, *Letuchie Listki 1812 goda*, ed. P.A. Kartakov, St Petersburg 1904
Rotenhan, *Denkwürdigkeiten eines württembergischen Offiziers aus dem Feldzuge im Jahre 1812*, Berlin 1892
Roy, J.J., *Les Français en Russie. Souvenirs de la Campagne de 1812 et de deux ans de captivité en Russie*, Tours 1856
Rumigny, Marie Théodore Gueilly, Comte de, *Souvenirs du Général Comte de Rumigny*, Paris 1921
Safanovich, V.I., *Vospominania Valeriana Ivanovicha Safanovicha*, in *Russkii Arkhiv*, Kniga 1, 1903

Saint-Chamans, Alfred, *Mémoires du Général Comte de Saint-Chamans, ancien aide de camp du Maréchal Soult*, Paris 1896
Saint-Denis, Louis Étienne, *Souvenirs du Mameluck Ali sur l'Empereur Napoléon*, Paris 1926
Sanguszko, Prince Eustachy, *Pamiętnik 1786–1815*, Krakow 1876
Savary, Anne Jean Marie, Duc de Rovigo, *Mémoires du duc de Rovigo pour servir à l'histoire de l'Empereur Napoléon*, Vol. V, Paris 1828
Sayve, Auguste de, *Souvenirs de Pologne et scènes de la campagne de 1812*, Paris 1833
Sbornik Istoricheskikh Materialov, izvlechennykh iz arkhiva sobstv. Evo Imp. Vielichestva Kantselarii, 15 vols, St Petersburg 1876–1917
Ségur, Philippe de, *Histoire et Mémoires*, Vols III, IV, V, VI, Paris 1873
Serang, Marquis de, *Les Prisonniers Français en Russie, Mémoires et Souvenirs, recueillis et publiés par M. de Puibusque*, 2 vols, Paris 1837
Seruzier, Baron, *Mémoires Militaires du Baron Seruzier, colonel d'artillerie légère*, Paris 1823
Shchukin, P.I., *Bumagi otnosiashchiasia do Otechestvennoi Voiny 1812 goda*, 9 vols, Moscow 1897–1905
Sheremetev, S., *Borodino*, in *Russkii Arkhiv*, Vol. XXIX, Kniga 2, 1891
Shilder, N., *Nakanune Erfurtskavo Svidania 1808 goda*, in *Russkaia Starina*, Vol. 98, 1899
Shilder, N., *Zapiska Fligel-Adiutanta Chernysheva o Sredstviakh k Preduprezh deniu Vtorzhenia Niepriatelia v 1812 godu*, in *Voenny Sbornik*, St Petersburg January 1902
Shishkin, Ivan, *Bunt Opolchenia v 1812 godu*, in *Zaria*, No. 8, St Petersburg 1869
Shishkov, A.S., *Zapiski, Mnenia i Perepiska Admirala A.S. Shishkova*, 2 vols, Berlin 1870
Shtaingel, V.I., *Zapiski kasatelno sostavlenia i samovo pokhoda Sanktpeterburg skavo opolchenia protiv vragov otechestva v 1812 i 1813 godakh*, St Petersburg 1814–15
Shugurov, M.F. (ed.), *Doklad o Evreiakh Imperatoru Aleksandru Pavlovichu 1812*, in *Russkii Arkhiv*, Kniga 2, 1903
Shuvalov, P.A., *Pismo general-adiutanta P.A. Shuvalova k Imperatoru Aleksandru Pavlovichu*, in *Sbornik Imperatorskavo Russkavo Istoricheskavo Obshchestva*, Vol. XXI, St Petersburg 1877
Simanskii, Luka Aleksandrovich, *Zhurnal Uchastnika Voiny 1812 goda*, in *Voenno-Istoricheskii Sbornik*, Nos 2, 3, 4 1912; Nos 1, 2, 3, 4 1913; Nos 1, 2 1914
Sołtyk, Count Roman, *Napoléon en 1812. Mémoires Historiques et Militaires sur la Campagne de Russie*, Paris 1836
Staël-Holstein, Anne Louise Germaine de, *Dix années d'éxil*, Paris 1966

Steininger, J., *Mémoires d'un vieux déserteur*, ed. P. de Pardiellan, Paris n.d.
Suckow, C.F.E. von, *D'Iéna à Moscou. Fragments de Ma Vie*, Paris 1901
Sukhanin, N.N., *Iz zhurnala uchastnika voiny 1812 goda*, in *Russkaia Starina*, February 1912
Surrugues, Abbé, *Lettres sur l'Incendie de Moscou*, Paris 1823
Sverbeev, D.I., *Zapiski*, 2 vols, Moscow 1899
Szymanowski, Józef, *Pamiętniki*, Lwów 1898
Talleyrand, Charles Maurice de, *Lettres Inédites de Talleyrand à Napoleon*, Paris 1889
Tascher, Maurice de, *Journal de Campagne d'un cousin de l'Impératrice*, Paris 1933
Tatistchev (Tatishchev), Serge, *Alexandre Ier et Napoléon d'après leur correspondance inédite, 1801–1812*, Paris 1891
Ternaux-Compans, M., *Le Général Compans (1769–1845) d'après ses notes de campagne et sa correspondance de 1812 à 1813*, Paris 1912
Thévenin, Maurice, *Mémoires d'un vieux de la vieille*, in *Bulletin de la Société des Sciences Historiques et Naturelles de l'Yonne*, 98, 1959–60
Thiebault, Baron, *Mémoires du Général Baron Thiebault*, Vol. IV, Paris 1895
Thirion, Auguste, *Souvenirs Militaires*, Paris 1892
Timofev, S.P. (ed.), *Opisanie, shto proiskhodilo vo vremia nashestvia niepriatelia v donskom Monastire 1812 goda*, in *Russkii Arkhiv*, Vol. 29, Kniga 10, 1891
Tolichev, T. (ed.), *Raskazy Ochevidtsev o Dvenadtsatom Gode*, Moscow 1912
Toll, Karl Friedrich von der, *Denkwürdigkeiten des russischen Generals von der Toll*, ed. Theodor von Bernhardi, Leipzig 1856
Trefcon, Toussaint-Jean, *Carnet de Campagne du Colonel Trefcon*, Paris 1914
Turgenev, A.M., *Iz Dnievnika Nieizvestanvo litsa*, in *Russkii Arkhiv*, Kniga 3, 1903
Turno, Karol, *Souvenirs d'un Officier Polonais*, in *Revue des Études Napoléoniennes*, Tome 33, Paris, August–September 1931
Uxküll, Boris von, *Arms and the Woman. The Diaries of Baron Boris von Uxküll 1812–1819*, trs. Joel Carmichael, London 1966
Vassilchikova, A.I., *Vospominania*, in *Russkii Arkhiv*, Kniga 3, 1912
Vaudoncourt, Frédéric Guillaume de, *Relation Impartiale du Passage de La Bérézina, par un témoin oculaire*, Paris 1814
Vaudoncourt, Frédéric Guillaume de, *Quinze Années d'un Proscrit*, 4 vols, Paris 1835
Venturini, Joseph Louis Auguste, *Carnets d'un Italien au service de la France*, in *Nouvelle Revue Retrospective*, Paris, Jan–June 1904
Vermeil de Conchard, Colonel (trs.), *Campagne et Déféction du corps Prussien de la Grande Armée, traduit du journal du Général de Seydlitz*, Paris 1903
Viazemskii, P.A., *Vospominanie o 1812 gode*, in *Polnoe Sobranie Sochinienii*, Vol. VII, St Petersburg 1882

Villamov, G.I., *Dnevnik Stats-sekretaria Grigoria Ivanovicha Villamova*, in *Russkaia Starina*, July 1912
Villemain, Abel François, *Souvenirs Contemporains d'Histoire et de Littérature*, Vol. I, Paris 1854
Vionnet de Maringoné, Lt Général Louis Joseph, *Campagnes de Russie et de Saxe*, Paris 1899
Vlijmen, B.R.F. van, *Vers la Bérésina*, Paris 1908
Voenskii, K. (ed.), *Sviashchennoi Pamiati Dvenadtsaty god. Istoricheskie ocherki, raskazy, vospominania i stati, otnosiashchiasia k epokhe otechestvennoi voiny*, St Petersburg n.d.
Volkonskii, S.G., *Zapiski*, St Petersburg 1902
Voltaire, François Marie Arouet, *Histoire de Charles XII Roi de Suède*, Paris 1832
Vossen, Anton, *Dnievnik Poruchika Vossena*, in *Russkii Arkhiv*, Kniga 3, 1903
Vossler, H.A., *With Napoleon in Russia 1812. The diary of Lt. H. A. Vossler a soldier of the Grand Army*, trs. Walter Wallich, London 1969
Wachsmuth, I.I., *Geschichte meiner Kriegsgefandenschaft in Russland in den Jahren 1812–1813; in gedrängter Kürze dargestellt von I. I. Wachsmuth Leutenant in der König. Westphalischen Armee*, Magdeburg 1910
Walter, Jakob, *The Diary of a Napoleonic Foot Soldier*, Moreton-in-Marsh 1991
Wesemann, J.H.C., *Kanonier des Kaisers. Kriegstagenbuch der Heinrich Wesemann, 1808–14*, Cologne 1971
Widemann, M., *Les Oceanocrates et leurs partisans, ou la guerre avec la Russie en 1812*, Paris 1812
Wilson, Sir Robert, *Brief Remarks on the Character and Composition of the Russian Army*, London 1810
Wilson, Sir Robert, *Narrative of Events during the Invasion of Russia*, London 1860
Wilson, Sir Robert, *Private Diary of Travels, Personal Services, and Public Events during Mission and Employment with the European Armies in the Campaigns of 1812, 1813 and 1814*, 2 vols, London 1861
Wolzogen, Ludwig, Freiherr von, *Memoiren des konigl. preuss. Generals Ludwig Freiherrn von Wolzogen (1807–1814 in russischen Diensten)*, Leipzig 1851
Württemberg, Prince Eugene of, *Mémoires*, in *Journal des Campagnes du Prince de Wurtemberg*, Paris 1907
Wybranowski, Roman, *Pamiętniki Jenerała Romana Wybranowskiego*, 2 vols, Lwów 1882
Ysarn, Chevalier François d', *Mémoires d'un habitant de Moscou pendant le séjour des Français en 1812*, Brussels 1871
Zaitsev, A., *O pokhodakh 1812 goda*, Moscow 1852
Załuski, Józef, *Wspomnienia*, Krakow 1976
Zotov, R.M., *Sochinienia*, Moscow 1996

Studies

Abalikhin, B.S., *Otechestvennaia Voina 1812 goda na Iugo-Zapade Rossii*, Volgograd 1987

Abalikhin, B.S., *Kontrnastuplienie russkikh voisk v 1812 g.*, in *Istoria SSSR*, 1987, No. 4

Anderson, M.S., *British Public Opinion and the Russian Campaign of 1812*, in *Slavonic and East European Review*, Vol. 34 (1956), pp. 408–25

Andolenko, S., *Histoire de l'Armée Russe*, Paris 1967

Angervo, J.M., *How Cold was 1812?*, in *The Times*, 8 February 1961

Askenazy, Szymon, *Książe Józef Poniatowski*, Warsaw 1974

Babenko, V.N., *Otechestvennaia Voina 1812: ukazatel sovietskoi literatury 1962–1987*, Moscow 1987

Babkin, V.I., *Narodnoe Opolchenie v Otechestvennoi Voinie 1812 g.*, Moscow 1962

Barclay de Tolly i Otechestvennaia Voina 1812g, in *Russkaia Starina*, August, September, October, December 1912

Barsukov, A., *Penzenskaia Starina*, in *Russkii Arkhiv*, No. 7, 1896

Belkovich, L., *Kniaz Piotr Ivanovich Bagration*, in *Russkaia Starina*, July 1912

Beskrovny, L.G., *Otechestvennaia Voina*, Moscow 1962

Beskrovny, L.G., *Borodinskoe Srazhenie*, Moscow 1971

Beskrovny, L.G. (ed.), *Polkovodets Kutuzov. Sbornik Statei*, Moscow 1955

Bogdanov, L.P., *Russkaia Armia v 1812 godu*, Moscow 1979

Bonnal, Général H., *La Manoeuvre de Vilna*, Paris 1905

Boppe, Paul Louis, *La Légion Portuguaise*, Paris 1897

Boppe, Paul Louis, *Les Espagnols à la Grande Armée*, Paris 1899

Boppe, Paul Louis, *La Croatie Militaire*, Paris 1900

Bourgoing, Paul, *Itinéraire de Napoléon Ier de Smorgoni à Paris*, Paris 1862

Brandys, M., *Kozietulski i Inni*, 2 vols, Warsaw 1967

Braquehay, A., *Le Général Baron Merle, 1766–1830. Notice Biografique*, Montreuil-sur-Mer 1892

Bruun, Geoffrey, *Europe and the French Imperium 1799–1814*, New York 1938

Buganov, A.V., *Russkaia Istoriia v pamiati krestian i natsionalnoe samosoznanie*, Moscow 1992

Bulgakov, A.Ya., *Razgovor Neapolitanskavo Korolia Miurata s Generalom Grafom Miloradovichom na Avanpostakh armii 14 oktiabria 1812 g*, Moscow 1843

Buturlin, D.P., *Kutuzov v 1812 godu*, in *Russkaia Starina*, Vol. 82, October–December 1894

Camp, F., *Un Espagnol Témoin de la Retraite de Napoléon en Russie (Don Raphael de Llanza)*, in *Revue des Études Napoléoniennes*, Vol. 30, Paris 1930

Cathcart, Col. the Hon. George, *Commentaries on the War in Russia and Germany in 1812 and 1813*, London 1850

Cavaignac, Godefroy, *La Formation de la Prusse Contemporaine*, 2 vols, Paris 1891

Chapuisat, E., *Les Suisses de l'Empereur*, in *Revue des Études Napoléoniennes*, Vol. 34, Paris 1932
Christian, R.F., *Tolstoy's War and Peace*, Oxford 1962
Chuquet, Arthur, *Études d'Histoire*, 8 vols, Paris n.d.
Creveld, Martin van, *Supplying War. Logistics from Wallenstein to Patton*, Cambridge 1977
Deutsch, Harold C., *The Genesis of Napoleonic Imperialism*, Harvard 1938
Driault, Edouard, *Tilsit. France et Russie sous le Premier Empire*, Paris 1917
Driault, Edouard, *Le Grand Empire (1809–1812)*, Paris 1924
Du Casse, A., *Le Général Vandamme et sa Correspondance*, 2 vols, Paris 1870
Du Casse, A., *Mémoires pour servir à l'histoire de la campagne de Russie*, Paris 1852
Duffy, Christopher, *Borodino and the War of 1812*, London 1972
Dundulis, Bronius, *Napoleon et la Lithuanie en 1812*, Paris 1940
Dwyer, Philip G. (ed.), *Napoleon and Europe*, London 2001
Dzhivelegov, A.K., *Aleksandr I i Napoleon. Istoricheskie Ocherki*, Moscow 1915
Dzhivelegov, A.K., Melgunov, S.P., & Pichet, V.I., *Otechestvennaia Voina i Russkoe Obshchestvo*, 7 vols, Moscow 1912
Elting, John R., *Swords Around a Throne. Napoleon's Grande Armée*, London 1989
Ernouf, Baron, *Les Français en Prusse, d'après des documents contemporains*, Paris 1872
Fabry, Lieutenant G., *Campagne de Russie (1812)*, 5 vols, Paris 1901
Fabry, Lieutenant G., *Campagne de 1812. Documents relatifs à l'aile gauche, 20 août–4 décembre, IIe, VIe, IXe corps*, Paris 1912
Feuer, Kathryn B., *Tolstoy and the Genesis of War and Peace*, Cornell 1996
Fontana, Biancamaria, *The Napoleonic Empire and the Europe of Nations*, in *The Idea of Europe from Antiquity to the European Union*, ed. Anthony Pagden, Cambridge 2002
Fregosi, Paul, *Dreams of Empire. Napoleon and the First World War 1792–1815*, London 1989
Garros, L.P., *Quel Roman que ma Vie! Itinéraire de Napoléon Bonaparte*, Paris 1947
Ginsburg, S, *Otechestvennaia Voina i Russkie Evrei*, St Petersburg 1912
Gotteri, Nicole, *Le Lorgne d'Ideville et le service de renseignements du Ministère des Relations Extérieures pendant la campagne de Russie*, in *Revue d'Histoire Diplomatique*, Nos 1–2, Paris 1989
Grunwald, Constantin de, *Baron Stein, Enemy of Napoleon*, London 1936
Grunwald, Constantin de, *Alexandre Ier, Le Tsar Mystique*, Paris 1955
Handelsman, Marceli, *Napoléon et la Pologne 1806–7*, Paris 1909
Hartley, Janet M., *Alexander I*, London 1994

Hartley, Janet M., *Russia in 1812*, in *Jahrbucher fur Geschichte Osteuropas*, Stuttgart

Hazlitt, William, *The Life of Napoleon Buonaparte*, 4 vols, London 1830

Henderson, E.F., *Blucher and the Uprising of Prussia against Napoleon 1806–1815*, London 1911

Higham, Robin, & Kagan, Frederick (eds), *The Military History of Tsarist Russia*, London 2002

Hosking, Geoffrey, *Russia and the Russians. A History*, London 2001

Hosking, Geoffrey, *Russia. People and Empire, 1552–1917*, London 1997

Iudin, P., *Ssylnie 1812 goda v Orenburgskom kraie*, in *Russkii Arkhiv*, Vol. 9, 1896

Jenkins, Michael, *Arakcheev. Grand-Vizir of the Russian Empire*, London 1969

Josselson, Michael & Diana, *The Commander. A Life of Barclay de Tolly*, Oxford 1980

Karabanov, N.V., *Otechestvennaia Voina v Izobrazhenii Russkikh Pisatelei*, Moscow 1912

Kartsov, Yu. & Voenskii, K., *Prichiny Voiny 1812 goda*, St Petersburg 1911

Keep, John, *Soldiers of the Tsar. Army and Society in Russia 1462–1874*, Oxford 1985

Kemble, J., *Napoleon Immortal. The Medical History and Private Life of Napoleon Bonaparte*, London 1959

Kharkievich, V., *Voina 1812 goda ot Nemana de Smolenska*, 2 vols, Vilna 1901

Kistorii Otechestvennoi Voiny 1812 goda. Poslyednyaia popytka Napoleona nachat' mirnie peregovory s Imperatorom Aleksandrom vo vremia zanyatia Moskvy frantsuzkimi voiskami, in *Russkaia Starina*, January 1912

Koliubakhin, B., *1812 god. Poslednie dni komandovania Barklayem 1oi i 2oi zapadnymi armiami*, in *Russkaia Starina*, June 1912

Koliubakhin, B., *1812 god. Izbranie Kutuzova glavnokomanduiushchym nad vsemi armiami, priezd ievo v armiu i pervie dni evo deiatelnosti*, in *Russkaia Starina*, July 1812

Koliubakhin, B., *1812 god. Borodinskoie srazhenie 26 avgusta*, in *Russkaia Starina*, August 1912

Kraehe, Enno, *Metternich's German Policy. The Contest against Napoleon 1799–1814*, Vol. I, Princeton 1963

Kukiel, M., *Les Polonais à la Moskowa*, in *Revue des Études Napoléoniennes*, Vol. 28, Paris 1929

Kukiel, M., *Wojna 1812 roku*, 2 vols, Kraków 1937

Kukiel, M., *Vues sur le Trône de Pologne en 1812*, in *Revue des Études Napoléoniennes*, Vol. 34, Paris 1932

Langsam, W.C., *The Napoleonic Wars and German Nationalism in Austria*, Columbia 1931

Ley, Francis, *Alexandre Ier et sa Sainte Alliance (1811–1825)*, Paris 1975

Lobanov-Rostovskii, A.A., *Russia and Europe 1789–1825*, Duke 1947
Madariaga, Isabel de, *Russia in the Age of Catherine the Great*, London 1981
Madariaga, Isabel de, *Spain and the Decembrists*, in *European Studies Review*, Vol. III, No. 2, April 1973
Mansuy, A., *Jérôme Napoléon et la Pologne en 1812*, Paris 1931
Marcel-Paon, A., *Du Niemen au Niemen*, in *Revue des Études Napoléoniennes*, Vol. 37, Paris 1933
Marco de Saint-Hilaire, Émile, *Histoire de la Campagne de Russie pendant l'année 1812 et de la captivité des prisonniers français en Sibérie et dans les autres provinces de l'empire*, 2 vols, Paris 1846–48
Margueron, L.J., *Campagne de Russie*, Part 1, 4 vols, Paris n.d.
Markham, F.M.H., *Napoleon and the Awakening of Europe*, London 1954
Martinien, Aristide, *Tableaux par corps et par batailles des officiers tués et blessés pendant les guerres de l'Empire, 1805–1815*, Paris 1899
Meynier, Albert, *Les Morts de la Grande Armée*, Paris 1930
Mikhailovskii-Danilevskii, A.I., *Opisanie Otechestvennoi Voiny 1812 goda*, 4 vols, St Petersburg 1839
Moreau, Jean, *Le Soldat Impérial 1800–1814*, 2 vols, Paris 1904
Morley, Charles, *Alexander I and Czartoryski. The Polish Question 1801–1813*, in *Slavonic and East European Review*, Vol. 25, London April 1947
Nafziger, George, *Napoleon's Invasion of Russia*, Novato 1998
Nagengast, W.E., *Moscow, the Stalingrad of 1812; the American Reaction to Napoleon's Retreat from Russia*, in *Russian Review*, Vol. 8, No. 4 (1949), pp. 302–15
Nechkina, M.V., *1812 god. K Stopiatidesiatiletiu Otechestvennoi Voiny. Sbornik Statei*, Moscow 1962
Nersisian, M.G., *Otechestvennaia Voina 1812 goda i Narodi Kavkaza*, Erevan 1965
Nikolai Mikhailovich, Grand Duke, *L'Empereur Alexandre Ier*, St Petersburg 1912
Norov, A.S., *Voina i Mir 1805–1812*, Moscow 1914
Okuniev, Colonel N., *Considérations sur les Grandes Operations, les Batailles et les Combats de la Campagne de 1812 en Russie*, Paris 1829
Olivier, Daria, *L'Incendie de Moscou*, 1964
Otechestvennaia Voina 1812 goda. Istochniki, Pamiatniki, Problemy. Materialy VI Miezhdunarodnoi Nauchnoi Konferentsii, Borodino 1998
Otechestvennaia Voina 1812 goda. Istochniki, Pamiatniki, Problemy. Materialy VII Vserossiiskoi Nauchnoi Konferentsii, Borodino 1999
Otechestvennaia Voina 1812 goda. Istochniki, Pamiatniki, Problemy. Materialy VIII Vserossiiskoi Nauchnoi Konferentsii, Borodino 2000
Otechestvennaia Voina 1812 goda. Istochniki, Pamiatniki, Problemy. Materialy IX Vserossiiskoi Nauchnoi Konferentsii, Moscow 2001

Otechestvennaia Voina v Kaluzhskoi Gubernii i Rossiiskoi Provintsii, Materiali Nauchnoi Konferentsii 21 Oktiabria 2000g, Maloyaroslavets 2001
Palitsyn, N.A., *Manifesty Napisannie Shishkovym v Otechestvennuiu Voinu i patrioticheskie ich znachenie*, in *Russkaia Starina*, June 1812
Palmer, Alan, *Napoleon in Russia*, London 1967
Palmer, Alan, *Alexander I. The Tsar of War and Peace*, London 1974
Parkinson, Roger, *The Fox of the North. The Life of Kutuzov, General of War and Peace*, London 1976
Pokrovskii, M.N., *Russkaia Istoria s Drevnieishykh Vremien*, 5 vols, Moscow 1910–13
Ponzio, Luigi, *L'Italia nella campagnia di Russia: narrazione popolare a riccordo del primo centenario*, Pavia 1912
Popov, A.I., *Velikaia Armia v Rossii. Pogonia za Mirazhem*, Samara 2002
Popov, A.N., *Dvizhenie Russkikh Voisk ot Moskvy do Krasnoi Pakhry*, in *Russkaia Starina*, June, July, August, September 1897
Popov, A.N., *Moskva v 1812g*, in *Russkii Arkhiv*, 1875
Popov, A.N., *Frantsuzy v Moskve w 1812g*, in *Russkii Arkhiv*, 1876
Porter, Sir Robert Ker, *A Narrative of the Campaign in Russia During the Year 1812*, London 1814
Priezd Imperatora Aleksandra I v Moskvu, in *Russkaia Starina*, July 1912
Raeff, Marc, *Michael Speransky. Statesman of Imperial Russia, 1772–1839*, The Hague 1969
Ragsdale, Hugh, *Détente in the Napoleonic Era. Bonaparte and the Russians*, Lawrence, Kansas, 1980
Ramm, Agatha, *Germany 1789–1919. A Political History*, London 1967
Ratchinski, André, *Napoléon et Alexandre Ier. La Guerre des Idées*, Paris 2002
Riley, J.P., *Napoleon and the World War of 1813*, London 2001
Rober, A., *L'Idée nationale autrichienne et les guerres de Napoléon*, Paris 1933
Roberts, Andrew, *Napoleon and Wellington*, London 2001
Rose, J. Holland, *The Napoleonic Empire at its Height*, in *Cambridge Modern History*, Vol. IX, Cambridge 1906
Sarrazin, M., *Histoire de la guerre de Russie*, Paris 1815
Sauvage, N.J., *Relation de la Campagne de Russie*, Paris n.d.
Sauzey, Jean Camille, *Les Allemands sous les Aigles Françaises*, 6 vols, Paris 1902–12
Schmitt, H.A., *Stein, Alexander and the Crusade against Napoleon*, in *Journal of Modern History*, Vol. 31 (December 1959)
Schmittlein, R., *Un district lithuanien sous l'occupation française (1812)*, Mainz 1952
Schuermans, Albert, *Itinéraire Général de Napoléon*, Paris 1911

Schwarzfuchs, S., *Napoleon, the Jews and the Sanhedrin*, London 1979
Scott, Franklin D., *Bernadotte and the Fall of Napoleon*, Harvard 1935
Senkowska-Gluck, Monika (ed.), *Europa i Świat w Epoce Napoleońskiej*, Warsaw 1988
Servières, G., *L'Allemagne Française sous Napoléon*, Paris 1904
Shanahan, W.O., *Prussian Military Reforms 1786–1813*, Columbia 1945
Shchegolev, P.E., *Dekabristy*, Moscow-Leningrad 1926
Shilder, N.K., *Imperator Aleksandr I*, 4 vols, St Petersburg 1904–5
Shishov, A.V., *Nieizvestny Kutuzov. Novoe Prochtenie Biografii*, Moscow 2001
Shtrange, M.M., *Russkoie Obshchestvo i Frantsuzkaia Revolutsia 1789–1794*, Moscow 1956
Shtrange, M.M., *Demokraticheskaia Inteligentsia Rossii v XVIII viekie*, Moscow 1965
Shvedov, S.V., *Komplektovanie, chislennost' i poteri russkoi armii v 1812 godu*, in *Istoria SSSR*, No. 4, 1987
Sirotkin, V.G., *Duel Dvukh Diplomatsii: Rossia i Frantsia v 1801–1812*, Moscow 1966
Sirotkin, V.G., *Napoleonskaia "Voina Nerev" protiv Rossii*, Moscow 1981
Sirotkin, V.G., *Otechestvennaia Voina 1812g.*, Moscow 1988
Sirotkin, V.G., *Napoleon i Rossia*, Moscow 2000
Sirotkin, V.G. & Kozlov, V.T., *Traditsi Borodina: Pamiat' i Pamiatniki*, Moscow 1989
Skallon, D.A. (ed.), *Istoria gosudarsvennoi svity*, in *Russkaia Starina*, June 1912
Smith, D.G., *Borodino*, Moreton-in-Marsh 1998
Sokolov, Oleg, *La Campagne de Russie*, in *Napoléon Ier*, Saint-Cloud November 2000–October 2001
Stchoupak, N., *L'Entrevue de I. Iakovlev avec Napoléon*, in *Revue des Études Napoléoniennes*, Vol. 33, Paris 1931
Strong, John W., *Russia's Plans for an Invasion of India in 1801*, in *Canadian Slavonic Papers*, Toronto, Vol. 7, 1965
Tarle, E.V., *Napoleon*, Moscow 1936
Tarle, E.V., *Nashestvie Napoleona na Rossiu 1812 god*, Moscow 1992
Tezisy Nauchnoi Konferentsii "Otechestvennaia Voina 1812 goda. Rossia i Evropa," Borodino 1992
Thiry, Jean, *La Campagne de Russie*, Paris 1969
Troitskii, N.A., *1812, Velikii god Rossii*, Moscow 1988
Troitskii, N.A., *Otechestvennaia Voina 1812 goda. Istoria Temy*, Saratov 1991
Troitskii, N.A., *Aleksandr I i Napoleon*, Moscow 1994
Troitskii, N.A., *Dien Borodina*, in *Znamia*, No. 8, 1987
Troitskii, N.A., *O dislokatsii i chislennosti russkikh voisk v nachale Otechestvennoi Voiny 1812g.*, in *Voenno-Istoricheskii Zhurnal*, No. 8, 1987

Troitskii, N.A., *O chislennosti russkikh armii v nachale Otechestvennoi Voiny 1812g.*, in *Voprosy Istorii*, No. 11, 1987
Troitskii, N.A., *Otechestvennaia Voina 1812 g i Russkaia Literatura XIX veka*, Moscow 1998
Tselorungo, D.G., *Ofitsery Russkoi armii-uchestniki borodinskovo srazhenia*, Moscow 2002
Tulard, Jean, *Bibliographie des Mémoires sur le Consulat et l'Empire, écrites ou traduites en Français*, Paris 1971
Tulard, Jean, *Le Grand Empire*, Paris 1982
Tulard, Jean, *Itinéraire de Napoléon au jour de jour 1769–1821*, Paris 1992
Tulard, Jean, *Le Dépôt de la Guerre et la Préparation de la Campagne de Russie*, in *Revue Historique de l'Armée*, 1939, No. 3
Turner, Wesley B., *The War of 1812*, Toronto 2000
Vandal, Albert, *Napoléon et Alexandre 1er. L'Alliance Russe sous le Premier Empire*, 3 vols, Paris 1891
Vaudoncourt, Frédéric-Guillaume de, *Mémoires pour servir à l'histoire de la guerre entre la France et la Russie en 1812*, London 1815
Vertlib, E., *1812 god u Pushkina i Zagoskina. K voprosu ob istokakh russkovo samosoznania*, New York 1990
Vigel, F.F., *Zapiski Filipa Filipovicha Vigelia*, 7 vols, Moscow 1892
Vilatte de Prugnes, Robert, *Les Efféctifs de la Grande Armée pour la Campagne de Russie de 1812*, in *Revue des Études Historiques*, Paris 1913
Vitberg, F., *O pamiatnikakh Otechestvennoi Voiny*, in *Russkaia Starina*, December 1912
Voenskii, K.A., *Bonapart i Russkie Plennie*, St Petersburg 1907
Voenskii, K.A., *Russkoe Dukhoventsvo i Otechestvennaia Voina 1812 goda*, Moscow 1912
Volkov, S.V., *Russkii ofitserskii korpus*, Moscow 1993
Volkovskoi, K., *Materialy dlia biografii kniazia M.I., Golenishcheva-Kutuzova*, in *Russkaia Starina*, September 1912
Voronov, P., *Kto upravlial russkimi voiskami v iunie 1812 goda, poslie perepravy armii Napoleona cherez Nieman*, in *Russkaia Starina*, July 1912
Voronovskii, V.M., *Otechestvennaia Voina 1812 g v predelakh Smolenskoi Gubernii*, St Petersburg 1912
Waliszewski, K., *Le Règne d'Alexandre 1er*, 3 vols, Paris 1924
Wirtschafter, Elise Kimerling, *From Serf to Russian Soldier*, Princeton 1989
Woolf, Stuart, *Napoleon's Integration of Europe*, London 1991
Zanoli, Alessandro, *Sulla Milizia Cisalpino-Italiana*, 2 vols, Milan 1845
Zhilin, P.A., *Feldmarshal Mikhail Ilarionovich Kutuzov*, Moscow 1987
Zhilin, P.A., *Otechestvennaia Voina 1812 goda*, Moscow 1988
Zhilin, P.A., *Gibel' Napoleonskoi Armii v Rossii*, Moscow 1968

Index